DNA Repair Protocols

METHODS IN MOLECULAR BIOLOGY™

John M. Walker, SERIES EDITOR

335. **Fluorescent Energy Transfer Nucleic Acid Probes:** *Designs and Protocols,* edited by *Vladimir V. Didenko, 2006*

334. **PRINS and *In Situ* PCR Protocols:** *Second Edition,* edited by *Franck Pellestor, 2006*

333. **Transplantation Immunology:** *Methods and Protocols,* edited by *Philip Hornick and Marlene Rose, 2006*

332. **Transmembrane Signaling Protocols:** *Second Edition,* edited by *Hydar ali and Haribabu Bodduluri, 2006*

331. **Human Embryonic Stem Cell Protocols,** edited by *Kursad Turksen, 2006*

330. **Nonhuman Embryonic Stem Cell Protocols, Vol. II:** *Differentiation Models,* edited by *Kursad Turksen, 2006*

329. **Nonhuman Embryonic Stem Cell Protocols, Vol. I:** *Isolation and Characterization,* edited by *Kursad Turksen, 2006*

328. **New and Emerging Proteomic Techniques,** edited by *Dobrin Nedelkov and Randall W. Nelson, 2006*

327. **Epidermal Growth Factor:** *Methods and Protocols,* edited by *Tarun B. Patel and Paul J. Bertics, 2006*

326. **In Situ Hybridization Protocols,** *Third Edition,* edited by *Ian A. Darby and Tim D. Hewitson, 2006*

325. **Nuclear Reprogramming:** *Methods and Protocols,* edited by *Steve Pells, 2006*

324. **Hormone Assays in Biological Fluids,** edited by *Michael J. Wheeler and J. S. Morley Hutchinson, 2006*

323. **Arabidopsis Protocols,** *Second Edition,* edited by *Julio Salinas and Jose J. Sanchez-Serrano, 2006*

322. ***Xenopus* Protocols:** *Cell Biology and Signal Transduction,* edited by *X. Johné Liu, 2006*

321. **Microfluidic Techniques:** *Reviews and Protocols,* edited by *Shelley D. Minteer, 2006*

320. **Cytochrome P450 Protocols,** *Second Edition,* edited by *Ian R. Phillips and Elizabeth A. Shephard, 2006*

319. **Cell Imaging Techniques,** *Methods and Protocols,* edited by *Douglas J. Taatjes and Brooke T. Mossman, 2006*

318. **Plant Cell Culture Protocols,** *Second Edition,* edited by *Victor M. Loyola-Vargas and Felipe Vázquez-Flota, 2005*

317. **Differential Display Methods and Protocols,** *Second Edition,* edited by *Peng Liang, Jonathan Meade, and Arthur B. Pardee, 2005*

316. **Bioinformatics and Drug Discovery,** edited by *Richard S. Larson, 2005*

315. **Mast Cells:** *Methods and Protocols,* edited by *Guha Krishnaswamy and David S. Chi, 2005*

314. **DNA Repair Protocols:** *Mammalian Systems, Second Edition,* edited by *Daryl S. Henderson, 2006*

313. **Yeast Protocols:** *Second Edition,* edited by *Wei Xiao, 2005*

312. **Calcium Signaling Protocols,** *Second Edition,* edited by *David G. Lambert, 2005*

311. **Pharmacogenomics:** *Methods and Protocols,* edited by *Federico Innocenti, 2005*

310. **Chemical Genomics:** *Reviews and Protocols,* edited by *Edward D. Zanders, 2005*

309. **RNA Silencing:** *Methods and Protocols,* edited by *Gordon Carmichael, 2005*

308. **Therapeutic Proteins:** *Methods and Protocols,* edited by *C. Mark Smales and David C. James, 2005*

307. **Phosphodiesterase Methods and Protocols,** edited by *Claire Lugnier, 2005*

306. **Receptor Binding Techniques:** *Second Edition,* edited by *Anthony P. Davenport, 2005*

305. **Protein–Ligand Interactions:** *Methods and Applications,* edited by *G. Ulrich Nienhaus, 2005*

304. **Human Retrovirus Protocols:** *Virology and Molecular Biology,* edited by *Tuofu Zhu, 2005*

303. **NanoBiotechnology Protocols,** edited by *Sandra J. Rosenthal and David W. Wright, 2005*

302. **Handbook of ELISPOT: Methods and Protocols,** edited by *Alexander E. Kalyuzhny, 2005*

301. **Ubiquitin–Proteasome Protocols,** edited by *Cam Patterson and Douglas M. Cyr, 2005*

300. **Protein Nanotechnology:** *Protocols, Instrumentation, and Applications,* edited by *Tuan Vo-Dinh, 2005*

299. **Amyloid Proteins:** *Methods and Protocols,* edited by *Einar M. Sigurdsson, 2005*

298. **Peptide Synthesis and Application,** edited by *John Howl, 2005*

297. **Forensic DNA Typing Protocols,** edited by *Angel Carracedo, 2005*

296. **Cell Cycle Control:** *Mechanisms and Protocols,* edited by *Tim Humphrey and Gavin Brooks, 2005*

295. **Immunochemical Protocols,** *Third Edition,* edited by *Robert Burns, 2005*

294. **Cell Migration:** *Developmental Methods and Protocols,* edited by *Jun-Lin Guan, 2005*

293. **Laser Capture Microdissection:** *Methods and Protocols,* edited by *Graeme I. Murray and Stephanie Curran, 2005*

292. **DNA Viruses:** *Methods and Protocols,* edited by *Paul M. Lieberman, 2005*

291. **Molecular Toxicology Protocols,** edited by *Phouthone Keohavong and Stephen G. Grant, 2005*

290. **Basic Cell Culture Protocols,** *Third Edition,* edited by *Cheryl D. Helgason and Cindy L. Miller, 2005*

METHODS IN MOLECULAR BIOLOGY™

DNA Repair Protocols

Mammalian Systems

SECOND EDITION

Edited by

Daryl S. Henderson

*Pharmacological Sciences, SUNY Stony Brook
Stony Brook, NY*

HUMANA PRESS ✳ TOTOWA, NEW JERSEY

© 2006 Humana Press Inc.
999 Riverview Drive, Suite 208
Totowa, New Jersey 07512

www.humanapress.com

All rights reserved. No part of this book may be reproduced, stored in a retrieval system, or transmitted in any form or by any means, electronic, mechanical, photocopying, microfilming, recording, or otherwise without written permission from the Publisher. Methods in Molecular Biology™ is a trademark of The Humana Press Inc.

All papers, comments, opinions, conclusions, or recommendations are those of the author(s), and do not necessarily reflect the views of the publisher.

This publication is printed on acid-free paper. ∞
ANSI Z39.48-1984 (American Standards Institute)
Permanence of Paper for Printed Library Materials.

Cover illustration: Artistic rendering of a confocal microscope image of human fibroblasts irradiated with ultraviolet light through a micropore filter, immunostained with antibodies to PCNA and CAF-1, and counterstained with Hoechst 33258 to reveal DNA. Original image courtesy of Dr. Ennio Prosperi. See Chapter 31, "Analysis of Proliferating Cell Nuclear Antigen (PCNA) Associated with DNA Excision Repair Sites in Mammalian Cells," by A. Ivana Scovassi and Ennio Prosperi.

Cover design by Patricia F. Cleary.

For additional copies, pricing for bulk purchases, and/or information about other Humana titles, contact Humana at the above address or at any of the following numbers: Tel.: 973-256-1699; Fax: 973-256-8341; E-mail: orders@humanapr.com; or visit our Website: www.humanapress.com

Photocopy Authorization Policy:
Authorization to photocopy items for internal or personal use, or the internal or personal use of specific clients, is granted by Humana Press Inc., provided that the base fee of US $30.00 per copy is paid directly to the Copyright Clearance Center at 222 Rosewood Drive, Danvers, MA 01923. For those organizations that have been granted a photocopy license from the CCC, a separate system of payment has been arranged and is acceptable to Humana Press Inc. The fee code for users of the Transactional Reporting Service is: [1-58829-513-3/06 $30.00].

Printed in the United States of America. 10 9 8 7 6 5 4 3 2 1

Library of Congress Cataloging in Publication Data
DNA repair protocols : mammalian systems / edited by Daryl S. Henderson.—2nd ed.
  p. ; cm. — (Methods in molecular biology ; 314)
  Includes bibliographical references and index.
  ISBN 1-58829-513-3 (alk. paper) EISBN 1-59259-973-7
  1. DNA repair—Laboratory manuals.
  [DNLM: 1. DNA Repair—Laboratory Manuals. 2. DNA Damage—Laboratory Manuals. 3. Mammals—genetics—Laboratory Manuals. QU 25 D755 2005] I. Henderson, Daryl S. II. Series: Methods in molecular biology (Clifton, N.J.); v. 314.
  QH467.D627 2005
  572.8'6—dc22   2005006250

Preface

The first edition of this book, published in 1999 and called *DNA Repair Protocols: Eukaryotic Systems*, brought together laboratory-based methods for studying DNA damage and repair in diverse eukaryotes: namely, two kinds of yeast, a nematode, a fruit fly, a toad, three different plants, and human and murine cells. This second edition of *DNA Repair Protocols* covers mammalian cells only and hence its new subtitle, *Mammalian Systems*. There are two reasons for this fresh emphasis, both of them pragmatic: to cater to the interests of what is now a largely mammalocentric DNA repair field, and to expedite editing and production of this volume.

Although *DNA Repair Protocols: Mammalian Systems* is a smaller book than its predecessor, it actually contains a greater variety of methods. Fourteen of the book's thirty-two chapters are entirely new and areas of redundancy present in the first edition have been eliminated here (for example, now just two chapters describe assays for nucleotide excision repair [NER], rather than seven). All eighteen returning chapters have been revised, many of them extensively. In order to maintain a coherent arrangement of topics, the four-part partitioning seen in the first edition was dispensed with and chapters concerned with ionizing radiation damage and DNA strand breakage and repair were relocated to near the front of the book. Finally, an abstract now heads each chapter.

I have aimed to make *DNA Repair Protocols: Mammalian Systems* a well-rounded book, intended to address a broad range of questions about practical mammalian DNA repair, including arcana such as "what is radioresistant DNA synthesis and how is it measured?" (see Chapter 5, by Jaspers and Zdzienicka). The final selection of topics was influenced by both the contents of the first edition (naturally) and the willingness of its authors to contribute again; by my desire to correct deficiencies in the first edition (e.g., it lacks a chapter on DNA helicase assays); and by what new methods have come into use since 1998, when the first edition went into production. Below I summarize and put into context some of what's new, as a way of illustrating the diversity and scope of *DNA Repair Protocols: Mammalian Systems*. (My apologies to authors of chapters not mentioned; no slight is intended.)

Cytogenetic analysis, a topic that had scant coverage in the first edition, is emphasized in Chapters 3 and 4. In Chapter 3, Au and Salama detail a cytogenetic challenge assay in which blood lymphocytes obtained from individuals of interest (e.g., smokers, chemical workers, chemotherapy patients) are irradiated—that is, challenged—with ionizing or UV radiation and then scored for chromosomal anomalies. Abnormally high levels of induced chromosomal aberrations may be indicative of a constitutional repair deficiency, as the authors and their colleagues recently described for certain variant alleles of two DNA repair genes, *XPD* and *XRCC1*.

The cellular response to ionizing radiation is complex. For example, the irradiated cell may be killed outright or it may repair whatever damage occurs and continue to grow and divide no worse for wear. A third outcome, only recently appreciated, goes something like this: the cell survives irradiation apparently unharmed and proliferates, but then its descendents go on to display "genomic instabilities" of various kinds. This intriguing transgenerational phenomenon (or phenomena) and cytological methods for studying it are described by Nagar, Corcoran, and Morgan in Chapter 4.

In Chapter 7, Huang and Darzynkiewicz present a new cytometric approach for detecting DNA double-strand breaks (DSBs). Their method is based on work by W. M. Bonner and colleagues who discovered that a phosphorylated form of histone H2AX, called γH2AX, accumulates in quantity at chromosomal DSBs. Specific antibodies against γH2AX are now available commercially, making immunofluorescence detection of γH2AX a convenient way to detect DSBs induced by clastogenic agents (with attendant caveats noted in the chapter).

Two very different methods for quantifying levels of DNA damage (Chapters 14 and 21) both make use of PicoGreen, an ultrasensitive nucleic acid stain that shows especially strong enhancement of fluorescence when bound to double-stranded DNA (dsDNA). In Chapter 14, Santos, Meyer, Mandavilli and Van Houten describe a quantitative PCR (QPCR) method that can be used to measure DNA damage in specific regions of either the mitochondrial or nuclear genome. The crux of this method is that DNA lesions able to impede movement of the polymerase during PCR will cause less product to be synthesized compared to control reactions using a lesion-free template; in both cases the amount of dsDNA product is quantified with PicoGreen. The QPCR technique combines aspects of Bohr's pioneering Southern blot method, described by Anson, Mason, and Bohr in Chapter 13 (both approaches quantify underrepresentation of a specific DNA fragment and use Poisson probabilities to calculate levels of damage), and Pfeifer's ligation-mediated PCR in Chapter 15 (the stopping of polymerases by lesions in DNA). (Note that Hays and Hoffman in the first edition described a QPCR approach for measuring activities of photolyases, although that method uses UV-damaged plasmids as template and radiolabel to quantify PCR product levels.)

In Chapter 21, Schröder, Batel, Schwertner, Boreiko, and Müller detail their sensitive microtiter plate assay ("Fast Micromethod") for estimating levels of single-strand breaks (nicks) in DNA. The single-strand breaks are identified indirectly by measuring the reduction in amount of dsDNA—quantified by

PicoGreen—in mutagen-treated and untreated samples following alkaline denaturation. Nicked duplex DNA denatures more readily and hence loses fluorescence more rapidly than does intact DNA. The Fast Micromethod shares some methodological features with the alkaline version of the comet assay described by Speit and Hartmann in Chapter 20 (e.g., denaturation of DNA under alkaline conditions). In addition, it can be adapted for high throughput screening and allows testing of both cells and tissues, including frozen material.

Chapter 22, contributed by Shibutani, Kim, and Suzuki, describes a vastly improved ^{32}P-postlabeling protocol, which is a very sensitive method for detecting the presence of "adducted" bases in DNA. ^{32}P-postlabeling protocols have traditionally used thin layer chromatography to separate modified deoxynucleotides from their normal counterparts. The new technique, developed in Shibutani's laboratory, uses polyacrylamide gel electrophoresis, which allows multiple samples to be run out and compared on a single gel. Other strengths of the method are its improved handling and high sensitivity.

In Chapter 24, Wang and Hays describe a new and efficient protocol for preparing mismatch repair (MMR) plasmid substrates. Their method makes use of nicking endonucleases that introduce a single-strand nick at specific sequences in duplex DNA. One nicking endonuclease is used to create a pair of nicks separated by tens of nucleotides on one strand of a plasmid specially engineered for this purpose. The oligonucleotide defined by the nicks is melted out by heating, the resulting gapped plasmid is purified by BND-cellulose chromatography, and the desired mismatch-containing synthetic oligomer is annealed into the gap. Following ligation and purification, the mismatch-containing plasmid substrate is treated with a different nicking endonuclease to generate *the nick* that is essential for initiation of MMR excision.

In the first edition, Matsumoto described a base excision repair (BER) assay using *Xenopus* oocyte extracts. Here, in Chapter 26, he details an updated protocol for studying BER in mammalian cells that includes several important technical refinements. For example, in the synthetic BER substrate that he described in the first edition, the abasic site was positioned opposite an adenine. In his new construct the abasic site is opposite a cytosine, thus mimicking depurination of guanine, which in vivo occurs much more frequently than loss of thymine. Two further improvements to his assay are the use of SYBR Green (rather than radiolabel) to detect repair products and a simplified method for preparing whole-cell extracts.

Missing from the first edition was a chapter on DNA helicase assays. This significant lacuna has now been filled by Brosh and Sharma's excellent contribution, Chapter 28. New authors Scovassi and Prosperi have written a considerably expanded chapter (Chapter 31) on proliferating cell nuclear antigen (PCNA), in keeping with that molecule's central role in DNA repair. Those two chapters, together with one by Mello, Moggs, and Almouzni addressing the role of chromatin in repair (Chapter 32), round out a strong collection of BER- and NER-related chapters.

I thank all the authors for their excellent chapters and much patience; John Walker, the series editor, for his advice; James Geronimo, who deftly guided the book through production, and his colleagues at Humana Press; and Nadine Henderson and other Hendersons and Tindalls for their interest and support.

Daryl S. Henderson

Contents

Preface v
Contributors xiii
Technical Notes xvii

 1 Isolation of Mutagen-Sensitive Chinese Hamster Cell Lines
    by Replica Plating
    Malgorzata Z. Zdzienicka 1
 2 Complementation Assays Adapted for DNA Repair–Deficient
    Keratinocytes
    **Mathilde Fréchet, Valérie Bergoglio, Odile Chevallier-Lagente,
    Alain Sarasin, and Thierry Magnaldo** 9
 3 Cytogenetic Challenge Assays for Assessment
    of DNA Repair Capacities
    William W. Au and Salama A. Salama 25
 4 Evaluating the Delayed Effects of Cellular Exposure
    to Ionizing Radiation
    Shruti Nagar, James J. Corcoran, and William F. Morgan 43
 5 Inhibition of DNA Synthesis by Ionizing Radiation:
    A Marker for an S-Phase Checkpoint
    Nicolaas G. J. Jaspers and Malgorzata Z. Zdzienicka 51
 6 Analysis of Inhibition of DNA Replication in Irradiated Cells
    Using the SV40-Based In Vitro Assay of DNA Replication
    George Iliakis, Ya Wang, and Hong Yan Wang 61
 7 Cytometric Assessment of Histone H2AX Phosphorylation:
    A Reporter of DNA Damage
    Xuan Huang and Zbigniew Darzynkiewicz 73
 8 Detection of DNA Strand Breaks by Flow and Laser Scanning
    Cytometry in Studies of Apoptosis and Cell Proliferation
    (DNA Replication)
    Zbigniew Darzynkiewicz, Xuan Huang, and Masaki Okafuji 81
 9 In Vitro Rejoining of Double-Strand Breaks in Genomic DNA
    George Iliakis and Nge Cheong 95
10 Detection of DNA Double-Strand Breaks and Chromosome
    Translocations Using Ligation-Mediated PCR and Inverse PCR
    **Michael J. Villalobos, Christopher J. Betti,
    and Andrew T. M. Vaughan** 109

11 Plasmid-Based Assays for DNA End-Joining In Vitro
    *George Iliakis, Bustanur Rosidi, Minli Wang,
    and Huichen Wang* 123

12 Use of Gene Targeting to Study Recombination
    in Mammalian Cell DNA Repair Mutants
    Rodney S. Nairn and Gerald M. Adair 133

13 Gene-Specific and Mitochondrial Repair
    of Oxidative DNA Damage
    R. Michael Anson, Penelope A. Mason, and Vilhelm A. Bohr 155

14 Quantitative PCR-Based Measurement of Nuclear
    and Mitochondrial DNA Damage
    and Repair in Mammalian Cells
    *Janine H. Santos, Joel N. Meyer, Bhaskar S. Mandavilli,
    and Bennett Van Houten* 183

15 Measuring the Formation and Repair of DNA Damage
    by Ligation-Mediated PCR
    Gerd P. Pfeifer 201

16 Immunochemical Detection of UV-Induced DNA
    Damage and Repair
    Marcus S. Cooke and Alistair Robson 215

17 A Dot-Blot Immunoassay for Measuring Repair
    of Ultraviolet Photoproducts
    Shirley McCready 229

18 Quantification of Photoproducts in Mammalian Cell DNA Using
    Radioimmunoassay
    David L. Mitchell 239

19 DNA Damage Quantitation by Alkaline Gel Electrophoresis
    *Betsy M. Sutherland, Paula V. Bennett,
    and John C. Sutherland* 251

20 The Comet Assay: *A Sensitive Genotoxicity Test
    for the Detection of DNA Damage and Repair*
    Günter Speit and Andreas Hartmann 275

21 Fast Micromethod DNA Single-Strand-Break Assay
    *Heinz C. Schröder, Renato Batel, Heiko Schwertner,
    Oleksandra Boreiko, and Werner E. G. Müller* 287

22 ^{32}P-Postlabeling DNA Damage Assays: *PAGE, TLC, and HPLC*
    Shinya Shibutani, Sung Yeon Kim, and Naomi Suzuki 307

23 Electrophoretic Mobility Shift Assays to Study Protein Binding
    to Damaged DNA
    Vaughn Smider, Byung Joon Hwang, and Gilbert Chu 323

24	Construction of MMR Plasmid Substrates and Analysis of MMR Error Correction and Excision
	Huixian Wang and John B. Hays 345
25	Characterization of Enzymes that Initiate Base Excision Repair at Abasic Sites
	Walter A. Deutsch and Vijay Hegde 355
26	Base Excision Repair in Mammalian Cells
	Yoshihiro Matsumoto 365
27	In Vitro Base Excision Repair Assay Using Mammalian Cell Extracts
	Guido Frosina, Enrico Cappelli, Monica Ropolo, Paola Fortini, Barbara Pascucci, and Eugenia Dogliotti 377
28	Biochemical Assays for the Characterization of DNA Helicases
	Robert M. Brosh, Jr. and Sudha Sharma 397
29	Repair Synthesis Assay for Nucleotide Excision Repair Activity Using Fractionated Cell Extracts and UV-Damaged Plasmid DNA
	Maureen Biggerstaff and Richard D. Wood 417
30	Assaying for the Dual Incisions of Nucleotide Excision Repair Using DNA with a Lesion at a Specific Site
	Mahmud K. K. Shivji, Jonathan G. Moggs, Isao Kuraoka, and Richard D. Wood 435
31	Analysis of Proliferating Cell Nuclear Antigen (PCNA) Associated with DNA Excision Repair Sites in Mammalian Cells
	A. Ivana Scovassi and Ennio Prosperi 457
32	Analysis of DNA Repair and Chromatin Assembly In Vitro Using Immobilized Damaged DNA Substrates
	Jill A. Mello, Jonathan G. Moggs, and Geneviève Almouzni 477
Index	 489

Contributors

GERALD M. ADAIR • *Department of Carcinogenesis, Science Park-Research Division, University of Texas M.D. Anderson Cancer Center, Smithville, TX*
GENEVIÈVE ALMOUZNI • *Research Section, Institut Curie, Paris, France*
R. MICHAEL ANSON • *Laboratory of Molecular Gerontology, National Institute on Aging, NIH, Baltimore, MD*
WILLIAM W. AU • *Department of Preventative Medicine and Community Health, The University of Texas Medical Branch, Galveston, TX*
RENATO BATEL • *Laboratory for Marine Molecular Toxicology, Center for Marine Research, Institute Ruder Boscovic, Rovinj, Croatia*
PAULA V. BENNETT • *Biology Department, Brookhaven National Laboratory, Upton, NY*
VALÉRIE BERGOGLIO • *Laboratory of Genetic Instability and Cancer, Institute Gustave Roussy, Villejuif, France*
CHRISTOPHER J. BETTI • *Program in Molecular Biology, Loyola University Chicago Medical Center, Maywood, IL*
MAUREEN BIGGERSTAFF • *Cancer Research UK, Clare Hall Laboratories, London Research Institute, South Mimms, United Kingdom*
VILHELM A. BOHR • *Laboratory of Molecular Gerontology, National Institute on Aging, NIH, Baltimore, MD*
OLEKSANDRA BOREIKO • *Abteilung Angewandte Molekularbiologie, Institut für Physiologische Chemie, Universität, Mainz, Germany*
ROBERT M. BROSH, JR. • *Laboratory of Molecular Gerontology, National Institute on Aging, NIH, Baltimore, MD*
ENRICO CAPPELLI • *Department of Aetiology and Epidemiology, Mutagenesis Laboratory, Istituto Nazionale Ricerca Cancro, Genova, Italy*
NGE CHEONG • *Department of Paediatrics, Clinical Research Centre, National University of Singapore, Singapore*
ODILE CHEVALLIER-LAGENTE • *Laboratory of Genetic Instability and Cancer, Institute Gustave Roussy, Villejuif, France*
GILBERT CHU • *Departments of Medicine and Biochemistry, Stanford University Medical Center, Stanford, CA*
MARCUS S. COOKE • *Genome Instability Group, Department of Cancer Studies and Molecular Medicine, University of Leicester, Leicester, UK*
JAMES J. CORCORAN • *Radiation Oncology Research Laboratory, University of Maryland, Baltimore, MD*

ZBIGNIEW DARZYNKIEWICZ • *Brander Cancer Research Institute, New York Medical College, Valhalla, NY*

WALTER A. DEUTSCH • *Pennington Biomedical Research Center, Louisiana State University, Baton Rouge, LA*

EUGENIA DOGLIOTTI • *Department of Environment and Primary Prevention, Istituto Superiore di Sanitá, Roma, Italy*

PAOLA FORTINI • *Department of Environment and Primary Prevention, Istituto Superiore di Sanitá, Roma, Italy*

MATHILDE FRÉCHET • *Laboratory of Genetic Instability and Cancer, Institute Gustave Roussy, Villejuif, France*

GUIDO FROSINA • *Department of Aetiology and Epidemiology, Mutagenesis Laboratory, Istituto Nazionale Ricerca Cancro, Genova, Italy*

ANDREAS HARTMANN • *Novartis Pharma AG, Basel, Switzerland*

JOHN B. HAYS • *Department of Environmental and Molecular Toxicology, Oregon State University, Corvallis, OR*

VIJAY HEGDE • *Pennington Biomedical Research Center, Louisiana State University, Baton Rouge, LA*

DARYL S. HENDERSON • *Pharmacological Sciences, State University of New York at Stony Brook, Stony Brook, NY*

XUAN HUANG • *Brander Cancer Research Institute, New York Medical College, Valhalla, NY*

BYUNG JOON HWANG • *Division of Biology, California Institute of Technology, Pasadena, CA*

GEORGE ILIAKIS • *Institute of Medical Radiation Biology, Medical School, University of Duisburg-Essen, Essen, Germany*

NICOLAAS G. J. JASPERS • *Department of Genetics, Erasmus Medical Center, Rotterdam, The Netherlands*

SUNG YEON KIM • *Laboratory of Chemical Biology, Department of Pharmacological Sciences, State University of New York at Stony Brook, Stony Brook, NY*

ISAO KURAOKA • *Laboratory for Organismal Biosystems, Graduate School of Frontier Biosciences, Osaka University, Osaka, Japan*

THIERRY MAGNALDO • *Laboratory of Genetic Instability and Cancer, Institute Gustave Roussy, Villejuif, France*

BHASKAR S. MANDAVILLI • *Laboratory of Molecular Genetics, National Institute of Environmental Health Sciences, NIH, Research Triangle Park, NC*

PENELOPE A. MASON • *Laboratory of Molecular Gerontology, National Institute on Aging, NIH, Baltimore, MD*

YOSHIHIRO MATSUMOTO • *Department of Pharmacology, Fox Chase Cancer Center, Philadelphia, PA*

SHIRLEY MCCREADY • *School of Biological and Molecular Sciences, Oxford Brookes University, Oxford, UK*
JILL A. MELLO • *Research Section, Institut Curie, Paris, France*
JOEL N. MEYER • *Laboratory of Molecular Genetics, National Institute of Environmental Health Sciences, NIH, Research Triangle Park, NC*
DAVID L. MITCHELL • *Department of Carcinogenesis, Science Park-Research Division, University of Texas M.D. Anderson Cancer Center, Smithville, TX*
JONATHAN G. MOGGS • *Syngenta Central Toxicology Laboratory, Macclesfield, United Kingdom*
WILLIAM F. MORGAN • *Radiation Oncology Research Laboratory and Greenebaum Cancer Center, University of Maryland, Baltimore, MD*
WERNER E. G. MÜLLER • *Abteilung Angewandte Molekularbiologie, Institut für Physiologische Chemie, Universität, Mainz, Germany*
SHRUTI NAGAR • *Radiation Oncology Research Laboratory, University of Maryland, Baltimore, MD*
RODNEY S. NAIRN • *Department of Carcinogenesis, Science Park-Research Division, University of Texas M.D. Anderson Cancer Center, Smithville, TX*
MASAKI OKAFUJI • *Brander Cancer Research Institute, New York Medical College, Valhalla, NY*
BARBARA PASCUCCI • *Department of Environment and Primary Prevention, Istituto Superiore di Sanitá and Istituto di Cristallografia, CNR, Sezione di Roma, Roma, Italy*
GERD P. PFEIFER • *Division of Biology, Beckman Research Institute of the City of Hope, Duarte, CA*
ENNIO PROSPERI • *Istituto di Genetica Molecolare del CNR, sez. Istochimica e Citometria, Dipartimento di Biologia Animale, Università di Pavia, Pavia, Italy*
ALISTAIR ROBSON • *Dermatopathology, St. Johns Institute of Dermatology, St. Thomas' Hospital, London, UK*
MONICA ROPOLO • *Mutagenesis Laboratory, Department of Aetiology and Epidemiology, Istituto Nazionale Ricerca Cancro, Genova, Italy*
BUSTANUR ROSIDI • *Institute of Medical Radiation Biology, Medical School, University of Duisburg-Essen, Essen, Germany*
SALAMA A. SALAMA • *Department of Obstetrics and Gynecology, The University of Texas Medical Branch, Galveston, TX*
JANINE H. SANTOS • *Laboratory of Molecular Genetics, National Institute of Environmental Health Sciences, NIH, Research Triangle Park, NC*
ALAIN SARASIN • *Laboratory of Genetic Instability and Cancer, Institute Gustave Roussy, Villejuif, France*
HEINZ C. SCHRÖDER • *Abteilung Angewandte Molekularbiologie, Institut für Physiologische Chemie, Universität, Mainz, Germany*

HEIKO SCHWERTNER • *BIOTECmarin GmbH, Mainz, Germany*
A. IVANA SCOVASSI • *Istituto di Genetica Molecolare del CNR, sez. Genetica e Genomica Umana, Pavia, Italy*
SUDHA SHARMA • *Laboratory of Molecular Gerontology, National Institute on Aging, NIH, Baltimore, MD*
SHINYA SHIBUTANI • *Laboratory of Chemical Biology, Department of Pharmacological Sciences, State University of New York at Stony Brook, Stony Brook, NY*
MAHMUD K. K. SHIVJI • *Hutchison/MRC Research Centre, MRC Cancer Cell Unit and University of Cambridge/Cancer Research UK, Cambridge, United Kingdom*
VAUGHN SMIDER • *Integrigen Inc., Novato, CA*
GÜNTER SPEIT • *Abteilung Humangenetik, Universitätsklinikum Ulm, Ulm, Germany*
BETSY M. SUTHERLAND • *Biology Department, Brookhaven National Laboratory, Upton, NY*
JOHN C. SUTHERLAND • *Department of Physics, East Carolina University, Greenville, NC and Biology Department, Brookhaven National Laboratory, Upton, NY*
NAOMI SUZUKI • *Laboratory of Chemical Biology, Department of Pharmacological Sciences, State University of New York at Stony Brook, Stony Brook, NY*
BENNETT VAN HOUTEN • *Laboratory of Molecular Genetics, National Institute of Environmental Health Sciences, NIH, Research Triangle Park, NC*
ANDREW T. M. VAUGHAN • *Department of Radiation Oncology, University of California Davis Medical Center, Sacramento, CA*
MICHAEL J. VILLALOBOS • *Purdue Research Foundation, West Lafayette, IN*
HONG YAN WANG • *Department of Radiation Oncology, Thomas Jefferson University, Philadelphia, PA*
HUICHEN WANG • *Center for Neurovirology and Cancer Biology, College of Science and Technology, Temple University, Philadelphia, PA*
HUIXIAN WANG • *Department of Environmental and Molecular Toxicology, Oregon State University, Corvallis, OR*
MINLI WANG • *Institute of Medical Radiation Biology, Medical School, University of Duisburg-Essen, Essen, Germany*
YA WANG • *Department of Radiation Oncology, Thomas Jefferson University, Philadelphia, PA*
RICHARD D. WOOD • *University of Pittsburgh Cancer Institute, Pittsburgh, PA*
MALGORZATA Z. ZDZIENICKA • *Department of Toxicogenetics, University of Leiden Medical Center, Leiden, The Netherlands*

Technical Notes

1. UV-A, UV-B, and *UV-C*: This terminology, which divides the ultraviolet (UV) light spectrum into three wave bands, was first proposed in 1932 by the American spectroscopist William Coblentz and his colleagues to begin to address the problem of standardizing the measurement of UV radiation used in medicine *(1,2)*. Each spectral band was defined "provisionally" and "approximately" by the absorption characteristics of specific glass filters as follows: UV-A, 400-315 nm; UV-B, 315-280 nm; and UV-C, <280 nm *(1)*. Although based on physical specifications, these definitions were influenced by knowledge of other UV phenomenology, including biological effects and physical properties. For example, wavelengths in the UV-B band were known to have potent erythemic effects, and wavelengths below 290 nm were known to be absent from sunlight reaching the Earth's surface (because they are absorbed by stratospheric ozone). Moreover, the germicidal effects of UV-C wavelengths from artificial UV sources had recently been described, with the greatest killing effect observed at about 266 nm *(3)*. Today, the spectral bands implied by these terms may be found to vary somewhat from Coblentz's original definitions, depending on the scientific discipline. Environmental photobiologists, for example, generally use the following definitions: UV-A, 400-320 nm; UV-B, 320–290 nm; and UV-C, 290–200 nm *(4)*.

2. Relative centrifugal forces: The g-forces listed in this book are calculated for the maximum radius of the rotor unless stated otherwise. For microcentrifuges similar to Eppendorf's 5415 and 5417 models, maximal rotational velocity (approx 14,000 rpm) corresponds to approx 16,000g and approx 25,000g, respectively.

Daryl S. Henderson

References

1. Coblentz, W. W. (1932) The Copenhagen meeting of the Second International Congress on Light. *Science* **76,** 412–415.
2. Coblentz, W. W. (1930) Instruments for measuring ultraviolet radiation and the unit of dosage in ultraviolet therapy. *Br. J. Radiol.* **3,** 354–363.
3. Gates, F. L. (1930) A study of the bactericidal action of ultra violet light. III. The absorption of ultra violet light by bacteria. *J. Gen. Physiol.* **14,** 31–42.
4. Diffey, B. L. (1991) Solar ultraviolet radiation effects on biological systems. *Phys. Med. Biol.* **36,** 299–328.

1

Isolation of Mutagen-Sensitive Chinese Hamster Cell Lines by Replica Plating

Malgorzata Z. Zdzienicka

Summary

Mutant rodent cell lines hypersensitive to DNA-damaging agents have provided a useful tool for the characterization of DNA repair pathways and have contributed to a better understanding of the mechanisms involved in the cellular responses to mutagenic treatment. Here we present a detailed description of how to isolate mutagen-sensitive mutants from hamster "wild-type" cell lines. First, cells are treated with ethyl nitrosourea, and then the mutagenized cell populations are screened for cells with an increased sensitivity to various mutagens using a replica-plating method. Mutagen-sensitive clones are identified and then characterized by assessing their stability, degree of sensitivity to various mutagens, and by genetic complementation analysis.

Key Words: EMS; ENU; hamster cell mutants; mutagenesis; mutant isolation; mutagen sensitivity; UV radiation; X-rays.

1. Introduction

Cell lines with an increased sensitivity to mutagens such as ultraviolet (UV) light, X-rays, alkylating compounds, and crosslinking agents are defective in a cellular response to these agents. These responses include mechanisms that process DNA lesions, scavenge free radicals, or regulate cell cycle progression. The molecular defects in cell lines derived from patients with inherited recessive disorders that combine cancer proneness with an abnormal response to DNA-damaging agents, such as xeroderma pigmentosum, ataxia telangiectasia, Nijmegen breakage syndrome, and Fanconi anemia, have been studied extensively. However, such human diseases do not identify all possible cellular responses to mutagenic treatments, because only those defects that are manifested at the clinical level, and not lethal in vivo, can be detected.

From: *Methods in Molecular Biology: DNA Repair Protocols: Mammalian Systems, Second Edition*
Edited by: D. S. Henderson © Humana Press Inc., Totowa, NJ

Therefore, in addition, many mutagen-sensitive mutants have been obtained in rodent cell lines. The availability of such mutants is essential to identify the genes involved, their products and functions, as well as to assess the biological consequences of their impact. It has also become evident that rodent cell mutants defective in DNA repair provide an important tool for the isolation of human genes complementing the defect in these mutants *(1)*. Therefore, to dissect the cellular response to a specific mutagen it is essential to have a comprehensive set of mutants showing increased sensitivity to the agent. In addition to human mutants, such rodent cell mutants were identified in several laboratories *(2)*.

To induce mutants, "wild-type" cells are treated with a strong mutagen, such as ethyl nitrosourea (ENU) or ethyl methanesulfonate (EMS), following which the mutagenized cell population is screened to identify clones having an increased sensitivity to one or more DNA-damaging agents. To test large numbers of clones, the replica-plating technique has been used successfully in our laboratory *(3)*, and many mutants with increased sensitivity to mutagenic agents have been isolated in Chinese hamster (CHO or V79) cell lines *(2–13)*.

The replica-plating technique is a standard method used in microbial genetics where many clones are screened. Cells from the mutagenized population are plated in plastic tissue-culture dishes. When the single cell-derived clones are visible to the eye they are transferred to 96-well master plates and cultured for several days. Cells from the master plate are then used to make replicas of the 96 lines, whereupon each replica is treated with a mutagen at a dose that is only marginally toxic to wild-type cells (giving approx 90% survival). Sensitive clones are identified by comparing the control replica with the mutagen-treated replica, because sensitive clones show growth retardation. The main advantage of this technique is that it enables cells derived from a single colony to be screened for hypersensitivity to a number of different mutagenic agents. Once such mutants are identified, they should be examined for their sensitivity to additional mutagenic agents, and characterized to determine whether they represent new complementation groups.

2. Materials

1. Cells: CHO or V79 cells (*see* **Note 1**).
2. Ham's F10 medium: modified by omission of hypoxanthine and thymidine; supplemented with 10% fetal calf serum, 100 U/mL of penicillin and 0.1 mg/mL of streptomycin. This is referred to as "standard" medium (*see* **Note 2**).
3. Standard medium supplemented with 20 mM N-2-hydroxyethylpiperazine-N'-2-ethanesulfonic acid (HEPES), pH 7.4.
4. Phosphate-buffered saline (PBS), pH 7.4.
5. Trypsin solution: 0.25% trypsin, 0.02 % EDTA in PBS.

6. Cryo Tubes, 1.8 mL (Nunc).
7. Tissue-culture dishes: 10- and 15-cm (e.g., Greiner).
8. 96-Well microtiter plates with flat bottoms (Costar).
9. Multichannel pipetter for dispensing liquids (20–200 µL) into 96-well plates.
10. Transtar-96 portable liquid handling system; 96-tip sterile, disposable cartridges (Costar). (*See* **Note 3**.)
11. 0.9% NaCl solution.
12. 0.2% Methylene blue (Sigma, St. Louis, MO) solution: Dissolve 2 g of methylene blue in a few milliliters of ethanol, and then add water to 1 L.
13. 0.4 M ENU (Pfaltz & Bauer, Waterbury, CT) freshly prepared in dimethyl sulfoxide (DMSO).
14. Ethyl methane sulfonate (EMS) (Kodak, Tramedico BV, Weesp, Holland), freshly dissolved in PBS (1% v/v; 94 mM). Vortex-mix to dissolve.
15. Methyl methanesulfonate (MMS) (Merck): Prepare as for EMS (1% v/v; 118 mM).
16. Mitomycin C (MMC) (Kyowa Hakko, Kogyo Co. Ltd. Tokyo, Japan), stock solution (4 mg/mL); Dissolve in sterile H_2O (or PBS). Keep at 0–4°C.
17. Bleomycin (BLM) (Lundbeck BV, Amsterdam, Holland), freshly dissolved in PBS.

3. Methods

Culture cells in 10- or 15-cm plastic dishes containing 10 mL or 25 mL of standard medium, respectively, at 37°C, in an atmosphere of 5% CO_2 in a humidified incubator.

3.1. ENU Mutagenesis

1. Treat a suspension of >10^7 cells with freshly prepared ENU (4 mM final concentration) in 10 mL of prewarmed medium supplemented with 20 mM HEPES. Incubate at 37°C for 1 h. This should give a surviving fraction of approx 1% (*see* **step 3**).
2. Collect the cells by low-speed centrifugation (135g), wash twice with PBS, and resuspend in 10 mL of standard medium.
3. To assess the level of killing by ENU, seed 100–1000 treated cells in 10-cm dishes. At the same time, seed 100 cells from the untreated population to serve as a control. Incubate the dishes (in duplicate) for 8–10 d. Rinse with 0.9% NaCl solution, air-dry, and stain with methylene blue solution. Count the visible colonies.
4. Seed the remaining treated cells in two or three 15-cm dishes. After 4 d of incubation, trypsinize the cells, and collect by low-speed centrifugation. Set aside approx 5×10^5 cells to assess the degree of mutagensis as described in **Subheading 3.2**. To the rest of the cells add DMSO to a final concentration of 6%. Aliquot 10^6 cells/Cryo tube, and transfer the tubes to a –100°C freezer. After 1 d, the tubes can be transferred to liquid nitrogen for long-term storage. Each plate will yield about 7–10 tubes.

3.2. Estimation of the Level of Mutagenesis

To evaluate the degree of mutagenesis, the frequency of induced $hprt^-$ mutants in the mutagenized population is determined by incubating the ENU-treated cells in standard medium containing 6-thioguanine (TG) (*see* **Note 4**).

1. Incubate 5×10^5 cells from **step 4** of **Subheading 3.1.** for an additional 4 d (8 d in total).
2. Trypsinize the cells, and collect by low-speed centrifugation.
3. Seed 10^5 cells in each of three to five 10-cm dishes containing medium supplemented with 7 µg/mL of TG. Seed 200 cells in each of three to five dishes containing standard medium (without TG) to determine the number of viable cells.
4. After 10 d of incubation, rinse the dishes with NaCl solution, air-dry, and stain with methylene blue. Count visible colonies and calculate the frequency of TG-resistant ($hprt^-$) mutants. There should be at least 1 $hprt^-$ mutant/10^3 viable cells.

3.3. Screening for Mutants by Replica Plating

1. For each mutant isolation, dilute an ampoule of frozen cells (from **step 4** of **Subheading 3.1.**) in standard medium and seed to obtain about 20–50 single cell-derived clones/ 10-cm dish.
2. After 8–12 d of incubation, remove the medium, pick up cells from each colony with the flat end of a sterile wooden toothpick and transfer them to a 96-well microtiter plate whose wells contain 25 µL of the trypsin solution. Trypsinize the cells for 3 min at 37° C, and then add 200 µL of standard medium (*see* **Note 5**). Return the plates to the incubator.
3. After about 1 wk, remove the medium, wash the cells with PBS, add trypsin solution for 3 min at 37°C (*see* **Note 5**), and add 200 µL of standard medium. Mix the cells by shaking the master plate.
4. Using a Transtar-96 pipetter, transfer cells from the master plate to three to six replica plates containing standard medium (*see* **Note 6**). One plate will serve as a control for retrieving mutants, a second control plate will be stained with methylene blue, and the remaining replica plates will be treated with mutagens.
5. Irradiate or treat with a chemical mutagen at the following doses, which allow 80–100% survival of wild-type cells:
    a. 254 nm UV light (6 J/m^2): 4 h after transfer, when the cells are attached, remove the medium, wash the cells with PBS, irradiate, and then add back standard medium.
    b. X-rays (2–3 Gy): Irradiate the cells in medium without waiting for attachment. There is no need to change the medium.
    c. MMC (5 ng/mL): At **step 4**, transfer the cells to standard medium already containing MMC. Cells receive a continuous treatment.
    d. EMS (2 mM): Treat as for MMC.
    e. MMS (0.3 mM): Treat as for MMC.
6. After more than 1 wk of growth, rinse all of the plates except one of the controls with NaCl solution, air-dry, and stain the cells with methylene blue.

7. Compare the stained control plate with the plates treated with mutagens. Those wells containing no cells, or a greatly reduced number of cells, in the treated plates compared to the control plate are considered putative mutants.
8. Pick the putative mutant cells from the untreated control replica plate, grow them, and retest for sensitivity.
9. Reclone the sensitive mutant cells by seeding approx 20 cells/10-cm dish (*see* **Note 7**). Isolate three subclones after 10 d, and test the sensitivity of these subclones.

3.4. Survival Experiment to Assess the Degree of Mutagen Sensitivity

1. Trypsinize the mutant cells in exponential growth, and plate approx 300 cells into 10-cm dishes in triplicate.
2. After 4 h, when cells are attached, treat the cells with a mutagen over a range of doses. Use as a guide the doses given in **step 5** of **Subheading 3.3.** For UV irradiation, remove the medium, wash the cells with PBS, irradiate, and add back standard medium. For X-rays, irradiate the cells in the standard medium without changing it. After chemical treatment, wash the cells twice with PBS, and add fresh medium. Incubate the cells for 8–10 d to obtain colonies visible by eye.
3. Rinse the dishes with NaCl solution, air-dry, stain with methylene blue, and count visible colonies.

3.5. Characterization of Mutants

1. To determine the stability of the isolated mutant, culture the mutant cells for 2–3 mo. After 1, 2, and 3 mo, re-examine the sensitivity of the mutant over a range of doses. A tailing curve may indicate the presence of revertants in the cell culture. In this case reclone and retest the mutant.
2. To assess the degree of sensitivity of the mutant to other mutagenic agents, perform cell-survival studies after treatment with mutagens from at least four different classes of DNA damaging agents:
   a. UV.
   b. X-rays, bleomycin.
   c. Monofunctional alkylating agents, such as, MMS, EMS, and ENU.
   d. Bifunctional alkylating (crosslinking) agents, such as MMC and *cis*-dichlorodiammine platinum (II) (cisplatin; *cis*-DDP). Compare the D_{10} values (i.e., the dose required to kill 90% of the cells) of the parental and mutant cells (*see* **Note 8**).
3. To determine whether the isolated mutant represents a new complementation group, perform genetic complementation analyses with representative mutants of different extant complementation groups as described in **refs. 6–8**.

4. Notes

1. Cell lines used for the isolation of mutants must be pseudodiploid. They cannot be strictly diploid, since the probability of inducing mutations in two wild-type

alleles is extremely low. Therefore, mutants can only be induced in cells that are functionally or structurally hemizygous. Different hamster cell lines, even of the same origin but growing for several years in different laboratories, may show different hemizygosities, thus allowing mutants defective in different genes to be isolated. To obtain new complementation groups of mutants, use cell lines that have not been used extensively for the isolation of such mutants. We have found that mutagen-sensitive cells were obtained with 10-fold higher frequency from V79 than CHO9 cells. Most probably this is owing to the different extent of hemizygosity in these two "wild-type" cell lines. V79 cell lines cultured for a long time in different laboratories have different hemizygosities.
2. Other types of culture media may be used. However, the cells should first be adapted to the medium. Medium lacking hypoxanthine and thymidine is required to select for 6-thioguanine-resistant mutants (*see* **Subheading 3.2.**).
3. When a Transtar-96 multipipetter is not available, use the sterile cartridges as "stamps" to transfer cells.
4. *hprt* is an X-linked gene encoding a nonessential purine salvage pathway enzyme, hypoxanthine phosphoribosyl transferase (HPRT). HPRT metabolizes TG to a cytotoxic nucleotide. *See* **ref. *14*** for a detailed discussion of *hprt* mutagenesis.
5. When preparing the master plate, transfer no more than 12–24 clones to a part of the 96-well plate, trypsinize them, and add standard medium before transferring the next 12–24 clones. Cells will be killed if kept in trypsin for too long. Do the transfers quickly and do not have too many clones in trypsin at any one time.
6. An alternative method of transferring cells from the master plate to the replica plate employs Cytodex-1 microcarrier beads (Amersham Biosciences) *(15)*. However, this method cannot be used to screen for UV-sensitive clones because of shielding from the beads. When cells in the replica plate form colonies, add medium containing 600 beads, which gives a monolayer of beads covering the bottom of each well. Incubate for 2 additional days to allow the cells to grow onto the microcarrier beads. Make replicas of each plate by transferring medium containing the suspended beads with attached cells.
7. All mutagen-sensitive lines should be recloned to avoid possible contamination with other cells.
8. For further studies, use cells with a significantly increased sensitivity to the given agent. To date, the isolated mutants show a 2- to 10-fold increased sensitivity toward X-ray or UV radiation, and more than 10-fold to MMC.

References

1. Thompson, L. H. (1999) Strategies for cloning mammalian DNA repair genes. *Methods Mol. Biol.* **113,** 57–85.
2. Collins, A. R. (1993) Mutant rodent cell lines sensitive to ultraviolet light, ionising radiation and cross-linking agents: a comprehensive survey of genetic and biochemical characteristics. *Mutat. Res.* **293,** 99–118.
3. Zdzienicka, M. Z. and Simons, J. W. I. M. (1987) Mutagen-sensitive cell lines are obtained with a high frequency in V79 Chinese hamster cells. *Mutat. Res.* 178, 235–244.

4. Zdzienicka, M. Z. (1996) Mammalian X-ray-sensitive mutants: a tool for the elucidation of the cellular response to ionizing radiation, in *Genetic Instability in Cancer*, (Lindahl, T., ed.), Cold Spring Harbor Laboratory Press, Plainview, NY, pp. 281–293.
5. Thacker, J. and Zdzienicka, M. Z. (2003) The mammalian XRCC genes: their roles in DNA repair and genetic stability. *DNA Repair* **2**, 665–672.
6. Zdzienicka, M. Z., van der Schans, G. P., Natarajan, A. T. Thompson, L. H., Neuteboom, I., and Simons, J. W. I. M. (1992) A CHO mutant (EM-C11) with sensitivity to simple alkylating agents and a very high level of sister chromatid exchanges. *Mutagenesis* **7**, 265–269.
7. Zdzienicka, M. Z., Tran, Q., van der Schans, G. P., and J. W. I. M. Simons. (1998) Characterization of an X-ray-hypersensitive mutant of V79 Chinese hamster cells. *Mutat. Res.* **194**, 239–249.
8. Telleman, P., Overkamp, W. J. I., van Wessel, N., et al. (1995) A new complementation group of mitomycin C-hypersensitive Chinese hamster cell mutants that closely resembles the phenotype of Fanconi anemia cells. *Cancer Res.* **55**, 3412–3416.
9. Errami, A., He, D. M., Friedl, A. A., et al. (1998) XR-C1, a new CHO cell mutant which is defective in DNA-PKcs, is impaired in both V(D)J coding and signal joint formation. *Nucleic Acids Res.* **26**, 3146–3153.
10. Shen, M. R., Zdzienicka, M. Z., Mohrenweiser, H., Thompson, L. H., and Thelen, M. P. (1998) Mutations in hamster single-strand break repair gene XRCC1 causing defective DNA repair. *Nucleic Acids Res.* **26**, 1032–1037.
11. Errami, A., Overkamp, W. J. I., He, D. M., et al. (2000) A new X-ray sensitive CHO cell mutant of ionizing radiation group 7, XR-C2, that is defective in DSB repair but has only a mild defect in V(D)J recombination. *Mutat. Res.* **461**, 59–69.
12. Kraakman-van der Zwet, M., Overkamp, W. J. I., van Lange, R. E. E., et al. (2002) Brca2 (XRCC11) deficiency results in radioresistant DNA synthesis and a higher frequency of spontaneous deletions. *Mol. Cell. Biol.* **22**, 669–679.
13. Godthelp, B. C., Wiegant, W. W., van Duijn-Goedhart, et al. (2002) Mammalian Rad51C contributes to DNA cross-link resistance, sister chromatid cohesion and genomic stability. *Nucleic Acids Res.* **30**, 2172–2182.
14. McCormick, J. J. and Maher, V. M. (1988) Measurement of colony-forming ability and mutagenesis in diploid human cells, in *DNA Repair: A Laboratory Manual of Research Procedures*, Vol. 1B (Friedberg, E. C. and Hanawalt, P. C., eds.) Marcel Dekker, New York, NY, pp. 501–521.
15. Stackhouse, M. A. and Bedford, J. S. (1993) An ionizing radiation-sensitive mutant of CHO cells: irs20. I. Isolation and characterization. *Radiat. Res.* **136**, 241–249.

2

Complementation Assays Adapted for DNA Repair–Deficient Keratinocytes

Mathilde Fréchet, Valérie Bergoglio,
Odile Chevallier-Lagente, Alain Sarasin, and Thierry Magnaldo

Summary

Genetic alterations affecting nucleotide excision repair, the most versatile DNA-repair mechanism responsible for removal of bulky DNA adducts including ultraviolet (UV) light-induced DNA lesions, may result in the rare, recessively inherited autosomal syndromes xeroderma pigmentosum (XP), Cockayne syndrome (CS), or trichothiodystrophy (TTD). Classical approaches such as somatic cell fusions or microinjection assays have formalized the genetic complexity of these related but clinically distinct syndromes, and contributed to the determination of seven, five, and three complementation groups for XP, CS, and TTD, respectively. XP patients are highly susceptible to photoinduced cutaneous cancers of epidermal origin. To better study the responses to UV irradiation of XP keratinocytes, and to objectively determine the extent to which cutaneous gene therapy may be realized, we set up experimental procedures adapted to ex vivo genetic complementation of keratinocytes from XP patients. We provide here detailed rationales and procedures for these approaches.

Key Words: Complementation; DNA repair; gene therapy, keratinocytes; ultraviolet radiation; xeroderma pigmentosum.

1. Introduction
1.1. Clinical Aspects and Rationale

Xeroderma pigmentosum (XP), Cockayne syndrome (CS), and trichothiodystrophy (TTD) are rare human disorders inherited as autosomal recessive traits with an estimated frequency of 1 per 250,000 and 1 per 40,000 newborns in North America/Europe and Japan/North Africa, respectively *(1)*. All XP and some CS and TTD patients exhibit some degree of photosensitivity, and addi-

From: *Methods in Molecular Biology: DNA Repair Protocols: Mammalian Systems, Second Edition*
Edited by: D. S. Henderson © Humana Press Inc., Totowa, NJ

tional clinical hallmarks are distinctive to each syndrome. Most dramatically, XP patients, but not CS or TTD patients, are highly susceptible to skin cancers on areas of the body exposed to sunlight. These diseases are life threatening and even for those XP forms exhibiting the simpler phenotype (restricted to skin) no curative treatment is available. For these reasons, XP and more specifically certain complementation groups of XP patients (*see* **Subheadings 1.7. and 1.8.**), are good candidates for cutaneous gene therapy. Together with the general purpose of assigning a patient to a specific complementation group, the focus of this chapter is the application of genetic complementation to epidermal keratinocytes. These skin cells are the primary targets of ultraviolet (UV) irradiation, leading to their mutagenic transformation and the development of basal and squamous cell carcinomas, which constitute about 30% of human cancers.

1.2. Molecular Aspects

XP and photosensitive CS and TTD patients exhibit impaired capacity in the most versatile DNA repair mechanism, nucleotide excision repair (NER) *(2)*. Following UV irradiation, residual NER capacity of a patient cell line can thus be determined by quantifying the extent of unscheduled DNA synthesis (UDS), as measured by incorporation of [^{3}H]thymidine into the DNA (**Fig. 1**). Somatic cell fusions followed by UDS have contributed to the definition of complementation groups into which each patient falls. Seven complementation groups called XP-A to XP-G have been identified for classical XP, five in the case of CS (CS-A, CS-B, XP-B, XP-D, XP-G), and three in the case of TTD (XP-B, XP-D and TTD-A). These findings have pointed out some genetic overlap between these syndromes since alterations to the same gene (e.g., *XPD*) can give rise to XP (XP-D), CS (CS/XP-D), or TTD (TTD/XP-D) *(3)*.

1.3. Significance of Complementation Group Determination

Determination of complementation group is obviously essential for refining clinical diagnosis. It may also aid in genetic counselling and perhaps eventually lead to improvements in patient treatment. In addition, determination of complementation group may contribute to:

1. Knowledge of genotype–phenotype relationships (which specific mutations result in which disease?).
2. Dissection of the molecular events underlying expression of the phenotypic traits characteristic of each disease (what function[s] of the mutant protein is impaired?).
3. Elaboration of targeted pharmacological treatments.
4. Furthering prospects for gene therapy.

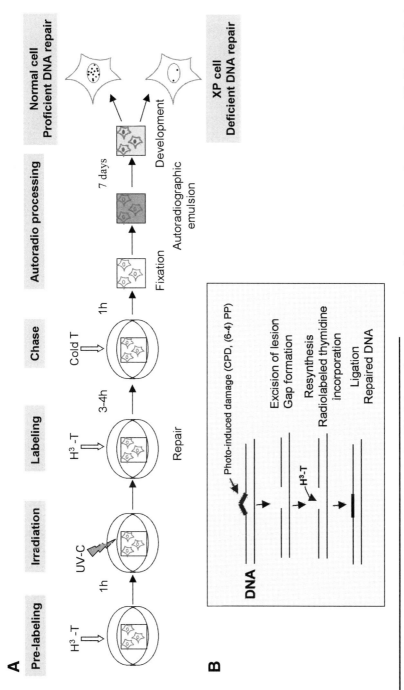

Fig. 1. Unscheduled DNA Synthesis (UDS). (**A**) Schematic representation of the experimental procedure. (**B**) Simplified view of molecular steps occurring upon UDS. *See* text and **ref. 4**.

1.4. Historical Methods for Determining the Complementation Group of NER-Deficient Cells

Molecular cloning of DNA repair genes involved in NER has led to considerable improvement of classical methods for complementation group determination. Reintroduction of the appropriate cloned DNA repair gene to a patient's cells using conventional expression vectors can be performed by either microinjection or classical transfection followed by assessment of repair capacity after UV irradiation and UDS. Microinjection, however, requires specific material and technical skills, and efficiency of transfection in primary cells remains quite poor. A fourth method, called host–cell reactivation, involves the cotransfection of a DNA repair gene together with a reporter vector (luciferase, chloramphenicol acetyltransferase) previously inactivated by exposure to UV light. In each case, the presence of the appropriate DNA repair gene will restore normal levels of either UDS or reporter gene activity. These methods are helpful for determining the complementation group of fibroblasts or transformed cells and have been described in detail previously *(4)*.

1.5. Specifications for Keratinocyte Complementation

The epidermis constitutes the external skin compartment. It is mainly composed of keratinocytes organized in stratified cell layers distinguishable by their proliferation and differentiation capacity. Some keratinocytes forming the basal (innermost) layer are stem cells *(5)*. In vivo, the presence of stem cells and transient amplifying cells (i.e., rapidly proliferating cells exhibiting a limited [<15 cell divisions] lifespan) sustains permanent epidermal replenishment according to a stepwise program of differentiation initiated in the basal layer and completed with the formation of dead, outermost keratinocytes that form horny layers. Under appropriate procedures, primary epidermal keratinocytes can be cultured in vitro from a very small (generally 2–5 mm diameter) punch skin biopsy (*see* **Subheading 3.**). Adequate culture conditions allow the maintenance of stem cells and the preservation of their growth and differentiation potential. Under these circumstances, it is possible to: (1) genetically (and phenotypically) modify stem cells and transient amplifying cells ex vivo; and (2) reconstruct skin *in vitro* from these cells *(6,7)* or regenerate genetically modified skin in vivo, for instance, after engraftment of epithelial sheets onto an athymic mouse *(8)*. These specifications depend on strict cell culture conditions as described in **Subheading 3.** Further refinements concerning control of expression (duration, level, regulation) of the corrective gene and the topology (i.e., stratum specificity) of its expression within the epidermal compartment then relies on the type and design of the vector utilized. Ideally, the gene of interest should be placed under transcriptional control of its own DNA regulatory elements, but many difficulties related to cargo capacity of the vectors and their specific efficiency of gene transfer into host DNA constitute limiting parameters.

1.6. New Complementation Methods Adapted to Keratinocyte Complementation

Besides classical retrovirus-based gene transfer, methods of nonviral transfer have been developed. Nonviral gene transfer methodologies such as direct injection or electroporation of naked DNA or topical gene delivery using a liposomal–DNA mixture have the advantage of being simple and easy to perform. However, they have the disadvantages of poor efficiency of epidermal gene delivery and generally drive only transitory expression of the transgene *(9–11)*. Integration of large DNA sequences is now possible using the C31 bacteriophage integrase-based gene transfer, but the quantity of cells showing stable, long-term expression of the transgene remains lower than with retroviral vectors *(12)*.

Virus-based methodologies rely on the use of adenoviral vectors, adenovirus-associated vectors, retrovirus and lentivirus (**Table 1**). Adenoviral vectors remain episomal leading to short-term expression of the gene of interest, because the viral construct is lost in proliferating cells. Adenovirus-associated vectors are able to integrate into genomic DNA but exhibit a rather small carrying capacity, which can be limiting for the transfer of large genes. Recombinant retroviral vectors (RRV) are highly efficient at gene transfer and gene integration, making them the most utilized vehicle in gene therapy approaches. One obstacle to transduction by retroviral vectors, however, is the requirement for replication of target cells at the time of infection *(13)*. Consequently, only ex vivo gene therapy can be realized using RRV. In other respects, lentiviral vectors have demonstrated efficient delivery and integration and stable expression of genes in nondividing cells. Consequently ex vivo as well as in vivo gene transfer can be envisaged with lentiviral vectors.

1.7. Retrovirus-Mediated Transduction of DNA Repair Genes into Keratinocytes

Among the vectors briefly described above, as a first attempt we used the backbone of a retroviral vector derived from Moloney murine leukemia virus (LXSN) encoding the bacterial *neomycin phosphotransferase* selection gene *neo*. As described previously, cDNAs of repair genes were inserted downstream of the 5′ long terminal repeat (LTR) of the proviral DNA which contains regulatory elements of transcription *(14)*. Production of infectious retroviral particles was performed using the packaging Crip cell line, a derivative of NIH-3T3 cells genetically engineered for the production of GAG, POL, and ENV retroviral proteins that are necessary for the production of infectious retroviral particles *(4)*.

Table 1
Comparison of Properties of Viral Vectors Used in Gene Transfer Approaches in Keratinocytes

	Vector capacity	Viral titer (cfu/mL)	Efficiency of transfer	Genomic integration	Long term expression in vivo	Disadvantages	Advantages	References
Adenoviral vector (AV)	>30 kb	10^{11}–10^{13}	Transduction of replicating and non-replicating cells.	No	No	Transient expression.	High cargo capacity.	*17, 18*
Adeno-associated viral vector (AAV)	5 kb	~ 10^{10}	Transduction of nonreplicating cells possible but transduction of replicating cells is more efficient.	Yes, when not coinfected with helper virus, at one specific site in human cells (19q13-qter).	Possible	Small cargo capacity.	Possibility of specific integration.	*19, 20*
Retroviral vector (RV)	10 kb	~ 10^6–10^7	Transduction of replicating cells uniquely.	Yes	>40 wk	Random integration: insertional mutation. Influence of LTR on the expression of the transgene.	High efficiency of transduction. Long term expression.	*13, 21–23*
Lentiviral vector (LV)	10 kb	>10^9	Transduction of replicating and non-replicating cells.	Yes	> 6 mo	Random integration: insertional mutation.	High efficiency of transduction. Long term expression. Adapted to in vivo transduction.	*24, 25*

Psi-Crip cells are transfected with the recombinant proviral DNA encoding the DNA repair gene of interest together with the *neo* selectable marker. Neomycin selection of transfected cells permits the isolation of clones exhibiting a high titer of infection ($\geq 10^6$ plaque-forming units per milliliter) of target cells. Alternatively, infectious retroviral particles can be produced by transient cotransfection of human embryonic cell line 293 using, on the one hand, two helper vectors, one encoding GAG and POL proteins and the other encoding ENV protein, and on the other hand, the proviral DNA encoding the DNA-repair gene *(15)*. Collected retroviral particles can then be used to transfer the DNA repair gene into patient cells. After G418 selection of transduced cells, production and function of the DNA repair protein is checked by challenging the transduced cells with UV light. Examples of recovery of normal levels of cell survival following UV-irradiation and UDS in *XPC*-transduced XP-C primary keratinocytes are shown in **Fig. 2**.

1.8. Contribution of Keratinocyte Complementation to Cutaneous Gene Therapy of XP

Except for strict sun avoidance and surgical resection of epidermal tumors, no effective curative treatment is available to XP patients. At least for some complementation groups of XP (in particular XP-C), cutaneous gene therapy may thus provide a promising approach for improving the quality of patient life. In this respect, it is necessary to ensure long term expression of the correcting gene and absence of immunogenicity of both the DNA repair protein and selection marker. Using a novel generation of RRV, our preclinical assays based on regeneration of genetically corrected skin onto laboratory animals are currently addressing some of these questions.

2. Materials
2.1. Cell Culture

1. Keratinocyte culture medium, CFAD: three volumes of Dulbecco's minimal essential medium (DMEM) (Gibco/Invitrogen), one volume of F10 (Ham's) (Gibco), 5 µg/mL of insulin (Sigma), 2×10^{-9} M triiodothyronin (T3, Sigma), 0.4 µg/mL of hydrocortisone (Calbiochem), 10^{-10} M cholera toxin (ICN), 1.8×10^{-5} M adenine (Sigma), 5 µg/mL of transferrin (Sigma), 1% nonessential amino acids (Gibco), 1% sodium pyruvate (Gibco), 1% penicillin–streptomycin (Gibco), 1% glutamine (Gibco), 0.2% fungizone (Gibco), 10% fetal bovine serum. CFAD supplemented with 10 ng/mL of epidermal growth factor (EGF; Chemicon).
2. Swiss 3T3 J2 mouse fibroblasts were a kind gift of Dr. James Rheinwald (Brigham and Women's Hospital, Harvard Institutes of Medicine, Boston, MA).
3. 3T3 J2 and Psi-Crip culture medium: DMEM (Gibco), 1% penicillin–streptomycin (Gibco), 0.2% fungizone (Gibco), 10% bovine calf serum (HyClone).
4. Phosphate-buffered saline (PBS), pH 7.4: 137 mM NaCl, 2.7 mM KCl, 4,3 mM Na$_2$HPO$_4$ (Gibco).

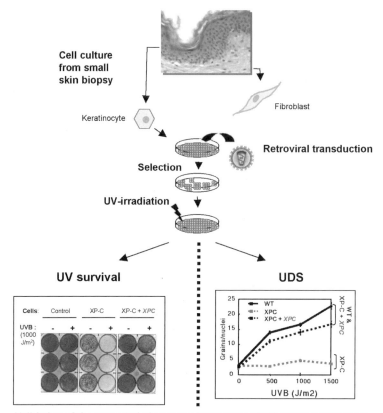
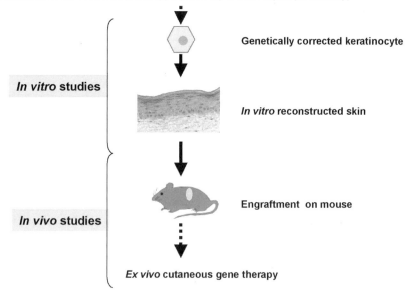

5. 0.1% Trypsin solution: 1X PBS, 0.1% glucose, 0.1% trypsin, 0.0015% phenol red, 1% penicillin–streptomycin (Gibco). Adjust the pH to 7.45 with 2 N NaOH, filter the solution and store at –20°C.
6. 0.02% EDTA in 1X PBS.
7. 35-, 60- and 100-mm tissue-culture dishes, 25-, 75-, and 150-cm^2 tissue-culture flasks (T25, 75, and 150).
8. 37°C, 10% CO_2 humidified incubators.
9. Sterile 15-mL polypropylene centrifuge tubes (Falcon).

2.2. Transduction of Cells

1. Psi-Crip cells producing the retrovirus of interest.
2. Psi-Crip culture medium (*see* **item 3**, **Subheading 2.1.**).
3. SFM-keratinocyte medium, containing 0.2 ng/mL of EGF, 0.25 µg/mL of bovine pituary extract (BPE; Gibco), 0.1 mM $CaCl_2$.
4. 10 mg/mL of polybrene (Sigma).
5. 0.45-µm filter.
6. Neomycin (G418) at 50 mg/mL (Gibco).

2.3. Measurement of DNA Repair

2.3.1. UDS

1. UV-B radiation source: Spectroline transilluminator, model TR-312 (Spectronics Corporation, Westbury, NY), equipped with a cutoff filter, WG 305 (Schott). Fluence is measured using a VLX312 radiophotometer (Vilbert-Lourma). A UV-C radiation source (254 nm) can also be used. (*See* **Note 1**.)
2. Fetal bovine serum dialyzed with PBS (DFBS) and sterilized by filtration.
3. F10 (Ham's) medium (Gibco).
4. [^{3}H]thymidine (SA of 50 Ci/mmol).
5. 1 M Unlabeled thymidine.
6. 10 µM Fluorodeoxyuridine.
7. 5% Trichloroacetic acid (TCA) diluted in water, kept at 4°C.
8. 100% Ethanol.

Fig. 2. *(Opposite)* From biopsy to cutaneous gene therapy. Fibroblasts and keratinocytes are isolated from small skin biopsies taken from patients suspected of having a DNA repair deficiency. Keratinocytes are transduced with retroviruses expressing a wild-type copy of the DNA repair gene (e.g., *XPC* or *XPD*) together with the selection gene, such as the G418 resistance gene *neo*. Cells are then selected using G418. Phenotypic correction is checked by both cell survival and UDS assays following UV irradiation. Only cells transduced by the appropriate DNA repair gene exhibit the corrected phenotype in both assays. In vitro studies can be done and it is possible to reconstruct skin in vitro using genetically corrected keratinocytes. Engraftment of genetically corrected skin onto a laboratory mouse constitutes a validation step for the principle and methodologies of cutaneous gene therapy.

9. PBS (*see* **item 4**, **Subheading 2.1.**).
10. Glass microscope slides (2.5 × 7.5 cm^2) and sterile glass coverslips (2 × 2 cm^2).
11. Mounting medium for microscopic preparations (Eukitt).
12. Kodak emulsion NTB2, Kodak D19 developer, Kodak 3000 fixer.
13. Mayer's hematoxylin staining solution (Merck).

2.3.2. Clonal Survival Following UV Irradiation

1. UV-B irradiation device and WG 305 Schott cutoff filter; VLX312 UV-B light radiophotometer (*see* **item 1**, **Subheading 2.3.1.**).
2. 60- and 100-mm tissue-culture dishes.
3. PBS (*see* **item 4**, **Subheading 2.1.**).
4. 0.02% EDTA in 1X PBS.
5. Trypsin (*see* **item 5**, **Subheading 2.1.**).
6. Rhodamine blue staining solution: one volume of 1% rhodamine and one volume of 1% Nile blue.
7. 3.7% Formaldehyde in PBS.

2.3.3. Mass Culture Survival Following UV Irradiation

1. Six-well Multidishes.
2. UV-B lamp at 312 nm and corresponding UV light dosimeter (*see* **item 1**, **Subheading 2.3.1.**).
3. Rhodamine blue staining solution (*see* **item 6**, **Subheading 2.3.2.**).
4. 3.7% Formaldehyde in PBS.
5. Fluor-S Max MultiImager (Bio-Rad).
6. Quantity One software (Bio-Rad).

3. Methods
3.1. Isolation of Keratinocytes from Skin Biopsy

1. Seed 3T3 J2 cells at 1500 cells/ cm^2 in a T150 flask (*see* **Notes 2** and **3**).
2. Two days later, 3T3 J2 cells are trypsinized and γ-irradiated at 45 Gy in CFAD medium (*see* **Note 4**). Seed the γ-irradiated 3T3 J2 cells in 35-mm plates at 15,000 cells/cm^2 in CFAD medium and incubate at 37°C, 10% CO$_2$ from 2 h (time necessary for the cells to adhere to tissue culture plastic) to 24 h.
3. Obtain skin biopsies from nonexposed (to sunlight) and normally pigmented sites from consenting patients. Disinfect the biopsy area with 0.1% chlorexidine, obtain the biopsy material and immediately place it in DMEM medium, 10% FCS, 1% penicillin–streptomycin.
4. Wash the biopsy three times with 25 mL of PBS for 3 min.
5. Place the biopsy in 0.1% trypsin solution and mince the tissue into 1-mm^2 fragments with a scalpel. Transfer to a 12.5-cm^2 flask.
6. Repeat **step 5** until the volume of trypsin reaches 1 mL.
7. Add 1 mL of 0.02% EDTA solution, resuspend and transfer the remaining tissue to the 12.5-cm^2 flask.
8. Incubate for 2 h at 37°C

Keratinocyte Gene Correction

9. Add 8 mL of CFAD to the 12.5-cm^2 flask, mix and settle the suspension for 5 min.
10. Transfer the supernatant to centrifugation tube 1.
11. Add 10 mL of CFAD to the 12.5-cm^2 flask containing the remaining skin fragments, mix vigorously again and let the suspension settle for 5 min.
12. Transfer the second supernatant to centrifugation tube 2.
13. Repeat **step 11** and transfer the supernatant to centrifugation tube 3.
14. Centrifuge tubes 1, 2, and 3 at 400g (1500 rpm) for 10 min.
15. Resuspend the pellets in CFAD and seed the cells in plates previously coated with γ-irradiated 3T3 J2 feeder cells.

3.2. Culture of Keratinocytes

1. D 1: Seed 3T3 J2 cells at 1500 cells/ cm^2 in a T150 flask.
2. D 4: Irradiate 3T3 J2 at 45 Gy and plate at 15,000 cells/cm^2. After 2–18 h, thaw primary keratinocytes and seed at 10,000 cells/cm^2 on the γ-irradiated 3T3 J2 feeder cell layer.
3. D 6: Change CFAD medium and replace with CFAD containing 10 ng/mL of EGF. From this date, change CFAD + EGF (10 ng/mL) every 2 d.
4. Preconfluent keratinocytes can be frozen or passed onto freshly γ-irradiated 3T3 J2 cells. Cocultures are treated with EDTA (0.02% in PBS) for 5 min at room temperature in order to remove 3T3 J2 feeder cells. Keratinocytes are then trypsinized and either frozen (*see* **Note 5**) or reseeded.

3.3. Transduction of Cells with Retrovirus

1. D 1: Seed keratinocytes to be infected at 10,000 cells/cm^2. Seed the stably retrovirus producing cell line (i.e., psi-crip) at 13,000 cells/cm^2 in a 75-cm^2 flask (*see* **Note 6**).
2. D 3: Place the keratinocytes and the producer cells overnight in SFM–keratinocytes medium supplemented with 0.2 ng/mL of EGF, 0.25 µg/mL BPE, 0.1 mM CaCl$_2$, 50,000 U of penicillin–streptomycin, and 250 U of fungizone (*see* **Note 7**).
3. D 4: Collect the virus-containing medium, filter through a 0.45-µm filter and add Polybrene at 5 µg/mL. Remove the keratinocytes medium. Refeed keratinocytes using virus-containing medium. Change the medium with CFAD + EGF 8–10 h following infection.
4. At 72 h after infection, trypsinize the infected keratinocytes and seed onto 3T3 J2 cells resistant to neomycin, that is, 3T3 J2 cells that have been transduced using retroviral particles expressing the *LXSN* proviral genome. Add neomycin at 200 µg/mL to select the transduced cells.

3.4. Unscheduled DNA Synthesis

1. D 1: Seed keratinocytes at 10,000 cells/cm^2 on sterile glass coverslips in a 35-mm dish.
2. D 4: When keratinocyte colonies contain about 20–30 cells, wash the cells once with serum-free DMEM, and refeed the cells with F10 (Ham's), 1% DFBS. Incubate keratinocytes for 16 h at 37°C in F10, 1 % DFBS.

3. D 5: a. Prelabeling: Remove the F10, 1% DFBS medium and refeed the keratinocytes with 1.5 mL of F10, 1% DFBS containing 10 µCi/mL of [^{3}H]thymidine and 10 µM of fluorodeoxyuridine. Incubate for 1 h at 37°C (put the plates in plastic boxes to avoid contaminating the incubator with ^{3}H). (*See* **Note 8**.)
   b. UV irradiation: Remove the prelabeling medium and wash the cells with PBS (37°C). Irradiate control and "of interest" keratinocytes using three UV-B doses (0, 1000, 1500, and 2000 J/m^2). Doses should be measured using a UV-B photoradiometer. (*See* **Note 1**.)
   c. Labeling: Refeed the keratinocytes with F10, 1% DFCS containing 10 µCi /mL of [^{3}H]thymidine and incubate at 37°C for 3 h.
   d. Chase: Remove the labeling medium and refeed the keratinocytes with 1.5 mL of F10, 1% DFCS containing 10^{-5} M of unlabeled thymidine. Incubate for 1 h at 37°C.
4. D 7: Remove the medium and wash the cells twice with PBS.
5. Fix the cells with 100% methanol under a fume hood for 10 min and TCA-precipitate nucleic acids using ice cold 5% TCA twice for 15 min.
6. Dehydrate the cells once in 70% ethanol for 5 min and twice in 100% ethanol for 5 min.
7. Air-dry and mount the cover slips on a glass microscope slide with mounting medium. Position the cover slips so that the cellular side is up. Air-dry overnight.
8. D 8: In the dark, prewarm two volumes of photographic emulsion at 42°C and dilute it in three volumes of prewarmed water. Plunge the slide with cover slips into the emulsion for 5 s and air-dry for 4 h. Put the slide in a dark box, wrap it with aluminum foil, and keep at 4°C for 5 d.
9. In the dark, develop the slides with Kodak D19 at 14°C for 5 min, stop the reaction with 2% acetic acid for 30 s and fix with diluted Kodak 3000 solution (1 part fix: 4 parts water) for 10 min.
10. Gently wash the slides in tap water for 10 min.
11. Stain the cells with Mayer's hematoxylin for 5 min and wash in tap water.
12. Dehydrate cells in successive ethanol solutions, 70%, 90%, 95%, and 100%, for 1 min each.
13. Air-dry the slides and mount another cover slip on the cells with mounting medium.
14. Using a microscope (×100 magnification), count the grain number in each nucleus. DNA repair synthesis rate is expressed by mean grain number of at least 30 nuclei.

3.5. Cell Survival Assay Following UV Irradiation
3.5.1. Clonal Analysis

1. D 1: For each cell line (control and complemented cell lines), seed four 60-mm dishes with keratinocytes at 10,000 cell/cm^2 on γ irradiated 3T3 J2 cells.
2. D 5: a. Irradiation: Wash the keratinocytes with PBS and immediately irradiate with UV-B. Usually, one dish of each cell strain previously pre-

pared is exposed to one of the following UV-B doses: 0, 500, 1000, and 1500 J/m^2.
b. Immediately after irradiation, trypsinize, count, and seed the cells at 35 and 70 cells/cm^2 in 100-mm dishes onto γ-irradiated 3T3 J2 feeder cells. Change the culture medium (CFAD + EGF) every 2 d.
3. D 17: Wash the keratinocytes with PBS and fix for 30 min in 3.7% formaldehyde in PBS. Stain for 1 h in 1% rhodamine blue staining solution.
4. Determine the numbers and types of colonies as described *(16)*.

3.5.2. Mass Culture

1. D 1: For each cell line (control and complemented cell lines), seed keratinocytes (15,000 cell/cm^2) in six-well multidishes on γ-irradiated 3T3 J2 fibroblasts in CFAD medium.
2. D 4: Irradiate the cells for clonal analysis and grow 6 additional days.
3. D 11: Fix the keratinocytes for 30 min in 3.7% formaldehyde in PBS and stain using 1% rhodamine blue staining solution.
4. Cell survival is determined after scanning the dishes (Fluor-S Max MultiImager, Bio-Rad) and image analysis using Quantity One software (Bio-Rad) *(7)*.

4. Notes

1. UV-C (254 nm) can also be used. We have shown previously that UV-C and UV-B irradiations result in equivalent UDS values *(16)*. Because UV-B wavelengths are more physiologically relevant, we perform analyses of cell survival after UV-B exposure.
2. Culture of primary keratinocytes from very small skin biopsies has been optimized for growth on γ-irradiated 3T3 J2 Swiss fibroblasts. In our hands, other methods of keratinocyte growth relying on the use of a "defined" medium (e.g., SFM keratinocyte medium, Invitrogen Corp.) are not adapted to very small skin biopsies from XP patients.
3. 3T3 J2 cells can be used as a feeder layer until they reach passage 15 if they have never been cultivated to confluence.
4. Irradiation of 3T3 J2 can be replaced by mitomycin C (MMC) treatment as follows: Prepare a stock solution of MMC 0.5 mg/mL in PBS. Dilute MMC in DMEM without serum to an intermediate concentration of 0.1 mg/mL. Filter through a 0.22-μm filter. Prepare enough medium + MMC to treat all the flasks to a final concentration of 0.01 mg/mL. Rinse cells with PBS twice and replace PBS with DMEM + MMC (0.01 mg/mL). Incubate cells for 2 h in the incubator. Rinse cells with PBS and trypsinize. After counting, seed cells at the same density used for γ-irradiated 3T3 J2 cells. Do this a few hours before seeding the keratinocytes.
5. Keratinocytes are frozen in one volume of freezing solution: DMEM, 25% serum, 10% DMSO.
6. The retrovirus producer cell line Psi-crip must be grown after confluency before the retrovirus containing medium is harvested.

7. Cultivation of keratinocytes in low calcium (0.1 mM) SFM–keratinocyte medium changes their morphology and impedes their capacity to attach to each other and form highly cohesive colonies.
8. This step is to label and identify replicative cells.

Acknowledgments

MF and VB are equal contributors. This work was supported by grants from the Association pour la Recherche contre le Cancer (ARC, contract 9500 to T. M.; ARC, contract 5765 to A. S.), the Association Française contre les Myopathies (AFM, to T. M.), the Fondation de l'Avenir and funds from Centre National de la Recherche Scientifique (CNRS). MF is supported by Lóreal. We gratefully acknowledge Dr. F. Bernard for expert advice concerning skin reconstruction in vitro.

References

1. Cleaver, J. E. and Kraemer, K. H. (1995) Xeroderma pigmentosum and Cockayne syndrome, in *The Metabolic and Molecular Bases of Inherited Disease*, 7th Ed. (Scriver, C. R., Beaudet, A. L., Sly, W. S., and Valle, D., eds.), McGraw-Hill, New York, NY, pp. 4393–4419.
2. Bootsma, D., Weeda, G., Vermeulen, W., et al. (1995) Nucleotide excision repair syndromes: molecular basis and clinical symptoms. *Philo. Trans. R. Soc. Lond. B Biol. Sci.* **347,** 75–81.
3. Lehmann, A. R. (2001) The xeroderma pigmentosum group D (XPD) gene: one gene, two functions, three diseases. *Genes Dev.* **15,** 15–23.
4. Zeng, L., Sarasin, A., and Mezzina, M. (1999) Novel complementation assays for DNA repair-deficient cells: transient and stable expression of DNA repair genes. *Methods Mol. Biol.* **113,** 87–100.
5. Alonso, L. and Fuchs, E. (2003) Stem cells of the skin epithelium. *Proc. Natl. Acad. Sci. USA* **100,** 11,830–11,835.
6. Bernerd, F., Asselineau, D., Vioux, C., et al. (2001) Clues to epidermal cancer proneness revealed by reconstruction of DNA repair-deficient xeroderma pigment-osum skin in vitro. *Proc. Natl. Acad. Sci. USA* **98,** 7817–7822.
7. Arnaudeau-Begard, C., Brellier, F., Chevallier-Lagente, O., et al. (2003) Genetic correction of DNA repair-deficient/cancer-prone xeroderma pigmentosum group C keratinocytes. *Hum. Gene Ther.* **14,** 983–996.
8. Del Rio, M., Larcher, F., Serrano, F., et al. (2002) A preclinical model for the analysis of genetically modified human skin in vivo. *Hum. Gene Ther.* **13,** 959–968.
9. Hengge, U. R., Chan, E. F., Foster, R. A., Walker, P. S., and Vogel, J. C. (1995) Cytokine gene expression in epidermis with biological effects following injection of naked DNA. *Nat. Genet.* **10,** 161–166.
10. Rols, M. P., Delteil, C., Golzio, M., Dumond, P., Cros, S., and Teissie, J. (1998) In vivo electrically mediated protein and gene transfer in murine melanoma. *Nat. Biotechnol.* **16,** 168–171.
11. Yarosh, D. B. (2001) Liposomes in investigative dermatology. *Photodermatol. Photoimmunol. Photomed.* **17,** 203–212.

12. Ortiz-Urda, S., Thyagarajan, B., Keene, D. R., Lin, Q., Fang, M., Calos, M. P., and Khavari, P. A. (2002) Stable nonviral genetic correction of inherited human skin disease. *Nat. Med.* **8,** 1166–1170.
13. Miller, D. G., Adam, M. A., and Miller, A. D. (1990) Gene transfer by retrovirus vectors occurs only in cells that are actively replicating at the time of infection. *Mol. Cell. Biol.* **10,** 4239–4242.
14. Zeng, L., Quilliet, X., Chevallier-Lagente, O., Eveno, E., Sarasin, A., and Mezzina, M. (1997) Retrovirus-mediated gene transfer corrects DNA repair defect of xeroderma pigmentosum cells of complementation groups A, B and C. *Gene Ther.* **4,** 1077–1084.
15. Merten, O. W., Cruz, P. E., Rochette, C., Geny-Fiamma, C., Bouquet, C., Goncalves, D., Danos, O., and Carrondo, M. J. (2001) Comparison of different bioreactor systems for the production of high titer retroviral vectors. *Biotechnol. Prog.* **17,** 326–335.
16. Otto, A., Riou, L., Marionnet, C., Mori, T., Sarasin, A., and Magnaldo, T. (1999) Differential behaviors toward utraviolet A and B radiations of fibroblasts and keratinocytes from normal and DNA-repair deficients individuals. *Cancer Res.* **59,** 1212–1218
17. Setoguchi, Y., Jaffe, H. A., Danel, C., and Crystal, R. G. (1994) Ex vivo and in vivo gene transfer to the skin using replication-deficient recombinant adenovirus vectors. *J. Invest. Dermatol.* **102,** 415–421.
18. Yeh, P. and Perricaudet, M. (1997) Advances in adenoviral vectors: from genetic engineering to their biology. *Faseb J.* **11,** 615–623.
19. Braun-Falco, M., Doenecke, A., Smola, H., and Hallek, M. (1999) Efficient gene transfer into human keratinocytes with recombinant adeno-associated virus vectors. *Gene Ther.* **6,** 432–441.
20. Kotin, R. M., Menninger, J. C., Ward, D. C., and Berns, K. I. (1991) Mapping and direct visualization of a region-specific viral DNA integration site on chromosome 19q13-qter. *Genomics* **10,** 831–834.
21. Mathor, M. B., Ferrari, G., Dellambra, E., Cilli, M., Mavilio, F., Cancedda, R., and De Luca, M. (1996) Clonal analysis of stably transduced human epidermal stem cells in culture. *Proc. Natl. Acad. Sci. USA* **93,** 10,371–10,376.
22. Kolodka, T. M., Garlick, J. A., and Taichman, L. B. (1998) Evidence for keratinocyte stem cells in vitro: long term engraftment and persistence of transgene expression from retrovirus-transduced keratinocytes. *Proc. Natl. Acad. Sci. USA* **95,** 4356–4361.
23. Choate, K. A. and Khavari, P. A. (1997) Sustainability of keratinocyte gene transfer and cell survival in vivo. *Hum. Gene Ther.* **8,** 895–901.
24. Kuhn, U., Terunuma, A., Pfutzner, W., Foster, R. A., and Vogel, J. C. (2002) In vivo assessment of gene delivery to keratinocytes by lentiviral vectors. *J. Virol.* **76,** 1496–1504.
25. Chen, M., Li, W., Fan, J., Kasahara, N., and Woodley, D. (2003) An efficient gene transduction system for studying gene function in primary human dermal fibroblasts and epidermal keratinocytes. *Clin. Exp. Dermatol.* **28,** 193–199.

3

Cytogenetic Challenge Assays for Assessment of DNA Repair Capacities

William W. Au and Salama A. Salama

Summary

Different challenge assays have been used to investigate cellular responses following exposure to DNA damaging agents. Our protocol uses X- or γ-rays or ultraviolet light to challenge cells to repair the induced damage, and chromosome aberrations as a biomarker to indicate DNA repair proficiency. The assay was used successfully to demonstrate base- and nucleotide-excision repair deficiency in certain polymorphic DNA repair genes, namely *XRCC1* 751Gln and *XPD* 312Asn, respectively. In addition, populations with elevated exposure to certain environmental mutagenic agents—cigarette smokers, pesticide sprayers, and residents who lived near uranium mining and milling sites—showed DNA repair deficiency. Because expression of chromosome aberrations is associated with a significantly increased incidence of both cancer morbidity and mortality, the challenge assay may be useful in predicting cancer risk. The protocol for the assay is straightforward and the data have practical applications.

Key Words: Challenge assays; DNA repair; mutagen sensitivity; population monitoring.

1. Introduction

The basic concept of challenge assays is to use the assays to investigate cellular activities after exposing cells to DNA damaging agents. The information is used to evaluate how exposed cells respond to different challenging conditions, what cellular functions are involved in the response, and whether the response can have long-term biological consequences. Different types of cellular responses are useful for these evaluations. Depending on the objectives of investigations, the responses may range from the measurement of cell cycle delay, expression of chromosome aberrations (CAs) and activation of certain cellular genes. Quantitation of CAs (cytogenetic assay) is frequently used as a measure of variations in DNA repair capacities. With this objective in mind, cytogenetic challenge assays have traditionally been applied for two types of investigations. One is to investigate whether patients who have cancer may have an inherent deficiency in DNA repair capacity. For example, lymphocytes from lung cancer and head and neck cancer patients were found to be highly sensitive to the induction of CAs after their cells were exposed in vitro to bleomycin *(1,2)*. In addition, the bleomycin sensitivity appeared to be inherited and to predispose the sensitive individuals to an increased risk for cancer *(3,4)*. Therefore, mutagen-sensitive individuals, including phenotypically normal people, are believed to have a DNA repair deficiency. Consequently, these individuals may be at increased risk for environmental cancer.

The second application of challenge assays is to investigate whether populations exposed to environmental hazardous agents may have exposure compromised DNA repair activities. In this approach, lymphocytes from workers and matched controls are exposed to physical agents such as X- or γ-rays in vitro and the induction of CAs is used as an indicator of repair proficiency. Cigarette smokers; workers exposed to butadiene, pesticides, and styrene; and residents exposed to uranium mining and milling waste were found to have significantly higher CAs than the respective matched controls *(5–7)*. On the other hand, mothers who had given birth to children with neural tube defects and workers who were exposed to very low levels of benzene did not show the defective repair capacity *(8,9)*. These studies show that prolonged exposure to environmental toxic substances, at high enough concentrations, may cause damage to cellular repair mechanisms that can increase the long-term risk of disease. In these studies, the observed abnormal challenge response was unlikely to have been caused by the inheritance of certain susceptibility genes because the sample size for each population was too small to have included a substantial number of genetically susceptible individuals and because there was no selection mechanism to enrich for such individuals.

The use of CAs in the two applications of the challenge assay is relevant to health risk assessment. This is based on the evidence from prospective studies that indicate that individuals having elevated CAs are at significantly higher risk for development of cancer than those with lower frequencies *(10,11)*.

Recently, challenge assays have found new applications in the genome program. For example, polymorphisms in DNA repair genes have recently been discovered and some variant genes have been found to be associated with the development of certain cancers *(12–14)*. However, the functional significance of the variant alleles has not been characterized yet. A recent study by Au et al. *(15)* has shed some light onto the role of polymorphisms in these genes on DNA repair capacity. They used their challenge assays to investigate associations between variant DNA repair genotypes and abnormal repair of γ-ray or ultraviolet (UV) light-induced CAs in lymphocytes from healthy nonsmokers. Among the variant genes studied, *XRCC1* 194Trp and *APE1* 148Glu were not associated with abnormal repair of the induced DNA damage; therefore, the authors concluded that these variant genotypes may not have a significant effect on base excision or nucleotide excision repair capacity (BER and NER, respectively). On the other hand, *XPD* 312Asn and *XPD* 751Gln were associated with significant increase of CAs after the cells were challenged with UV light but not with γ-rays. The authors therefore concluded that these two variant genotypes are defective in NER but not BER. In addition, *XRCC1* 751Gln and *XRCC3* 241Met were associated with a significant increase of CAs after the cells were challenged with γ-rays but not with UV light, suggesting that the two variant genotypes are defective in BER but not NER. Knowledge of such variant genotypes can make a significant contribution to understanding genetic susceptibility to cancer in human populations.

The challenge assay as used by Au et al. *(5,6,15,16)* has features that offer special advantages in revealing DNA repair deficiencies. In the assay, lymphocytes are irradiated with X- or γ-rays or UV light during the G_1 phase of the cell cycle and the frequency of CAs is determined in the metaphase stage of the same cell cycle. With this protocol, the induced DNA damage is subjected to modification by DNA repair activities that are present at different stages of the same cell cycle and converted into observable CAs. If the duration between irradiation and observation is shortened too much, certain cell cycle stage-dependent DNA repair activities may not become involved and the assessment of DNA repair capacity may therefore be inadequate. On the other hand, if the duration is prolonged beyond one cell cycle, abnormal cells may be eliminated from the cell division process. This will cause an underestimation of DNA repair deficiency.

Besides CAs, other biomarkers can be incorporated into studies that use the challenge approach for investigation. One is the use of the single cell gel electrophoresis assay to document the repair of DNA strand breaks after the challenge *(17,18)*.

This brief introduction shows that a variety of challenge assays are useful for different types of investigations. In this chapter, the focus is on how to conduct the cytogenetic challenge assays to assess variations in DNA repair capacities, especially in the use of X- or γ-rays and UV light as the challenging agents to induce CAs. The advantages of using physical instead of chemical agents are several. First, the dosage to the target cells can be precisely delivered. Second, physical agents are not affected by variations in cellular uptake and capacities for metabolic activation or detoxification from one individual to another. Thus, the use of physical agents can minimize some potentially confounding factors that can affect the outcome of investigations.

2. Materials
2.1. Laboratory Supplies
1. 15-mL sterile conical centrifuge tubes.
2. Sterile pipets (1-, 5-, 10-, and 25-mL capacities).
3. Sterile syringes (1-, 5-, and 10-mL capacities).
4. Sterile flasks (100- and 500-mL capacities).
5. Sterile Petri dishes (100-mm diameter; for bacterial, not mammalian, cell cultures; Fisher Scientific, Pittsburgh, PA)
6. Sterile 15-mL Vacutainer™ CPT™ tubes containing sodium heparin as the anticoagulating agent (green top tubes) for the collection of peripheral blood from donors (Becton, Dickinson & Co., Sparks, MD).
7. Portable and hand-held vacuum pumps for pipetting fluids (Fisher Scientific).
8. Glass Pasteur pipets (5.7 in. long) and rubber bulbs that can fit on top of the pipets (Corning Incorporated, Corning, NY).
9. Microscope slides with frosting at one end for writing labels (Fisher Scientific).
10. Coverslips for the microscope slides (Fisher Scientific).
11. Coplin jars (Fisher Scientific).
12. Other general laboratory supplies such as beakers, flasks, sterile and non-sterile gloves, nonsterile pipets, and so forth.

2.2. Cell Culture
1. Incomplete medium (sterile, in 97 mL quantity and stored frozen at $<-20°C$; multiples of 97 mL incomplete medium can also be prepared).
    a. RPMI 1640 medium, 85 mL (Gibco Laboratory, Grand Island, NY).
    b. Heat inactivated fetal bovine serum, 10 mL (Irvin Scientific, Santa Ana, CA).
    c. Penicillin and streptomycin, 1 mL (stock concentrations at 10,000 U and 10,000 µg/mL, respectively, Gibco Laboratory).
    d. Sodium heparin, 1 mL (stock concentration at 1000 usp U, Gibco Laboratory).

   Store the 97 mL of incomplete medium frozen in one bottle if a large amount of culture medium is used in each experiment. Otherwise, store the medium in 50-mL aliquots.
2. Phytohemagglutinin (M form, lyophilized PHA; Gibco Laboratory).
3. L-Glutamine (200 mM liquid, Gibco Laboratory).
4. Colcemid, KaryoMAX® Colcemid® Solution, liquid (10 µg/mL), in phosphate-buffered saline (PBS; Life Technologies, Inc., Grand Island, NY.).
5. PBS (Sigma Chemical Co., St. Louis, MO).

2.3. Reagents
1. Histopaque 1077 (Sigma Chemical Co.).
2. Potassium chloride.
3. Methanol.
4. Glacial acetic acid.
5. Gurr Giemsa stain (Biomedical Specialties, Santa Monica, CA).

6. Pro-texx as slide mounting medium (American Scientific Products, McGraw Park, IL).
7. Xylene (EM Science, Gibbston, NJ).

2.4. Equipment

1. Laminar flow hoods for handling sterile procedures such as blood culture and UV irradiation.
2. X-ray or γ-ray machine that can deliver 100 cGy ionizing radiation to cell cultures in approx 1–2 min and with a capacity to irradiate multiple cultures simultaneously is preferable.
3. Phase contrast microscopes that have a ×400 magnification.
4. Binocular research light microscopes that have a minimum of ×100 and ×1000 magnification.
5. GS-6R centrifuge (Beckman Instruments Inc., Palo Alto, CA).
6. CO_2 incubator, water jacketed and maintained at 37°C with 95% humidity (Forma Scientific Co., Marietta, OH).

2.5. UV Irradiation Equipment

1. A short-wave 8-W germicidal lamp in a housing that emits primarily 254 nm UV light at a peak intensity of 1100 µW/cm^2 (Spectronics Corp., NY).
2. Glass attenuators that are custom-made to reduce the UV irradiation dose to the target cells to around 1 J/m^2/s from a distance of 10 in. The attenuators will be placed underneath the lamp.
3. The setup with stands to support the lamp, lamp housing, and attenuators is placed in the laminar flow hood for irradiation. The set up is arranged so that Petri dishes of cell culture can be irradiated, one at a time, under sterile conditions. The lamp is set at a distance from the target (Petri dish of cells) to deliver a dose of 1 J/m^2/s. The usual UV-dose to cells is 4 J/m^2, that is, requiring 4 s of irradiation.
4. UVX digital radiometer (San Gabriel, CA). The dosimeter will be used to verify the UV light dose before each irradiation procedure.

2.6. Cell Counting

1. Bright-Line hemacytometer and cover glass (Hausser Science, Horsham, PA).
2. Pasteur pipets.
3. PBS.
4. Trypan blue, 0.4% in PBS.
5. Microscope.
6. Sterile 15-mL conical tubes.
7. Hand counter.

3. Methods
3.1. Recruitment of Volunteers

Depending on the objectives of an investigation, different types of volunteers will be recruited. These may be certain types of cancer patients or exposed populations, certain susceptible subpopulations and healthy individuals. It is important, however, to document each recruited population carefully. The documentation may involve interviews using questionnaires and/or laboratory analyses. Questionnaires can be developed to collect confidential data regarding personal information, health history, exposure history, and family history. Individuals exposed to factors that may affect the frequency of CAs, such as intake of therapeutic drugs, should be excluded. Laboratory analyses may include the analysis of chemicals in body fluids to verify exposure. (*See* **Note 1**.)

3.2. Collection of Specimens for Laboratory Analysis

For the challenge assay, the most frequently used tissue is blood and the cells are lymphocytes. The main reason is that these cells are readily available and they can be grown in culture in a relatively synchronous cell cycle progression mode. Other cell types such as buccal mucosal and sputum cells can also be used *(19)*. Since some of these alternative cells are not actively dividing, they are not useful in cytogenetic challenge assays but may be in other challenge assays such as those using DNA strand breaks as biomarkers *(17,18)*.

Blood samples will be collected from peripheral veins in volunteers and into vacutainer tubes that contain sodium heparin as the anticoagulating agent. As soon as blood samples are drawn into tubes, the tubes are rocked back and forth gently to mix the sodium heparin with the blood. These 15-mL tubes usually collect only 12 mL of blood. Usually, 30 mL of blood samples are collected from each donor to provide adequate samples for a variety of studies. The blood tubes from each volunteer are labeled with a unique code number and maintained at room temperature. These tubes are transported to the laboratory as soon as possible for processing (*see* **Notes 2** and **3**).

3.3. Isolation of Blood Lymphocytes

Use sterile procedures, sterile tubes, pipets, and solutions in the following steps.

1. Place an appropriate volume of Histopaque 1077 (e.g., 22 mL, that is equal to the volume of the blood used for lymphocyte isolation) at room temperature into a 50-mL centrifuge tube.
2. Layer carefully the blood from a donor on top of the Histopaque solution, with the formation of two layers: blood and Histopaque (the remaining whole blood sample will be used for other assays). Additional tubes will be used if more samples need to be processed.
3. Centrifuge the tubes at the speed of 300–400g (1500 rpm) for 30 min at room temperature, with the brake of the centrifuge turned off.
4. Carefully remove the tubes from the centrifuge without shaking and tilting that would disturb the separated cells inside the tubes. The tube should contain three distinct layers of content. The top yellowish layer is made up of plasma. The thin

Cytogenetic Challenge Assays

middle layer is opaque and it contains nucleated cells, such as lymphocytes. The bottom layer contains Histopaque on top of the red blood cells.

5. Remove the top layer, leaving a small amount of the plasma with the opaque layer in the tube. The plasma can be saved and used for other purposes, for example, as a medium supplement to cell cultures and for analysis for metabolites, and so forth.
6. Use a 10-mL pipet to transfer the layer of lymphocytes together with a small amount of the Histopaque into another 50-mL centrifuge tube. This can be done a few times to remove all the cells but not picking up any of the red blood cells that are located at the lower layer in the tube. Cells from the same donor from multiple tubes are placed into the same tube.
7. Add approx 40 mL of prewarmed (37°C) PBS (or plain RPMI medium) into each of the tubes that contain the isolated cells. Mix the saline with the cells and centrifuge the tubes at 300g for 10 min. Make sure that the cell pellets are formed after the centrifugation.
8. Use pipets to remove or pour off the supernatant and add 10 mL of prewarmed plain RPMI 1640 medium to the cell pellet in each tube. Mix the cells and use the suspension for a viability cell count.

3.4. Cell and Viability Count

1. Dilute 0.2 mL of trypan blue with 0.8 mL of PBS.
2. Place cover glass over hemacytometer chamber.
3. Transfer 0.5 mL of the well-mixed cell suspension to a 15-mL tube and add 0.5 mL of the diluted trypan blue solution to give a dilution factor of 2.
4. Use a Pasteur pipet to mix the cell suspension and fill both chambers of the hemacytometer (without overflow) by capillary action.
5. Use the microscope with a ×10 ocular (and a ×10 objective) to count the cells in each of 10 squares (1 mm^2 each). If more than 10% of the cells are in clumps, repeat the entire sequence. If fewer than 200 or more than 500 cells are present in the 10 squares, repeat with a more suitable dilution factor.
6. Count only the nucleated cells that have not picked up the blue dye indicating that they are viable.
7. Calculate the number of cells per milliliter, and the total number of cells, in the original culture as follows:

   Cells/mL = Average count per square × 10^4

   Total cells = Cells/mL × any dilution factor (2 in this case) × total volume of cell suspension from which the sample was taken.

3.5. Whole Blood Cultures for the X-Ray Challenge Assay

Use sterile procedures throughout.

1. Thaw and warm the incomplete cell culture medium to 37°C, and add 2 mL of PHA and 1 mL of L-glutamine to make 100 mL of complete medium. Proportionally less complete medium can be made than is needed for use in one experiment, for example, 50 mL.

2. Add 5 mL of complete medium to each 15-mL centrifuge tube.
3. To each tube, add 0.5 mL of whole blood and mix the blood with the medium.
4. Place the tubes in an incubator that is set at 37°C, 95% humidity, and 5% CO_2. The caps of the tubes are not tightened to allow gaseous exchange in the incubator. In addition, the tubes are placed at a 30° angle to maximize the air-liquid surface for the exchange.
5. Initiate the cell cultures in the morning, usually before 11 am to provide convenience in the subsequent laboratory procedures. Usually only two cultures are set up from each donor for the challenge assay (one irradiated and one unirradiated control).
6. Incubate the cell cultures for 50–52 h before the cells are harvested for chromosome analysis.

3.6. Ionizing Irradiation Procedure

1. We use a Mark I Cs137 irradiator (J. L. Shepherd, Glendale, CA). The machine can be set up to generate a 100 cGy dose to target cells in approx 1 min and eight culture tubes can be irradiated at the same time. Depending on the availability of irradiators, X- or γ-rays can be used and they produce similar biological effects. (*See* **Note 4**.)
2. At 24 h after initiation of cultures, remove one tube from each donor from the incubator and place it in an insulated container, for example, a small cooler. The second tube from each donor stays in the incubator and is used as an unirradiated control for the determination of background chromosome aberration frequencies.
3. Bring the culture tubes in the cooler to the facility for irradiation. For us, the entire process takes approx 25 min. Cultures are returned to the incubator immediately after irradiation. In cases when the procedure takes longer than 30 min, it is recommended that heated insulating boxes be used to keep the blood cultures warm. Any prolonged cooling may interfere with cellular repair processes and will delay cell cycle progression.

3.7. Isolated Lymphocyte Cultures for the UV Light Challenge Assay

Use sterile procedures throughout.

1. Prepare the complete medium as described in **Subheading 3.5.** and place 5 mL of complete medium into each 15-mL cell culture tube.
2. Initiate cell cultures with a lymphocyte concentration of 5×10^5 cells/mL by adding an appropriate volume of the isolated lymphocytes to the culture medium.
3. Place the cell cultures in the incubator as described in **Subheading 3.5., step 4**.

3.8. UV Light Irradiation Procedures

Use sterile procedures throughout. (*See* **Note 4**.)

1. At about 1 h before irradiation of cells, the UV light needs to be turned on to warm up the irradiation source. Irradiation doses need to be checked with the dosimeter several times until the dose is stabilized before irradiation of cells can proceed.

2. At 24 h after culture initiation, centrifuge one tube from each donor at 300g (1500 rpm) for 10 min to pack the cells. The packed cells should not have any red color, that is, there should be as few red blood cells as possible.
3. Remove the supernatant cell culture medium using sterile technique, label with the code from the donor and store for later reuse.
4. Add 1 mL of sterile saline and resuspend the packed cells.
5. Transfer the cell suspension to a sterile Petri dish (*see* **Note 5**).
6. Place the dish with the lid under the UV-irradiation equipment. At this time, the personnel must have UV-protective eye wear, gloves and clothing because the UV light has already been turned on. With one hand on the lid of the Petri dish and another hand on a timer, remove the lid and turn on the timer simultaneously. As soon as the irradiation time is completed, put the lid back on the dish and remove the dish from the irradiation source.
7. After the irradiation, dispense 2 mL of sterile culture medium into each dish and pipet the suspension all around the dish to remove cells from the dish (*see* **Note 6**). Transfer the cell suspension back to the original culture tube. Centrifuge the culture tubes from several donors at 300g for 10 min to pack the cells. Remove the supernatant medium and add back the previously saved medium from each donor to the cells of the same donor tube. Resuspend the packed cells and put the culture tubes back into the incubator.

3.9. Cell Culture Harvest

1. Harvest cultures from both X-ray and UV light–irradiated cultures in a similar manner. Since the X-ray irradiated cultures are manipulated while the unirradiated cultures are not, the former cultures are given an extra hour of culture time due to the delay in cell cycle progression caused by the manipulation. In addition, since the UV light irradiated cultures involve more manipulation than the others, the former cultures are given an additional hour. Under these conditions, all unirradiated cultures are harvested at the 50th h after the initiation of cell culture. The X-irradiated cultures are harvested at the 51st h and the UV-irradiated cultures at the 52nd h. This arrangement also reduces complications due to the harvesting of too many cultures from several donors at once.
2. Before the cultures are actually harvested, add colcemid to each culture to block and accumulate mitotic cells at the metaphase stage of the cell cycle. Colcemid (0.1 mL) is added to give a final concentration of 0.2 µg/mL in the cultures. Incubate the cells for 1.5 h before harvesting.
3. Centrifuge the culture tubes at 300g for 10 min to pack the cells.
4. Remove the supernatant medium using Pasteur pipets.
5. Add 10 mL of prewarmed (37°C) KCl hypotonic solution (0.075 M) to each tube that contains packed cells. Resuspend the cells using one pipet for each culture tube. If necessary, pipet the suspension to completely break up the pellet. Incubate the tubes at 37°C for 10 min. Do the same for all the other tubes. The same pipet for each donor can be used for the remaining steps.
6. Add 1 mL of ice-cold fixative (3:1 ratio of methanol and glacial acetic acid) onto the top of each tube at the end of the hypotonic treatment and mix the suspension

using the pipet. At this time, the tubes containing cultures from whole blood will turn from red to black, indicating lysis of red blood cells.
7. Centrifuge the tubes at 300g for 10 min to pack the cells. The liquid in the tubes that contain cells from whole blood culture may be so dark that the size of the cell button may be difficult to see. Remove all but 0.5 mL of the supernatant.
8. Add 4 mL (2 Pasteur pipetfuls) of fixative that has been chilled in an ice bucket. Mix immediately and rapidly using the Pasteur pipet so that all the cells are fixed evenly; if mixed too slowly, clumps may form. Incubate on ice for 15–20 min.
9. Centrifuge the tubes to pack the cells again, remove the supernatant and add 5 mL of fixative to resuspend the cells. Repeat **steps 8** and **9** several times until the pellet is clean and white. The culture is now ready to make slides. For best results, slides should be made the same day as the harvest. However, the cells can be stored in the refrigerator with the caps on tightly until the next day before making the slides.

3.10. Slide Preparation

1. Remove the tubes from the refrigerator and centrifuge at 300g for 10 min.
2. Remove the supernatant fixative. Add newly prepared and pre-chilled fixative to the cell button, approx 0.3 mL per cell button to make a concentrated cell suspension.
3. Prepare a 60% glacial acetic acid solution with water and place the solution in a small beaker that is deep enough to dip slides inside. The solution should cover the entire slide up to the frosted end that is used to write the slide label.
4. Label at least two slides for each donor for each experimental condition (irradiated and unirradiated). The label should contain the specimen code, the irradiation condition, the experiment number, the date of the experiment and the slide number (one or two).
5. Dip each slide into the 60% glacial acetic acid solution and pull the slides out. The slide should contain a thin film of the liquid. If not, the slides may be dirty or may be coated with oil.
6. Hold the wet slide at a 45° angle; dispense a drop of the cell suspension from the Pasteur pipet to a location near the label side of the slide that is being held by the fingers. The cell suspension in fixative should be seen to react vigorously with the water-based solution on the slide. In addition, the cell suspension should spread downward on the slide and away from the fingers. Place the slides with cells near a wall at a 75° angle to facilitate spreading and drying. Paper towels should be placed underneath the drying slides to absorb the methanol and acid. In addition, one should avoid excessive contact with the solution on the skin or via inhalation. Wearing of gloves and conducting the procedures in a well-ventilated hood are highly recommended.
7. Approximately 1 h after the preparation of slides, view the slides under a phase contrast microscope. With ×400 magnification, one should see about 25–100 cells with about one metaphase cell per field of view from several fields of observation. This preliminary observation is used to determine several conditions: are there too many or too few cells on the slide and are the chromosomes in the

metaphase cells well spread out or not? The former can be changed by adjusting the concentration in the cell suspension. This can be achieved by repacking the cells and adding more or less fixative for slide preparation. The latter can be improved by subjecting the cells to more fixative treatment by adding 5 mL of fixative to the cells and keeping the cells for another day before making slides.

8. Place the slides containing cells onto trays for drying for 24 h in an airy and nondusty environment.
9. Stain the air-dried slides in a 15% Gurr Giemsa stain (15 mL of Gurr Giemsa stain + 85 mL of sterile water and filter through Whatman Filter paper) for 10 min. Rinse the stained slides in water and then air dry at a 45° angle for 24 h.
10. At locations with high humidity, it is necessary to put the slide on a slide warmer at 37°C for 30 min to remove the moisture. Mount the stained slides with Protexx medium and cover slips. Lay the mounted slides horizontally and let dry for 24 h.

3.11. Microscopic Evaluation

1. The staining on the slides should provide an indication of where the cells are located. A decision should be made on how to scan the slide and this should be consistent from slide to slide. For example, scan the slides from left to right and from top to bottom. Place the mounted slides on the stage of a light microscope to start the evaluation. Scanning is usually performed using a ×10 objective to give ×100 magnification (with the ×10 eye piece). When a metaphase appears in the field of view, place the cell in the center of the field for an in-depth evaluation.
2. The in-depth evaluation is performed using the ×100 objective. The first evaluation criterion is to count the number of chromosomes in the cell. It should have 46 chromosomes, in normal or abnormal morphology. Cells having incomplete chromosome numbers are not useful because some chromosomes are lost from the cells and the cells cannot be used to quantitate abnormal chromosomes with confidence. Experience in chromosome analysis is needed to determine whether overlapping chromosomes contain normal or abnormal chromosomes. A chromosome score sheet is used to record the analysis (**Table 1**). In general, 100 metaphase cells should be scored per experimental condition. The criteria that determine the number of cells to be analyzed has been described earlier *(20)*.
3. All normal chromosomes should have one constriction, the centromere, that is lightly stained. The centromere can be located near the center or at one end of a chromosome. The frequently observed abnormal chromosomes are described as follows:
    a. Chromosome deletions are chromosomes with two arms but without centromeres (**Fig. 1**; refs. *21,22*). They are frequently induced by exposure to ionizing radiation. With a dose of 100 cGy of X- or γ-rays, one may find approx 25 deletions per 100 metaphase cells.
    b. Chromosome translocations are exchanges between two abnormal chromosomes. Exposure to ionizing radiation induces a characteristic translocation known as a dicentric (**Fig. 1**). Each dicentric is formed by the joining of two chromosomes, each having a double strand break in the DNA molecule.

Table 1
An Example of a Score Sheet for Analysis of Chromosome Aberrations

Experiment No.: _____ ; Experiment Date: _____ ; Microscope No.: _____

Slide Code: _____ ; Date of Scoring: _____ ; Scorer: _____

Cell no.	Chromosome count	Chromatid abnormalities		Chromosome abnormalities		Unusual chromosomes	Coordinates
		Breaks	Exchanges	Deletions	Translocations		
1							
2							
3							
4							

Cytogenetic Challenge Assays

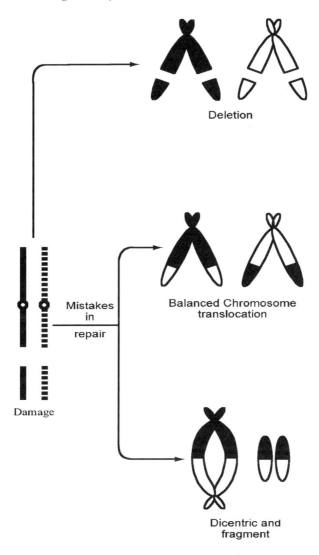

Fig. 1. Diagrammatic illustration of the formation of chromosome-type abnormalities. (Reprinted with permission from Wiley-Liss.)

The damaged chromosomes that contain centromeres are joined to form one dicentric chromosome whereas the broken DNA molecules (without the centromeres) are joined to form the deletion abnormality. Therefore, each dicentric has two centromeres and the deletion has none. Since the two abnormalities (dicentric and deletion) are derived from the same abnormal event, they are counted together as one event: dicentric plus a deletion.

The abnormality is entered under the translocation category in **Table 1**. The deletion is not entered under the deletion category. From irradiation with 100 cGy of X- or γ-rays, one may find approx 10 dicentrics per 100 metaphase cells. Occasionally, there are dicentrics without deletions. This can be caused by the loss of the deletion in the cell or loss of the deletion when the cell progressed from the first to the second cell division. The former may occur once in 200 metaphase cells. The latter occurs frequently if cells progress through the cell cycle more rapidly than expected. Therefore, if dicentrics without deletions occur more than once per 200 metaphase cells, there is a possibility of many second division metaphase cells in the preparation. In this case, the slides from that particular experimental condition should not be used because the frequency of abnormal cells will be underestimated. An alternative form of chromosome translocation is known as a balanced translocation (**Fig. 1**). In this case, the broken arms are joined to the wrong chromosome without the formation of a deletion. As a result there is no abnormal looking chromosome, for example, a chromosome with two centromeres. In this case, the abnormal chromosome is not detectable using the traditional cytogenetic assay. On the other hand, the balanced and unbalanced chromosome translocations are considered to be reciprocal events.

   c. Chromatid breaks are chromosomes with one instead of two broken arms as in deletions (**Fig. 2**). The broken fragment is usually found near the chromosome where the abnormality occurs. This abnormality is frequently induced after exposure to UV light. The frequency is about 10 per 100 metaphase cells after exposure to 4 J/m^2 of UV light.

   d. Chromatid exchanges are exchanges between two abnormal chromosomes and only one arm from each chromosome is involved (**Fig. 2**). In chromatid exchanges, the broken arms without centromeres that result from the exchange event are usually lost. The frequency of exchanges is approx 1 per 100 metaphase cells after exposure to 4 J/m^2 of UV light.

   e. Unusual chromosomes are abnormal chromosomes that do not fall into any of the four categories described earlier. Unusual chromosomes take many forms and characteristics. They may include chromosomes with three centromeres (tricentrics), unusually long chromosomes and chromosomes with unique shapes. These chromosomes, in aggregate, are low in frequency, fewer than one per 100 metaphase cells. Their low frequencies and the many different mechanisms involved in their formation make them unsuitable for statistical analysis as a group.

4. Use the coordinate column (**Table 1**) to record the location of the abnormal cells on the slide as indicated by the coordinates on the microscope stage.
5. Cells that contain any type of chromosome abnormality, for example, dicentrics, chromatid breaks and unusual chromosomes, and so forth, are categorized as aberrant cells. Therefore, this category is calculated from the collected data and used to indicate the overall response of cells to a challenge condition.

Cytogenetic Challenge Assays

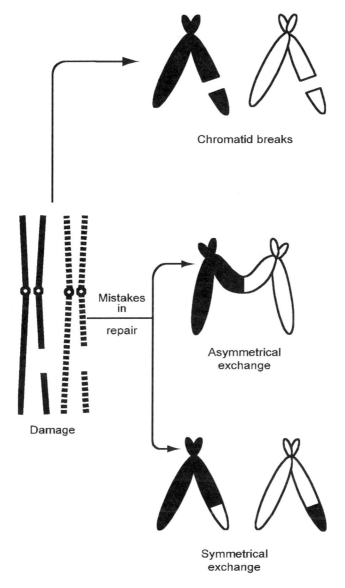

Fig. 2. Diagrammatic illustration of the formation of chromatid-type abnormalities. (Reprinted with permission from Wiley-Liss.)

6. The six categories of chromosome abnormalities from each donor are entered into a data base for statistical analysis. Data from several donors from one experimental condition, for example, irradiated with 100 cGy of X- or γ-rays, is entered as mean chromosome abnormality per 100 cells ± standard errors of the mean.

7. A biostatistician is consulted to conduct the statistical analysis. Typically, the software SPSS for Windows (version 10: SPSS Inc., Chicago, IL) is used to determine statistical significance using analysis of variance (ANOVA). The analysis can be performed further using stratified analysis for other experimental conditions, for example, genotypes, exposure to environmental toxic substances, and so forth. In general, an α error (p) of less than 0.05 is the criterion of significance and two-sided tests are used. Examples of the statistical analysis are described elsewhere *(23)*.

4. Notes

1. It is often emphasized that the crucial factor that influences the success or failure of an investigation is the validity of the experimental design to meet the objectives of an investigation. In designing population studies, some major areas to consider are the sample size, the type of tissues to be collected and the assays to be conducted because failure to consider these factors is frequently the cause for discrepancies in experimental results *(20,24)*.
2. To minimize experimental variations from different populations, for example, workers and controls, analyses for both groups need to be conducted simultaneously. An excellent tracking system needs to be designed such that the donors and the experimental data can be uniquely identified. For example, a coding system should uniquely identify the experiment, date, donor, tissue type, assay, and so forth. The same code is used to build the database for entering the collected data.
3. The experimental logistics should facilitate the efficient use of personnel, facilities, and other resources to complete the study. Several tasks can be done before and during the experiment. For example, tubes for cell culture and harvest can be labeled ahead of time. Stations can be set up for the addition of hypotonic and of fixative solutions. These stations should be equipped with automatic pipeters and/or dispensers. An automatic suction apparatus can also be set up for rapid removal of supernatants and fluids. Microscope slides can also be labeled ahead of time.
4. The X-ray and UV light irradiation sources need to be calibrated frequently to make sure that the irradiation doses are consistent with expectation. For laboratories using the challenge assays for the first time, pilot runs of the assays should be conducted in order to set up the conditions for the experiments. The pilot runs should use tissue samples from at least two donors. Since the irradiation sources and culture conditions vary from one laboratory to another, cells may have different growth rates that can affect the quality of the harvest. Therefore, it is advisable to make several harvests after the irradiation, for example, 50, 52, and 54 h. The microscope slides from these harvests can be compared with that of the control for the availability of mitotic cells and the frequencies of chromosome aberrations.
5. Use Petri dishes meant for bacterial culture rather than for tissue culture because cells will stick to the latter dishes.

6. Collection of cells from the Petri dishes after UV irradiation can be a little tricky. After the irradiation, the cell suspension is pipeted around the entire dish to remove as many cells from the dish as possible. One rule of thumb is that if the cell button is invisible after centrifugation of the cell suspension, a lot of cells were not recovered from the dish. In this case, more consistent effort needs to be made.

Acknowledgments

Many colleagues have made significant contributions to the development of the challenge assay. Although we are not able to provide a comprehensive list of these colleagues we would like to show our appreciation to the late Dr. T. C. Hsu, Dr. Julian Preston, Dr. Marvin Legator, Dr. Jonathan Ward, Jr., Dr. Moon Heo, and Mrs. Sylvia Szucs.

References

1. Spitz, M. R., Fueger, J. J., Halabi, S., Schantz, S. P., Sampe, D., and Hsu, T.C. (1993) Mutagen sensitivity in upper aerodigestive tract cancer: a case-control analysis. *Cancer Epidemiol. Biom. Prev.* **2,** 329–333.
2. Zheng, Y. L., Loffredo, C. A., Zhipeng, Y., et al. (2003) Bleomycin-induced chromosome breaks as a risk marker for lung cancer: a case-control study with population and hospital controls. *Carcinogenesis* **24,** 269–274.
3. Cloos, J., Nieuwenhuis, E. J., Boomsma, D. I., et al. (1999) Inherited susceptibility to bleomycin-induced chromatid breaks in cultured peripheral blood lymphocytes. *J. Natl. Cancer Inst.* **91,** 1125–1130.
4. Wu, X., Lippman, S. M., Lee, J. J., et al. (2002) Chromosome instability in lymphocytes: a potential indicator of predisposition to oral premalignant lesions. *Cancer Res.* **62,** 2813–2818.
5. Au, W. W., Bechtold, W. E., Whorton, E. B. J., and Legator, M. S. (1995) Chromosome aberrations and response to gamma-ray challenge in lymphocytes of workers exposed to 1,3 butadiene. *Mutat. Res.* **334,** 125–130.
6. Au, W. W., Lane R. G., Legator, M. S., Whorton, E. B., Wilkinson, G. S., and Gabehart, G. J. (1995) Biomarker monitoring of a population residing near uranium mining activities. *Environ. Health Perspect.* **103,** 466–470.
7. Oberheitmann, B., Frentzel-Beyme, R., and Hoffmann, W. (2001) An application of the challenge assay in boat builders exposed to low levels of styrene—a feasibility study of a possible biomarker for acquired susceptibility. *Int. J. Hygiene Environ. Health* **204,** 54–60.
8. Au, W. W., Rodriguez, G., Rocco, C., Legator, M. S., and Wilkinson, G. S (1996) Chromosomal damage and DNA repair response in lymphocytes of women who had children with neural tube defects. *Mutat. Res.* **361,** 17–21.
9. Hallberg, L. M., El Zein, R., Grossman, L., and Au, W. W.(1996) Measurement of DNA repair deficiency in workers exposed to benzene. *Environ. Health Perspect.* **104 (suppl 3),** 529–534.

10. Hagmar, L., Bonassi,S., Strömberg, U., et al. (1998) Chromosomal aberrations in lymphocytes predict human cancer—A report from the European Study Group on Cytogenetic Biomarkers and Health (ESCH). *Cancer Res.* **58,** 4117–4121.
11. Bonassi, S., Hagmar, L., Strömberg, U., et al, and the European Study Group on Cytogenetic Biomarkers and Health (ESCH). (2000) Chromosomal aberrations in lymphocytes predict human cancer independently from exposure to carcinogens. *Cancer Res.* **60,** 1619–1625.
12. Sturgis, E. M., Dahlstrom, K. R., Spitz, M. R., and Wei, Q. (2002) DNA repair gene ERCC1 and ERCC2/XPD polymorphisms and risk of squamous cell carcinoma of head and neck. *Arch. Otolaryngol. Head Neck Surg.* **128,** 1084–1088.
13. Zhou, W., Liu, G., Miller, D. P., Thurston, S. W., Xu, L., and Wain, J. C. (2003) Polymorphisms in the DNA repair genes XRCC1 and ERCC2, smoking and lung cancer risk. *Cancer Epid. Biomarkers Prev.* **12,** 359–365.
14. Harms, C., Salama, S. A., Sierra-Torres, C. H., Cajas-Salazar, N., and Au, W. W. (2004) Polymorphisms in DNA repair genes, chromosome aberrations and lung cancer. *Environ. Mol. Mutagen.* **44,** 74–82.
15. Au, W. W., Salama, S. A., and Sierra-Torres, C. (2003) Functional characterization of polymorphisms in DNA repair genes using cytogenetic challenge assays. *Environ. Health Perspect.* **111,** 1843–1850.
16. Au, W. W., Sierra-Torres, C. H., Cajas-Salazar, N., Shipp, B. K., and Legator, M. S. (1999) Cytogenetic effects from exposure to mixed pesticides and the influence from genetic susceptibility. *Environ. Health Persp.* **107,** 501–505.
17. Cebulska-Wasilewska, A. (2003) Response to challenging dose of X-rays as a predictive assay for molecular epidemiology. *Mutat. Res.* **544,** 289–298.
18. Cebulska-Wasilewska, A., Panek, A., Zabinski, Z., Moszczynski, P., and Au, W.W. (2005) Influence of mercury vapours on lymphocytes in vivo and on their susceptibility to UV-C and X-rays, and repair efficiency in vitro. *Mutat. Res.*, in press.
19. Salama, S. A., Serrana, M., and Au, W.W. (1999) Biomonitoring using accessible human cells for exposure and health risk assessment. *Mutat. Res.* **436,** 99–112.
20. Au, W. W., Cajas-Salazar, N., and Salama, S. (1998) Factors contributing to discrepancies in population monitoring studies. *Mutat. Res.* **400,** 467–478.
21. Au, W. W. (1991) Monitoring human population for the effects of radiation and chemical exposures using cytogenetic techniques, in *Occupational Medicine: State of the Art Reviews* (Wilkinson, G., ed.), Hanley and Belfus, Philadelphia, PA, pp. 597–611.
22. Au, W. W., Badary, O., and Heo, M. Y. (2001) Cytogenetic assays for monitoring populations exposed to environmental mutagens, in *Occupational Medicine: State of the Art Reviews* (Wilkinson, G. ed.), Vol. 16, Hanley and Belfus, Inc., pp. 345–357.
23. Au, W. W., Oh, H. Y., Grady, J., Salama, S., and Heo, M. Y. (2001) Usefulness of genetic susceptibility and biomarkers for evaluation of environmental health risk. *Environ. Mol. Mutagen.* **37,** 215–225.
24. Au, W. W., (2000) Strategies for conducting human population monitoring studies. *NATO Sci. A Life Sci.* **313,** 86–93.

4

Evaluating the Delayed Effects of Cellular Exposure to Ionizing Radiation

Shruti Nagar, James J. Corcoran, and William F. Morgan

Summary

A number of ongoing delayed effects have now been described in the progeny of an irradiated cell. These are grouped under the rubric of radiation induced genomic instability. Perhaps the best characterized is the dynamic production of chromosomal rearrangements in some clonally expanded cells surviving irradiation. In this chapter we provide the protocols for irradiation, cell culture, chromosome analysis, and characterization of the status of genomic stability in the context of delayed radiation effects.

Key Words: Chromosomal aberrations; clonogenic survival; fluorescence *in situ* hybridization; genomic instability; ionizing radiation.

1. Introduction

For many years it was thought that the biological effects of ionizing radiation were solely the result of cellular processes acting on DNA damage after the deposition of energy in the cell nucleus. For instance, a misrepaired DNA double strand break could lead to a mutation, or two breaks on different chromosomes could misrejoin to form either a dicentric chromosome and its associated fragment or a reciprocal translocation. If these were not lethal events then all the progeny of that damaged cell would contain either the mutation or the chromosomal rearrangement. Although this can certainly account for much of the damage associated with radiation exposure, there is increasing evidence for the dynamic, ongoing production of nonclonal chromosome changes occurring in the progeny of cells surviving exposure to ionizing radiation (reviewed in **refs. *1,2***).

This so called radiation induced genomic instability has been described in normal human lymphocytes (*3*), normal human and murine bone marrow cells

From: *Methods in Molecular Biology: DNA Repair Protocols: Mammalian Systems, Second Edition*
Edited by: D. S. Henderson © Humana Press Inc., Totowa, NJ

(4,5), and a host of different cell lines in in vitro tissue culture systems *(6)*. Chromosomal instability appears to be independent of the p53 status of the cell *(7)*, although this is somewhat controversial *(8)*. While chromosomal changes are perhaps the best characterized end point of radiation-induced genomic instability a number of other end points including delayed mutation, gene amplification, recombination, mini- and microsatellite instability, transformation, and cell killing have also been described (reviewed in **ref. 2**). Not surprisingly, chromosomal instability is frequently associated with delayed reproductive cell death as measured by persistently reduced plating efficiency in the chromosomally unstable cell clones *(9,10)*. This is likely to be because a fraction of the chromosomal rearrangements produced are not compatible with cell survival. Generally, however, there is no a relationship between the different phenotypes associated with radiation induced genomic instability. That is, cells showing a particular instability end point usually do not show evidence of other end points *(10)*, for example, chromosomal unstable clones do not show increased gene amplification or delayed mutations; thus they do not exhibit a true mutator phenotype.

The molecular, biochemical, and cellular mechanisms responsible for these delayed effects are currently unclear. While dicentric chromosome formation followed by bridge–breakage–fusion cycles can account for some of the observed chromosomal instability, there is intriguing evidence for a role for soluble secreted factor or factors produced either directly by the radiation, or as a consequence of induced instability perpetuating the dynamic production of chromosomal change over time *(11)*. Genomic instability is a well-characterized hallmark of the cancer cell *(12)*. Whatever the mechanism(s) for radiation-induced instability, understanding how exposure to DNA damaging agents induces and perpetuates ongoing genomic destabilization could have important implications for understanding radiation carcinogenesis. In this chapter we provide the protocols for irradiation, cell culture, chromosome analysis, and characterization of the status of genomic stability in the context of delayed radiation effects.

2. Materials
2.1. Cell Culture and Irradiation

1. Phosphate-buffered sucrose (PBS): 7 mM KH_2PO_4, pH 7.4, 1 mM $MgCl_2$, 272 mM sucrose.
2. Culture medium: GM10115 Chinese hamster ovary/human chromosome 4 hybrid cells (Human Genetic Mutant Cell Repository [line HHW416]; Institute for Medical Research, Camden, NJ) are cultured as monolayers in Dulbecco's minimal essential medium supplemented with 10% fetal bovine serum (FBS), 2 mM L-glutamine, 100 U/mL of penicillin, 100 µg/mL of streptomycin, and 0.2 mM L-proline. Cells are cultured at 34°C in humidified incubators containing 5% CO_2.
3. Trypsin–EDTA: Stock solutions of 0.5 g/L of trypsin, 0.2 g/L of EDTA, 1.0 g/L of glucose and 0.58 g/L of $NaHCO_3$ in 8% NaCl and 0.4% KCl in H_2O.

4. 2% Crystal violet in 30% methanol.
5. X-ray source (Agfa NDT Inc., Lewistown, PA).

2.2. Preparation of Metaphase Spreads

1. Colcemid (Calbiochem, La Jolla, CA): Stock solution 1×10^{-5} M in H_2O.
2. Hypotonic solution: 0.075 M KCl in HPLC-grade H_2O.
3. Methanol.
4. Acetic acid.

2.3. Giemsa Staining

1. Giemsa solution (5% v/v Giemsa stain in distilled water).
2. Permount mounting medium (Fisher Scientific, Pittsburgh, PA).

2.4. *Fluorescence* In Situ *Hybridization*

1. QiaFilter Plasmid Maxi kit (Quiagen, Valencia, CA).
2. BioNick DNA Labeling System (Invitrogen, Carlsbad, CA).
3. Rapid polymerase chain reaction (PCR) Purification System (Marligen Biosciences, Ijamsville, MD).
4. 2X Saline sodium citrate (SSC). (20X SSC stock solution is: 3 M NaCl, 0.3 M sodium citrate, pH 7.0.)
5. 70%, 85%, and 100% ethanol.
6. 50% Formamide–2X SSC.
7. 70% Formamide–2X SSC.
8. Hybridization mix (stock): 80% (v/v) formamide, 15% (w/v) Dextran sulfate, 2.8X SSC stored indefinitely at –20°C.
9. Sheared herring sperm DNA, 1 mg/mL (Invitrogen).
10. PN buffer: 0.1 M phosphate buffer, pH 8.0, 0.5% Igepal CA-630 (Sigma-Aldrich, St. Louis, MO).
11. Rubber cement (Starkey Chemical Process Co., La Grange, IL).
12. Fluorescent conjugates: fluorescein avidin D, biotinylated anti-avidin D (Vector Laboratories, Inc., Burlingame, CA).
13. Blocking solution for fluorescent conjugates: 5% nonfat dry milk, 0.1% Tween-20 (Sigma-Aldrich), 4X SSC.
14. 0.05 µg/mL of 4′,6-diamidino-2-phenylindole (DAPI) or 0.30 µg/mL of propidium iodide (PI) (Sigma-Aldrich) in Vectashield Mounting Medium (Vector Laboratories, Inc.).
15. Plastic cover slips, cut to size (Ber Plastics Incorporated, Riverdale, NJ).
16. Fluorescence microscope (Nikon Eclipse E600; Image Systems, Inc., Columbia, MD).

3. Methods
3.1. Cellular Irradiation and Isolation of Clones

1. Clonally expand a single cell to obtain a homogeneous cell population (*see* **Note 1**).

2. Based on survival data for the cell type being used, seed sufficient cells into a 25-cm^2 tissue culture flask so as to yield 5–10 surviving colonies after radiation exposure. At 1 h after plating cells, irradiate the flask with the desired radiation dose (*see* **Note 2**).
3. Culture the cells for 10–15 d to form colonies. When surviving colonies are visible to the naked eye they are ready for isolation. Using the tip of a sterile Dacron swab dipped in trypsin–EDTA solution, individually pick 5–10 well spaced single colonies off the bottom of the culture flask by swiping the swab across the colony. Briefly dip the tip of the swab containing cells in a 25-cm^2 tissue culture flask containing fresh culture medium, and agitate so that the cells are dispersed.
4. Incubate the culture flask to allow the cells to clonally expand. Once the cells have reached approx 90% confluence, split the culture in half, and grow both cultures again to approx 90% confluence. Freeze one of the two flasks for future use, if necessary, and use the other flask to prepare metaphase chromosomes for analysis of chromosomal instability.
5. Stain the remaining colonies from the irradiated 25-cm^2 tissue culture flask with 2% crystal violet in 30% methanol to determine surviving fraction for future reference (*see* **Note 3**).

3.2. Cell Freezing

1. Remove the cells from the flask using trypsin–EDTA and pellet by centrifugation. Discard the supernatant and resuspend the cell pellet in 10% dimethyl sulfoxide (DMSO), 20% cell culture medium, and 70% FBS to a final volume of 0.5 mL (*see* **Note 4**). Store the vials at –70°C for approx 1 wk, then move to liquid nitrogen for long-term storage.

3.3. Preparation of Metaphase Chromosome Spreads

1. Add Colcemid to exponentially growing cells at a final concentration of $2 \times 10^{-7} M$.
2. After 3 h, gently shake the flask to dislodge mitotic cells and collect the medium (*see* **Note 5**).
3. Centrifuge the cells at approx 1500 rpm (Beckman Coulter Allegra 6 centrifuge with swinging bucket rotor), discard the supernatant, and resuspend the pellet in 10 mL of hypotonic 0.075 M KCl solution prewarmed to 37°C.
4. Incubate the cells at 37°C for 20 min, recentrifuge, discard the supernatant, and resuspend the pellet by adding methanol dropwise while gently shaking the centrifuge tube. Add 5 mL of methanol.
5. Centrifuge the cells, discard the supernatant, and resuspend the pellet by adding methanol–acetic acid (3:1 v/v) dropwise while gently shaking the centrifuge tube. Add 5 mL of methanol–acetic acid.
6. Centrifuge the cells and repeat **step 5** above.
7. Resuspend the pellet in three to five drops of methanol–acetic acid and, using a Pasteur pipet, drop the metaphase cells onto clean glass microscope slides and allow them to air-dry.

3.4. Giemsa Staining for Conventional Cytogenetics Analysis

1. Age metaphase slides overnight (at least) on a 70°C hot plate.
2. Make up a fresh 5% Giemsa solution in water (*see* **Note 6**).
3. Stain slides for 5 min, wash thoroughly and air-dry.
4. Permanently mount a cover slip.
5. Analyze using brightfield microscopy.

3.5. Fluorescence In Situ Hybridization

1. Obtain desired plasmid vector(s) containing chromosome-specific DNA sequences (*see* **Note 7**) and amplify in *Escherichia coli*, then purify with a QiaFilter Plasmid Maxi kit (*see* manufacturer's instructions).
2. Label 1 µg of purified plasmid with biotin using nick translation system. Separate the labeled DNA from unincorporated nucleotides using a Concert PCR purification kit (*see* manufacturer's instructions).
3. Incubate metaphase spread slides in 2X SSC at 37°C for 30 min to increase probe penetration.
4. Dehydrate the slides in ice-cold 70% ethanol for 2 min followed by 2 min in 85% ethanol and 2 min in 100% ethanol.
5. Denature the slides in 70% formamide–2X SSC solution for 3 min at 70°C.
6. Following denaturation, dehydrate the slides once again in 70%, 85%, and 100% ice-cold ethanol for 2 min each.
7. Air-dry the slides and place on a 42°C slide warmer.
8. Vortex hybridization solution containing 1 µg of labeled probe, 3 µg/mL of herring sperm DNA, 25 µL of hybridization mix, and 2 µL of water to a total volume of 35 µL. Centrifuge briefly.
9. Denature the hybridization solution at 72°C for 5 min and centrifuge briefly.
10. Add hybridization solution to the metaphase slide on the slide warmer. Add a glass cover slip and seal the edges of the cover slip with rubber cement.
11. Incubate 24–96 h at 37°C in a humidified chamber.
12. Gently take off glass cover slip. Wash the slide with 50% formamide–2X SSC at 45°C for 5 min.
13. Wash the slide in PN buffer for 2 min.
14. Add 35 µL of fluorescein avidin D and incubate at 37°C in a humidified chamber for 20 min.
15. Wash the slides in PN buffer for 2 min.
16. Add 35 µL of biotinylated anti-avidin D antibody and incubate at 37°C in a humidified chamber for 20 min.
17. Wash the slide in PN buffer for 2 min.
18. Add 35 µL of fluorescein avidin D and incubate at 37°C in a humidified chamber for 20 min.
19. Wash the slide in PN buffer for 2 min.
20. Add 35 µL of DAPI in Vectashield, place a glass cover slip on the slide, and observe under a fluorescence microscope using a filter set for fluorescein at 465–495 nm. Chromosome spreads that are either stained with Giemsa (*see* **Note 8**)

or are fluorescently labeled with chromosome-specific probes (*see* **Note 9**) can be scored to determine whether the clones are genomically unstable (*see* **Note 10** for criteria for genomic instability).

Notes

1. Prior to beginning any experiment(s) on induced instability we recommend clonally expanding a single cell to form a homogeneous population of cells for the proposed experiments. This eliminates inherent instability associated with many cell lines as much as practically possible.
2. The dose of X-ray typically used to obtain 10–15 colonies per dish is dependent on the survival fraction based on cell survival curves for the cell type being used.

$$\text{Surviving fraction} = \frac{\text{Colonies counted}}{(\text{Colonies seeded} \times \text{PE})}$$

   where PE is plating efficiency, which is the number of cells seeded that grow into colonies. Estimates of survival are obtained for a range of X-ray doses typically comprising 2, 4, 6, 8, 10, and 12 Gy. The number of cells seeded per dish is adjusted so that a countable number of colonies result after exposure to radiation. In our hands, using the GM10115 human–hamster hybrid cells, we find chromosomal instability induced in approx 3% of surviving clones per Gy of X-rays.
3. Crystal violet is a positively charged (cationic) dye that binds to negatively charged cellular components such as nucleic acids and acidic polysaccharides present in the cell membrane. Most staining procedures including that of crystal violet are performed on fixed cells, which kills the cells and causes them to stick to the culture dish. The methanol present in the crystal violet solution acts as a fixative and thus all cells that are stained with the dye are essentially dead cells. Simple stains such as crystal violet allow one to better visualize the cells present in the culture dish, which aids in the subsequent counting of colonies. Other simple stains such as Giemsa may also be used.
4. The conventional composition of freezing medium has typically consisted of 10% DMSO and 10% FBS in 80% culture medium. However, we have found that freezing cells in medium with a large percentage of FBS (70%) causes not only a significant reduction in the time it takes cells to recover from the freeze-thaw process but also results in more robust cell growth and morphology. Approximately 24 h after thawing, the freezing medium is replaced with fresh culture medium and cells are incubated until confluence.
5. If the metaphase cells from the cell line being investigated do not round up and detach at mitosis, trypsinize the cell population and process as you would if mitotic shake off were feasible.
6. Giemsa stain is specific for the phosphate groups of DNA and solid staining with Giemsa has been used extensively to visualize whole chromosomes. Solid staining with Giemsa however, does not distinguish between chromosomes (unless

chromsome banding techniques such as G-banding are performed). Fluorescent *in situ* hybridization (FISH), on the other hand, is a technique to visualize specific chromosomes and chromosome regions. The advantages of FISH over Giemsa staining are that not only are fluorescence techniques more sensitive and precise, but they also allow the simultaneous analysis of multiple chromosomes or chromosome fragments. Another advantage of using FISH over Giemsa is the ability to detect chromosomal aberrations such as balanced translocations and insertions that cannot be visualized with Giemsa staining. Unlike Giemsa staining, however, FISH is more time and labor intensive, and is also a more expensive technique.

7. The decision on which chromosome(s) to paint in a metaphase spread is largely dependent on the type of cells being used and the number of chromosome-specific probes available. In our cell system, the human–hamster hybrid GM10115 cells, there is one copy of human chromosome 4 present in a background of 20–24 hamster chromosomes. Owing to the unavailability of hamster chromosome-specific probes, we paint only the human chromosome present in these cells. Multiple chromosomes in a metaphase spread may be painted using different fluorescent dyes, thus increasing the percentage of the genome that may be analyzed for chromosomal aberrations *(13)*.

8. To cytogenetically characterize surviving clones as to their stability/instability status, those metaphase chromosomes stained with Giemsa can be analyzed for chromosome- and chromatid-type chromosomal aberrations. At least 200 well spread, sequential metaphase cells per clone should be analyzed. If a statistically significant increase in aberrations is observed over the unirradited control clones, the clonally expanded population derived from an irradiated cell can be described as unstable.

9. For those metaphase chromosomes analyzed by FISH, at least 200 metaphase spreads should be scored for rearrangements involving the "painted" chromosomes(s). In our hands, a clone is classified as unstable if it contains three or more distinct metaphase subpopulations involving the labeled chromosome and these subpopulations represent at least 5% of the 200 spreads scored.

10. The criterion for judging a clone unstable is arbitrary and varies from laboratory to laboratory. Those criteria presented above are used in our laboratory and are presented as guidelines only. Given the inherent instability of many cell lines in culture this should be decided based on the cell line used and the hypothesis to be tested.

References

1. Morgan, W. F. (2003) Non-targeted and delayed effects of exposure to ionizing radiation: II. Radiation-induced genomic instability and bystander effects in vivo, clastogenic factors and transgenerational effects. *Radiat. Res.* **159,** 581–596.
2. Morgan, W. F. (2003) Non-targeted and delayed effects of exposure to ionizing radiation: I. Radiation-induced genomic instability and bystander effects in vitro. *Radiat. Res.* **159,** 567–580.

3. Holmberg, K., Meijer, A. E., Harms-Ringdahl, M., and Lambert, B. (1998) Chromosomal instability in human lymphocytes after low dose rate gamma-irradiation and delayed mitogen stimulation. *Int. J. Radiat. Biol.* **73,** 21–34.
4. Kadhim, M. A., Lorimore, S. A., Hepburn, M. D., Goodhead, D. T., Buckle, V. J., and Wright, E. G. (1994) Alpha-particle-induced chromosomal instability in human bone marrow cells. *Lancet* **344,** 987–988.
5. Kadhim, M. A., Lorimore, S. A., Townsend, K. M., Goodhead, D. T., Buckle, V. J., and Wright, E.G. (1995) Radiation-induced genomic instability: delayed cytogenetic aberrations and apoptosis in primary human bone marrow cells. *Int. J. Radiat. Biol.* **67,** 287–293.
6. Morgan, W. F., Day, J. P., Kaplan, M. I., McGhee, E. M., and Limoli, C. L. (1996) Genomic instability induced by ionizing radiation. *Radiat. Res.* **146,** 247–258.
7. Kadhim, M. A., Walker, C. A., Plumb, M. A., and Wright, E. G. (1996) No association between p53 status and alpha-particle-induced chromosomal instability in human lymphoblastoid cells. *Int. J. Radiat. Biol.* **69,** 167–174.
8. Schwartz, J. L., Jordan, R., Evans, H. H., Lenarczyk, M., and Liber, H. (2003) The TP53 dependence of radiation-induced chromosome instability in human lymphoblastoid cells. *Radiat. Res.* **159,** 730–736.
9. Marder, B. A. and Morgan, W. F. (1993) Delayed chromosomal instability induced by DNA damage. *Mol. Cell. Biol.* **13,** 6667–6677.
10. Limoli, C. L., Kaplan, M. I., Corcoran, J., Meyers, M., Boothman, D. A. and Morgan, W. F. (1997) Chromosomal instability and its relationship to other end points of genomic instability. *Cancer Res.* **57,** 5557–5563.
11. Morgan, W. F. (2003) Is there a common mechanism underlying genomic instability, bystander effects and other nontargeted effects of exposure to ionizing radiation. *Oncogene* **22,** 7094–7099.
12. Lengauer, C., Kinzler, K. W., and Vogelstein, B. (1998) Genetic instabilities in human cancers. *Nature* **396,** 643–649.
13. Huang, L., Grim, S., Smith, L. E., Kim, P. M., Nickoloff, J. A., Goloubeva, O. G., and Morgan, W. F. (2004) Ionizing radiation induces delayed hyperrecombination in mammalian cells. *Mol. Cell. Biol.* **24,** 5060–5068.

5

Inhibition of DNA Synthesis by Ionizing Radiation
A Marker for an S-Phase Checkpoint

Nicolaas G. J. Jaspers and Malgorzata Z. Zdzienicka

Summary

Inhibition of replicative DNA synthesis by ionizing radiation is partly caused by an active, signal-mediated response termed the "S-phase checkpoint." Defects in this checkpoint were first discovered in the human inherited disorder ataxia-telangiectasia (AT). γ-Irradiated cells from AT patients consistently display a diminished inhibition of DNA synthesis, a feature called "radioresistant DNA synthesis" (RDS). RDS has been widely used as a diagnostic marker for AT, in postnatal as well as prenatal material. The regulation and control of the S-phase checkpoint is complex and multifaceted; it is not restricted to ionizing radiation, but can occur after many genotoxic stressors. Defects in both upstream control functions, such as ATM, NBS1, and MRE11, as well as downstream modulators can provoke an RDS phenotype. Here a simple, accurate and highly reproducible experimental protocol is presented for the generation of DNA synthesis inhibition curves from cells in culture.

Key Words: Ataxia-telangiectasia; γ-irradiation; genotoxic stress; MRE11; Nijmegen breakage syndrome; radioresistant DNA synthesis; S-phase checkpoint.

1. Introduction

In mammalian cells, the rate of DNA synthesis decreases after X- or γ-ray exposure. The dose-response curve indicates biphasic kinetics of inhibition (*see* **Fig. 1**). The initial, steep component of the curve represents inhibition of initiation of new replicons, whereas the shallow component is a manifestation of chain growth failure. The latter is probably the result of direct interference of DNA lesions with the replication machinery; on the other hand, slowdown of replicon initiation is one of the manifestations of an active signal-mediated response. The first evidence for this was obtained from studies on cells derived from patients with the autosomal recessive human diseases

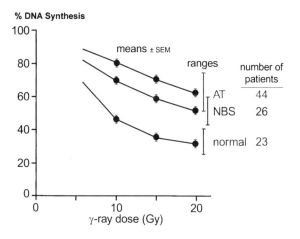

Fig. 1. Inhibition of DNA replication by ionizing radiation in primary human fibroblasts and its use in diagnosis of AT and NBS. Note that the RDS phenotype is somewhat less pronounced in NBS. (Unpublished data by N. G. J. Jaspers and W. J. Kleijer.)

ataxia-telangiectasia (AT) and Nijmegen breakage syndrome (NBS), as well as some laboratory-generated radiosensitive rodent cell lines *(1,2)*. AT and NBS cells have virtually lost the steep component of the inhibition curve. This property is usually referred to as "radioresistant DNA synthesis" (RDS), a phrase originally coined by Painter *(3)*. In the rodent mutants, diminished replicon initiation is also evident, but the degree of DNA synthesis inhibition is generally lower in the parental hamster cell lines as well *(4,5)*.

The RDS feature is widely considered to be a marker of the so-called radiation-induced "S-phase checkpoint." Upstream mediators of this thoroughly investigated checkpoint are ATM and the MRE11/RAD50/NBS1-complex (MRN). More downstream components are the CHK1 and CHK2 kinases that act on the regular cyclin-based cell cycle control machinery. So, characteristically, the RDS feature is encountered in AT patients having a defective *ATM* gene *(3)*, as well as in patients with NBS *(6)* or ataxia-telangiectasia-like disorder (ATLD) *(7)*, who carry mutations in one of the MRN subunits. However, regulation of the radiation-induced S-phase checkpoint is highly complex, and tampering with any of its modulators may provoke an RDS phenotype as well, as has been observed (for instance) in the case of Trp53 *(8)*, CDC25A *(9)*, BRCA2 *(10)*, or even HSP70 *(11)*. Moreover, a challenge with other types of genotoxic stressors (such as interstrand crosslinks, UV and DNA synthesis inhibitors) can provoke an alternative S-phase checkpoint route governed by the kinase ATR and a complementary set of coactors and modulators, which probably include the Fanconi anemia complex, the helicases BLM and WRN *(12)*, and the kinase BCR/ABL *(13)*. In normal cells, RDS can also be provoked by caffeine, an inhibitor of both ATR and ATM kinases *(14)*.

RDS is a typical feature of AT and NBS cells (*see* **Fig.1**). To date, more than 250 AT and NBS patients have been tested in this respect with almost invariably similar results. In very rare cases, DNA replication inhibition at intermediate or normal levels has been found. However, such cases always showed a nonstandard (mostly milder) clinical course of the disease, which was related either to a leaky mutation in the *ATM* gene or to a different genetic entity resembling AT in some respect *(15,16)*. AT and NBS heterozygotes, claimed to be slightly radiosensitive, cannot be distinguished from normal individuals with the RDS assay *(17)*. Although screening of the large *ATM* gene for the wide range of mutations found in AT families is feasible *(18,19)*, DNA synthesis rate measurements have remained an attractive first-line diagnostic tool for AT. In contrast, direct sequencing methods are preferred in the case of NBS where >90% of the cases carry a typical homozygous 5-base pair deletion in the *NBS1* gene.

Genetic heterogeneity in AT and NBS has been suggested on the basis of complementation of RDS phenotype in somatic cell hybrids *(20–22)*. However, the same *ATM* and *NBS1* genes were found mutated in all of the established complementation groups. Obviously, RDS is not a useful marker in cell-fusion– complementation studies, reflecting complicated and subtle interactions of the multifaceted S-phase checkpoint players *(5,23)*.

Traditionally, overall DNA replication is best measured by incorporation of radioactively labeled thymidine, followed by some quantification method such as liquid scintillation counting or autoradiography. The fact that the results of both techniques are very similar *(20)*, one reflecting whole cultures and the other single cells, indicates that over the labeling period, effects of G_1 arrest do not play a major role. What is measured here is clearly a checkpoint governing S-phase dependent DNA synthesis itself. Alternative DNA precursors, such as bromodeoxyuridine, have also come into use, after sensitive immunofluorimetric detection methods became available.

This chapter describes the most straightforward and reproducible experimental protocol, making use of double-labeling with radioactive thymidine. The procedure is highly accurate and reproducible within a few percent, and therefore also useful for clinical purposes (*see* **Note 1**).

2. Materials

1. A thymidine-free media formulation, with moderate buffering capacity, such as RPMI1640 or thymidine-free Ham's F10. (*See* **Note 2**.)

2. [^{14}C]Thymidine, specific activity 0.05 Ci/mmol (1 Ci = 37 GBq).
3. [*methyl*-^{3}H]Thymidine, specific activity from 2–50 Ci/mmol (or higher).
4. 1 m*M* Thymidine stock: Filter-sterilize, store at 4°C.
5. Phosphate-buffered saline (PBS).
6. 1 *M* Na–*N*-2-hydroxyethylpiperazine-*N'*-2-ethanesulfonic acid (HEPES), pH 7.35: Filter-sterilize and store at room temperature.
7. Freshly prepared 0.25 *M* NaOH.
8. Scintillation cocktail suitable for alkaline solutions (e.g., Hionic Fluor from Packard).
9. Cell culture dishes 30-mm or six-well clusters.

3. Methods
3.1. Cell Prelabeling

1. Seed cells (*well-dispersed!*) into 3-cm dishes; 5–7×10^4 primary human fibroblasts, or 1×10^4 transformed HeLa or hamster cells (e.g., Chinese hamster ovary) per dish (unirradiated controls four dishes, irradiated samples in duplicate) and culture for 20–30 h. Include standard (e.g., AT and normal) strains in each test. (*See* **Note 3**.)
2. Incubate cells overnight (15–17 h) in the presence of [^{14}C]thymidine, 0.05 µCi/mL (molar concentration = 1 µ*M*). Add HEPES to a concentration of 20 m*M* in the medium, from a 1 *M* stock, pH 7.35. (*See* **Note 4**.)

3.2. Radiation Exposure

Take the cells to the radiation source, preferably without changing the ^{14}C-medium. If medium change is required for safety reasons, use standard medium at room temperature with additional HEPES for strict pH control. Cooling down to 0–5°C is undesirable (the cold shock compromises the control of DNA synthesis) and unnecessary, since rejoining of most of the single- and double-strand breaks is fast (a few and 30 min, respectively) in comparison to the labeling period. (*See* **Note 4**.) With most ionizing radiation sources, dose rates are in the order of 1–2 Gy/min (*see also* **Notes 5** and **6**).

3.3. Second Labeling

1. Remove the ^{14}C-medium and immediately change with tritiated HEPES medium. Use [*methyl*-^{3}H]thymidine at 2.0 µCi/mL. With higher specific activities, add unlabeled thymidine to a final molar concentration of 1 µ*M* to reduce the effective specific activity and produce a predictable ^{3}H/^{14}C ratio of approx 10 (*see* **Subheading 3.5.** and **Note 7**).
2. Incubate for 3–4 h. (*See* **Note 6**.)

3.4. Collection of Labeled Cells

Any measurement of incorporation requires disposal of free labeled precursors present in the cells. The classic way of doing this is acid (trichloroacetic

acid [TCA]) precipitation (*see* **Note 8**). This tedious work can be replaced by a greatly simplified procedure:
1. Remove the medium and rinse the cells with warm PBS (37°C).
2. Incubate at 37°C with unlabeled medium for 20–30 min, to chase the labeled precursor pool.
3. Remove the medium and rinse once with PBS.
4. Squirt 500 μL of fresh 0.25 *M* NaOH over the monolayer to lyse the cells.
5. Pipet-mix the lysate and transfer to a scintillation vial.
6. Mix well with 7.5 mL of a scintillation cocktail. The solution should be clear.

3.5. Scintillation Counting

Make use of a dual-label counting program with in-line standard quenching curves, that discriminates between ^{3}H and ^{14}C and calculates the corrected dpm values for you. Checking proper action of this program with standard amounts of labeled mixtures is strongly advised. Counting time per vial is 5 min at least. The dpm calculations are accurate only within certain limits: ^{3}H/^{14}C dpm ratios should always be within 3 and 15 for all unirradiated samples. Values outside this range should be seriously mistrusted, because of this calculation problem and for other reasons (*see* **Note 7**).

3.6. Processing Data into Inhibition Curves

Use the ^{3}H/^{14}C ratios as a measure for the overall rate of DNA synthesis. Duplicate values are usually within 5%. Any spreadsheet program is suitable to process the data into standard inhibition curves. If standard errors (SEs) are calculated from the duplicates and represented in curves, the SE obtained in the unexposed controls should be transported into the data points (*see also* **Note 3**).

4. Notes

1. Clinical application. The RDS test has been routinely used for prenatal diagnosis of AT and NBS *(24,25)*. If the mutations in the index patient are known, direct sequencing methods are preferred of course *(26)*. Use of polymorphic genetic markers located near the ATM gene has also been advocated *(27)*. Sequencing methods have the significant advantage of direct processing of chorionic villus samples without culture. But since in most AT cases mutational information is not available, DNA synthesis measurement in cultured chorionic villus or amniotic fluid cells is usually the easiest and least costly alternative.
2. Media. Do not use DMEM. It is not possible to keep the pH stable within the required narrow range, even with added HEPES, and experimental variability will result. RPMI1640 and Ham's F10 or F12 maintain a stable pH. With Ham's F10, a special thymidine-free batch should be ordered. Dialysis of the serum supplement to remove thymidine (usually present at 0.10–0.15 μ*M*) is not required with the labeling protocol used here.

3. Strain variability. The degree of inhibition of DNA replication by ionizing radiation is widely different among cell lines, with human cells showing, in general, much more pronounced inhibition of DNA synthesis than rodent cells such as Chinese hamster V79 and CHO9 cells, although mouse A9 cells respond similarly to human cells *(4,5,28)*. The hamster cell mutants defective in Ku86 show a more pronounced inhibition than their parental cell lines *(4)*. Since transformed and tumor cells are characterized by seriously compromised growth control, diverse responses are to be expected in the steep component of the inhibition curve, which is in fact what is found. In addition, commonly used transformed strains may have very different average replicon sizes, which affect the shallow component of the inhibition curve.
4. Reproducibility. A DNA synthesis experiment is not at all complicated, however many investigators have reported difficulties with reproducibility. To obtain reproducible results:
   a. Minimize handling and temperature changes.
   b. Add HEPES buffer for strict pH control during all labeling periods.
   c. Include a prelabeling to have an internal standard, facilitating experimental handling.

   The first and second points are particularly vital, since the overall rate of DNA replication in cultures responds immediately to sudden changes in many physical parameters including temperature, pH, and osmotic pressure, even if they are small.
5. Other DNA damaging agents. Incubations with other agents such as bleomycin, adriamycin, cisplatin or mitomycin C (usually for 30–60 min) should always be performed immediately after prelabeling, and in presence of supplementary HEPES, to strictly control the pH.
6. Radiation doses. With primary human fibroblasts inhibition is largely reflecting the fast component in a dose range below 10–15 Gy. This dose range is therefore sufficient to study the effects on replicon initiation, which reflects the action of the S-phase checkpoint. After a dose of 10–20 Gy the minimal rate of overall replication is reached between 1.5 and 3 h, which implies that a good picture of DNA synthesis inhibition is best obtained in a pulse-labeling protocol of 2–4 h. Shorter times tend to produce more technical variation, without much gain in information.
7. Inappropriate ratios of ^{3}H to ^{14}C. Very low ratios may indicate that dishes were too full, and cells have started to reach confluence: in this case, discard the experiment. Very high ratios indicate serious problems in the initial phase of the experiment or point to microbial contamination (mycoplasma!). The amount of ^{14}C dpm per dish should be $2–7 \times 10^3$ with primary fibroblasts; with transformed cells 10 times higher values are usual. Incorporation levels of less than 1000 dpm per dish are unacceptably low—discard such an experiment.
8. Alternative protocol. TCA precipitation was traditionally used to harvest cells for scintillation counting, as follows:
   a. Remove the medium and rinse the cells with ice-cold PBS + 1 mM "cold" thymidine.

b. Scrape the cells into 1 mL of ice-cold PBS with a rubber policeman.
   c. Pipet the cells onto a glass filter disc (e.g., Whatman GF/C) soaked in 10% TCA (w/v), with gentle vacuum suction.
   d. Rinse the filter once with ice-cold 10% TCA and twice with ice-cold ethanol.
   e. Dry the filter in a scintillation vial (placed in an oven) and add 0.5 mL of solubilizer, for example, Soluene-350 (Perkin-Elmer).
   f. After 15–30 min, add scintillation fluid, mix well, and count.

References

1. Young, B. R., and Painter, R. B. (1989) Radioresistant DNA synthesis and human genetic diseases. *Hum. Genet.* **82,** 113–117.
2. Zdzienicka, M. Z. (1996) Mammalian X-ray-sensitive mutants: A tool for the elucidation of the cellular response to ionizing radiation, in *Genetic Instability and Cancer* (Lindahl, T., ed.), Cold Spring Harbor Laboratory Press, Plainview, NY, pp. 281–293.
3. Painter, R.B. (1981) Radioresistant DNA synthesis: an intrinsic feature of ataxia telangiectasia. *Mutat. Res.* 84, 183–190.
4. Verhaegh, G. W. C. T., Jaspers, N. G. J., Lohman, P. H. M, and Zdzienicka, M. Z. (1994) The relation between radiosensitivity and radioresistant DNA synthesis in ionizing radiation-sensitive rodent cell mutants, in *Molecular Mechanisms in Radiation Mutagenesis and Carcinogenesis* (Chadwick, K. H., Cox, R., Leenhouts, H. P., and Thacker, J., eds.), European Commission, European Atomic Energy Community, Brussels, Luxembourg, pp. 23–28.
5. Verhaegh, G. W. C. T., Jaspers, N. G. J., Lohman, P. H. M., and Zdzienicka, M. Z. (1993) Co-dominance of radioresistant DNA synthesis in a group of AT-like hamster cell mutants. *Cytogenet. Cell Genet.* **63,** 176–180.
6. Taalman, R. D, Jaspers, N. G. J. , Scheres, J. M., de Wit, J., and Hustinx, T. W. (1983) Hypersensitivity to ionizing radiation, in vitro, in a new chromosomal breakage disorder, the Nijmegen Breakage Syndrome. *Mutat. Res.* **112,** 23–32.
7. Stewart, G. S., Maser, R. S., Stankovic, T., et al. (1999) The DNA double-strand break repair gene hMRE11 is mutated in individuals with an ataxia-telangiectasia-like disorder. *Cell* **99,** 577–587.
8. Nasrin, N., Mimish, L. A., Manogaran, P. S., et al. (1997) Cellular radiosensitivity, radioresistant DNA synthesis, and defect in radioinduction of p53 in fibroblasts from atherosclerosis patients. *Arterioscler. Thromb. Vasc. Biol.* **17,** 947–953.
9. Falck, J., Mailand, N., Syljuasen, R. G., Bartek, J., and Lukas, J. (2001) The ATM-Chk2-Cdc25A checkpoint pathway guards against radioresistant DNA synthesis. *Nature* **410,** 842–847.
10. Kraakman-van der Zwet, M., Overkamp, W. J., van Lange, R. E., et al. (2002) Brca2 (XRCC11) deficiency results in radioresistant DNA synthesis and a higher frequency of spontaneous deletions. *Mol. Cell. Biol.* **22,** 669–679.
11. Hunt, C. R., Dix, D. J., Sharma, G. G., et al. (2004) Genomic instability and enhanced radiosensitivity in Hsp70.1- and Hsp70.3-deficient mice. *Mol. Cell. Biol.* **24,** 899–911.

12. Pichierri, P. and Rosselli, F. (2004) Fanconi anemia proteins and the S phase checkpoint. *Cell Cycle* **3,** 698–700.
13. Dierov, J., Dierova, R., and Carroll, M. (2004) BCR/ABL translocates to the nucleus and disrupts an ATR-dependent intra-S phase checkpoint. *Cancer Cell* **5,** 275–285.
14. Sarkaria, J. N., Busby, E. C., Tibbetts, R. S., et al. (1999) Inhibition of ATM and ATR kinase activities by the radiosensitizing agent, caffeine. *Cancer Res.* **59,** 4375–4382.
15. Morgan, S. E., Lovly, C., Pandita, T., Shiloh, Y., and Kastan M. B. (1997) Fragments of ATM which have dominant-negative or complementing activity. *Mol. Cell. Biol.* **17,** 2020–2029.
16. Taylor, A. M. R., McConville, C. M., Rotman, G., Shiloh, Y., and Byrd, P. J. (1994) A haplotype common to intermediate radiosensitivity variants of ataxia-telangiectasia in the UK. *Int. J. Radiat. Biol.* **66,** S35–41.
17. Jaspers, N. G. J., Scheres, J. M., De Wit, J., and Bootsma, D. (1981) Rapid diagnostic test for ataxia telangiectasia. *Lancet* **8244,** 473.
18. Broeks, A., de Klein, A., Floore, A. N., et al. (1998) *ATM* germline mutations in classical ataxia-telangiectasia patients in the Dutch population. *Hum Mutat.* **12,** 330–337.
19. Hellani, A., Lauge, A., Ozand, P., Jaroudi, K., and Coskun, S. (2002) Pregnancy after preimplantation genetic diagnosis for ataxia telangiectasia. *Mol. Hum. Reprod.* **8,** 785–788.
20. Jaspers, N. G. J. and Bootsma, D. (1982) Genetic heterogeneity in ataxia-telangiectasia studied by somatic cell fusion. *Proc. Natl. Acad. Sci. USA* **79,** 2641–2644.
21. Murnane, J. P. and Painter, R. B. (1982) Complementation of the defects in DNA synthesis in irradiated and unirradiated ataxia-telangiectasia cells. *Proc. Natl. Acad. Sci. USA* **79,** 1960–1963.
22. Jaspers, N. G. J., Gatti, R. A., Baan, C., Linssen, P. C., and Bootsma, D. (1988) Genetic complementation analysis of ataxia telangiectasia and Nijmegen breakage syndrome: a survey of 50 patients. *Cytogenet. Cell Genet.* **49,** 259–263.
23. Kraakman-van der Zwet, M., Overkamp, W. J., Jaspers, N. G. J., Natarajan, A. T., Lohman, P. H., and Zdzienicka, M. Z. (2001) Complementation of chromosomal aberrations in AT/NBS hybrids: inadequacy of RDS as an endpoint in complementation studies with immortal NBS cells. *Mutat. Res.* **485,** 177–185.
24. Jaspers, N. G. J., van der Kraan, M., Linssen, P. C., Macek, M., Seemanova, E., and Kleijer, W. J. (1990) First-trimester prenatal diagnosis of the Nijmegen Breakage Syndrome. *Prenat. Diagn.* **10,** 667–674.
25. Kleijer, W. J., van der Kraan, M., Los, F. J., and Jaspers, N. G. J. (1994) Prenatal diagnosis of ataxia-telangiectasia and Nijmegen breakage syndrome by the assay of radioresistant DNA synthesis. *Int. J. Radiat. Biol.* **66,** S167–174.
26. Chessa, L., Piane, M., Prudente, S., et al. (1999) Molecular prenatal diagnosis of ataxia-telangiectasia heterozygosity by direct mutational assays. *Prenat. Diagn.* **19,** 542–545.

27. Gatti, R. A., Peterson, K. L., Novak, J., et al. (1993) Prenatal genotyping of ataxia-telangiectasia. *Lancet* **342,** 376.
28. Verhaegh, G. W. C. T., Jongmans, W., Jaspers, N. G. J., et al. (1993) A gene which regulates DNA replication in response to DNA damage is located on human chromosome 4q. *Am. J. Hum. Genet.* **57,** 1095–1103.

6

Analysis of Inhibition of DNA Replication in Irradiated Cells Using the SV40-Based In Vitro Assay of DNA Replication

George Iliakis, Ya Wang, and Hong Yan Wang

Summary

The mechanisms of inhibition of DNA replication after DNA damage through the activation of the S-phase checkpoint have been the focus of several investigations over the last 40 yr. Recent studies have identified several components of this checkpoint response and there is strong interest in its biochemical characterization. Helpful for the delineation of the mechanism of the S-phase checkpoint is the observation that factors inhibiting DNA replication in vivo can be found in active form in extracts prepared from irradiated cells, when these are tested using the simian virus 40 (SV40) assay for in vitro DNA replication. In this assay, replication of plasmids carrying the minimal origin of SV40 DNA replication is achieved in vitro using cytoplasmic cell extracts and SV40 large tumor antigen (TAg) as the only noncellular protein. Here, we describe protocols developed to measure in vitro DNA replication with the purpose of analyzing its regulation after exposure to DNA damage. The procedures include the preparation of components of the in vitro DNA replication reaction including cytoplasmic extracts from cells that have sustained DNA damage. The assay is powerful but is limited by the fact that initiation steps carried out by the TAg in vitro may have different cellular determinants.

Key Words: Cell extract; DNA replication; in vitro assay; S-phase checkpoint; SV40; T-antigen.

1. Introduction

Inhibition of DNA replication in eukaryotic cells was one of the earliest effects of radiation to be reported and quantified. Elucidation of the mechanism causing this inhibition has been the focus of several laboratories for at least four decades. A significant development from these efforts was the recognition that the mechanism of inhibition has a direct (*cis*-acting) and an indirect (*trans*-acting) component (for reviews *see* **refs. 1,2**). Whereas the direct component is thought to derive from radiation-induced DNA damage that alters chromatin structure and inhibits DNA replication in *cis*, the indirect component is attributed to the activation by DNA damage of regulatory processes that inhibit DNA replication in *trans*. The latter is equivalent to the activation of a checkpoint during the S-phase *(2–4)*.

The realization that a checkpoint is activated in S-phase after induction of DNA damage has led to intensive studies aiming at its genetic and biochemical characterization. The best documented genetic alteration that affects the regulation of DNA replication in response to radiation exposure is found in individuals with the hereditary genetic disorder ataxia-telangiectasia (AT). AT cells fail to inhibit DNA replication in response to DNA damage suggesting that *ATM*, the gene mutated in these cells, is normally involved in the regulation of DNA replication (for reviews *see* **refs. 5,6**). Recent studies have identified several additional components of this checkpoint response and there is strong interest in its biochemical characterization.

Promising for the delineation of the mechanism of the S-phase checkpoint is the observation that factors that inhibit DNA replication in vivo can be found in active form in extracts prepared from irradiated cells, when these are tested for replication activity using the simian virus 40 (SV40) assay for in vitro DNA replication *(7–9)*. In this assay, replication of plasmids carrying the minimal origin of SV40 DNA replication is achieved in vitro using cytoplasmic cell extracts and SV40 large tumor antigen (TAg) as the only noncellular protein *(10–13)*. It is thought that cellular proteins function in this assay in the same manner as in vivo.

The assay has been extremely successful in the field of DNA replication and has led to the characterization of a number of factors involved in eukaryotic DNA replication. In a similar way, the assay could also help in the biochemical characterization of important components of the regulatory pathway activated in response to DNA damage (for example *see* **refs. 14,15**). While the assay is limited by the fact that a single noncellular protein, TAg, carries out functions assigned to different families of proteins in eukaryotic cells *(16)*, its potential for the biochemical characterization of checkpoint responses has not been exhausted and modifications mitigating some limitations may be possible.

Here, we describe protocols developed to measure in vitro DNA replication with the purpose of analyzing its regulation after exposure to DNA damage. The required procedures include: (1) preparation of cytoplasmic extract from cells that have sustained DNA damage; (2) preparation of the SV40 TAg; (3) Preparation of supercoiled plasmid DNA carrying the SV40 origin of DNA replication; (4) Assembly of in vitro replication reactions; and (5) assay of DNA replication using incorporation of radioactive precursors, and of the DNA

Inhibition Analysis With SV40 Assay

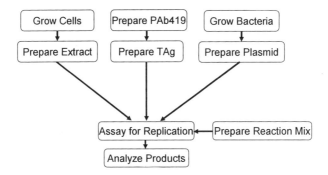

Fig. 1. Outline of the individual steps and preparations required for assembling DNA replication reactions using the SV40 system for in vitro DNA replication and for evaluating their outcome.

replication products using gel electrophoresis. **Figure 1** shows a graphic outline of these steps.

2. Materials
2.1. Preparation of HeLa Cell Extract

1. Minimum essential medium (MEM) modified for suspension cultures (S-MEM), supplemented with 5% iron-supplemented bovine serum and antibiotics (100 U/mL of penicillin, 100 µg/mL of streptomycin).
2. Hypotonic buffer solution: 10 mM N-2hydroxyethylpiperazine-N'-2-ethanesulfonic acid (HEPES), pH 7.5 (0.6 M stock, pH 7.5 at room temperature [RT]), 1.5 mM MgCl$_2$ (1 M stock), 5 mM KCl (3 M stock). Immediately before use add 0.2 mM phenylmethylsufonyl fluoride (PMSF) (100 mM stock in ethanol), and 0.5 mM dithiothreitol (DTT) (1 M stock in H$_2$O; store at –20°C), 20 mM β-glycerophosphate.
3. 10X cytoplasmic buffer: 10 mM HEPES, pH 7.5, 1.4 M KCl, and 1.5 mM MgCl$_2$.
4. Dialysis buffer: 25 mM Tris-HCl, pH 7.5 at 4°C (1 M stock, pH 7.5 at 4°C), 10% glycerol, 50 mM NaCl (5 M stock), 1 mM EDTA (0.5 M stock, pH 8.0). Immediately before use add 0.2 mM PMSF, 0.5 mM DTT, and 20 mM β-glycerophosphate.
5. Microcarrier spinner flasks of 30 L nominal volume (Bellco Glass, Inc.).
6. Microcarrier magnetic stirrers (Bellco Glass, Inc.).
7. 100-mm tissue culture dishes.
8. 50-mL Dounce homogenizer with B pestle.

2.2. Preparation of TAg

1. Hybridoma cell line PAb419. It is the L19 clone generated by Harlow et al. (17). It produces a monoclonal antibody that recognizes the amino terminal region of

TAg and is used for the preparation of immunoaffinity columns employed in the purification of TAg. The cell line can be requested from Dr. Harlow, and is used to produce antibody that can be purified using standard procedures *(18)*.
2. Insect cells *(Sf9)*. Available from ATCC.
3. Baculovirus *Autographa californica* expressing TAg. This virus was constructed by Lanford (941T) *(19)* using a cDNA copy of TAg mRNA and can be requested from the author. The procedures used to prepare stocks of the virus and to measure its infectivity in pfu/mL are described in specialized protocols and the reader is referred to these publications for more information, for example **ref. 20**.
4. TD buffer: 25 mM Tris-HCl, (1 M stock, pH 7.4), 136 mM NaCl (5 M stock), 5.7 mM KCl (3 M stock), 0.7 mM Na$_2$HPO$_4$ (0.2 M stock).
5. Buffer B: 50 mM Tris-HCl, (1 M stock, pH 8.0), 150 mM NaCl (5 M stock), 1 mM EDTA (0.5 M stock, pH 8.0), 10% glycerol, 1 mM PMSF (0.1 M stock), 1 mM DTT (1 M stock).
6. Buffer C: 50 mM Tris-HCl (1 M stock, pH 8.0), 500 mM LiCl (1 M stock), 1 mM EDTA (0.5 M stock, pH 8.0), 10% glycerol, 1 mM PMSF (0.1 M stock), 1 mM DTT (1 M stock).
7. Buffer D: 10 mM piperazine-1,4-*bis*(2-ethanesulfonic acid) (PIPES), (1 M stock, pH 7.4, dissolved in 1 M NaOH), 5 mM NaCl (5 M stock), 1 mM EDTA (0.5 M stock, pH 8.0), 10% glycerol, 1 mM PMSF (0.1 M stock), 1 mM DTT (1 M stock).
8. Buffer E: 20 mM triethylamine, 10% glycerol, pH 10.8. Make just before using.
9. Buffer F: 10 mM PIPES, (1 M stock, pH 7.0), 5 mM NaCl (5 M stock), 0.1 mM EDTA (stock 0.5 M, pH 8.0), 10% glycerol, 1 mM PMSF (0.1 M stock), 1 mM DTT (1 M stock).
10. Reagents: sodium borate, dimethylpipelimidate, ethanolamine, triethylamine, merthiolate, Nonidet P-40 (NP-40).
11. Chromatography supplies: Protein A agarose, Sepharose 4B-Cl (Invitrogen, Amersham Biosciences, etc.), 5-mL syringes or EconoColumns with ID of 0.75 cm (Bio-Rad).
12. Supplies and equipment for sodium dodecyl sulfate-polyacrylamide gel electrophoresis (SDS-PAGE).
13. Two 1-L and one 250-mL microcarrier spinner flasks (Bellco).
14. Material for protein determination using the Bradford assay (a ready-to-use solution is available from Bio-Rad).

2.3. Preparation of Supercoiled Plasmid DNA Carrying the SV40 Origin of DNA Replication

Several plasmids that carry the mimimum origin of SV40 DNA replication are available and can be used for this purpose, for example, pSV01ΔEP *(21)*, pSV010 *(22)*, and pJLO *(23)*. Large quantities of these plasmids can be prepared using cesium chloride ethidium bromide gradients. The description of these methods are beyond the scope of this chapter and can be found in publications describing molecular biology protocols *(20,24)*.

2.4. Assembly of In Vitro Replication Reactions and Product Analysis

1. Replication reaction solution (5X): 200 mM HEPES–KOH, pH 7.5, 40 mM MgCl$_2$, 2.5 mM DTT, 200 mM phosphocreatine, 15 mM ATP, 1 mM CTP, 1 mM GTP, 1 mM UTP, 0.5 mM dATP, 0.125 mM dCTP, 0.5 mM dGTP, 0.5 mM dTTP. Prepare 10 mL of replication reaction solution and freeze in small aliquots.
2. Creatine phosphokinase stock: 2.5 mg/mL prepared in 50% v/v glycerol. Phosphokinase together with phosphocreatine present in the replication reaction solution form an ATP regeneration system that is required for DNA replication.
3. Salmon sperm DNA.
4. Reagents: EDTA, trichloroacetic acid, SDS.
5. Enzymes: Proteinase K, RNase A.
6. Scintillation counter.
7. Gel electrophoresis equipment.
8. Glass fiber filters (GF/C, Whatman).

3. Methods

3.1. Preparation of Cell Extract (see Notes 1 and 2)

The method described here is a modification of a method originally developed by Dignam et al. *(25)*, and allows the preparation of cell extract from 10 L of cell suspension. Higher or lower amounts of extract can be prepared by appropriate scaling. Ionizing radiation or radiomimetic chemicals can be used for the generation of damage in the DNA. We obtained satisfactory results using a 3-h treatment with 0.5 µg/mL of camptothecin. We also use routinely 10–40 Gy of 25 MV X-rays from a linear accelerator. The details for the latter treatment depend heavily upon the type of equipment used. It is therefore impractical to describe them here. Investigators with access to such equipment are advised to request the assistance of radiation physicists for dosimetry and other details on the irradiation protocol. Standard X-ray-producing equipment (50–250 kV) cannot be used to irradiate volumes of the magnitude described here owing to their low penetration characteristics. Whatever the solution for the irradiation problem, it is important to keep in mind that disturbance in the cell culture has to be kept to a minimum before and after irradiation for reproducible results. Significant reductions in temperature, changes of the growth vessels, centrifugations, and so on, should be avoided. Extracts can be processed for repair at a specific time after irradiation. The precise timing will depend upon the type of experiments and its specific goals. We prepare extracts 0–3 h after the treatment to induce DNA damage.

1. Grow HeLa cells at 37°C for 3 d in 25 100-mm tissue culture dishes prepared at an initial density of 6×10^6 cells/dish in 20 mL of S-MEM supplemented with serum and antibiotics. The final density after 3 d of growth should be approx 20 $\times 10^6$ cells/dish, giving a total of 5×10^8 cells in 25 dishes.

2. Trypsinize cells from all dishes and resuspend in 10 L of prewarmed complete growth medium in a 30-L nominal volume microcarrier flask, thoroughly pregassed with 5% CO_2 in air. The initial cell concentration should be approx 5×10^4 cells/mL. Place in a warm room at 37°C; provide adequate stirring (approx 40–60 rpm).
3. Allow cells to grow for four days, to a final concentration of $4–6 \times 10^5$ cells/mL. Do not exceed this concentration.
4. Collect cells by centrifugation (8 min at $2500g$). Collection should be fast and is best done using a refrigerated centrifuge that can accept 1-L bottles (e.g., Beckman J6-MI). All further processing should be carried out at 0–4°C.
5. Rinse twice in phosphate-buffered saline (PBS) and centrifuge (5 min at $500g$). Determine the packed cell volume (pcv) (approx 12–15 mL total).
6. Resuspend the cell pellet in 5 pcv of hypotonic buffer solution and centrifuge quickly (5 min, $500g$). Cells swell and the pcv approx doubles.
7. Determine the new pcv. Resuspend the cell pellet in 3 pcv of hypotonic buffer and disrupt in a Dounce homogenizer (20 strokes, pestle B). It is advisable to test cell disruption using a phase contrast microscope.
8. Add 0.11 volumes of high-salt buffer and centrifuge at $3000g$ for 20 min.
9. Carefully remove the supernatant and centrifuge at $100,000g$ for 1 h.
10. Place the resulting extract (S100) in dialysis tubing with a molecular mass cutoff of 10–14 kDa and dialyze overnight against 50–100 volumes of dialysis buffer.
11. Collect the extract. Centrifuge at $15,000g$ for 20 min to remove precipitated protein. Aliquot, and snap freeze. Store at –70°C. Keep a small aliquot for determining protein concentration using the Bradford assay (Protein assay, Bio-Rad).

3.2. Preparation of TAg

Good quality TAg is essential for efficient replication in vitro of plasmids carrying the SV40 origin of DNA replication. Investigators can either obtain this protein from commercially available sources, or can prepare it in the laboratory using available reagents. The method of preparation described here is essentially the one described by Simanis and Lane *(26)*, and can be conveniently separated into two parts: (1) preparation of the TAg immunoaffinity column and (2) purification of TAg from extracts of *sf*9 cells infected with the baculovirus 941T which expresses Tag.

3.2.1. Preparation of TAg Immunoaffinity Column

1. Mix 5 mg of PAb419 antibody with 2 mL of wet Protein A beads. Incubate at room temperature for 1 h with gentle rocking.
2. Wash the beads twice with 20 mL of 0.2 M sodium borate, pH 9.0, by centrifugation at $1000g$ for 5 min.
3. Resuspend the beads in 20 mL of 0.2 M sodium borate (pH 9.0), mix, and remove 100 µL of bead suspension for assaying coupling efficiency. Add solid dimethylpipelimidate to bring the final concentration to 20 mM.

4. Mix for 30 min at room temperature on a rocker and remove a 100-µL suspension of the coupled beads.
5. Stop the coupling reaction by washing the beads with 20 mL of 0.2 M ethanolamine, pH 8.0, and incubate for 2 h at room temperature in 0.2 M ethanolamine with gentle rocking.
6. Spin down and resuspend the beads in PBS. Add 0.01% merthiolate if extensive storage is anticipated. At this point the beads are ready for use in the purification of TAg (if the quality control performed next is positive).
7. Check the efficiency of coupling by boiling in Laemmli sample buffer the samples of beads taken before and after coupling. Run the equivalent of 1-µL and 9-µL beads from both samples on a 10% SDS-PAGE gel and stain with Coomassie blue. Good coupling is indicated by heavy chain bands (55 kDa) in the samples obtained before but not in the samples obtained after coupling.
8. Prepare the immunoaffinity column by pouring beads into a 5-mL syringe, or a 0.75-cm diameter EconoColumn.

3.2.2. Purification of TAg From Extracts of sf9 Cells

1. Grow enough *sf*9 cells to prepare 1 L of cell suspension at 2×10^5 cells/mL. Distribute the cell suspension in two 1-L microcarrier spinner flasks (500 mL of cell suspension per spinner) and incubate at 27°C under gentle spinning (1–1.5 revolutions per second). Allow the cells to grow until they reach a concentration of 2×10^6 cells/mL (3–4 d). (*See* **Note 3**.)
2. Centrifuge the cells for 5 min at 500g and carefully return the supernatant to spinner flasks.
3. Resuspend the cells in enough volume of virus stock to reach a multiplicity of infection equal to 10 plaque-forming units (pfu) per cell. Place the cell suspension in a 250-mL spinner flask and allow attachment of virus to the cells by gentle stirring at 27°C for 2 h.
4. Return the cell suspension to the original spinner flasks and incubate for 48 h at 27°C under gentle stirring to allow for protein expression. We have recently noted that good protein expression is also achieved by adding the viral stock directly into the cell culture and incubating for 48 h.
5. Harvest the cells by centrifugation at 500g for 5 min and resuspend in 25 mL of TD buffer.
6. Centrifuge at 500g for 5 min and resuspend in 30 mL of buffer B. Add 10% NP-40 to a final concentration of 0.5%.
7. Place in ice for 30 min. Invert the tube several times every 10 min.
8. Centrifuge for 10 min at 10,000g in a Corex tube. At this stage extract can be removed and quickly frozen at –70°C for use at a later time to purify TAg. When needed, this extract is thawed quickly by immersing in warm water. Do not allow extract to warm up above 4°C. All subsequent steps should be carried out in a cold room.
9. Save a 50-µL aliquot of the cell extract and load the remaining material onto a 2 mL of 4B-Cl Sepharose column equilibrated with buffer B. Elution can be

achieved either by gravity, or with the help of a peristaltic pump giving a flow rate of approx 1 mL/min. This step retains proteins binding nonspecifically to agarose.
10. Allow the flowthrough of **step 9** to pass through a 2-mL Protein A–Sepharose column equilibrated with buffer B. This can be achieved either by gravity, or with the help of a peristaltic pump giving a flow rate of approx 1 mL/min. This step retains material binding nonspecifically to Protein A.
11. Allow the flowthrough of **step 10** to pass through a 2-mL immunoaffinity column of Protein A–agarose coupled with antibody PAb419. This can be achieved either by gravity, or with the help of a peristaltic pump at a flow rate of approx 0.5 mL/min. This step retains TAg from the cell extract.
12. Repeat **step 11**. Save the flowthrough.
13. Wash the immunoaffinity column with 200 mL of buffer C. This is best achieved with a peristaltic pump giving a flow rate of approx 2 mL/min.
14. Wash the immunoaffinity column with 100 mL of buffer D. This is best achieved with a peristaltic pump giving a flow rate of approx 2 mL/min.
15. Elute with buffer E (*see* **Note 4**). Collect 0.5 mL fractions in tubes containing 25 µL of 1.0 M PIPES, pH 7.0. Place the fractions in ice.
16. Wash the immunoaffinity column with 20 mL of buffer E, and then with 40 mL of buffer B. Column can be reused four or five times.
17. Measure protein concentration using the Bradford assay.
18. Combine fractions with protein. Dialyze overnight against 2 L of buffer F.
19. Test purity by SDS-PAGE followed by silver staining. Test activity for in vitro SV40 DNA replication. The procedure yields 1–2 mg TAg per liter of *sf*9 cell culture.

3.3. Preparation of Supercoiled Plasmid DNA Carrying the SV40 Origin of DNA Replication

Standard procedures can be used for the preparation of supercoiled plasmid DNA. It is preferable to purify the DNA using a two-step purification on a CsCl/ethidium bromide gradient. Detailed protocols for this purpose can be found in **refs. 20,24**.

3.4. Assembly and Analysis of In Vitro Replication Reactions

3.4.1. Assembly of Reactions and Evaluation of Replication Activity

1. Assemble 50-µL reactions by mixing in an Eppendorf tube kept in ice, 5 µL of reaction buffer, 2.5 µL of creatine phosphate, 200–400 µg of extract protein, 1 µg of TAg, 0.3 µg of superhelical plasmid DNA, 0.001 µCi/mL [α-^{32}P]dCTP; add H$_2$O to 50 µL. Extracts from untreated and treated cells should be used in parallel so that the results obtained can be directly compared.
2. Incubate reactions at 37°C for 1 h. Longer or shorter incubations can also be used if information on the kinetics of replication is desired.
3. Terminate reactions by adding EDTA to a final concentration of 20 m*M*.

4. Add 25 µg of denatured salmon sperm DNA and mix well.
5. Add 1 mL of cold 10% trichloroacetic acid to precipitate acid-insoluble material. Mix well.
6. Collect the precipitate onto Whatman GF-C glass fiber filters. Wash three times with 10 mL of cold 10% trichloroacetic acid. Wash four times with 10 mL of deionized water.
7. Add 5 mL of scintillation fluid. Measure incorporated activity in a scintillation counter.

3.4.2. Analysis of the DNA Replication Products

1. To analyze the DNA replication products by gel electrophoresis, add 0.1% SDS to the stopped replication reactions.
2. Digest with RNase (20 µg/mL) for 15 min at 37°C.
3. Add Proteinase K (200 µg/mL) and incubate at 37°C for 30 min.
4. Purify DNA either by extraction in phenol–chloroform followed by precipitation in ethanol, or by using commercially available DNA purification systems.
5. Separate in 1% agarose at 6.5 V/cm for 2 h. Electrophoretic conditions may need to be modified if larger plasmids are used. For optimal resolution reduce field intensity to 1 V/cm.

4. Notes

1. The preparation of a good extract depends strongly on the quality of the cells used. When cell growth is not optimal, or when cells overgrow, low replication activity may result and the inhibition in extracts of treated cells may be suboptimal. To ensure optimal growth we routinely take the following measures:
   a. Carefully test different batches of serum to find one with good growth characteristics. HeLa cells have a generation time of less than 20 h when grown as a monolayer, and less than 24 h when grown in suspension, under optimal growth conditions.
   b. Use cells grown in dishes to start the suspension cultures for extract preparation. This helps to reduce clumping after extensive growth in suspension.
   c. We follow cell growth daily, and collect cells for extract preparation when they reach a concentration of 4–6 × 10^5 cells/mL.
   d. We measure cell cycle distribution by flow cytometry. A high percentage of S-phase cells (approx 25% for HeLa cells), suggests that the cell culture is still in an active state of growth.
2. The effect of the DNA damage inducing agent on DNA replication in vitro can be variable. We found that this is usually due to the growth conditions (overgrown cultures) or to the absence of phosphatase inhibitors. We routinely add β-glycerophosphate since we found it to significantly improve the reproducibility. Other phosphatase inhibitors as well as the use of protease inhibitors should be considered if reproducibility problems persist.
3. Optimal cell growth is also a prerequisite of a successful preparation of TAg. We find that *sf*9 cells grow more consistently if kept in suspension. Transfer from a

monolayer state to a suspension state is usually associated with a shock that takes the cells some time to overcome.
4. We have observed that 20 mM triethylamine may not elute all bound TAg from the immunoaffinity column. If this proves to be the case, increasing the triethylamine concentration (up to 100 mM), or alternative eluting methods (*see* **ref. *18***) should be considered. However, it should be kept in mind that such alternatives may reduce TAg activity.

References

1. Iliakis, G. (1997) Cell cycle regulation in irradiated and nonirradiated cells. *Semi. Oncol.* **24,** 602–615.
2. Iliakis, G., Wang, Y., Guan, J., and Wang, H. (2003) DNA damage checkpoint control in cells exposed to ionizing radiation. *Oncogene* **22,** 5834–5847.
3. Zhou, B.-B. S. and Elledge, S. J. (2000) The DNA damage response: putting checkpoints in perspective. *Nature* **408,** 433–439.
4. Nyberg, K. A., Michelson, R. J., Putnam, C. W., and Weinert, T. A. (2002) Toward maintaining the genome: DNA damage and replication checkpoints. *Annu. Rev. Genet.* **36,** 617–656.
5. Shiloh, Y. (2001) ATM and ATR: networking cellular responses to DNA damage. *Curr. Opin. Genet. Dev.* **11,** 71–77.
6. Abraham, R. T. (2001) Cell cycle checkpoint signaling through the ATM and ATR kinases. *Genes Dev.* **15,** 2177–2196.
7. Wang, Y., Huq, M. S., Cheng, X., and Iliakis, G. (1995) Regulation of DNA replication in irradiated cells by *trans*-acting factors. *Radiat. Res.* **142,** 169–175.
8. Wang, Y., Huq, M. S., and Iliakis, G. (1996) Evidence for activities inhibiting *in trans* initiation of DNA replication in extract prepared from irradiated cells. *Radiat. Res.* **145,** 408–418.
9. Wang, Y., Perrault, A. R., and Iliakis, G. (1997) Down-regulation of DNA replication in extracts of camptothecin-treated cells: activation of an S-phase checkpoint? *Cancer Res.* **57,** 1654–1659.
10. Challberg, M. D. and Kelly, T. J. (1989) Animal virus DNA replication. *Annu. Rev. Biochem.* **58,** 671–717.
11. Hurwitz, J., Dean, F. B., Kwong, A. D., and Lee, S.K. (1990) The *in vitro* replication of DNA containing the SV40 origin. *J. Biol. Chem.* **265,** 18,043–18,046.
12. Kelly, T. J. (1988) SV40 DNA replication. *J. Biol. Chem.* **263,** 17,889–17,892.
13. Stillman, B. (1989) Initiation of eukaryotic DNA replication in vitro. *Annu. Rev. Cell Biol.* **5,** 197–245.
14. Wang, Y., Zhou, X. Y., Wang, H.-Y., and Iliakis, G. (1999) Roles of replication protein A and DNA-dependent protein kinase in the regulation of DNA replication following DNA damage. *J. Biol. Chem.* **274,** 22,060–22,064.
15. Wang, Y., Guan, J., Wang, H., Wang, Y., Leeper, D.B., and Iliakis, G. (2001) Regulation of DNA replication after heat shock by RPA-nucleolin interactions. *J. Biol. Chem.* **276,** 20,579–20,588.

16. Mendez, J. and Stillman, B. (2003) Perpetuating the double helix: molecular machines at eukaryotic DNA replication origins. *BioEssays* **25,** 1158–1167.
17. Harlow, E., Crawford, L. V., Pim, D. C., and Williamson, N. M. (1981) Monoclonal antibodies specific for simian virus 40 tumor antigens. *J. Virol.* **39,** 861–869.
18. Harlow, E. and Lane, D. (1999) *Using Antibodies*, Cold Spring Harbor Laboratory Press, Cold Spring Harbor, NY.
19. Lanford, R. E. (1988) Expression of simian virus 40 T antigen in insect cells using a baculovirus expression vector. *Virology* **167,** 72–81.
20. Ausubel, F. M., Brent, R., Kingston, R. E., et al. (1994) *Current Protocols in Molecular Biology.* Greene Publishing Associates and Wiley-Interscience, New York, NY.
21. Wobbe, C. R., Dean, F., Weissbach, L., and Hurwitz, J. (1985) *In vitro* replication of duplex circular DNA containing the simian virus 40 DNA origin site. *Proc. Natl. Acad. Sci. USA* **82,** 5710–5714.
22. Prelich, G. and Stillman, B. (1988) Coordinated leading and lagging strand synthesis during SV40 DNA replication in vitro requires PCNA. *Cell* **53,** 117–126.
23. Li, J. J. and Kelly, T. J. (1984) Simian virus 40 DNA replication in vitro. *Proc. Natl. Acad. Sci. USA* **81,** 6973–6977.
24. Sambrook, J. and Russell, D. W. (2001) *Molecular Cloning.* 3rd Ed. Cold Spring Harbor Laboratory Press, Cold Spring Harbor, NY.
25. Dignam, J. D., Lebovitz, R. M., and Roeder, R. G. (1983) Accurate transcription initiation by RNA polymerase II in a soluble extract from isolated mammalian nuclei. *Nucleic Acids Res.* **11,** 1475–1489.
26. Simanis, V. and Lane, D. P. (1985) An immunoaffinity purification procedure for SV40 large T antigen. *Virology* **144,** 88–100.

7

Cytometric Assessment of Histone H2AX Phosphorylation

A Reporter of DNA Damage

Xuan Huang and Zbigniew Darzynkiewicz

Summary

DNA damage that leads to formation of DNA double-strand breaks (DSBs) induces phosphorylation of histone H2AX on Ser-139 at sites flanking the breakage. Immunocytochemical detection of phosphorylated H2AX (denoted as γH2AX) thus provides a marker of DSBs. The method presented in this chapter describes the detection of γH2AX for revealing the presence of DSBs, combined with differential staining of cellular DNA for revealing the cell cycle phase. The detection of γH2AX is based on indirect immunofluorescence using secondary antibody tagged with fluorescein isothiocyanate (FITC) while DNA is counterstained with propidium iodide (PI). Intensity of cellular green (FITC) and red (PI) fluorescence is measured by flow cytometry and bivariate analysis of the data is used to correlate the presence of DSBs with the cell cycle phase.

Key Words: Antitumor drugs; cell cycle phase; DNA damage; double-strand DNA breaks; flow cytometry; histone H2AX phosphorylation; immunofluorescence; ionizing radiation.

Introduction

DNA damage that involves formation of DNA double-strand breaks (DSBs) triggers phosphorylation of histone H2AX *(1)* which is one of several variants of the nucleosome core histone H2A family *(2)*. The phosphorylation, mediated either by ataxia-telangiectasia mutated (ATM), ataxia telangiectasia related (ATR) or DNA-dependent protein kinase (DNA-PK) *(3–5)*, occurs on Ser-139 at the C-terminus of H2AX molecules flanking the DSBs in chromatin *(2)*. The phosphorylated form of H2AX is defined as γH2AX *(6)*. Antibodies that detect γH2AX have recently become commercially available. The appearance of γH2AX in chromatin in the form of discrete nuclear foci, each focus represent-

From: *Methods in Molecular Biology: DNA Repair Protocols: Mammalian Systems, Second Edition*
Edited by: D. S. Henderson © Humana Press Inc., Totowa, NJ

ing a single DSB, can be detected immunocytochemically shortly after induction of DSBs *(7)*. Intensity of γH2AX immunofluorescence (IF) of the individual cell thus reports the extent of DNA damage (frequency of DSBs) in its nucleus. Flow or laser scanning cytometry offers the possibility to rapidly quantify γH2AX IF in large cell populations and multiparameter analysis of the cytometric data makes it possible to correlate DNA damage with other attributes of the cell, for example, cell cycle phase *(8–13)*. This approach to measure DNA damage is rapid, more sensitive, and less cumbersome compared with the alternative, commonly used method, the comet assay *(14)*.

The method presented in this chapter is designed to immunocytochemically detect γH2AX for revealing the presence of DSBs, combined with differential staining of DNA, to define the cell cycle phase of the cells in which DSBs are being detected (*see* **Note 1**). The detection of γH2AX is based on indirect immunofluorescence using the secondary antibody tagged with fluorescein isothiocyanate (FITC) while DNA is counterstained with propidium iodide (PI). The cells are briefly fixed in methanol-free formaldehyde and then transferred into 70% ethanol in which they can be stored at –20°C at least for 2 wk, perhaps longer. Ethanol treatment makes the plasma membrane permeable to the γH2AX antibody; further permeabilization is achieved by including the detergent Triton X-100 into a solution used to incubate cells with the antibody. After incubation with the primary γH2AX antibody, the cells are incubated with FITC-labeled secondary antibody and their DNA is then counterstained with PI in the presence of RNase A to remove RNA, which otherwise may also be stained with PI. Intensity of cellular green (FITC) and red (PI) fluorescence is measured by flow cytometry.

It should be noted that DSBs can also be intrinsic, occurring in healthy, nontreated cells, for example in the course of V(D)J and class-switch recombination during immune system development *(15,16)* or during DNA replication *(8,9)*. Furthermore, DSBs are formed in the course of DNA fragmentation in apoptotic cells *(12)*. Strategies are presented in this chapter to differentiate between DSBs induced by DNA damaging agents vs the intrinsically formed DSBs in untreated cells (*see* **Note 2**), or vs apoptosis-associated DSBs (*see* **Note 3**).

2. Materials

1. Cells to be analyzed: $10^6 - 5 \times 10^6$ cells, untreated (control) and treated with the DSB inducing agent(s), suspended in 1 mL of tissue culture medium.
2. 70% Ethanol.
3. Phosphate-buffered saline (PBS).
4. Methanol-free formaldehyde fixative: Prepare 1% (v/v) solution of methanol-free formaldehyde (Polysciences, Warrington, PA) in PBS. This solution may be stored at 4°C for up to 2 wk.

Assessment of Histone H2AX Phosphorylation

5. BSA–T–PBS: Dissolve bovine serum albumin (BSA; Sigma) in PBS to obtain a 1% (w/v) BSA solution. Add Triton X-100 (Sigma) to obtain 0.2% (v/v) of its concentration. This solution may be stored at 4°C for up to 2 wk.
6. PI (Molecular Probes, Eugene, OR) stock solution: Dissolve PI in distilled water to obtain 1 mg/mL of solution. This solution can be stored at 4°C in the dark (e.g., in the tube wrapped in aluminum foil) for several months.
7. PI staining solution: Dissolve RNase A (DNase-free; Sigma) in PBS to obtain 0.1% (w/v; 100 mg/mL) solution. Add an appropriate aliquot of PI stock solution (e.g., 5 µL per 1 mL) to obtain its 5 µg/mL final concentration. Store the PI staining solution in the dark. This solution may be stored at 4°C for up to 2 wk.
8. Unconjugated primary antibody: Histone γH2AX antibody (murine monoclonal, available from Upstate Biotechnology, Lake Placid, NY; alternatively, rabbit polyclonal, available from Trevigen, Gaithersburg, MD).
9. FITC-conjugated secondary antibody, for example, either polyclonal goat anti-mouse, or antirabbit-F(ab′)^2, depending on the source of the primary antibody, appropriately titered.
10. 12 × 75 mm polypropylene tubes.
11. Centrifuge and rotor capable of 300g.
12. Flow cytometers of different types, offered by several manufacturers, can be used to measure cell fluorescence following staining according to the protocol given below. The most common flow cytometers are from Coulter Corporation (Miami, FL), Becton Dickinson Immunocytometry Systems (San Jose, CA), Cytomation/DAKO (Fort Collins, CO) and PARTEC (Zurich, Switzerland).

3. Methods

1. Centrifuge cells collected from tissue culture (suspended in culture medium) at 300g for 4 min at room temperature. Suspend the cell pellet (1–2 × 10^6 cells) in 0.5 mL of PBS.
2. With a Pasteur pipet transfer this cell suspension into a 6-mL polypropylene tube (*see* **Note 4**) containing 4.5 mL of ice-cold 1% methanol-free formaldehyde solution in PBS. Keep on ice for 15 min.
3. Centrifuge at 300g for 4 min at room temperature and suspend the cell pellet in 4.5 mL of PBS. Centrifuge again as in **step 1** above and suspend the cell pellet in 0.5 mL of PBS. With a Pasteur pipet, transfer the suspension to a tube containing 4.5 mL of ice-cold 70% ethanol. The cells should be maintained in 70% ethanol at −20°C for at least 2 h, but may be stored under these conditions for up to 2 wk.
4. Centrifuge at 200g for 4 min at room temperature, remove the ethanol and suspend the cell pellet in 2 mL of BSA–T–PBS solution.
5. Centrifuge at 300g for 4 min at room temperature and suspend the cells again in 2 mL of BSA–T–PBS. Keep at room temperature for 5 min.
6. Centrifuge at 300g for 4 min at room temperature and suspend the cells in 100 µL of BSA–T–PBS containing 1 µg of the primary γH2AX antibody (*see* **Note 5**).
7. Cap the tubes to prevent drying and incubate them overnight at 4°C (*see* **Note 6**).
8. Add 2 mL of BSA–T–PBS and centrifuge at 300g for 4 min at room temperature.

9. Suspend the cells in 2 mL of BSA–T–PBS and centrifuge at 300g for 4 min at room temperature.
10. Suspend the cell pellet in 100 µL of BSA–T–PBS containing the appropriate (antimouse or antirabbit, depending on the source of the primary antibody) FITC-tagged secondary antibody (see **Note 5**).
11. Incubate for 1 h at room temperature, occasionally gently shaking. Add 5 mL of BSA–T–PBS and after 2 min centrifuge at 300g for 4 min at room temperature.
12. Suspend the cells in 1 mL of the PI staining solution. Incubate at room temperature for 30 min in the dark.
13. Set up and adjust the flow cytometer for excitation with light at blue wavelength (488-nm laser line or BG-12 excitation filter).
14. Measure the intensity of green (530 ± 20 nm) and red (>600 nm) fluorescence of the cells by flow cytometry. Record the data.

4. Notes

1. On the bivariate distributions (scatterplots), subpopulations of cells in G_1 vs S vs G_2/M are distinguished based on differences in their DNA content (intensity of PI fluorescence; see **Fig. 1**). To assess the mean extent of DNA damage (frequency of DSBs) for cells at a particular phase of the cycle, the mean values of γH2AX IF are calculated separately for G_1, S, and G_2/M cells by the computer-interactive "gating" analysis. It should be noted, however, that along with the doubling of DNA, histone content also doubles during the cell cycle. In fact the histone:DNA content ratio remains invariable throughout the cell cycle *(17)*. Therefore, cells in S and G_2/M with the same "degree" of H2AX phosphorylation (i.e., the "same percentage of phosphorylated H2AX molecules per total H2AX") as cells in G_1, will have (because of their higher histone content) 1.5 and 2.0 times higher γH2AX IF compared to cells in G_1 phase. To assess the degree of H2AX phosphorylation, that is, to make γH2AX IF independent of histone doubling during the cycle, the data may be normalized by presenting per unit of DNA (histone) by dividing the mean γH2AX IF of S- and G_2/M-phase cells by 1.5 and 2.0, respectively. After such normalization the intensity of γH2AX IF reflects the frequency of DSBs per unit of DNA rather than per nucleus, making this parameter independent of the cell cycle phase.
2. A low level of γH2AX IF seen in the cells that have not been treated with inducers of DSBs represents an intrinsic ("programmed") H2AX phosphorylation, primarily associated with DNA replication *(8)*. The level of intrinsic γH2AX IF varies between different cell lines. To quantify the γH2AX IF induced by external factors that damage DNA, this intrinsic component of γH2AX IF has to be subtracted. Towards this end the means of γH2AX IF of G_1, S, and G_2/M-phase untreated cells are subtracted from the respective means of the G_1, S, and G_2/M subpopulations of the radiation-, drug- or carcinogen-treated cells, respectively (**ref. *11***; and see **Fig. 1**). After the subtraction the extent of increase in intensity of γH2AX IF (Δ γH2AX IF) over the untreated sample represents the "treatment-induced phosphorylation" of this protein. There is no need, therefore, to use the

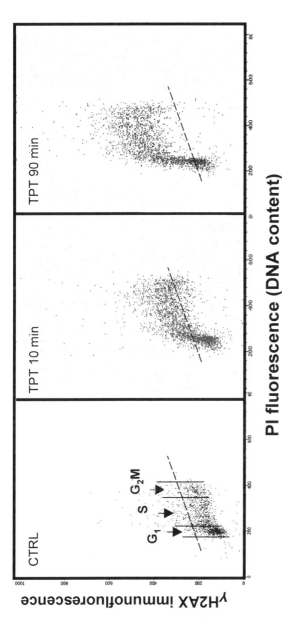

PI fluorescence (DNA content)

Fig. 1. Detection of histone H2AX phosphorylation in relation to the cell position in the cell cycle. The bivariate (γH2AX IF vs DNA content) distributions (scatterplots) of the untreated cells (CTRL) and, to induce DSBs, of the cells treated with 0.15 μM of the DNA topisomerase I inhibitor topotecan (TPT) for 10 or 90 min (11). The cells were processed as described in **Subheading 3.**, their fluorescence was measured by flow cytometry (FACScan; Becton Dickinson Immunochemistry, San Jose, CA). Notice increased γH2AX IF of TPT-treated cells. Based on differences in DNA content, subpopulations of cells in G_1 vs S vs G_2M phases of the cycle may be distinguished and gated as shown in the CTRL sample. The mean γH2AX IF may then be calculated for each subpopulation. To estimate the extent of treatment-induced H2AX phosphorylation the γH2AX IF means of the untreated cells have to be subtracted from the respective G_1, S, and G_2M means of the treated samples. The dashed line (– – –) represents the γH2AX IF level below which 95% of cells from the untreated culture (CTRL) express γH2AX.

isotype control to estimate the nonspecific antibody binding component, because it is expected that this component is similar for the untreated and treated cells, and thus is being subtracted while calculating Δ γH2AX IF.

DNA undergoes extensive fragmentation in apoptotic cells *(18)*. It is often desirable, therefore, to distinguish between primary DSBs induced by DNA damaging agents vs DSBs generated during apoptosis. The following attributes of γH2AX IF allow one to distinguish cells with radiation-, drug-, or carcinogen-induced H2AX phosphorylation from cells that have phosphorylation of this histone triggered by apoptosis-associated (AA) DNA fragmentation:

a. The γH2AX IF induced by external DNA damaging agents is seen rather early during the treatment (10 min–2 h) whereas AA γH2AX IF is seen later (>3 h) *(11,12)*.
b. The intensity AA γH2AX IF is generally much higher than that of DNA damage-induced γH2AX IF, unless the cells are at a very late stage of apoptosis *(11)*.
c. The induction of AA γH2AX IF is prevented by cell treatment with the caspase inhibitor z-VAD-FMK, which precludes activation of endonuclease responsible for DNA fragmentation. In its presence the AA-γH2AX IF is suppressed *(11)*.
d. AA H2AX phosphorylation occurs concurrently with activation of caspase-3 in the same cells. Multiparameter analysis (active caspase-3 vs γH2AX IF) thus provides the direct approach to distinguish cells in which DSBs were caused by inducers of DNA damage (active caspase-3 is undetectable) from the cells that have H2AX phosphorylation additionally triggered in response to apoptotic DNA fragmentation (active caspase-3 is present).

These strategies are discussed in more detail elsewhere *(11)*.

4. If the sample initially contains a small number of cells, they may be lost during repeated centrifugations. To minimize cell loss, polypropylene or siliconized glass tubes are recommended. Since transferring cells from one tube to another causes electrostatic attachment of a large fraction of cells to the surface of each new tube, all steps of the procedure (including fixation) preferably should be done in the same tube. Addition of 1% BSA to rinsing solutions also decreases cell loss. When the sample contains very few cells, carrier cells (e.g., chick erythrocytes) may be included; they can be recognized during analysis based on differences in DNA content (intensity of PI fluorescence).
5. Quality of the primary and secondary antibodies is of particular importance. Their ability to detect γH2AX is often lost during improper transport or storage conditions. Also of importance is their use at optimal concentration. It is recommended that with the first use of every new batch of primary or secondary antibody they be tested at serial dilution (e.g., within the range between 0.2 and 2.0 µg/100 µL) to find their optimal titer for detection of γH2AX. The titer recommended by vendor is not always the optimal one.
6. Alternatively, incubate for 1 h at 22–24°C. The overnight incubation at 4°C, however, appears to yield somewhat higher intensity of γH2AX IF compared to a 1-h incubation.

Acknowledgments

This work was supported by NCI grant RO1 28704.

References

1. Rogakou, E. P., Pilch, D. R., Orr, A. H., Ivanova, V. S., and Bonner, W. M. (1998) DNA double-stranded breaks induce histone H2AX phosphorylation on serine 139. *J. Biol. Chem.* **273,** 5858–5868.
2. West, M. H. and Bonner, W. M. (1980) Histone 2A, a heteromorphous family of eight protein species. *Biochemistry* **19,** 3238–3245.
3. Burma, S., Chen, B. P., Murphy, M., Kurimasa, A., and Chen, D. J. (2001) ATM phosphorylates histone H2AX in response to DNA double-strand breaks. *J. Biol. Chem.* **276,** 42,462–42,467.
4. Furuta, T., Takemura, H., Liao, Z.-Y., et al. (2003) Phosphorylation of histone H2AX and activation of Mre11, Rad50, and Nbs1 in response to replication-dependent DNA double-strand breaks induced by mammalian topoisomerase I cleavage complexes. *J. Biol. Chem.* **278,** 20,303–20,312.
5. Park, E. J., Chan, D. W., Park, J. H., Oettinger, M. A., and Kwon, J. (2003) DNA-PK is activated by nucleosomes and phosphorylated H2AX within the nucleosomes in an acetylation-dependent manner. *Nucleic Acids Res.* **31,** 6819–6827.
6. Rogakou, E. P., Boon, C., Redon, C., and Bonner W. M. (1999) Megabase chromatin domains involved in DNA double-strand breaks *in vivo*. *J. Cell Biol.* **146,** 905–916.
7. Sedelnikova, O. A., Rogakou, E. P., Panuytin, I. G., and Bonner, W. (2002) Quantitive detection of ^{125}IUdr-induced DNA double-strand breaks with γ-H2AX antibody. *Radiation Res.* **158,** 486–492.
8. MacPhail, S. H., Banath, J. P., Yu, Y., Chu, E., and Olive, P. L. (2003) Cell cycle-dependent expression of phosphorylated histone H2AX: reduced expression in unirradiated but not X-irradiated G1-phase cells. *Radiat. Res.* **159,** 759–767.
9. MacPhail, S. H., Banath, J. P., Yu, T. Y., Chu, E. H., Lambur, H., and Olive, P. L. (2003) Expression of phosphorylated histone H2AX in cultured cell lines following exposure to X-rays. *Int. J. Radiat. Biol.* **79,** 351–358.
10. Yoshida, K., Yoshida, S. H., Shimoda, C., and Morita, T. (2003) Expression and radiation-induced phosphorylation of H2AX in mammalian cells. *J. Radiat. Res. (Tokyo)* **44,** 47–51.
11. Huang, X., Okafuji, M., Traganos, F., Luther, E., Holden, E., and Darzynkiewicz, Z. (2004) Assessment of histone H2AX phosphorylation induced by DNA topoisomerase I and II inhibitors topotecan and mitoxantrone and by DNA cross-linking agent cisplatin. *Cytometry* **58A,** 99–110.
12. Huang, X., Traganos, F., and Darzynkiewicz, Z. (2003) DNA damage induced by DNA topoisomerase I- or topoisomerase II- inhibitors detected by histone H2AX phosphorylation in relation to the cell cycle phase and apoptosis. *Cell Cycle* **2,** 614–619.
13. Albino, A. P., Huang, X., Jorgensen, E., et al. (2004) Induction of histone H2AX phosphorylation in A549 human pulmonary adenocarcinoma cells by tobacco

smoke and in human bronchial epithelial cells by smoke condensate: a new assay to detect the presence of potential carcinogens in tobacco. *Cell Cycle* **3,** 1062–1068.
14. Speit, G. and Hartmann, A. (1999) The comet assay (single-cell gel test): a sensitive genotoxicity test for detection of DNA damage and repair. *Methods Mol. Biol.* **113,** 203–212.
15. Downs, J. A., Lowndes, N. F., and Jackson, S. P. (2000) A role for *Saccharomyces cerevisiae* histone H2A in DNA repair. *Nature* **408,** 1001–1004.
16. Jackson, S. P. (2001) DNA damage signaling and apoptosis. *Biochem. Soc. Transactions* **29,** 655–661.
17. Gorczyca, W., Bruno, S., Darzynkiewicz, R. J., Gong, J., and Darzynkiewicz, Z. (1992) DNA strand breaks occurring during apoptosis: their early *in situ* detection by the terminal-deoxynucleotidyl transferase and nick translation and prevention by serine protease inhibitors. *Int. J. Oncol.* **1,** 639–648.
18. Marzluff, W. F. and Duronio, R. J. (2002) Histone mRNA expression: multiple levels of cell cycle regulation and important developmental consequences. *Curr. Opin. Cell Biol.* **14,** 692–699.

8

Detection of DNA Strand Breaks by Flow and Laser Scanning Cytometry in Studies of Apoptosis and Cell Proliferation (DNA Replication)

Zbigniew Darzynkiewicz, Xuan Huang, and Masaki Okafuji

Summary

Extensive fragmentation of nuclear DNA occurs during apoptosis, and the presence of DNA strand breaks is considered to be a marker of the apoptotic mode of cell death. This chapter describes methods to label *in situ* DNA strand breaks with fluorochromes for detection by flow or laser scanning cytometry. By staining DNA with a fluorochrome of another color, cellular DNA content is measured concurrently and the bivariate analysis of such a data reveals DNA ploidy and cell-cycle phase position of apoptotic cells. The DNA strand break-labeling methodology is also used for detecting the incorporation of halogenated DNA precursors in studies of the cell cycle, proliferation, and DNA replication. In this application, termed "strand breaks induced by photolysis" (SBIP), the cells are incubated with 5-bromo-2′-deoxyuridine (BrdU) to incorporate it into DNA and sensitize the DNA to ultraviolet (UV) light. DNA strand breaks are then photolytically generated by exposing the cells to UV light. The DNA strand breaks resulting from UV-photolysis are subsequently fluorochrome-labeled as for labeling apoptotic-DNA breaks. Because SBIP, unlike the alternative method of detection of BrdU incorporation, does not require subjecting cells to harsh conditions (strong acid or heat) of DNA denaturation, it is compatible with concurrent detection of intracellular or cell surface antigens by immunocytochemical means.

Key Words: Apoptosis; BrdU; cell cycle; DNA fragmentation; DNA photolysis; flow cytometry; laser-scanning cytometry; SBIP; TUNEL; UV light.

1. Introduction

DNA fragmentation that leads to an abundance of strand breaks in nuclear DNA is considered to be a hallmark of apoptosis (*1–3*). One of the most widely used methods to identify apoptotic cells thus relies on labeling DNA strand breaks *in situ* either with fluorochromes (*4–6*) or absorption dyes (*7,8*). The

From: *Methods in Molecular Biology: DNA Repair Protocols: Mammalian Systems, Second Edition*
Edited by: D. S. Henderson © Humana Press Inc., Totowa, NJ

advantage of DNA strand break labeling with fluorochromes is that such cells can be rapidly analyzed by flow or laser scanning cytometry. Furthermore, when cellular DNA content is measured concurrently, after being stained with a fluorochrome of another color, the bivariate analysis of such data reveals DNA ploidy and cell-cycle phase position of cells undergoing apoptosis *(4,5,9)*.

Some methods presented in this chapter can be applied to cells in suspension whose fluorescence is measured by flow cytometry (**Subheadings 3.1.1.** and **3.2.1.**). However, a simple modification of these methods, also presented here (**Subheadings 3.1.3.** and **3.2.2.**), permits the analysis of cells attached to microscope slides. The fluorescence of cells attached to microscope slides can be measured by the laser scanning cytometer (LSC), or by its more recent version, the automatic imaging cytometer, iCyte. These are the microscope-based cytofluorimeters that allow one to measure rapidly, with high sensitivity and accuracy, the fluorescence of individual cells *(10,11)*. The instruments combine advantages of both flow and image cytometry. The staining of cells on slides prevents their loss, which otherwise occurs during the repeated centrifugations of samples during preparation for flow cytometry. Another advantage stems from the possibility of spatially localizing particular cells on the slide for their visual examination or morphometric imaging analysis after the initial measurement of a large population and electronic selection (gating) of the cells of interest. The measured cells can also be bleached and restained with another set of dyes. This permits the cell attributes measured after restaining to be correlated with the attributes measured before, on a cell-by-cell basis *(11–13)*.

Cell fixation and permeabilization are essential steps to label DNA strand breaks successfully. Cells are briefly fixed with a crosslinking fixative, such as formaldehyde, and then permeabilized by suspending them in ethanol. By crosslinking low molecular weight DNA fragments to other cell constituents, formaldehyde prevents extraction of the fragmented DNA, that otherwise occurs during the repeated centrifugations and rinses required by this procedure. The 3′-OH termini of the fragmented DNA serve as primers and become labeled with 5-bromo-2′-deoxyuridine (BrdU) in a reaction catalyzed by exogenous terminal deoxynucleotidyl transferase (TdT) and BrdUTP *(14)*. The incorporated BrdU is immunocytochemically detected by anti-BrdU antibody conjugated to fluorescein isothiocyanate (FITC) *(14,15)*. This reagent is widely used in studies of cell proliferation to detect BrdU incorporated during DNA replication *(16)*. The overall cost of reagents is markedly lower and the sensitivity of DNA strand break detection is higher when BrdUTP is used as a marker, compared to labeling with biotin-, digoxygenin-, or directly fluorochrome-tagged deoxynucleotides (**Fig. 1**) *(14)*. However, because certain applications may require the use of multiple fluorochromes, alternative procedures utilizing the latter three reagents are also described in this chapter.

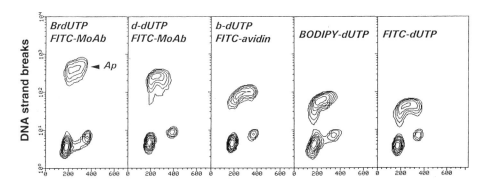

Fig. 1. Detection of apoptotic cells by flow cytometry using different methods of DNA strand break labeling. Bivariate distributions represented by isometric contour maps (DNA content vs DNA strand break labeling) of HL-60 cells incubated for 3 h with 0.15 µM camptothecin, which preferentially induces apoptosis (Ap) of DNA replicating cells *(4,9)*. The first three panels to the left represent indirect labeling of DNA strand breaks, utilizing either BrdUTP, digoxygenin-conjugated dUTP (d-dUTP), or biotinylated dUTP (b-dUTP). The two right panels show cell distributions following a direct, single-step DNA strand break labeling, either with BODIPY- or FITC-conjugated dUTP. Note the exponential scale of the ordinate. As is evident, the greatest difference is achieved following DNA strand break labeling with BrdUTP *(14)*.

The DNA strand break labeling methodology also can be used for detecting the incorporation of halogenated DNA precursors in studies of the cell cycle, proliferation and DNA replication. In this application, termed "strand breaks induced by photolysis" (SBIP; **Subheading 3.2.**), the cells are incubated with BrdU or 5-iodo-2′-deoxyuridine (IdU) to incorporate these base analogs into DNA, and DNA strand breaks are photolytically generated by exposure of the cells to ultraviolet (UV) light; the incorporated precursor sensitizes DNA to UV light *(14,15)*. DNA breaks resulting from UV photolysis, which in this case are markers of precursor incorporation (i.e., DNA replication), are subsequently fluorochrome-labeled in the same way as described for labeling apoptotic DNA breaks. Because the SBIP method does not require subjecting cells to the harsh conditions (strong acid or heat) of DNA denaturation, as the alternative approach does *(16)*, it is compatible with the concurrent detection of intracellular or cell surface antigens by immunocytochemical means *(14)*.

2. Materials
2.1. DNA Strand Break Labeling
1. Phosphate-buffered saline (PBS), pH 7.4.

2. 1% Formaldehyde (methanol-free, ultrapure; Polysciences, Warrington, PA) in PBS, pH 7.4.
3. 70% Ethanol.
4. TdT (Roche). TdT 5X reaction buffer: 1 M potassium (or sodium) cacodylate, 125 mM HCl, pH 6.6, 1.25 mg/mL of bovine serum albumin (BSA; Sigma, St. Louis, MO). This 5X reaction buffer can be purchased from Roche.
5. 5-Bromo-2′-deoxyuridine-5′-triphosphate (BrdUTP) stock solution (50 µL): 2 mM BrdUTP (Sigma) in 50 mM Tris-HCI, pH 7.5.
6. 10 mM CoCl$_2$ (Roche).
7. Rinsing buffer: 0.1% Triton X-100 and 5 mg/mL of BSA dissolved in PBS.
8. FITC-conjugated anti-BrdU monoclonal antibody (MAb): Dissolve 0.3 µg of FITC-conjugated anti-BrdU MAb (Becton Dickinson, Immunocytometry Systems, San Jose, CA) in 100 µL of PBS containing 0.3% Triton X-100 (Sigma) and 1% BSA.
9. Propidium iodide (PI) staining buffer: 5 µg/mL of PI (Molecular Probes, Eugene, OR), 100 µg/mL of RNase A (DNase-free; Sigma) in PBS.
10. Cytospin centrifuge (Shandon, Pittsburgh, PA): for use in conjunction with laser scanning cytometry.

Kits for labeling DNA strand breaks that utilize the BrdUTP–TdT methodology similar to that described in this chapter are commercially available from Phoenix Flow Systems (San Diego, CA) and several other vendors (APO-BRDU™). Some kits contain positive-control cell samples that have a subpopulation of cells with abundant DNA breakage. Such controls are helpful to reveal whether the inability to detect cells with DNA strand breaks in the experimental sample is due to their absence or technical problems with the kit (e.g., inactive TdT).

2.2. Additional Reagents and Equipment for SBIP

1. BrdU (Sigma).
2. 60 × 15 mm polystyrene Petri dishes (Corning, Corning, NY).
3. UV light illumination source: Fotodyne UV 300 analytic DNA transilluminator containing four 15-W bulbs (Fotodyne Inc., New Berlin, WI).
4. UV light photometer: UVX-25 Sensor (UVP, Upland, CA).

A kit based on the SBIP methodology (APO-DIRECT) is commercially available from Phoenix Flow Systems and from several other suppliers of reagents for cytometry.

2.3. Instrumentation

Flow cytometers of different types, offered by several manufacturers, can be used to measure cell fluorescence following staining according to the procedures described below. The manufacturers of the most common flow cytometers are Coulter Corporation (Miami, FL), Becton Dickinson Immunocytometry Systems (San Jose, CA), Cytomation/DAKO (Fort Collins, CO) and PARTEC (Zurich, Switzerland). The multiparameter LSC and iCyte are available from CompuCyte (Cambridge, MA).

The software to deconvolute the DNA content frequency histograms to analyze the cell cycle distributions is available from Phoenix Flow Systems (San Diego, CA) and Verity Software House (Topham, MA).

3. Methods

3.1. Detection of Apoptotic Cells by DNA Strand Break Labeling (see **Note 1**)

3.1.1. DNA Strand Break Labeling With BrdUTP for Analysis by Flow Cytometry

1. Suspend $1–2 \times 10^6$ cells in 0.5 mL of PBS. With a Pasteur pipet, transfer this suspension to a 6-mL polypropylene tube (*see* **Note 2**) containing 4.5 mL of ice-cold 1% formaldehyde (*see* **Note 3**). Incubate on ice for 15 min.
2. Centrifuge at 300g for 5 min, and resuspend the cell pellet in 5 mL of PBS. Centrifuge again, and resuspend the cells in 0.5 mL of PBS. With a Pasteur pipet, transfer the suspension to a tube containing 4.5 mL of ice-cold 70% ethanol. The cells can be stored in ethanol at –20°C for several weeks.
3. Centrifuge at 200g for 3 min, remove the ethanol, resuspend the cells in 5 mL of PBS, and centrifuge at 300g for 5 min.
4. Resuspend the pellet in 50 µL of a solution containing:
   a. 10 µL of TdT 5X reaction buffer.
   b. 2.0 µL of BrdUTP stock solution.
   c. 0.5 µL (12.5 U) of TdT.
   d. 5 µL of $CoCl_2$ solution.
   e. 33.5 µL of distilled H_2O.
5. Incubate the cells in this solution for 40 min at 37°C (*see* **Notes 4** and **5**).
6. Add 1.5 mL of the rinsing buffer, and centrifuge at 300g for 5 min.
7. Resuspend the cell pellet in 100 µL of FITC-conjugated anti-BrdU Ab solution.
8. Incubate at room temperature for 1 h.
9. Add 1 mL of PI staining solution.
10. Incubate for 30 min at room temperature, or 20 min at 37°C, in the dark.
11. Analyze the cells by flow cytometry.
    a. Illuminate with blue light (488-nm laser line or BG12 excitation filter).
    b. Measure green fluorescence of FITC at 530 ± 20 nm.
    c. Measure red fluorescence of PI at >600 nm.

    Figure 1 (left-most panel) shows apoptotic cells detected by this method (*see* **Notes 6** and **7**).

3.1.2. DNA Strand Break Labeling With Other Markers for Analysis by Flow Cytometry

As mentioned in **Subheading 1.**, DNA strand breaks can be labeled with deoxynucleotides tagged with a variety of other fluorochromes. For example, the Molecular Probes (Eugene, OR) catalog lists seven types of dUTP conjugates, including three BODIPY dyes (e.g., BODIPY-FL-X-dUTP), fluorescein, cascade blue, Texas red, and dinitrophenol. Several cyanine dyes conjugates (e.g., CY-3-dCTP) are available from Biological Detection Systems (Pittsburgh, PA). Indirect labeling via biotinylated or digoxygenin-conjugated deoxynucleotides offers a multiplicity of commercially available fluorochromes (fluorochrome-conjugated avidin or streptavidin, as well as digoxygenin antibodies) with different excitation and emission characteristics. DNA strand breaks, thus, can be labeled with a dye of any desired fluorescence color and excitation wavelength (**Fig. 1**).

The procedure described in **Subheading 3.1.1.** can be adapted to utilize each of these fluorochromes. In the case of direct labeling, the fluorochrome-conjugated deoxynucleotide is included in the reaction solution (0.25–0.5 nmol/50 μL) instead of BrdUTP, as described in **step 4** of **Subheading 3.1.1.** Following the incubation step (**step 5**), omit **steps 6–8** and stain the cells directly with PI (**step 9**). In the case of indirect labeling, digoxygenin- or biotin-conjugated deoxynucleotides are included in the reaction buffer (0.25–0.5 nmol/50 μL) instead of BrdUTP at **step 4**. The cells are then incubated either with fluorochrome-conjugated antidigoxigenin MAb (0.2–0.5 μg/100 μL of PBS containing 0.1% Triton X-100 and 1% BSA) or with fluorochrome-conjugated avidin or streptavidin (0.2–0.5 μg/100 μL, as above) at **step 7** and then processed through **steps 8–10** as described in the protocol. Cytometric analysis is performed with the excitation and emission wavelength appropriate to the applied fluorochrome.

3.1.3. DNA Strand Break Labeling for Analysis by Laser Scanning Cytometry

1. Transfer 300 μL of cell suspension (in tissue-culture medium with serum) containing approx 20,000 cells to a cytospin chamber. Cytocentrifuge at 1000 rpm for 5 min.
2. Without allowing the cytocentrifuged cells to dry completely, prefix them in 1% formaldehyde in PBS (in a Coplin jar) for 15 min on ice.
3. Transfer the slides to 70% ethanol, and fix for at least 1 h. The cells can be stored in ethanol for several weeks at –20°C.
4. Follow **steps 4–8** of **Subheading 3.1.1.** Carefully layer small volumes (approx 100 μL) of the respective buffers, rinses, or staining solutions on the cytospin area of the horizontally placed slides. At appropriate times, remove these solutions with a Pasteur pipet (or vacuum suction pipet). To prevent drying, place 2 × 4 cm^2 pieces of thin polyethylene foil on the slides over the cytospins, atop the drops of the solutions used for cell incubations (*see* **Note 8**).
5. Mount the cells under a cover slip in a drop of the PI staining solution. Seal the edges of the cover slip with melted paraffin or a gelatin-based sealer.
6. Measure cell fluorescence by laser scanning cytometry.
    a. Excite fluorescence with a 488-nm laser line.
    b. Measure green fluorescence of FITC at 530 ± 20 nm.

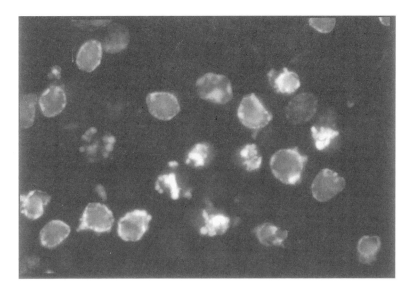

Fig. 2. Detection of apoptosis-associated DNA strand breaks. HL-60 cells were incubated with 0.15 µM camptothecin for 2.5 h, cytocentrifuged, fixed, and the DNA strand breaks were labeled with BrdUTP. The incorporated BrdU was then detected by FITC-conjugated anti-BrdU MAb, as described in **Subheading 3.1.3.**; however, the cells were not counterstained with PI. Note the predominance of DNA strand breaks in early apoptotic cells (prior to nuclear fragmentation) at the nuclear periphery, and strong labeling of the fragmented nuclei of late apoptotic cells.

c. Measure red fluorescence of PI at >600 nm.
Apoptotic cells detected by this method are shown in **Fig 2**.

3.1.4. Controls

The procedure of DNA strand break labeling is rather complex and involves many reagents. Negative results, therefore, may not necessarily mean the absence of DNA strand breaks, but may be due to methodological problems, such as loss of TdT activity, degradation of BrdUTP, and so forth. It is necessary, therefore, to include both positive and negative controls. An excellent control is to use HL-60 cells treated (during their exponential growth) for 3–4 h with 0.2 µM of the DNA topoisomerase I inhibitor camptothecin (CPT). Because CPT induces apoptosis selectively during S phase, cells in G_1, and G_2/M may serve as negative control populations, whereas the S phase cells in the same sample represent the positive control. As mentioned, some commercial kits provide such control cells. Another negative control consists of cells processed as described in **Subheading 3.1.1.**, except that TdT is excluded from **step 4**.

3.2. Detection of Cells Incorporating BrdU by the SBIP Method

The method of DNA strand break labeling described above for the identification of apoptotic cells also can be used to detect the presence of BrdU or IdU incorporated into DNA. A variety of different schemes may be used to label cells with these precursors. Pulse labeling, for example, is used to detect S-phase cells. A pulse-chase labeling strategy is used to follow a cohort of labeled cells progressing through various phases of the cycle for kinetic studies. Continuous labeling allows one to detect all proliferating cells in a culture or tumor to estimate the cell growth fraction. The scope of this chapter does not allow us to present technical details of cell labeling in cultures or in vivo, which are available elsewhere *(17)*. In general, 10–30 μ*M* BrdU is used for in vitro cell labeling, and the time of incubation for pulse-labeling varies between 10 and 60 min. It is important to maintain lightproof conditions (e.g., the cultures should be wrapped in aluminum foil) during and after cell labeling with BrdU, to prevent DNA photolysis.

Following incorporation of BrdU (or IdU), the cells are fixed, subjected to UV light illumination to photolyze the DNA at sites of the incorporated precursor, and the resulting DNA strand breaks are labeled identically to the DNA strand breaks of apoptotic cells in **Subheadings 3.1.1.** and **3.1.2.** To distinguish between DNA strand breaks in apoptotic cells and photolytically generated (BrdU-associated) breaks, the apoptotic DNA strand breaks may initially be labeled with a fluorochrome of one color, the cells then subjected to UV light illumination, and the photolytically generated breaks subsequently labeled with a fluorochrome of another color *(15)* (*see* **Subheading 3.1.2.**). The method of DNA photolysis presented below can be applied to any type of cells that have been labeled with BrdU or IdU.

3.2.1. SBIP Procedure for Cell Analysis by Flow Cytometry

1. Suspend $1-2 \times 10^6$ cells, previously incubated with BrdU, in 2 mL of ice-cold PBS.
2. Transfer the cell suspension to 60×15 mm polystyrene Petri dishes.
3. Place the dishes directly on the glass surface of a Fotodyne UV 300 analytic DNA gel transilluminator, which provides maximal illumination at 300-nm wavelength. Check the intensity of UV light by using a UV light photometer placed on the surface of the transilluminator. With relatively new UV bulbs, the intensity is expected to be 4–5 mW/cm^2. Other sources of UV light may be used provided that maximal intensity is at a wavelength close to 300 nm and the geometry of cell illumination favors uniform exposure of all cells (*see* **Note 9**).
4. Expose the cells to UV light for 5–10 min.
5. Transfer the cells to polypropylene tubes, and centrifuge at 300*g* for 5 min.
6. Suspend the cell pellet in 0.5 mL of PBS.
7. Transfer the cell suspension with a Pasteur pipet to a 6-mL polypropylene tube containing 4.5 mL of 70% ethanol, on ice. The cells can be stored in ethanol at –20°C for months.

8. Label the strand breaks, and process the cells for flow cytometry as described in **Subheading 3.1.1., steps 3–11**.

Controls should include cells incubated in the absence of BrdU (or IdU), as well as cells not illuminated with UV light.

3.2.2. SBIP Procedure for Cell Analysis by Laser Scanning Cytometry

1. Transfer 300 µL of cell suspension in tissue culture medium (with serum) containing approx 20,000 cells into a cytospin chamber. Cytocentrifuge at 1000 rpm for 6 min.
2. Without allowing the cytocentrifuged cells to dry completely, fix the slides in 70% ethanol, in Coplin jars, on ice, for at least 2 h. The slides can be stored in ethanol for months at –20°C.
3. Rinse the slides in PBS.
4. To photolyze the DNA, remove the slides from PBS and place (while still wet) on the glass surface of the transilluminator. The cytospinned cells should be placed face down, with the slide supported on both sides (e.g., with two other microscope slides) to prevent contact between the cells and the transilluminator glass surface (*see* **Note 10**).
5. Expose the cells to UV light for 5–10 min.
6. Process the cells as described in **steps 4** and **5** of **Subheading 3.1.3.**
7. Measure cell fluorescence by laser scanning cytometry as described in **step 6** of **Subheading 3.1.3.** (*See* **Figs. 3** and **4.**)

3.2.3. Controls for SBIP

As a negative control, analyze cells that were not incubated with BrdU (or IdU). Such a control is preferred over using an isotypic IgG (as a control for anti-BrdU Ab), since the latter does not always allow accurate discrimination between BrdU-labeled and unlabeled cells. Two types of positive controls are suggested. As a positive control for the DNA strand break labeling procedure alone, apoptotic cells prepared using CPT as described in **Subheading 3.1.4.** should be used. As another positive control, exponentially growing cells incubated with 30 µ*M* BrdU for 1 h, and then processed as described in **Subheading 3.2.1.** or **3.2.2.** should be used. In this control, one expects S-phase cells, that is, cells with a DNA content between 1.0 and 2.0 DNA index (DI), to show BrdU incorporation, and G_1 (DI = 1.0) and G_2/M cells (DI = 2.0) to be negative.

4. Notes

1. This method is useful for clinical material, such as obtained from leukemias, lymphomas, and solid tumors *(18,19)*, and can be combined with surface immunophenotyping. The cells are first immunophenotyped, then fixed with formaldehyde (which stabilizes the antibody bound on the cell surface), and subsequently subjected to the DNA strand break detection assay using different color fluorochromes (*see* **Subheading 3.1.2.**) than those used for immunophenotyping.

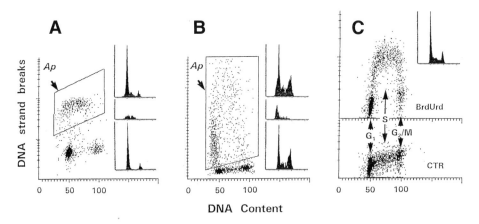

Fig. 3. Detection of apoptosis and DNA replication by differential labeling of DNA strand breaks and fluorescence measurement by laser scanning cytometry. Bivariate distributions (scattergrams) representing intensity of DNA strand break labeling with different fluorochromes vs cellular DNA content, identifying apoptotic, and BrdU-incorporating cells. (**A**) HL-60 cells were incubated with 0.15 µM camptothecin for 3 h. DNA strand breaks were directly labeled with dUTP conjugated to BODIPY. DNA histograms (insets) represent all cells (top), apoptotic cells located within the gating window (middle), and nonapoptotic cells (bottom). Notice that apoptosis is specific to S-phase cells. Ordinate, exponential scale. (**B**) Cells were subjected to hyperthermia (43.5°C, 30 min) and then incubated for 3 h at 37°C. DNA strand breaks in apoptotic cells were indirectly labeled with d-dUTP and detected by fluoresceinated antidigoxygenin antibody. Top DNA histogram, all cells; middle histogram, apoptotic cells (within the gating window); bottom histogram, nonapoptotic cells. Ordinate, linear scale. (**C**) Detection of BrdU incorporation (1-h pulse) by SBIP using indirect labeling with d-dUTP and detection by fluoresceinated antidigoxygenin antibody (top). Bottom, the cells were incubated in the absence of BrdU (control). DNA histogram represents all cells. Ordinate, exponential scale *(15)*.

2. When the sample initially contains a small number of cells, cell loss during repeated centrifugations is a problem. To minimize cell loss, polypropylene or siliconized glass tubes are recommended. Since transferring cells from one tube to another results in irreversible electrostatic attachment of a large fraction of cells to the surface of each new tube, all steps of the procedure (including fixation) should be done in the same tube. Addition of 1% BSA to rinsing solutions also decreases cell loss. When the sample contains very few cells, carrier cells (e.g., chick erythrocytes) may be included, which later can be recognized based on differences in DNA content. Cell analysis by LSC or iCyte, of course, has no such problem.
3. Cell prefixation with a crosslinking agent, such as formaldehyde, is required to prevent extraction of the fragmented DNA from apoptotic cells. This ensures that

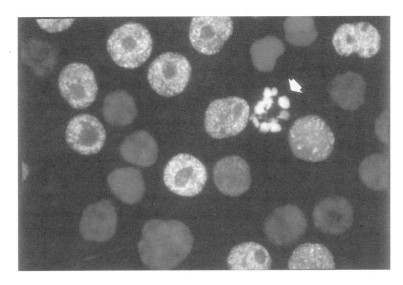

Fig. 4. Detection of photolysis-associated DNA strand breaks. HL-60 cells were pulse labeled (1 h) with BrdU. Their DNA was photolyzed by exposure to UV light, and DNA strand breaks were labeled with BrdUTP as described in **Subheading 3.2.2.** Cellular DNA was counterstained with 7-aminoactinomycin D. The sites of DNA replication ("replication factories") have a characteristic distribution, and nucleoli are unlabeled. A single apoptotic cell with a fragmented nucleus (arrow) is also labeled, but the labeling is diffuse, not granular.

despite repeated cell washings, the DNA content of apoptotic cells (and with it, the number of DNA strand breaks) is not markedly diminished. No prefixation with formaldehyde is required to detect DNA strand breaks induced by photolysis (**Subheading 3.2.**).
4. Alternatively, incubate at 22–24°C overnight.
5. Control cells may be incubated in the same solution, but without TdT.
6. It is generally easy to identify apoptotic cells owing to their intense labeling with FITC conjugated anti-BrdU MAb. Their high fluorescence intensity often requires the use of the exponential scale (logarithmic amplifiers of the flow cytometer) for data acquisition and display (**Fig. 1**). As is evident in **Fig. 1**, because cellular DNA content of both apoptotic and nonapoptotic cell populations is measured, the cell-cycle distribution and/or DNA ploidy of these populations can be estimated.
7. While strong fluorescence, which indicates the presence of extensive DNA breakage, is a characteristic feature of apoptosis, weak fluorescence does not necessarily mean the lack of apoptosis. In some cell systems, DNA cleavage generates DNA fragments 50–300 kb in size and does not progress into internucleosomal (spacer) sections *(20)*.

8. It is essential that the incubations are carried out in a moist atmosphere to prevent drying at any step of the reaction. Even minor drying produces severe artifacts.
9. In the SBIP procedure, to detect incorporated BrdU or IdU, the critical step is to expose the cells to an optimal and uniform dose of UV light. During the exposure, therefore, the layer of cell suspension should be thin and the Petri dishes should be exposed while in a horizontal position. Local cell crowding at the edges of the dish should be avoided, since it introduces undesired heterogeneity during illumination. Because the intensity of UV light at the surface of the transilluminator is uneven, depending very much on the position of the UV bulb underneath the glass, the "sweet spot" of relatively uniform intensity has to be found with a UV photometer. The cells should then be placed at this position for irradiation. Overexposure induces photolysis of native DNA, which has no incorporated BrdU. The signal-to-noise ratio in the detection of BrdU is then decreased owing to a high fluorescence background of the BrdU-unlabeled cells. Illumination of cells in the presence of Hoechst 33258, a dye that via a resonance energy transfer mechanism additionally photosensitizes BrdU, increases labeling of the DNA that contains incorporated BrdU *(21)*.
10. Alternatively, the cells maybe photolyzed in suspension, prior to fixation, as described in the procedure for flow cytometry (**steps 1–4** of **Subheading 3.2.1.**), then cytocentrifuged, fixed in ethanol, and processed for DNA strand break labeling.

Acknowledgments

This work was supported by NCI Grant R01 28704 and "This Close" Foundation for Cancer Research. Dr. Masaki Okafuji was on leave from the Department of Oral and Maxillofacial Surgery, Yamaguchi University School of Medicine, Ube, Japan.

References

1. Arends, M. J., Morris, R. G., and Wyllie, A. H. (1990) Apoptosis: the role of endonuclease. *Am. J. Pathol.* **136,** 593–608.
2. Compton, M. M. (1992) A biochemical hallmark of apoptosis: Internucleosomal degradation of the genome. *Cancer Metastasis Rev.* **11,** 105–119.
3. Nagata, S., Nagase, H., Mukae, N., and Fukuyama, H. (2003) Degradation of chromosomal DNA during apoptosis. *Cell Death Differ.* **10,** 108–116.
4. Gorczyca, W., Bruno, S., Darzynkiewicz, R. J., Gong, J., and Darzynkiewicz, Z. (1992) DNA strand breaks occurring during apoptosis: their early *in situ* detection by the terminal deoxynucleotidyl transferase and nick translation assays and prevention by serine protease inhibitors. *Int. J. Oncol.* **1,** 639–648.
5. Darzynkiewicz, Z., Juan, G., Li, X., Gorczyca, W., Murakami, T., and Traganos. F. (1997) Cytometry in cell necrobiology: analysis of apoptosis and accidental cell death (necrosis). *Cytometry* **27,** 1–20.

6. Ehemann, V., Sykora, J., Vera-Delgado, J., Lange, A., and Otto, H. F. (2003) Flow cytometric detection of spontaneous apoptosis in human breast cancer using the TUNEL-technique. *Cancer Lett.* **194,** 125–131.
7. Gold, R., Schmied, M., Rothe, G., et al. (1993) Detection of DNA fragmentation in apoptosis: Application of *in situ* nick translation to cell culture systems and tissue sections. *J. Histochem. Cytochem.* **41,** 1023–1030.
8. Wijsman, J. H., Jonker, R. R., Keijzer, R., Van De Velde, C. J. H., Cornelisse, C. J., and VanDierendonck, J. H. (1993) A new method to detect apoptosis in paraffin sections: *in situ* end-labeling of fragmented DNA. *J. Histochem. Cytochem.* **41,** 7–12.
9. Gorczyca, W., Gong, J., Ardelt, B., Traganos, F., and Darzynkiewicz, Z. (1993) The cell cycle related differences in susceptibility of HL-60 cells to apoptosis induced by various antitumor drugs. *Cancer Res.* **53,** 3186–3192.
10. Kamentsky, L. A. (2001) Laser scanning cytometry. *Methods Cell Biol.* **63,** 51–87.
11. Darzynkiewicz, Z., Bedner, E., Gorczyca, W., and Melamed, M. R. (1999) Laser scanning cytometry. A new instrumentation with many applications. *Exp. Cell Res.* **249,** 1–12.
12. Smolewski, P., Grabarek, J., Kamentsky, L. A., and Darzynkiewicz, Z. (2001) Bivariate analysis of cellular DNA vs RNA content by laser scanning cytometry using the product of signal subtraction (differential fluorescence; DF) as a separate parameter. *Cytometry* **45,** 73–78.
13. Haider, A. S., Grabarek, J., Eng, B., et al. (2003) *In vitro* wound healing analyzed by laser scanning cytometry. Accelerated healing of epithelial cell monolayers in the presence of hyaluronate. *Cytometry* **53A,** 1–8.
14. Li, X. and Darzynkiewicz, Z. (1995) Labelling DNA strand breaks with BrdUTP. Detection of apoptosis and cell proliferation. *Cell Prolif.* **28,** 571–579.
15. Li, X., Melamed, M. R., and Darzynkiewicz, Z. (1996) Detection of apoptosis and DNA replication by differential labeling of DNA strand breaks with fluorochromes of different color. *Exp. Cell Res.* **222,** 28–37.
16. Dolbeare, F. and Selden, J. R. (1994) Immunochemical quantitation of bromodeoxyuridine: application to cell cycle kinetics. *Methods Cell Biol.* **41,** 297–310.
17. Gray, J. W. and Darzynkiewicz, Z. (1987) *Techniques in Cell Cycle Analysis.* Humana Press, Totowa, NJ.
18. Halicka, H. D., Seiter, K., Feldman, E. J., et al. (1997) Cell cycle specificity during treatment of leukemias. *Apoptosis* **2,** 25–39.
19. Li, X., Gong, J., Feldman, E., Seiter, K., Traganos, F., and Darzynkiewicz, Z. (1994) Apoptotic cell death during treatment of leukemias. *Leuk. Lymphoma* **13,** 65–72.
20. Oberhammer, F., Wilson, J. W., Dive, C., et al. (1993) Apoptotic death in epithelial cells: cleavage of DNA to 300 and/or 50 kb fragments prior to or in the absence of internucleosomal fragmentation. *EMBO J.* **12,** 3679–3684.
21. Li, X., Traganos, F., Melamed, M. R., and Darzynkiewicz, Z. (1995) Single-step procedure for labeling DNA strand breaks with fluorescein- or BODIPY-conjugated deoxynucleotides: detection of apoptosis and bromodeoxyuridine incorporation. *Cytometry* **20,** 172–180.

9

In Vitro Rejoining of Double-Strand Breaks in Genomic DNA

George Iliakis and Nge Cheong

Summary

Recent genetic and biochemical studies have provided important insights into the mechanism of nonhomologous end-joining (NHEJ) in higher eukaryotes, and have helped to characterize several components including DNA-PKcs, Ku, DNA ligase IV, and XRCC4. There is evidence, however, that additional factors involved in NHEJ remain to be characterized. The biochemical characterization of NHEJ in higher eukaryotes has benefited significantly from in vitro plasmid-based end-joining assays. However, because of differences in the organization and sequence of genomic and plasmid DNA, and because multiple pathways of NHEJ are operational, it is possible that different factors are preferred for the rejoining of double-strand breaks (DSBs) induced in plasmid vs genomic DNA organized in chromatin. Here, we describe an in vitro assay that allows the study of DSB rejoining in genomic DNA. The assay utilizes as a substrate DSBs induced by various means in genomic DNA prepared from agarose-embedded cells after appropriate lysis. Two extremes in terms of state of DNA organization are described: "naked" DNA and DNA organized in chromatin. Here, we describe the protocols developed to carry out and analyze these in vitro reactions, including procedures for preparation of cell extract and the preparation of the substrate DNA ("naked" DNA or nuclei).

Key Words: Chromatin structure; DNA repair; double strand breaks; in vitro assay; nonhomologous end-joining.

1. Introduction

A large number of studies suggests that double-strand breaks (DSBs) induced in DNA by ionizing radiation or chemical agents are critical lesions that if unrepaired, or if misrepaired, may kill a cell, or cause its transformation to a cancer cell. DSBs can also be induced by endogenous oxidative stress, by interference of the replication fork with single-strand breaks, as well as during meiosis and V(D)J recombination. Cells have therefore developed efficient repair mechanisms for dealing with this type of lesion and restoring DNA integrity.

Two enzymatically distinct processes are utilized to repair DNA DSBs in eukaryotic cells: homology-directed repair (HDR) and nonhomologous end-joining (NHEJ) *(1–3)*. HDR removes DSBs by utilizing homologous DNA segments, on the same or different DNA molecules, and restores faithfully the original DNA sequence around the break. NHEJ removes DSBs from the genome by simply joining the DNA ends without homology requirements and without ensuring sequence restoration around the break. While homologous recombination is the preferred mechanism of DSB repair in prokaryotes and in lower eukaryotes such as yeast, nonhomologous end-joining is extensively used in higher eukaryotes *(4)*.

Recent genetic and biochemical studies have provided important insights into the mechanisms of NHEJ in higher eukaryotes, and have helped to characterize several important components of the NHEJ apparatus including DNA-PKcs, Ku, DNA ligase IV, and XRCC4 *(3)*. Evidence exists however that additional factors are required for the successful removal in vivo of DNA DSBs by the DNA-PK–dependent pathway of end-joining *(5–7)*. In addition, there is evidence that cells deficient in the DNA-PK–dependent pathway of NHEJ do not compensate by increasing utilization of HDR, but rely instead on backup pathways of NHEJ to remove ionizing radiation-induced DNA DSBs *(4,8–10)*. It is likely, therefore, that the characterization of components of the DNA-PK–dependent pathway of NHEJ, as well as the biochemical characterization of backup pathways of NHEJ, will benefit from in vitro assays of DNA end-joining.

The biochemical characterization of NHEJ in higher eukaryotes has significantly benefited from in vitro plasmid-based end-joining assays *(11)*. These assays are very useful and have generated information essential for our understanding of the mechanisms utilized by the cell to rejoin DNA ends. However, because of differences in the organization and sequence of genomic and plasmid DNA, and because multiple pathways are available for end-joining in mammalian cells, it is possible that different factors are preferred for the rejoining of DSBs induced in plasmid vs genomic DNA. For this reason, we developed and describe here an in vitro assay that allows the study of DSB rejoining in genomic DNA. The assay utilizes as a substrate DSBs induced by various means (e.g., ionizing radiation, restriction endonucleases, bleomycin, etc.) in genomic DNA prepared from agarose-embedded cells after appropriate lysis. At present, two extremes in terms of state of DNA organization have been tested: "naked" DNA and DNA organized in chromatin *(12–14)*. The former state is generated by complete lysis of cells embedded in agarose using detergents and proteases, while the latter state is generated by gentle lysis that

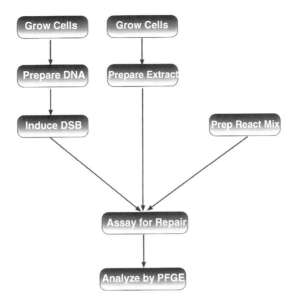

Fig. 1. Outline of the steps required to analyze repair of genomic DSBs in vitro.

leaves nuclei intact. Although not yet fully tested, intermediate states of chromatin organization can also be generated, by gradually extracting nuclear proteins and histones with increasing salt concentrations.

The unique feature of the assay lies in the fact that the agarose fiber network that encapsulates the intact cell generates a "cage" that protects and restricts the expansion and mobility of the cellular DNA, or chromatin, particularly when freed on lysis from the nucleus. This "cage" is expected to preserve essential features of DNA organization, especially after treatment to induce DSBs, and also to prevent extensive entangling. In this way, essential elements of the in vivo DNA repair process are maintained, including the large molecular weight DNA, the low number of DSBs (ends) present per mega-base pairs, the high local DNA concentration, the sequence context and, for nuclei-based or derivative assays, the possibility of retaining elements of chromatin structure.

Here, we describe the protocols developed to carry out these in vitro reactions for DSB rejoining. The required procedures include: (1) preparation of HeLa-cell extract; (2) preparation of the DNA ("naked" DNA or nuclei); (3) generation of DSBs in the DNA; (4) assembly of in vitro repair reactions; and (5) assay for DNA damage using pulsed-field gel electrophoresis. **Figure 1** shows a graphic outline of these steps.

2. Materials

2.1. Preparation of HeLa-Cell Extract (see Notes 1 and 2)

1. Minimum essential medium (MEM) modified for suspension cultures (S-MEM), supplemented with 5% fetal calf or iron-supplemented bovine serum and antibiotics (penicillin 100 U/mL, streptomycin 100 µg/mL).
2. Hypotonic buffer solution: 10 mM N-2-hydroxyethylpiperazine-N'-2-ethanesulfonic acid (HEPES), pH 7.5, at 4°C (0.6 M stock, pH 7.5 at room temperature [RT]), 1.5 mM MgCl$_2$ (0.5 M stock), 5 mM KCl (2 M stock). Immediately before use add 0.2 mM PMSF (100 mM stock in ethanol; can be stored at –20°C for more than a month), and 0.5 mM DTT (1 M stock in H$_2$O; store at –20°C).
3. High salt buffer: 10 mM HEPES, pH 7.5 at 4°C, 1.4 M KCl, and 1.5 mM MgCl$_2$.
4. Dialysis buffer: 25 mM Tris-HCl, pH 7.5 at 4°C (1 M stock, pH 7.5, at 4°C), 10% glycerol, 50 mM NaCl (5 M stock), 1 mM EDTA (0.5 M stock, pH 8.0). Immediately before use add 0.2 mM phenylmethylsulfonyl fluoride (PMSF), and 0.5 mM dithiothreitol (DTT).
5. Microcarrier spinner flasks of 30 L nominal volume (Bellco Glass, Inc.).
6. Microcarrier magnetic stirrers (Bellco Glass, Inc.).
7. 100-mm Tissue culture dishes.
8. 50-mL Dounce homogenizer with B pestle.

2.2. Preparation of DNA and Nuclei

2.2.1. Preparation of "Naked" DNA From Agarose-Embedded Cells

1. McCoys 5A growth medium (Gibco) supplemented with 5% bovine serum, 100 U/mL penicillin, and 100 µg/mL streptomycin.
2. HEPES-buffered McCoy's 5A growth medium. Prepared from McCoy's 5A medium by reducing the NaHCO$_3$ concentration to 5 mM and by adding HEPES to a final concentration of 20 mM.
3. [^{14}C]Thymidine, specific activity approx 50 mCi/mmol (see Note 3).
4. 500 µM thymidine in water.
5. 1% Solution of InCert agarose (FMC) in serum-free HEPES-buffered McCoy's 5A growth medium (no serum added). Incubate at 42°C in a water bath to prevent solidification.
6. 3.5-mm-diameter glass tubes cut in 15-to-20 cm lengths. Seal one end with tape. Precool by immersing into ice.
7. Lysis solution: 10 mM Tris-HCl, pH 7.5 at 50°C, 100 mM EDTA, 50 mM NaCl, 1% N-lauryl sarcosyl ([NLS] 10% stock in H$_2$O). Just before use add 0.2 mg/mL of Proteinase (Sigma cat. no. P6911).
8. Washing buffer: 10 mM Tris-HCl, pH 7.5 at 4°C, 100 mM EDTA and 50 mM NaCl.
9. RNase solution: 0.1 mg/mL of RNase A in 10 mM Tris-HCl, pH 7.5 at 37°C, 100 mM EDTA, 50 mM NaCl.

2.2.2. Preparation of Nuclei From Agarose-Embedded Cells

1. Saponin lysis solution: 50 mM Tris-HCl, pH 8.0 at RT, 100 mM EDTA, 100 mM KCl, 0.05% saponin, 10% glycerol. Add 200 mM β-mercaptoethanol immediately before use.
2. Nuclei washing solution: 10 mM Tris-HCl, pH 8.0 at RT, 100 mM EDTA. Add 1 mM PMSF immediately before use.
3. Nuclei prereaction washing solution: 10 mM Tris-HCl, pH 8.0 at RT. Add 1 mM PMSF just before use.

2.3. Generation of Double-Strand Breaks in DNA

The requirements will vary depending on the experimental protocol. *See* **Subheading 3.3.** for more details.

2.4. In Vitro Repair Reactions

2.4.1. Assembly of In Vitro Repair Reactions: "Naked" DNA

1. Prereaction washing buffer: 40 mM HEPES–KOH, pH 7.4 at 37°C, 1 mM EDTA, 20 mM KCl, 3 mM MgCl$_2$.
2. Reaction buffer: 40 mM HEPES-KOH, pH 7.4 at 37°C, 20 mM KCl, 3 mM MgCl$_2$, 10 µM (each) deoxyribonucleoside triphosphates (dNTPs) (2 mM stock each), 1.5 mM ATP (100 mM stock), 1 mM β-mercaptoethanol. To assemble reactions, prepare a 10X solution (*see* **Note 4**). Dilute with water to the final volume. Account for the volume of the extract (and of other components that may be added in the reaction) by reducing proportionally the amount of water added.
3. Lysis solution: 10 mM Tris-HCl, pH 7.5 at 50°C, 100 mM EDTA, 50 mM NaCl, 2% NLS, 0.1 mg/mL of Proteinase.
4. Washing solution: 10 mM Tris-HCl, pH 7.5 at 37°C, 100 mM EDTA, 50 mM NaCl.
5. RNase treatment solution: 0.1 mg/mL RNase in 10 mM Tris-HCl, pH 7.5 at 37°C, 100 mM EDTA, 50 mM NaCl.

2.4.2. Assembly of In Vitro Repair Reactions: Nuclei

1. Nuclei prereaction buffer: 20 mM Tris-HCl, pH 7.5 at RT, 2 mM EDTA, 50 mM KCl, 5 mM MgCl$_2$.
2. Nuclei reaction buffer: 20 mM Tris-HCl, pH 7.5 at RT, 2 mM EDTA, 50 mM KCl, 5 mM MgCl$_2$, 10 µM dNTPs, 1.5 mM ATP, 1 mM β-mercaptoethanol. *See* **item 2** of **Subheading 2.4.1.** for details in the use of this solution to assemble reactions.
3. Lysis solution: 10 mM Tris-HCl, pH 8.0 at RT, 100 mM EDTA, 2% NLS, 0.1 mg/mL of Proteinase.
4. Washing solution: 10 mM Tris-HCl, pH 8.0 at RT, 100 mM EDTA, 50 mM NaCl.
5. RNase treatment solution: 0.1 mg/mL RNase in 10 mM Tris-HCl, pH 7.5 at 37°C, 100 mM EDTA, 50 mM NaCl.

2.5. Assay for DNA Damage Using Pulsed-Field Gel Electrophoresis

1. 0.5X TBE: 45 mM Tris, 45 mM boric acid, 1 mM EDTA, pH 8.2. Prepare a 5X stock solution.
2. Agarose, Seakem, FMC.

3. Methods

3.1. Preparation of Cell Extract

The method described here allows the preparation of cell extract from 10 L of cell suspension. Higher or lower amounts of extract can be prepared by appropriate scaling. (*See* **Note 2**.)

1. Grow HeLa cells at 37°C for 3 d in 25 100-mm tissue culture dishes prepared at an initial density of 6×10^6 cells/dish in 20 mL of S-MEM supplemented with serum and antibiotics. The final density after 3 d of growth should be approx 20×10^6 cells/dish, giving a total of 5×10^8 cells in 25 dishes.
2. Trypsinize cells from all dishes and resuspend in 10 L of prewarmed complete growth medium in a 30-L nominal volume microcarrier flask, thoroughly pregassed with 5% CO_2 in air. The initial cell concentration should be approx 5×10^4 cells/mL. Place in a warm room at 37°C; provide adequate stirring (approx 60 rpm).
3. Allow cells to grow for 4 d, to a final concentration of $4-6 \times 10^5$ cells/mL. Do not allow cells to exceed this concentration.
4. Collect cells by centrifugation (7 min at 2500g). Collection should be fast and is best done using a refrigerated centrifuge that can accept 1-L bottles (e.g., Beckman J6-MI). All further processing should be carried out at 0–4°C.
5. Rinse twice in PBS and centrifuge (5 min at 500g). Determine the packed cell volume (pcv) (approx 12 mL total).
6. Resuspend cell pellet in five pcv volumes of hypotonic buffer solution and centrifuge quickly (10 min, 1200g). Cells swell and the pcv approximately doubles.
7. Determine the new pcv. Resuspend the cell pellet in an equal volume of hypotonic buffer and disrupt in a Dounce homogenizer (20 strokes, pestle B). It is advisable to test cell disruption using a phase-contrast microscope.
8. Add 0.11 volumes of high-salt buffer and centrifuge at 3000g for 20 min.
9. Carefully remove the supernatant and centrifuge at 100,000g for 1 h.
10. Place the resulting extract (S100) in dialysis tubing with a molecular weight cut-off of 10–14 kDa and dialyze overnight against 50–100 volumes of dialysis buffer.
11. Collect the extract. Centrifuge at 10,000g for 20 min to remove precipitated protein. Aliquot and snap freeze. Store at –70°C. Keep a small aliquot for determining protein concentration using the Bradford assay (Protein assay, Bio-Rad).

3.2. Preparation of DNA and Nuclei From Agarose-Embedded Cells

3.2.1. Preparation of "Naked" DNA

The procedure described here produces "naked" DNA from agarose-embedded cells with no detectable impurities of protein and RNA *(15)*. The protocol is for the preparation of DNA from the human lung carcinoma cell line A549, but it can be modified to prepare DNA from other cell lines (*see* **Note 5**).

1. Plate 2×10^6 A549 cells per 100 mm dish for 6 d in 20 mL of McCoy's 5A supplemented with serum, antibiotics, as well as with 0.01 µCi/mL of [^{14}C]thymidine ([methyl-^{14}C]thymidine) and 2.5 µ*M* unlabeled thymidine (*see* **Note 3**). Cells reach a plateau after this period of growth, with more than 95% accumulated in G_1/G_0 phase (if possible verify by flow cytometry). There will be approx 2×10^7 cells/dish.
2. Trypsinize the cells, collect by centrifugation (5 min at 500*g*) and wash once in serum-free HEPES-buffered McCoy's 5A growth medium.
3. Resuspend the cells in serum-free medium at a concentration of 6×10^6 cells/mL. Make sure no clumps are present at this stage. If clumps are visible under the microscope, they should be carefully disrupted using a narrow-bore Pasteur pipet, or a syringe with a 19-gage needle.
4. Mix with an equal volume of 1% InCert agarose. The final cell concentration is 3×10^6 cells/mL, and the final agarose concentration is 0.5%.
5. Quickly pipet the suspension into precooled 3.5-mm diameter glass tubes and incubate in ice until solidification (2–5 min).
6. Gently remove solidified agarose from the tubes and cut 5-mm-long blocks (approx 50 µL per block; approx 1.4×10^5 cells per block; approx 1.0 µg of DNA per block, assuming 5 pg of DNA per diploid human G_1 cell—A549 cells have 1.4-fold more DNA than human diploid cells; local DNA concentration in the "cage" generated by a cell with a diameter of approx 15 µm: approx 4 mg/mL). This operation should be carried out in a cold room. Immerse the agarose blocks into an excess of serum-free HEPES-buffered McCoy's 5A growth medium, enough to completely cover them.
7. Transfer the agarose blocks to lysis solution (five blocks per milliliter) and incubate at 4°C for 45 min.
8. Transfer to 50°C and incubate for 12 h in a moderately shaking water bath (longer incubation is also possible if desired, or if more convenient).
9. Carefully remove the lysis solution and rinse once in washing solution (five blocks per milliliter).
10. Remove the washing solution, add fresh washing solution, and repeat the washing for a further 6 h with three changes in a cold room with moderate shaking using an orbital shaker.
11. Carefully remove the washing solution and treat with RNase for 1 h at 37°C in a moderately shaking water bath (five blocks per milliliter).
12. Rinse the agarose blocks in washing solution and transfer to fresh washing solution. Incubate in washing solution at 4°C for 4 more hours under continuous shaking. Change the washing solution every 2 h.
13. Transfer the agarose blocks into fresh lysis solution and incubate at 50°C for 12 h in a moderately shaking water bath (Proteinase concentration can be reduced to 0.1 mg/mL at this stage).

14. Wash as described previously in **steps 9–11**. After completion of these steps, RNA and protein are not detectable in the agarose blocks. Agarose blocks are now ready to be used in in vitro repair reactions; they can be stored in washing buffer at 4°C in the dark for at least 12 mo (*see* **Note 6**).

3.2.2. Preparation of Nuclei

1. Grow A549 cells to a plateau phase as described in **Subheading 3.2.1., step 1**. (*See* **Note 5**).
2. Trypsinize, wash, and embed the cells in agarose as described in **Subheading 3.2.1., steps 2–6**.
3. Carefully transfer the agarose blocks to saponin lysis solution and incubate for 45 min at 4°C; then transfer to 50°C for 2 h in a moderately shaking water bath.
4. Remove the lysis solution and rinse the agarose blocks in nuclei washing solution.
5. Add fresh nuclei washing solution and incubate for 1 h at 37°C in a shaking water bath under moderate shaking. Repeat the washing procedure, if necessary. At this stage, agarose blocks are nearly free of cytoplasm, and are ready to be used in repair reactions. Agarose blocks so treated should be used immediately. If necessary, they can be stored overnight at 4°C. Longer storage is not advisable and, if adopted, it should be tested.

3.3. Generation of Double-Strand Breaks in DNA

Double-strand breaks can be induced, either in "naked" DNA or in nuclei embedded in agarose, by different agents. The choice of agent will be determined by the studies to be performed. Ionizing radiation can be used to randomly induce DSBs with nonligatable ends. If ligatable ends are desired, treatment with rare-cutting restriction endonucleases can be utilized. Aspects of these treatments are described next.

Exposure to X-rays or γ-rays can be carried out after embedding cells in agarose as described in **Subheading 3.2.1., steps 1–6**, and before lysis to generate "naked" DNA or nuclei. Doses in the range of 20–40 Gy should be used with the detection assays outlined here. Alternative detection assays should be utilized if experiments requiring lower doses must be performed. When intact, G_1-phase, A549 cells are exposed to radiation, 36 DSBs are produced per cell per Gy of 250 kVp X-rays (2-mm Al filter). Radiation exposure is carried out with the agarose blocks immersed in serum-free HEPES-buffered McCoy's 5A medium, on ice. After irradiation the agarose blocks are lysed for the preparation of either nuclei or "naked" DNA, as outlined in the corresponding protocols.

When experiments are performed with "naked" DNA it is more convenient to expose the agarose blocks to radiation after completion of the lysis procedure, shortly before the actual experiment. Because "naked" DNA is more sensitive to breakage by ionizing radiation than DNA organized in chromatin, approx threefold lower doses should be used (5–15 Gy) to generate a number of DSBs equivalent to that produced in intact cells exposed to doses between 10–50 Gy *(15)*.

When rare-cutting restriction endonucleases are used to generate DSBs, digestions should be carried out in the buffer recommended by the supplier of the enzyme (frequently cutting restriction endonucleases generate extensive DNA fragmentation that is only partly amenable to ligation under the conditions of the assay). It is advisable to wash the plugs in the enzyme buffer for 2–4 h with at least one change before proceeding with the actual digestion. The agarose barrier makes it necessary to increase the units of enzyme, as well as the treatment time employed. In general, increasing the treatment time by 4 h and the enzyme activity units 5- to 10-fold above what would have been used to digest DNA in solution, provide satisfactory results.

3.4. Assembly of In Vitro Repair Reactions

3.4.1. Assembly of Repair Reactions Using "Naked" DNA

Once cellular extracts and agarose blocks with DSBs are prepared, in vitro reactions to assay rejoining can be assembled.

1. Wash the required number of plugs in prereaction buffer (five agarose blocks per milliliter) for 2 h at 4°C under continuous, moderate shaking using an orbital shaker. Repeat the washing procedure two to three times.
2. Transfer the agarose blocks (one per 50-µL reaction) to 1.5-mL Eppendorf tubes containing 50 µL (total volume) reaction buffer supplemented with cell extract (50–100 µg/reaction). Maintain reactions at 4°C during these preparation steps.
3. Incubate at 37°C in a shaking water bath for various periods of time, as required by the experimental protocol.
4. After the required incubation period for repair, transfer the agarose blocks into cold lysis solution (0.5 mL per agarose block, in 10-mL round bottom tubes) for 30 min and then to 50°C for 2 h in a moderately shaking water bath.
5. Wash the agarose blocks for 1 h at 37°C in washing buffer (2.5 mL per agarose block). Remove the washing buffer and treat with RNase A solution (0.5 mL per agarose block) for 1 h at 37°C in a moderately shaking water bath.
6. Remove the RNase solution and wash the agarose blocks in washing buffer for 2 h at 4°C in a moderately shaking water bath (optional). Repeat the wash step with fresh washing buffer (optional). At this point, agarose blocks are ready to be loaded for pulsed-field gel electrophoresis.

3.4.2. Assembly of Repair Reactions Using Nuclei

1. Wash the required number of plugs in nuclei prereaction buffer for 2 h at 4°C.
2. Transfer the agarose blocks to 1.5-mL Eppendorf tubes containing 50 µL of nuclei reaction buffer supplemented with cell extract (50–100 µg/reaction). Keep the reactions at 4°C during these preparation steps.

3. Incubate at 37°C for various periods of time, as required by the experimental protocol.
4. After the incubation period for repair, transfer the agarose blocks into 1 mL of cold lysis solution (in 10-mL round-bottom tubes) for 30 min and then to 50°C for 12 h in a moderately shaking water bath.
5. Wash the agarose blocks for 1 h at 37°C in washing buffer. Repeat the wash step with fresh washing buffer. Remove the washing buffer and treat with 1 mL of RNase A solution for 1 h at 37°C in a moderately shaking water bath.
6. Remove the RNase solution and wash the agarose blocks in washing buffer for 2 h at 4°C in a moderately shaking water bath (optional). Repeat the wash step with fresh washing buffer (optional). At this point the agarose blocks are ready to be loaded for pulsed-field gel electrophoresis.

3.5. Assay for DNA Damage Using Pulsed-Field Gel Electrophoresis

A variety of pulsed-field gel electrophoresis techniques have been developed to quantitate DSBs in cellular DNA. Any one of these methods can, in principle, be used to evaluate in vitro repair in the reactions described previously. The choice will depend on equipment availability and the requirements of the individual experiment. We routinely use asymmetric field inversion gel electrophoresis (AFIGE) when a large number of samples needs to be evaluated and precise sizing of DNA is not required. When sizing of DNA fragments is considered important, we use clamped homogeneous electric field (CHEF) gel electrophoresis *(15)*. A detailed description of these methods is beyond the scope of this protocol, and only the parameters routinely used in our laboratory for the quantification of DSBs using the above assays are presented next.

3.5.1. AFIGE

1. Cast a 0.5% agarose (GTC, FMC) gel (250 mL, 20 × 25 × 0.5 cm) with 0.5 μg/mL of ethidium bromide in 0.5X TBE using the appropriate combs (20 wells, 3.5 × 6 mm). Allow the gel to solidify at RT, transfer to 4°C for 1 h for further solidification. Remove the combs just before loading.
2. Load the gel by inserting the agarose blocks into the wells. Close wells with 0.5% agarose.
3. Place the gel into the electrophoresis box (horizontal gel electrophoresis system, model H4, Gibco/BRL), containing 2.5 L of precooled (10°C) 0.5X TBE. Cooling is achieved by a refrigerated water bath and circulating pump.
4. Run at 10°C for 40 h by applying cycles of 1.25 V/cm for 900 s in the direction of DNA migration, and 5.0 V/cm for 75 s in the reverse direction.
5. After completion of electrophoresis, place gel under a UV table and take a photograph for documentation.
6. Dry the gel by carefully placing it on filter paper (model 583 gel dryer filter paper, 34 × 45 cm, Bio-Rad) in a gel dryer (1 h at 80°C), under vacuum (20–25 mmHg).

7. Expose a PhosphorImager screen to the dried gel for 24–48 h.
8. Quantitate DSBs present in the samples by evaluating the amount of activity present in the entire sample over that present in the lane (fraction of activity released [FAR]). This parameter is directly related to the number of DSBs present in the DNA.
9. Plot results as a function of time. A typical gel and its quantification are shown in **Fig. 2**.

3.5.2. CHEF Gel Electrophoresis

CHEF gel electrophoresis is performed in a CHEF DRII apparatus (Bio-Rad) using a 0.8% Seakem Agarose (BRL) gel in 0.5X TBE at 10°C for 68 h at 45 V with 60-min pulses. After completion of electrophoresis, the gel is stained with 0.5 µg/mL of ethidium bromide. Other details are as described for AFIGE.

4. Notes

1. Accurate pH control is essential for the preparation of extracts with high repair and low nuclease activity. Unless otherwise stated, the pH values given in the protocol are for the indicated temperatures. It is important to keep in mind that the pH of Tris-buffered solutions changes significantly with temperature (it increases as the temperature decreases). Thus, a solution adjusted to a pH of 7.0 at RT will reach a pH of nearly 7.5 at 4°C. We found it very helpful to adjust the pH at the temperature under which each solution will eventually be used.
2. The preparation of a good extract depends strongly on the quality of the cells used. When cell growth is not optimal, or when cells overgrow, high nuclease activity may severely compromise the use of the extract in repair reactions. To ensure optimal growth we routinely take the following measures:
   a. Carefully test different batches of serum to find one with good growth characteristics. HeLa cells have a generation time of less than 20 h when grown as a monolayer, and less than 24 h when grown in suspension, under optimal growth conditions.
   b. Use cells grown in dishes to start the suspension cultures for extract preparation. This helps reduce clumping occurring after extensive growth in suspension.
   c. We follow cell growth daily, and collect cells for extract preparation when they reach a concentration of 4–6 × 10^5 cells/mL.
   d. We measure cell cycle distribution by flow cytometry. A high percentage of S-phase cells (approx 25% for HeLa cells), suggests that the cell culture is still in an active state of growth.
3. In the method described here, cells are radioactively labeled with [^{14}C]thymidine for quantification. Quantification of radioactively labeled DNA can be carried out either using a scintillation counter or a PhosphorImager. We prefer the latter instrument as it is less labor intensive. It is also possible to use fluorescence detection methods for quantification (e.g., a FluorImager). If extreme sensitivity

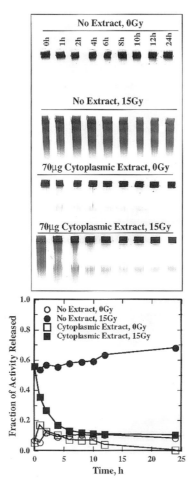

Fig. 2. Analysis of DSB rejoining in genomic DNA. The upper panel shows a series of four gels stained with ethidium bromide. The lower panel shows the results of their quantitation using a PhosphorImager. Open symbols represent nonirradiated samples, and closed symbols represent samples exposed to 15 Gy of X-rays just before assembling the repair reactions. A reduction in FAR indicates an increase in genomic DNA size as expected after DSB repair. Notice that the FAR is very low for samples not exposed to radiation, suggesting low nuclease activity. Notice also that in irradiated samples the FAR decreases only when extract is present in the reaction.

of detection is required, Southern blotting can also be considered. These alternative detection methods can easily be fitted in the described protocol. However, care should be exercised to ensure linear response for the parameter measured within the range of interest.

4. Since usually a large number of reactions need to be assembled, we prepare a 10X stock of the reaction buffer by mixing stocks of the individual components. Subsequently, we set up reactions by adding the appropriate amount of 10X reaction buffer (5 μL for a 50-μL reaction), the desired amount of extract, and enough H$_2$O to reach a final volume of 50 μL.
5. The choice of A549 cells for substrate preparation is arbitrary and is guided by their property to produce cultures with high percentage of G$_0$/G$_1$ cells when reaching the plateau phase of growth. Other cell lines can also be used if desired, but the protocol may require some modifications to accommodate the specific properties of these cells. When optimizing for the preparation of "naked" genomic DNA, the parameter to watch is degradation of nondamaged DNA during incubation under standard reaction conditions, but in the absence of extract; the FAR should be <10%. Relatively high FAR values under these conditions indicate the action of residual nucleases, which should be removed by further lysis.
6. A large number of agarose blocks with "naked" DNA can be prepared as a set and used, as needed, over a period of more than one year. This gives a great degree of flexibility in the planning of this type of repair experiment. Long-term storage of nuclei embedded in agarose is not possible.

References

1. Thompson, L. H. and Limoli, C. L. (2003) Origin, recognition, signaling, and repair of DNA double-strand breaks in mammalian cells. Eurekah.com and Kluwer Academic/Plenum, New York, NY.
2. Jackson, S. P. (2002) Sensing and repairing DNA double-strand breaks. *Carcinogenesis* **23**, 687–696.
3. Valerie, K. and Povirk, L. F. (2003) Regulation and mechanisms of mammalian double-strand break repair. *Oncogene* **22**, 5792–5812.
4. Iliakis, G., Wang, H., Perrault, A. R., et al. (2004) Mechanisms of DNA double strand break repair and chromosome aberration formation. *Cytogenet. Genome Res.* **104**, 14–20.
5. Baumann, P. and West, S. C. (1998) DNA end-joining catalyzed by human cell-free extracts. *Proc. Natl. Acad. Sci. USA* **95**, 14,066–14,070.
6. Udayakumar, D., Bladen, C. L., Hudson, F. Z., and Dynan, W. S. (2003) Distinct pathways of nonhomologous end-joining that are differentially regulated by DNA-dependent protein kinase-mediated phosphorylation. *J. Biol. Chem.* **278**, 41,631–41,635.
7. Huang, J. and Dynan, W. S. (2002) Reconstruction of the mammalian DNA double-strand break end-joining reaction reveals a requirement for an Mre11/Rad50/NBS1-containing fraction. *Nucleic Acids Res.* **30**, 1–8.
8. Wang, H., Zeng, Z.-C., Bui, T.-A., et al. (2001) Efficient rejoining of radiation-induced DNA double-strand breaks in vertebrate cells deficient in genes of the RAD52 epistasis group. *Oncogene* **20**, 2212–2224.
9. Wang, H., Perrault, A. R., Takeda, Y., Qin, W., Wang, H. and Iliakis, G. (2003) Biochemical evidence for Ku-independent backup pathways of NHEJ. *Nucleic Acids Res.* **31**, 5377–5388.

10. Perrault, R., Wang, H., Wang, M., Rosidi, B., and Iliakis, G. (2004) Backup pathways of NHEJ are suppressed by DNA-PK. *J. Cell. Biochem.* **92,** 781–794.
11. Labhart, P. (1999) Nonhomologous DNA end joining in cell-free systems. *Eur. J. Biochem.* **265,** 849–861.
12. Ganguly, T. and Iliakis, G. (1995) A cell-free assay using cytoplasmic cell extracts to study rejoining of radiation-induced DNA double-strand breaks in human cell nuclei. *Int. J. Radiat. Biol.* **68,** 447–457.
13. Cheong, N. and Iliakis, G. (1997) In vitro rejoining of double strand breaks induced in cellular DNA by bleomycin and restriction endonucleases. *Int. J. Radiat. Biol.* **71,** 365–375.
14. Cheong, N., Perrault, R., and Iliakis, G. (1998) *In vitro* rejoining of DNA double strand breaks: A comparison of genomic-DNA with plasmid-DNA-based assays. *Int. J. Radiat. Biol.* **73,** 481–493.
15. Cheong, N., Okayasu, R., Shah, S., Ganguly, T., Mammen, P., and Iliakis, G. (1996) In vitro rejoining of double-strand breaks in cellular DNA by factors present in extracts of HeLa cells. *Int. J. Radiat. Biol.* **69,** 665–677.

10

Detection of DNA Double-Strand Breaks and Chromosome Translocations Using Ligation-Mediated PCR and Inverse PCR

Michael J. Villalobos, Christopher J. Betti, and Andrew T. M. Vaughan

Summary

Current techniques for examining the global creation and repair of DNA double-strand breaks are restricted in their sensitivity, and such techniques mask any site-dependent variations in breakage and repair rate or fidelity. We present here a system for analyzing the fate of documented DNA breaks, using the *MLL* gene as an example, through application of ligation-mediated PCR. Here, a simple asymmetric double-stranded DNA adapter molecule is ligated to experimentally induced DNA breaks and subjected to seminested PCR using adapter and gene-specific primers. The rate of appearance and loss of specific PCR products allows detection of both the break and its repair. Using the additional technique of inverse PCR, the presence of misrepaired products (translocations) can be detected at the same site, providing information on the fidelity of the ligation reaction in intact cells. Such techniques may be adapted for the analysis of DNA breaks introduced into any identifiable genomic location.

Key Words: LM-PCR; IPCR; translocation; DNA; double-strand break repair; apoptosis; *MLL*.

1. Introduction

Chromosome translocations involving the mixed-lineage leukemia (*MLL*) gene are a frequent finding in infant, adult, and therapy-related leukemias *(1,2)*. Although the mechanism(s) responsible for the formation of a translocation is unknown, two models are beginning to evolve. The two mechanisms, illegitimate V(D)J recombination and apoptosis, share one element in common, in that both involve the presence of DNA double-strand breaks (DSBs) *(3–7)*.

Reprinted from *Molecular Toxicology Protocols*, edited by P. Keohavong and S. G. Grant, Humana Press, Totowa, NJ, 2005.

From: *Methods in Molecular Biology: DNA Repair Protocols: Mammalian Systems, Second Edition*
Edited by: D. S. Henderson © Humana Press Inc., Totowa, NJ

DNA DSBs have been shown to be potent inducers of chromosome translocations and can be produced by a multitude of agents, including ionizing radiation, genotoxic chemicals, and cellular processes such as apoptosis *(8,9)*. Traditional Southern blot-based techniques have been used in the past to visualize the presence of DNA breaks in a gene, but such methods have two major drawbacks: they require large amounts of starting material and are rather insensitive. The technique of ligation-mediated polymerase chain reaction (LM-PCR) coupled with nested PCR allows for greater sensitivity to detect DNA breaks and for the use of small amounts of DNA template.

The detection of chromosome translocations has previously employed both the Southern blot technique and *in situ* hybridization. However, these techniques are only capable of detecting chromosome translocations when both partner genes are known. Inverse PCR (IPCR) is performed using a gene-specific primer set oriented in opposing directions. Therefore, IPCR allows for the detection of chromosome translocations when only one of the translocation partners is known.

1.1. Ligation-Mediated PCR

This technique takes advantage of the extreme sensitivity of seminested PCR coupled with the specificity of LM-PCR to amplify gene fragments produced during apoptosis. LM-PCR has been used to amplify DNA adjacent to internucleosomal breaks induced during apoptosis, as well as breaks introduced during V(D)J recombination *(5,6)*. Although DNA lesions introduced by apoptotic nucleases produce blunt-end double-strand DNA breaks, it is possible to ligate a blunt-end linker molecule to the break site. This allows for the specific amplification of DNA sequences ligated to the linker molecule, using primers to the linker and to the gene of interest. In the second round of PCR, the use of nested primers exponentially increases the sensitivity of the assay to detect DNA breaks.

Prior to PCR amplification, the double-stranded linker molecule must be constructed and ligated to the genomic DNA. The linker is made by incubating two homologous oligonucleotides under a gradually decreasing temperature gradient. This is most easily achieved using the thermocycling file type on a Perkin Elmer 480 thermocycler. After the linker is made, it will have a staggered and a blunt-end terminus (**Fig. 1A**). Next, the linker is ligated to isolated genomic DNA using T4 DNA ligase. Just prior to LM-PCR, the reaction is heated to 72°C, causing the 11-mer oligomer to dissociate from the ligated linker molecule, leaving a 5′ 25-mer overhang. The 25-mer remains ligated to the DNA because the genomic DNA contributed its 5′ phosphate to the ligation reaction, and the 11-mer dissociates because the ligation lacked a 5′ phosphate, resulting in incomplete ligation (**Fig. 1B**). During PCR, *Taq* polymerase elongates the

Detection of DNA DSBs and Translocations

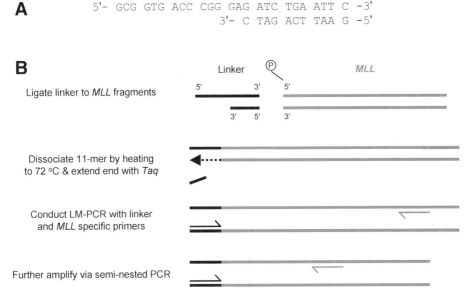

Fig. 1. Ligation-mediated polymerase chain reaction (LM-PCR) amplification of apoptotically cleaved *MLL*. (**A**) Depiction of the double-stranded blunt-end linker molecule used for LM-PCR. (**B**) Schematic of the LM-PCR technique used to amplify apoptotically cleaved *MLL*. The linker and linker-specific primers are black. *MLL* and its primers are in gray.

staggered 25-mer end to create the homologous strand. Now the DNA end has a double-stranded 25-mer linker molecule ligated to its terminus. LM-PCR is conducted using the linker-ligated DNA as a template, the 25-mer oligomer as a primer, and a primer specific to the gene or target of interest, in this case *MLL*. Seminested PCR is conducted using the PCR products from the first-round reactions. It is seminested because only the *MLL* primer is nested, whereas the same 25-mer oligomer is used as a primer. The second-round PCR products are analyzed by Southern blot using a 0.75-kb cDNA probe to the *MLL* breakpoint cluster region *(10)*. Amplification of cleaved *MLL* fragments with these primers results in a product of approx 290 bp (**Fig. 2**).

1.2. Inverse PCR

During apoptosis, the *MLL* breakpoint cluster region (bcr) is subjected to cleavage, creating a DNA DSB. One possible consequence of nonhomologous end joining (NHEJ) repair operating at the *MLL* cleavage site is incorrect repair of the break, leading to the formation of a chromosomal translocation. The identification of such misrepair events within early apoptotic cells would provide

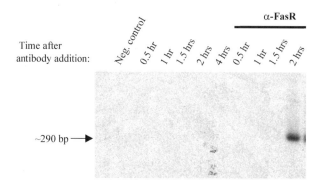

Fig. 2. Detection of *MLL* cleavage by LM-PCR. TK6 cells were induced to apoptosis by treatment with anti-Fas receptor antibody. Cell aliquots were removed at the stated times and DNA prepared. *MLL* cleavage was analyzed by LM-PCR. Control cells were not induced to undergo apoptosis. Negative control contains sfH$_2$O instead of DNA.

specific evidence that a break–rejoining cycle had occurred. To determine whether the apoptotic program is capable of generating a chromosomal translocation, an IPCR strategy was employed (**Fig. 3**). This procedure is able to amplify both the germline *MLL* sequence at the location of the apoptotic cut site and any rearranged fragments that contain a *MLL* translocation.

IPCR was first described by Ochman et al. *(11)* as a means of chromosome walking and identifying bacterial genes that were inactivated by insertional mutagenesis. Since its advent, this technique has been further tailored to identify DNA translocations when the partner gene is unknown *(12,13)*. The template for IPCR is a circularized DNA molecule created by restriction enzyme digestion followed by ligation. A restriction enzyme is chosen such that it has a high probability of cutting the unknown DNA sequence. This is achieved through the use of an enzyme with a short, 4-bp recognition sequence. Statistically, such an enzyme cuts random DNA sequences every 256 bp and thus would be very likely to cut an unknown DNA sequence.

The protocol described here utilizes a second infrequent cutting restriction enzyme with a 6-bp recognition sequence cutter that is predicted to cut a DNA sequence once every 4096 bp. This second restriction enzyme specifically allows for the detection of a chromosome translocation within a population of germline DNA sequences. The infrequently cutting restriction enzyme is chosen such that it lies on the 3′ side of the presumed chromosomal break site. The infrequent cutter also prevents the circularization of germline DNA molecules containing only the known DNA sequence (frequent cutter to frequent cutter) when both restriction enzymes are used. An infrequent restriction site

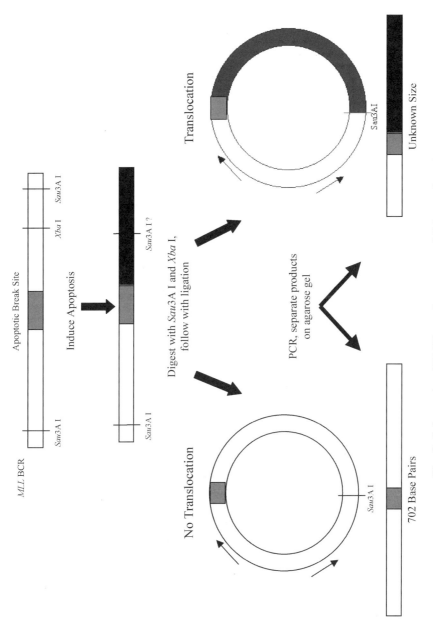

Fig. 3. Schematic of IPCR used to capture *MLL* translocations.

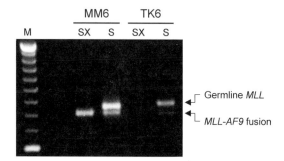

Fig. 4. IPCR detection of native and translocated *MLL*. In an effort to confirm the feasibility of detecting a chromosomal translocation, the IPCR technique was implemented using known positive and negative controls. The TK6 cell line contains one copy of the native (untranslocated) *MLL* gene, and the MM6 cell line is pseudotetraploid with respect to *MLL*, containing two native and two translocated copies of *MLL*. Using DNA cut only with the frequent cutter, a parental 782 bp is seen in both cell lines. Additionally, a 687-bp band is seen under the parental 782-bp band in the MM6 cell line, representing the *MLL-AF9* translocation. Both parental bands disappear when the DNA is digested with both the frequent (*Sau*3A I) and infrequent (*Xba*I) cutting enzymes. However, the 687-bp band under the MM6 parental band remains as expected since no *Xba*I site lies within this region of the *MLL-AF9* translocation.

would have a low probability of lying between two frequent cutting restriction enzyme sites. Therefore, circular germline molecules can be removed from the reaction mixture, thereby favoring the PCR amplification of molecules containing a DNA translocation (**Fig. 3**). However, any DNA translocations that contain this infrequent cutter site lying between the two frequent cutter sites will be eliminated from the pool. PCR is conducted from the circular templates using a pair of primers oriented in opposite directions that are specific to the known DNA sequence. This allows the PCR to proceed from the known sequence through the unknown sequence. An example of the IPCR technique's ability to detect *MLL* translocations is shown in **Fig. 4**. Here, DNA from both the TK6 cell line (not containing an *MLL* translocation) and the MM6 cell line (containing an *MLL-AF9* translocation) were both subjected to the IPCR assay.

2. Materials
2.1. Ligation-Mediated PCR
2.1.1. Construction of the Asymmetric Double-Stranded Linker
1. 0.22-µm filtered and autoclaved H_2O (sfH_2O).
2. Oligonucleotides (MWG Biotech, High Point, NC):

a. Linker 11 (5'-GAA TTC AGA TC-3').
b. Linker 25 (5'-GCG GTG ACC CGG GAG ATC TGA ATT C-3').
3. Oligonucleotide resuspension buffer: 75 mM Tris-HCl, pH 8.8 (Sigma, St. Louis, MO).
4. 1X T4 DNA ligase buffer: dilute from 10X with sfH$_2$0, keep at –20°C (Promega, Madison, WI).

2.1.2. Purification and Preparation of Template DNA

1. Puregene DNA isolation kit (Gentra Systems, Minneapolis, MN).
2. Asymmetric double-stranded linker.
3. 10X T4 DNA ligase buffer (Promega).
4. T4 DNA ligase (Promega).
5. 70°C Heating block.

2.1.3. Seminested PCR Amplification of Linker-Ligated DNA

1. *Taq* DNA polymerase (MBI Fermentas, Hanover, MD).
2. 10X and 1X (NH$_4$)$_2$SO$_4$ buffer: 10X stock comes with *Taq* polymerase; dilute to 1X with sfH$_2$0.
3. 500 μM Oligonucleotides (MWG Biotech).
   a. in12.2F (5'-ATG CCC AAG TCC CTA GAC AAA ATG GTG-3').
   b. Nin12.3F (5'-GTC TGT TCA CAT AGA GTA CAG AGG CAA CTA-3').
4. Oligonucleotide resuspension buffer: 75 mM Tris-HCl, pH 8.8.
5. 50 μM Stock solutions of Linker 11, Linker 25, in12.2F, and Nin12.3F.
6. 25 mM MgCl$_2$ (Sigma).
7. 25 mM dNTP stock: equal volumes of 100 mM dCTP, dTTP, dATP, and dGTP.
8. Thermocycler (e.g., Perkin Elmer 480, Foster City, CA).

2.1.4. Southern Blot Analysis of PCR-Amplified DNA

1. ZetaProbe GT nylon membrane (Bio-Rad, Hercules, CA).
2. 0.4 M NaOH buffer (Sigma).
3. Prime-a-Gene kit (Promega, Madison, WI).
4. 6000 Ci/mmol [α-^{32}P]dCTP (Amersham Pharmacia, Piscataway, NJ).
5. 0.75-kb cDNA fragment to *MLL* exons 8–14.

2.2. Inverse PCR

2.2.1. Preparation and Purification of the DNA Template

1. TK6 cells (American Type Culture Collection [ATCC]; Manassas, VA).
2. 25- and 75-cm^2 Tissue culture flasks (e.g., Corning, Corning, NY).
3. RPMI cell culture medium (e.g., Sigma), or medium appropriate for cells of interest.
4. Fetal bovine serum (FBS; e.g., Biologos, Naperville, IL).
5. Apoptotic stimulus: anti-Fas antibody (Kamiya Biomedical, Seattle, WA), radiation source (e.g., dual-head Cs-137 Gammacell irradiator, Nordion International, Ontario, Canada, or equivalent).

6. Centrifuge (e.g., Jouan, Winchester, VA).
7. Phosphate-buffered saline (PBS; Cambrex BioScience, Walkersville, MD).
8. Nuclei lysis solution (Promega).
9. 10 mg/mL of RNase (Sigma).
10. Protein precipitation solution (Promega).
11. 100% Isopropanol (Sigma).
12. Vortex mixer.
13. Ice.
14. 70% Ethanol (Sigma).
15. 0.22-μm Filtered and autoclaved H_2O.

2.2.2. Restriction Enzyme Digestion

1. *Sau*3AI and *Xba*I (Promega).
2. 3 *M* Sodium acetate, pH 5.2 (Sigma).

2.2.3. Ligation Reaction

1. T4 DNA ligase (Promega).
2. 0.1 *M* Spermidine (Sigma).

2.2.4. Inverse PCR

1. For intron 11, use the following primers:
   Forward 1 (5'-ATG TCC ATG ACA TAT CAC TG-3').
   Reverse 1 (5'-TAG GAC TTC ATA TTT GCC A-3').
   Forward 2 (5'-AGC ACA ATC CCA TCT TAG T-3').
   Reverse 2 (5'-TGA GAC GGA GTC TTG CT-3').

   For exon 12, use the following primers:
   Forward 1 (5'-CTT TGT TTA TAC CAC TC-3').
   Reverse 1 (5'-TAT GGG AAT ATA AAG GAG TGG G-3').
   Forward 2 (5'-TTA GGT CAC TTA GCA TGT TCT G-3').
   Reverse 2 (5'-CAG TTG TAA GGT CTG GTT TGT C-3') (Invitrogen, Carlsbad, CA).
2. *Taq* DNA polymerase (MBI Fermentas).
3. Thermocycler (e.g., Perkin Elmer 480).

2.2.5. Cloning and Sequencing of IPCR Products

1. Qiaquick gel extraction kit (Qiagen, Valencia, CA).
2. TOPO TA Cloning® Kit (Invitrogen).
3. Wizard Miniprep kit (Promega).
4. DNA sequencer (e.g., model 377 Prism DNA Sequencer®, Applied Biosystems, Foster City, CA).

2.2.6. Translocation Sequence Analysis

1. Computer with internet access.

3. Methods
3.1. Ligation-Mediated PCR
3.1.1. Construction of the Asymmetric Double-Stranded Linker

1. Resuspend Linker 11 and Linker 25 to 500 µM with oligonucleotide resuspension buffer by vortexing.
2. Mix 500 pmol of Linker 11 and 500 pmol Linker 25 by pipeting in the presence of 1X T4 DNA ligase buffer and sfH$_2$O, to a total volume of 5 µL.
3. Incubate the reaction in the thermocycler at 95°C for 5 min, 70°C for 1 s, and then lower the temperature from 70°C to 25°C over 60 min.
4. Continue to incubate the reaction at 25°C for 60 min, and subsequently lower the temperature from 25°C to 4°C over 60 min.
5. Store the double-stranded linker at –20°C.

3.1.2. Purification and Preparation of Template DNA

1. Isolate genomic DNA from approx 1–2 × 10^6 cells using the Puregene DNA isolation kit according to manufacturer's instruction (*see* **Note 1**).
2. To 1.0 µg of genomic DNA, add 100 pmol (1.0 µL) of double-strand linker, 6 µL of 10X T4 DNA ligase buffer, 1.0 µL (3 U) of T4 DNA ligase, and add sfH$_2$O to a final volume of 60 µL (*see* **Note 2**). Flick-mix the tubes.
3. Incubate the reaction at 15°C for 14 h, 70°C for 10 min in a heating block and then at 4°C (*see* **Note 3**).
4. Dilute the reaction to 5 ng/µL and then 0.5 ng/µL with sfH$_2$O (*see* **Note 4**).

3.1.3. Seminested PCR Amplification of Linker-Ligated DNA

1. Dilute *Taq* DNA polymerase 1:1 with 1X (NH$_4$)$_2$SO$_4$ buffer, for a final concentration of 2.5 U/µL.
2. Dilute 5 µL of 500 µM primers to 50 µM with 45 µL of oligonucleotide resuspension buffer.
3. To 2.0 µL (1.0 ng) of linker-ligated DNA add: 5.0 µL of 10X (NH$_4$)$_2$SO$_4$ buffer, 6.0 µL of 25 mM MgCl$_2$, 0.5 µL of 50 µM Linker 25 primer, 0.5 µL of 50 µM in 12.2F primer, 0.5 µL of 25 mM dNTP mix, and 34.5 µL of sfH$_2$O (*see* **Notes 5** and **6**).
4. Incubate the reactions at 72°C for 3 min, and then add 1.0 µL of diluted *Taq* DNA polymerase. Mix the reactions by vortexing and centrifuge briefly (*see* **Note 7**).
5. Process the PCRs at 72°C for 5 min, then increase to 95°C for 4 min, then 30 cycles of 95°C for 45 s, 66°C for 60 s, 72°C for 45 s, followed by 72°C for 10 min, and reduce to 4°C until removed from the machine.
6. To 1.0 µL of first-round PCR product add: 2.5 µL of 10X (NH$_4$)$_2$SO$_4$ buffer, 3.0 µL of 25 mM MgCl$_2$, 0.25 µL of 50 µM Linker 25 primer, 0.25 µL of 50 µM in 12.3F primer, 0.25 µL of 25 mM dNTP mix, 0.25 µL (1.25 U) of *Taq* polymerase, and 17.5 µL of sfH$_2$O (*see* **Note 6**).
7. Process the reactions at 95°C for 4 min, then 25 cycles at 95°C for 45 s, 66°C for 60 s, 72°C for 45 s, then 72°C for 5 min, and leave at 4°C.

8. Size-fractionate the second-round PCR products on a 2.0% agarose gel, and transfer the DNA to a ZetaProbe GT membrane with 0.4 M NaOH buffer according to the manufacturer's instructions.
9. Construct an *MLL* cDNA probe using the 0.75-kb fragment *(10,14)*, the Prime-a-Gene kit, and [α-^{32}P]dCTP.
10. Perform Southern blot analysis by standard methods using the *MLL* cDNA probe *(14)*.

3.2. Inverse PCR

3.2.1. Preparation and Purification of the DNA Template

1. Resuspend 1×10^7 TK6 cells in 20 mL of fresh RPMI medium supplemented with 10% FBS by vortexing. Treat with anti-Fas antibody (0.5 µg/mL final concentration) or 8 Gy γ-radiation.
2. Remove aliquots containing 3×10^6 cells at 0, 2, 4, 6, 8, and 24 h after addition of the apoptotic stimulus. Pellet cells by centrifugation at 700g for 5 min.
3. Wash cell pellets once in 500 µL of PBS and centrifuge again at 700g.
4. Resuspend pellets in 600 µL nuclei lysis solution, and mix by pipeting to ensure cell lysis.
5. Next, add 3 µL of RNase (stock 10 mg/mL) to each sample, and incubate at 37°C for 30 min.
6. Allow the samples to cool to room temperature.
7. Add 200 µL of protein precipitation solution and vortex on high for 20 s.
8. Allow the protein to precipitate for 5 min on ice.
9. Centrifuge at 14,000g to precipitate the protein.
10. Remove the liquid phase, leaving behind the white protein pellet, and transfer it to a clean microcentrifuge tube containing 600 µL of 100% cold isopropanol.
11. Repeatedly invert the tube until the white thread-like strands of DNA appear.
12. Pellet the DNA by centrifugation at 14,000g for 1 min, wash the pellet in 70% ethanol, and resuspend in 50 µL sfH$_2$O.

3.2.2. Restriction Enzyme Digestion

1. Set up a restriction enzyme digestion in a 1.5-mL microcentrifuge tube containing 6 µg of TK6 DNA (prepared as in **Subheading 3.2.1.**). Digest with 10 U *Sau*3AI or a combination of both *Sau*3AI (10 U) and *Xba*I (10 U) in a reaction volume of 50 µL according to the supplier's instructions with the restriction buffer provided.
2. Incubate the reaction at 37°C for 18 h.
3. Inactivate the restriction enzyme(s) by heating the reaction mixture to 70°C for 15 min.
4. Precipitate the DNA by adding 2 vol of ice-cold 100% ethanol and 1/10 vol of 3 M NaOAc, pH 5.2. Incubate at –80°C for 20 min, and then pellet at 14,000g in a microcentrifuge for 20 min at 4°C. Wash the pellet once in 70% ethanol, dry the pellet, and resuspend it in 172 µL of sfH$_2$O.

3.2.3. Ligation Reaction

1. Ligate cleaved DNA from **Subheading 3.2.2.** with 0.05 U of T4 ligase in a 200 µL vol with the addition of 5 µL of 0.1 M spermidine to a final concentration of 2.5 mM (*see* **Note 8**).
2. Allow reaction to proceed for 48 h at 14°C (e.g., using a water bath).
3. Inactivate T4 ligase by heating to 70°C for 10 min.
4. Precipitate DNA by adding 2 vol of ice-cold 100% ethanol and 1/10 vol of 3 M NaOAc, pH 5.2. Incubate at –80°C for 20 min, then pellet at 14,000g in a microcentrifuge for 20 min at 4°C. Wash the pellet once in 70% ethanol, dry the pellet, and resuspend it in 40 µL of sfH$_2$O (*see* **Note 9**).

3.2.4. Inverse PCR

1. Use 100 ng of the ligated DNA from **Subheading 3.2.3.** as a template for nested PCR analysis of the *MLL* gene. PCR is conducted using a nested set of primers to examine both the intron 11 control region and the exon 12 apoptotic cleavage site. Both sets were used identically for 28 cycles in the first round and 18 cycles in the second round at PCR temperatures of 95°C/55°C/72°C for 1 min/step using reaction conditions stated by Perkin Elmer (*see* **Note 10**).
2. Following PCR, analyze the samples by electrophoresis on a 1.2% agarose gel.

3.2.5. Cloning of Inverse PCR Products

1. Stain agarose gels with ethidium bromide and visualize under UV light. Excise bands present in lanes digested with both *Sau*3AI and *Xba*I, and gel-purify using a kit according to the manufacturer's instructions.
2. Clone gel-purified PCR products into pCR4-TOPO and select on plates containing 50 mg/mL ampicillin, according to the manufacturer's instructions.
3. Select and expand resistant colonies in 5 mL each of liquid LB and purify the cloned PCR product by performing a miniprep according to the manufacturer's protocol.
4. Sequence the cloned PCR product using either the T7 or T3 primers.

3.2.6. Translocation Sequence Analysis

1. Search obtained DNA sequences against the NCBI database (http://www.ncbi.nlm.nih.gov/BLAST/). Select accession numbers providing a match to the input DNA sequence, and analyze the region of DNA extending beyond the point at which it breaks off at *MLL*.

4. Notes

1. This protocol has been successfully conducted on the lymphoblastoid cell lines TK6 and WIL2-NS and the glioblastoma cell lines M059K and M059J. The genomic DNA must be fully resuspended. This can be achieved by resuspending the DNA with 200–350 µL of DNA hydration buffer and incubating it at 65°C for 2–3 h. During this incubation, flick-mix the tubes frequently.

2. The reaction stated is for one sample. It is best to make a master mix of double-stranded linker, T4 DNA ligase buffer, and T4 DNA ligase that can be added to each sample of DNA diluted in the required amount of sfH$_2$O. The master mix provides for greater accuracy in measuring small volumes and uniformity between samples.
3. The 70°C incubation heat-inactivates the T4 DNA ligase.
4. Vortex the DNA dilutions vigorously to ensure uniform solutions.
5. The amount of linker-ligated DNA used for amplification must be titrated. This is done to minimize the signal produced by basal levels of apoptosis in culture. Titrate the linker-ligated DNA to the point at which the break of interest is undetectable in control cells. A good titration range is 1.0–5.0 ng of linker-ligated DNA as template for LM-PCR.
6. Make a master mix of PCR components to ensure PCR uniformity.
7. Use of *Taq* DNA polymerase from Fermentas eliminates the need to optimize MgCl$_2$ concentrations in PCRs.
8. The volume of the ligation reaction should not be less than 200 µL. A large volume is necessary to favor intramolecular ligation kinetically.
9. Use of a master mix for PCR ensures equal concentration of reagents in each reaction.
10. DNA pellets resuspended in water can be incubated in a 50°C water bath to solubilize the DNA more readily.

References

1. Super, H. J., McCabe, N. R., Thirman, M. J., et al. (1993) Rearrangements of the *MLL* gene in therapy-related acute myeloid leukemia in patients previously treated with agents targeting DNA-topoisomerase II. *Blood* **82,** 3705–3711.
2. Rowley, J. D. (1993) Rearrangements involving chromosome band 11q23 in acute leukaemia. *Semin. Cancer Biol.* **4,** 377–385.
3. Aplan, P. D., Lombardi, D. P., Ginsberg, A. M., Cossman, J., Bertness, V. L., and Kirsch, I. R. (1990) Disruption of the human SCL locus by "illegitimate" V-(D)-J recombinase activity. *Science* **250,** 1426–1429.
4. Tycko, B. and Sklar, J. (1990) Chromosomal translocations in lymphoid neoplasia: a reappraisal of the recombinase model. *Cancer Cells* **2,** 1–8.
5. Schlissel, M. S. (1998) Structure of nonhairpin coding-end DNA breaks in cells undergoing V(D)J recombination. *Mol. Cell. Biol.* **18,** 2029–2037.
6. Staley, K., Blaschke, A. J., and Chun, J. (1997) Apoptotic DNA fragmentation is detected by a semi-quantitative ligation-mediated PCR of blunt DNA ends. *Cell Death Differ.* **4,** 66–75.
7. Wyllie, A. H. (1980) Glucocorticoid-induced thymocyte apoptosis is associated with endogenous endonuclease activation. *Nature* **284,** 555–556.
8. Richardson, C. and Jasin, M. (2000) Frequent chromosomal translocations induced by DNA double-strand breaks. *Nature* **405,** 697–700.
9. Vaux, D. L. and Strasser, A. (1996) The molecular biology of apoptosis. *Proc. Natl. Acad. Sci. USA* **93,** 2239–2244.

10. Betti, C. J., Villalobos, M. J., Diaz, M. O., and Vaughan, A. T. (2001) Apoptotic triggers initiate translocations within the *MLL* gene involving the nonhomologous end joining repair system. *Cancer Res.* **61,** 4550–4555.
11. Ochman, H., Gerber, A. S., and Hartl, D. L. (1988) Genetic applications of an inverse polymerase chain reaction. *Genetics* **120,** 621–623.
12. Forrester, H. B., Yeh, R. F., and Dewey, W. C. (1999) A dose response for radiation-induced intrachromosomal DNA rearrangements detected by inverse polymerase chain reaction. *Radiat. Res.* **152,** 232–238.
13. Forrester, H. B. and Radford, I. R. (1998) Detection and sequencing of ionizing radiation-induced DNA rearrangements using the inverse polymerase chain reaction. *Int. J. Radiat. Biol.* **74,** 1–15.
14. Ausubel, F. M., Brent, R., Kingston, R. E., et al. (eds.) (1997) *Short Protocols in Molecular Biology*. John Wiley & Sons, New York, NY.

11

Plasmid-Based Assays for DNA End-Joining In Vitro

George Iliakis, Bustanur Rosidi, Minli Wang, and Huichen Wang

Summary

Double-strand breaks (DSBs) disrupt DNA integrity and cause genomic instability and cancer, mutations, or cell death. Among the pathways utilized by cells of higher eukaryotes to repair this lesion, nonhomologous end-joining (NHEJ) is the most dominant. The biochemical characterization of NHEJ has significantly benefited from in vitro plasmid end-joining assays that can complement and extend information obtained from genetic studies. There is evidence that several factors involved in DNA-PK–dependent NHEJ remain to be identified. In addition, under certain circumstances, cells utilize backup pathways of NHEJ that depend on unknown factors to remove DSBs. Characterization of these putative factors will benefit from plasmid-based assays of DNA end-joining. Here, we describe a protocol for in vitro end-joining using plasmid DNA as substrate. The required procedures include: (1) preparation of HeLa-cell nuclear extract; (2) preparation of plasmid substrate DNA; (3) assembly of in vitro DNA repair reactions; and (4) product analysis by gel electrophoresis. The assay is powerful and easy to perform, but one should be aware that it represents an oversimplification, as it does not consider the in vivo organization of DNA into chromatin.

Key Words: Cell extract; DNA repair; double-strand breaks; in vitro assays; ionizing radiation; plasmid; restriction endonuclease.

1. Introduction

DNA double-strand breaks (DSBs) can be induced in the genome of eukaryotic cells by endogenous processes associated with oxidative metabolism, errors during DNA replication, and various forms of site-specific DNA recombination, as well as by exogenous agents such as ionizing radiation (IR) and chemicals. DNA DSBs disrupt the integrity of the genome and are therefore severe lesions that if unrepaired or if misrepaired can cause genomic instability and cancer, mutations, or cell death.

Cells utilize two enzymatically distinct processes to repair DNA DSBs: homology directed repair (HDR) and nonhomologous end-joining (NHEJ) *(1–3)*. HDR removes DSBs by utilizing homologous DNA segments, on the same or different DNA molecules, and restores faithfully the original DNA sequence around the break. NHEJ removes DSBs from the genome by simply joining the DNA ends without homology requirements and without ensuring sequence restoration around the break. While homologous recombination is the preferred mechanism of DSB repair in prokaryotes and in lower eukaryotes such as yeast, nonhomologous end-joining is extensively used in higher eukaryotes *(4)*.

The biochemical characterization of NHEJ in higher eukaryotes has significantly benefited from in vitro plasmid end-joining assays *(5)*. Such assays complement genetic studies and have confirmed the important roles of DNA-PKcs, Ku, DNA ligase IV, and XRCC4 in this repair process. Evidence exists however that additional factors are required for the successful removal in vivo of DNA DSBs by the DNA-PK–dependent pathway of end-joining *(6–8)*. In addition, there is evidence that cells deficient in the DNA-PK dependent pathway of NHEJ do not compensate by increasing utilization of HDR, but rely instead on backup pathways of NHEJ to remove ionizing radiation–induced DNA DSBs *(4,9–11)*. It is likely that the characterization of components of the DNA-PK–dependent pathway of NHEJ, as well as the biochemical characterization of backup pathways of NHEJ will continue to benefit from plasmid-based assays of DNA end-joining.

Here, we describe a protocol for in vitro end-joining using as substrate plasmid DNA. The required procedures include: (1) preparation of HeLa-cell nuclear extract; (2) preparation of plasmid substrate DNA; (3) assembly of in vitro DNA repair reactions; and (4) product analysis by gel electrophoresis. **Figure 1** shows a graphic outline of these steps.

2. Materials

2.1. Preparation of HeLa Nuclear Extract

0 2. Hypotonic buffer solution: 10 mM N-2-hydroxyethylpiperazine-N'-2-ethanesulfonic acid (HEPES), pH 7.5 at 4°C (1 M stock, pH 7.5 at room temperature [RT]), 1.5 mM MgCl$_2$ (1 M stock), 5 mM KCl (3 M stock). Immediately before use add 0.2 mM phenylmethylsulfonyl fluoride (PMSF) (200 mM stock in ethanol; can be stored at –20°C), and 0.5 mM dithiothreitol (DTT) (1 M stock in H$_2$O; store at –20°C).
3. High-salt buffer: 10 mM HEPES, pH 7.5 at 4°C, 1.6 M KCl, and 1.5 mM MgCl$_2$.
4. Dialysis buffer: 20 mM HEPES, pH 7.9 at 4°C, 10% glycerol, 100 mM KCl (3 M stock), 0.2 mM ethylenediamine tetraacetic acid (EDTA) (0.5 M stock, pH 8.0). Immediately before use add 0.2 mM PMSF, and 0.5 mM DTT.

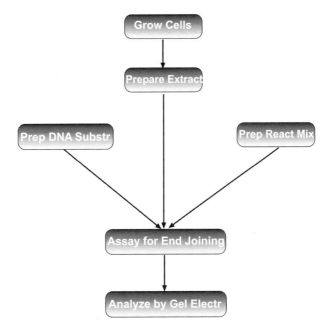

Fig. 1. Outline of steps required to analyze DNA end-joining using linearized plasmid as substrate.

5. Low salt buffer: 20 mM HEPES, pH 7.9 at 4°C, 1.5 mM MgCl$_2$, 20 mM KCl, 0.2 mM EDTA (0.5 M stock, pH 8.0). Immediately before use add 0.2 mM PMSF, and 0.5 mM DTT.
6. Microcarrier spinner flasks of 30 L nominal volume (Bellco Glass, Inc.).
7. Microcarrier magnetic stirrers (Bellco Glass, Inc.).
8. 100-mm Tissue culture dishes.
9. 50-mL Dounce homogenizer with B pestle.

2.2. Preparation of Substrate DNA

The choice of plasmid will depend on the specific requirements of the experiment (*see* **Note 1**). We have used extensively plasmid pSP65 (3 kb; Promega). After preparation of supercoiled plasmid DNA using standard procedures (*see* **Subheading 3.2.**), restriction endonucleases are needed for the linearization of the plasmid (*see* **Note 2**). This linearized plasmid is used as substrate in the end-joining reactions.

2.3. Assembly of In Vitro DNA Repair Reactions

1. 10X NHEJ buffer: 200 mM HEPES–KOH, pH 7.5 at RT, 800 mM KCl, 100 mM MgCl$_2$.

2. 1 mM ATP (20 mM stock stored at –20°C), 1 mM DTT (20 mM stock stored at –20°C), added into the reaction mixture just before use.
3. Stop solutions: 0.5 M EDTA, 0.5% sodium dodecylsulfate (SDS), 10 mg/mL of Protease (EC no. 232-909-5; Sigma cat. no. P6911).

2.4. Gel Electrophoresis

1. 6X DNA gel loading buffer: 30% glycerol, 50 mM EDTA, 100 mM Tris-HCl, pH 8.0, 0.01% bromophenol blue.
2. 0.5X TBE: 45 mM Tris base, 45 mM boric acid, 1 mM EDTA, pH 8.0. Prepare a 5X stock solution, store at room temperature.
3. Agarose, Seakem (FMC), or equivalent.
4. SYBR Gold (Invitrogen) DNA dye.
5. FluorImager (Typhoon, Amersham Biosciences), or other means for gel visualization.

3. Methods
3.1. Preparation of HeLa Nuclear Extract

The method described here allows the preparation of nuclear extract from a 10-L HeLa cell suspension. Higher or lower amounts of extract can be prepared by appropriate scaling. (*See* **Note 3**.)

1. Grow HeLa cells at 37°C for 3 d in 25 100-mm tissue culture dishes prepared at an initial density of 6×10^6 cells/dish in 20 mL of S-MEM supplemented with serum and antibiotics. (*See* **Chapter 9**.) The final density after 3 d of growth should be approx 20×10^6 cells/dish, giving a total of 5×10^8 cells in 25 dishes.
2. Trypsinize the cells from all dishes and resuspend in 10 L of prewarmed complete growth medium in a 30-L nominal volume microcarrier flask, thoroughly pregassed with 5% CO_2 in air. The initial cell concentration should be approx 5×10^4 cells/mL. Place in a warm room at 37°C; provide adequate stirring (~60 rpm).
3. Allow cells to grow for 4 d, to a final concentration of 4–6×10^5 cells/mL. Do not allow cells to exceed this concentration.
4. Collect the cells by centrifugation (7 min at approx 2000g). Quick collection is essential and is best done with a refrigerated centrifuge accepting 1-L bottles (e.g., Beckman J6-MI). All further processing should be carried out at 0–4°C.
5. Rinse twice in phosphate-buffered saline (PBS) by centrifugation (5 min at 500g). Use a marked tube of appropriate size to help determine the packed cell volume (pcv) (approx 12 mL total).
6. Resuspend the cell pellet in 5 pcv of hypotonic buffer solution, keep in ice for 10 min, and centrifuge (10 min, 1200g). Cells swell and the pcv approximately doubles.
7. Determine the new pcv. Resuspend the cell pellet in an equal volume of hypotonic buffer and disrupt in a Dounce homogenizer (20 strokes, pestle B). It is advisable to test cell disruption using a phase-contrast microscope.
8. Slowly add 3 M KCl to a final concentration of 50 mM KCl. Keep on ice for 10 min. (*See* **Note 4**.)

9. Centrifuge (3000g, 20 min) to precipitate the nuclei.
10. Carefully remove the supernatant, which is the cytoplasmic extract (CE), and, if needed, dialyze, aliquot and store at –80°C.
11. Resuspend the nuclear pellet in 2 packed nuclear volumes (pnv) of low salt buffer. Slowly add 1 pnv of high salt buffer (400 mM KCl final concentration). Keep at 4°C for 30 min under gentle agitation to extract proteins.
12. Centrifuge at 40,000g for 30 min at 4°C. Carefully remove the supernatant.
13. Add DEAE Sepharose (or DEAE Cellulose) to the extract (0.1 mL of DEAE Sepharose preequilibrated in dialysis buffer per mL of extract) and incubate at 4°C for 30 min under gentle shaking. This treatment removes DNA from the extract and is optional (*see* **Note 5**).
14. Centrifuge at 3000g for 15 min at 4°C to remove DEAE Sepharose.
15. Place the supernatant in dialysis tubing with a molecular weight cutoff of 10–14 kDa and dialyze overnight against 50–100 volumes of dialysis buffer. Alternatively, dialyze for 4 h with one change at 1 h. Dialysis under these conditions can cause protein precipitation and may lead to a significant loss in DNA end-joining activity (*see* **Note 6**).
16. Centrifuge at 10,000g for 20 min to remove precipitated protein, aliquot, and snap freeze. Store at –80°C. Keep a small aliquot for determining protein concentration using the Bradford assay (protein assay; Bio-Rad).

3.2. Preparation of Substrate DNA

Standard procedures can be used for the preparation of supercoiled plasmid DNA. It is preferable to purify the DNA using two-step purification on CsCl–ethidium bromide gradient. Detailed protocols for this purpose are beyond the scope of the present protocol and can be found in other sources of protocols *(12,13)*.

3.3. Digestion of pSP65 with SalI

1. In a 500-µL reaction prepared with the appropriate buffer depending on the restriction endonuclease used, add to 25 µg of pSP65 plasmid the appropriate units of the restriction endonuclease (50 U of *Sal* I) and incubate at 37°C under gentle shaking. At 1 h add 50 U more of enzyme and continue incubation. Repeat enzyme addition at 2 h. The plasmid should be practically completely digested after 3 h (times may vary depending on the type of enzyme used).
2. Use gel electrophoresis to test the degree of digestion. Continue digestion, if necessary, until practically complete linearization has been achieved.
3. Purify the completely digested plasmid with phenol–chloroform and precipitate with isopropanol.
4. Wash the precipitate twice with 70% ethanol.
5. Resuspend air-dried plasmid DNA in TE buffer aiming at a concentration of about 1 mg/mL.
6. Dialyze against TE and determine the concentration.

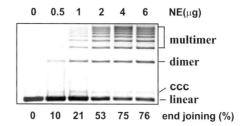

Fig. 2. Representative gel showing products of end-joining reactions carried out under the conditions described here. Reactions were assembled with the indicated amount of nuclear HeLa cell extract and were incubated at 25°C for 1 h. After gel electrophoresis, products were visualized in the Typhoon. Note the original linear substrate, as well as the dimers and other higher order multimers formed as a result of the end-joining activities present in the extract. The percentage end-joining shown in the lower part of the figure indicates the fraction of the input substrate that was found in rejoined products. Covalently closed circles are indicated by ccc.

3.4. Assembly of DNA End-Joining Reactions

1. Set up reactions (20 µL) by mixing 2 µL of 10X NHEJ buffer, 1 µL of ATP solution, 1 µL DTT solution, 50–250 ng of substrate DNA, 1–10 µg of nuclear extract, and H_2O up to 20 µL. (*See* **Note 7**.)
2. Incubate at 25°C for 60 min (end-joining is detectable after less than 5 min).
3. Terminate the end-joining reaction by adding 2 µL of 0.5% SDS, 2 µL of 0.5 M EDTA and 1 µL of protease (10 mg/mL). Incubate at 37°C for 30 min.
4. Add 5 µL of DNA gel loading buffer per reaction.
5. Load half the reaction on a 0.7% agarose gel. Run the gel for 4–5 h at 2 V/cm in 0.5X TBE buffer.
6. Stain the gel for 1 h at RT by diluting SYBR Gold 1:20,000 in 0.5X TBE under gentle shaking. Destain for 1 h in water. (*See* **Note 8**.)
7. Scan the gel in a FluorImager (Typhoon, Amersham Biosciences), or any other available gel documentation equipment. Analyze results quantitatively to calculate the percentage of plasmid found in dimers and other higher order polymers (**Fig. 2**).

Notes

1. Any plasmid with a polycloning site to allow digestion with a range of restriction endonucleases can be used as substrate in the reaction. The choice of plasmid depends on the specific question addressed.
2. The choice of restriction endonucleases is determined by the types of ends desired in the end-joining reaction. Restriction endonucleases can be used generating blunt or cohesive (with 3′ or 5′ protruding single strands) ends of varying lengths. Combinations of endonucleases can also be used to generate incompatible ends.

Substrates modeling the ends generated in cells exposed to ionizing radiation may also be considered *(14)*.

3. The preparation of a good extract depends strongly on the quality of the cells used. When cell growth is not optimal, or when cells overgrow, high nuclease activity may compromise the use of the extract in repair reactions. To ensure optimal growth we routinely take the following measures:

    a. Carefully test different batches of serum to find one with good growth characteristics. HeLa cells have a generation time of <20 h when grown as a monolayer, and less than 24 h when grown in suspension, under optimal growth conditions.
    b. Use cells grown in dishes to start the suspension cultures for extract preparation. This helps to reduce clumping occurring after extensive growth in suspension.
    c. We follow cell growth daily, and collect cells for extract preparation when they reach a concentration of 4–6×10^5 cells/mL.
    d. We measure cell cycle distribution by flow cytometry to confirm normal growth.

4. Whole-cell extract can be conveniently prepared, if desired, by adding high-salt buffer, to a final concentration of 400–500 mM KCl, at **step 8** of the preparation protocol and incubating for 30 min in ice. Preparation continues with **step 12** (**steps 9–11** are omitted). Whole cell extract contains nuclear proteins leaching into the cytoplasmic fraction during the cell disruption step of the nuclear extract preparation protocol and may, therefore, offer advantages under certain circumstances. On the other hand, whole cell extract contains a lot more proteins and has a lower specific activity. Since DNA repair reactions take place in the nucleus, nuclear extracts are considered by some investigators as the preferable starting material.

5. Although treatment with DEAE Sepharose removes DNA from the extract, which can interfere with end-joining, we also observed that a considerable amount of activity remains bound on DEAE Sepharose under the elution conditions employed. We, therefore, omit this step when preservation of activity is important and traces of DNA of no concern.

6. We have repeatedly observed that dialysis against 100 mM KCl leads to activity loss caused by protein precipitation. This loss of activity can be substantial. We found that end-joining activities are much better preserved if dialysis is carried out in a buffer containing 400 mM KCl.

7. The protocol described here uses a final concentration of 10 mM Mg^{2+} in the reaction buffer. End-joining activity is only slightly lower at 5 mM Mg^{2+}. These concentrations are close to the physiological range of 2–4 mM *(15)*. Occasionally, Mg^{2+} concentration is reduced to 0.5 mM to render the reaction DNA ligase IV dependent *(6,8)*. Under conditions of low Mg^{2+} concentration, end-joining activity is reduced by more than 90%. Mg^{2+} concentration should therefore be carefully adjusted depending on the specific question addressed.

8. We use SYBRGold and fluorimaging for DNA detection and quantification of end-joining. Under these conditions, reactions assembled with as little as 25 ng

of DNA can be analyzed. End labeling of the linearized DNA can further increase the detection limit and allow the analysis of reactions assembled with as little as 1–10 ng of plasmid DNA *(6)*. Similar sensitivity can also be achieved by detection with Southern blotting *(16)*. When sensitivity of detection is not an issue, ethidium bromide can also be used to stain the gel.

References

1. Thompson, L. H. and Limoli, C. L. (2003) Origin, recognition, signaling, and repair of DNA double-strand breaks in mammalian cells. Web-site: Eurekah.com and Kluwer Academic/Plenum, New York, NY.
2. Jackson, S. P. (2002) Sensing and repairing DNA double-strand breaks. *Carcinogenesis* **23,** 687–696.
3. Valerie, K. and Povirk, L. F. (2003) Regulation and mechanisms of mammalian double-strand break repair. *Oncogene* **22,** 5792–5812.
4. Iliakis, G., Wang, H., Perrault, A. R., et al. (2004) Mechanisms of DNA double-strand break repair and chromosome aberration formation. *Cytogenet. Genome Res.* **104,** 14–20.
5. Labhart, P. (1999) Nonhomologous DNA end-joining in cell-free systems. *Eur. J. Biochem.* **265,** 849–861.
6. Baumann, P. and West, S. C. (1998) DNA end-joining catalyzed by human cell-free extracts. *Proc. Natl. Acad. Sci. USA* **95,** 14,066–14,070.
7. Udayakumar, D., Bladen, C. L., Hudson, F. Z., and Dynan, W. S. (2003) Distinct pathways of nonhomologous end-joining that are differentially regulated by DNA-dependent protein kinase-mediated phosphorylation. *J. Biol. Chem.* **278,** 41,631–41,635.
8. Huang, J. and Dynan, W. S. (2002) Reconstruction of the mammalian DNA double-strand break end-joining reaction reveals a requirement for an Mre11/Rad50/NBS1-containing fraction. *Nucleic Acids Res.* **30,** 1–8.
9. Wang, H., Zeng, Z.-C., Bui, T.-A., Sonoda, E., Takata, M., Takeda, S., and Iliakis, G. (2001) Efficient rejoining of radiation-induced DNA double-strand breaks in vertebrate cells deficient in genes of the RAD52 epistasis group. *Oncogene* **20,** 2212–2224.
10. Wang, H., Perrault, A. R., Takeda, Y., Qin, W., Wang, H., and Iliakis, G. (2003) Biochemical evidence for Ku-independent backup pathways of NHEJ. *Nucleic Acids Res.* **31,** 5377–5388.
11. Perrault, R., Wang, H., Wang, M., Rosidi, B., and Iliakis, G. (2004) Backup Pathways of NHEJ Are Suppressed by DNA-PK. *J. Cell. Biochem.* **92,** 781–794.
12. Sambrook, J. and Russell, D. W. (2001) *Molecular Cloning*, 3rd Ed. Cold Spring Harbor Laboratory Press, Cold Spring Harbor, NY.
13. Ausubel, F. M., Brent, R., Kingston, R. E., et al. (1994) *Current Protocols in Molecular Biology*. Greene Publishing Associates and Wiley-Interscience, New York, NY.
14. Chen, S., Inamdar, K. V., Pfeiffer, P., et al. (2001) Accurate in vitro end joining of a DNA double-strand break with partially cohesive 3′-overhangs and 3′-phospho

glycolate termini: effect of Ku on repair fidelity. *J. Biol. Chem.* **276,** 24,323–24,330.
15. Strick, R., Strissel, P. L., Gavrilov, K., and Levi-Setti, R. (2001) Cation-chromatin binding as shown by ion microscopy is essential for the structural integrity of chromosomes. *J. Cell Biol.* **155,** 899–910.
16. Feldmann, E., Schmiemann, V., Goedecke, W., Reichenberger, S., and Pfeiffer, P. (2000) DNA double-strand break repair in cell-free extracts from Ku80-deficient cells: implications for Ku serving as an alignment factor in non-homologous DNA end joining. *Nucleic Acids Res.* **28,** 2585–2596.

12

Use of Gene Targeting to Study Recombination in Mammalian Cell DNA Repair Mutants

Rodney S. Nairn and Gerald M. Adair

Summary

Gene targeting by homologous recombination in mammalian cells is an important tool for generating genetically modified mice used for modeling human diseases. Gene targeting approaches are also useful for studying the mechanisms of homologous recombination. We have developed gene targeting methods that we have specifically used to investigate the mechanisms of recombination in cultured mammalian cells. In this chapter, we describe the generation of Chinese hamster ovary (CHO) cell gene disruption ("knockout") mutants in the repair/recombination gene *ERCC1*. Using this approach, we have constructed pairs of isogenic *ERCC1*-proficient and -deficient (null) CHO cell lines and used them as recipients for gene targeting assays in which a hemizygous mutant hamster adenine phosphoribosyltransferase (*APRT*) locus is corrected by homologous recombination with plasmid vectors containing hamster *APRT* DNA sequence homologous to the target gene in each cell line. The configuration of the targeting vector leads to experimental outcomes in which certain classes of *APRT* recombinants are over- or under-represented depending on the repair gene status of the transfection recipient. We describe methods both for targeted gene knockout of *ERCC1*, and for *APRT* targeted gene correction by homologous recombination, and some of our experimental results using these approaches.

Key Words: *APRT*; CHO cells; DNA repair; *ERCC1*; recombination.

1. Introduction

Gene targeting is defined as homologous recombination between a transfected DNA sequence and an endogenous chromosomal gene locus (the "target"), and has proven a powerful approach for genetic manipulation both in cultured cells and organisms. Gene targeting strategies and procedures have been developed both for yeast and higher eukaryotic cells, and enable target gene correction, disruption, deletion, replacement, or site-directed modification of virtually any gene or chromosomal locus for which a cloned DNA sequence is available *(1–4)*. Many of the first mammalian gene targeting studies were directed toward disruption ("knockout") of a selected target gene locus in mouse ES (embryo stem) cells, with the primary objective of obtaining the desired mutant mouse as quickly as possible *(4,5)*. Subsequently, approaches to introduce more subtle genetic alterations were developed and applied to the generation of mutant mouse disease models *(5–7)*. Recently, chicken DT40 cells, which exhibit a very high capacity for efficient gene targeting, have been developed and used as gene targeting recipients to study the roles of genes involved in DNA repair *(8)*. Gene targeting in yeast, DT40 and ES cells has been thoroughly reviewed by a number of workers (e.g., *see* **refs.** *1–5,9*). It is not within the scope of this chapter to review the general area of gene targeting; instead, here we describe some specific methods and the use of gene targeting as an approach to study mechanisms of recombination in cultured mammalian cells. Our particular focus is on the development and use of gene targeting to generate and analyze DNA repair-deficient Chinese hamster ovary (CHO) cell lines to reveal interactions between DNA repair and recombination pathways in mammalian cells.

In budding yeast, *Saccharomyces cerevisiae*, both the nucleotide excision repair (NER) and mismatch repair (MMR) pathways have components that are also involved in specialized recombination pathways. For example, in single-strand annealing (SSA) the Rad1–Rad10 structure-specific endonuclease which is responsible for making the 5′-side incision in the complex NER reaction is required to remove 3′ single-stranded DNA "tails" from SSA intermediates *(10–13)*. A component of the MMR pathway, MutS-β, consisting of the Msh2–Msh3 heterodimer, is also required in SSA *(14)*. We developed and used a variety of gene targeting approaches to generate CHO cell lines useful for investigating the roles of some of the homologous genes in mammalian cells (i.e., *XPF, ERCC1,* and cg*MSH3*, homologous to *RAD1, RAD10*, and sc*MSH3*, respectively) in such specialized recombination pathways. We and our colleagues developed the first gene targeting methods used to modify the endogenous hamster adenine phosphoribosyltransferase (*APRT*) locus, and subsequently used this system to investigate mechanisms of homologous gene targeting at this locus in CHO cells *(15–21)*. We have also used targeted gene disruption to construct *ERCC1* knockout mutants in DNA repair-proficient parental CHO cell lines *(22–25)* as well as in other CHO cell DNA repair mutants (our unpublished results), and used two-step gene replacement targeting strategies to modify endogenous target loci in conventionally derived CHO DNA repair mutants, creating isogenic pairs of cell lines for subsequent studies of targeted and mitotic recombination at the endogenous (*APRT*) gene locus *(23,25)*.

In the following subsections, we describe some of the underlying principles of gene targeting, disruption, and modification applied specifically to CHO cells, and the utility of the hemizygous CHO *APRT* gene as a model target locus for gene targeting studies designed to investigate mechanisms of homologous recombination in somatic mammalian cells. The advantages of using CHO cells for such studies include:

1. CHO cells retain a relatively stable, pseudodiploid karyotype, are easily grown and maintained, and have a high cloning efficiency.
2. CHO cells are widely used in somatic cell genetics, and a large database exists in the literature describing mutagenic and cytotoxic responses of CHO cells to a variety of physical and chemical agents *(26,27)*.
3. A significant number of gene loci are hemizygous in CHO cell lines, including several DNA repair genes *(28,29)*; hemizygous gene loci greatly simplify both the generation of targeted gene knockouts and the molecular analysis of gene alterations generated from gene targeting.

As an example of the use of gene targeting to create DNA repair-deficient knockout mutants, targeted gene disruption of the *ERCC1* DNA repair gene in CHO cells is described; however, other DNA repair genes from both the NER and DSBR (double-strand break repair) pathways should also be suitable candidates for knockout in CHO cells by gene targeting approaches, and the approaches we describe for knockout of *ERCC1* should be generally applicable to other DNA repair gene loci. In addition, we have used these approaches to construct isogenic pairs of DNA repair-proficient and -deficient CHO cell lines in which the endogenous *APRT* locus had been modified with a two-step gene targeting/replacement strategy *(16,30)*; the resulting, modified *APRT* allele can be further targeted using site-specific recombination systems (e.g., FLP/FRT) to create a variety of mitotic recombination substrates for comparison in isogenic repair-proficient and –deficient backgrounds (e.g., *see* **refs. 23,25**).

1.1. Targeted Knockout of a DNA Repair Gene Locus in Cultured Mammalian Cells

It has been shown that isogenic sequence (i.e., DNA isolated from the particular cell line to be used for gene targeting) is necessary for efficient targeted recombination in mammalian cells *(31,32)*. Therefore, for targeted gene disruption of the CHO *ERCC1* gene, a human cDNA *ERCC1* probe *(33)* was used to screen a CHO cell genomic library constructed in λ FIXII (Stratagene, La Jolla, CA), resulting in recovery of the CHO *ERCC1* gene. A fragment containing exons 4 and 5 was subcloned and used to construct a targeting vector (*see* **Fig. 1**); this vector has been used for targeted *ERCC1* knockout in several wild-type CHO cell lines (for example, CHO-K1, CHO–ATS49tg, RMP41; *see* **refs. 22–25**), as well as two DNA repair-deficient CHO cell mutants (cg*XPD* and

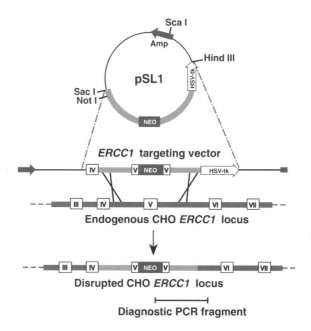

Fig. 1. Targeted gene knockout strategy for gene disruption of the *ERCC1* locus in CHO cells.

cg*MSH3*), generating double mutants defective in more than one DNA repair gene (our unpublished results).

Several parameters of gene targeting should be taken into consideration in designing a knockout vector. A study of gene targeting at the *HPRT* locus in mouse ES cells *(32)* showed that the use of isogenic targeting vector sequence/ target gene homology is more important for efficient targeted gene disruption than targeting vector configuration (i.e., integration vector configuration, having a double-strand break within target gene sequence homology in plasmid DNA, vs replacement vector configuration, with a double-strand break in the targeting vector outside target gene sequence homology in plasmid DNA). In CHO cells, we have found that although the overall efficiency of gene targeting is similar for both integration-type and replacement-type vectors *(18,34)*, the distribution of types of gene targeting events can be dramatically affected by targeting vector configuration. Use of a replacement-type targeting vector greatly reduces targeted integration recombinants compared to gene replacement/conversion recombinants (*see* **Fig. 2** for structures of integration and conversion recombinants at the model *APRT* locus). Therefore, integration-type targeting vectors are most efficient for two-step gene replacement or modifica-

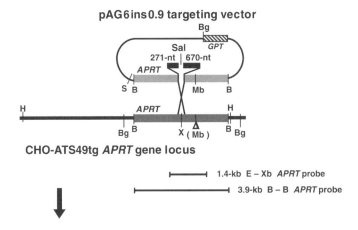

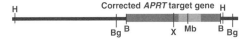

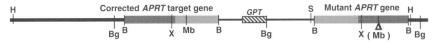

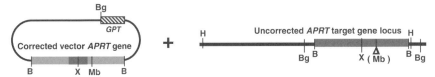

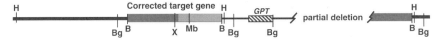

Fig. 2. Targeted recombination at the CHO–ATS49tg *APRT* locus.

tion strategies *(34–36)*, which require targeted integration in the first step, whereas replacement-type vectors may be preferable for gene disruption, or knockout, gene targeting approaches.

Another factor critical for efficient gene targeting is the length of target gene sequence homology in the targeting vector. A minimum length of sequence homology flanking each side of the positive selection marker (e.g., *NEO*) is required in a replacement-type targeting vector (*see* **Fig. 1**). Targeting efficiency decreases rapidly with decreasing length of sequence homology below about 500 base pairs (bp) *(19)*, suggesting that a minimum of 500–600 bp of homologous sequence flanking each side of *NEO* (or whatever selectable marker is used) should be designed into the knockout vector; typically, several kilobases of homologous sequence are used to flank each side of the selection marker *(37)*. Negative selection markers, such as *HSV-tk*, are used in positive–negative selection protocols to enrich the population of transformed cells for targeted homologous recombinants *(38)*. Selection against *HSV-tk* using gancyclovir or FIAU applied to the transfected cell population kills nonhomologous recombinants that have randomly integrated the targeting vector into their genomic DNA; gene replacement events that have lost the *HSV-tk* marker during recombination are thus enriched in a positive–negative selection protocol using *NEO* and gancyclovir or FIAU (*see* **Fig. 1**). Although very large enrichment ratios using positive–negative selection (up to three orders of magnitude) have been reported *(38)*, the degree of enrichment attainable depends on the target locus, the cell type, and the targeting vector; in practice, we have consistently achieved approx 6- to 12-fold enrichment for targeting the CHO *ERCC1* locus with the positive–negative gene disruption vector shown in **Fig. 1**.

1.2. Use of APRT Gene Targeting Assays to Study Recombination Deficiencies in Mammalian Cells

We have developed practical mammalian gene targeting assay systems utilizing hemizygous, CHO *APRT* deletion mutants, such as CHO–ATS49tg, for studying homologous recombination in cultured mammalian cells *(15–20,24,34)*. Our own work, and work from other laboratories, has demonstrated the utility of the hemizygous CHO *APRT* gene as a model target locus for studies designed to probe the mechanisms of mammalian recombination *(15–21,23-25,30,34,39–47)*. Targeted recombination/*APRT* gene correction assays, employing either "ends-in" integration-type or "ends-out" replacement-type targeting vector configurations, allow direct selection of APRT$^+$ recombinants. Use of a hemizygous, nonrevertible, deletion-inactivated CHO *APRT* gene as a target locus ensures that APRT$^+$ cells can arise only by targeted homologous recombination at the endogenous *APRT* locus. The small size of the CHO *APRT* gene (which spans only 2.2 kb from promoter to polyadenylation signal), absence of *APRT* pseudogenes, and hemizygosity for the target gene locus, greatly facilitate molecular analysis of APRT$^+$ recombinants.

In CHO *APRT* targeting experiments employing conventional "ends-in" integration vectors, with two unblocked arms of *APRT* targeting homology flanking a DSB, and a hemizygous APRT target locus with a 3-bp exon 5 MboII site deletion *(15,20,24,34)*, we have consistently recovered three distinct classes of APRT⁺ recombinants, reflecting at least three different types of targeted homologous recombination events:

1. Target gene conversions.
2. Targeted integrations at the chromosomal *APRT* locus.
3. Targeting vector correction recombinants, (which still retain an uncorrected *APRT* gene at the targeted chromosomal locus).

These three recombinant classes are readily distinguishable on the basis of their distinctive and diagnostic restriction fragment patterns. Target gene convertants (**Fig. 2A**), in which the 3-bp target gene deletion has been corrected by unidirectional transfer of sequence information from the targeting vector *APRT*, result in precise replacement of the three deleted bases and restoration of the exon 5 MboII restriction site. Targeted integration of the *APRT* targeting vector at the chromosomal gene locus by a single-reciprocal exchange/crossover event (**Fig. 2B**) results in a duplication of *APRT* gene sequences at the target gene locus (including at least one wild-type copy of the *APRT* gene). Targeting vector correction recombinants (**Fig. 2C**) retain an uncorrected *APRT* gene (with a 3-bp MboII site deletion) at the targeted chromosomal locus, but have also acquired an ectopically integrated wild-type *APRT* gene. In this class of recombinants, the targeting vector *APRT* sequence has been corrected by unidirectional transfer of sequence information from the target gene locus, by one-sided strand invasion and target gene-templated extension of the vector *APRT* sequence, either with or without gap-repair/vector recircularization. The corrected targeting vector, now carrying a wild-type *APRT* gene, is then randomly integrated at an ectopic site.

Comparisons of gene targeting frequencies and recombinant class distributions for DNA repair-proficient CHO–ATS49tg cells, and either conventionally derived CHO DNA repair-deficient mutants or isogenic, repair gene knockout cell lines, can provide valuable insights into the possible roles of specific mammalian repair gene products in various recombinational pathways. Although we have carried out *APRT* targeting experiments using several conventionally-derived *ERCC1* or *XPD* mutant, NER-deficient CHO cells lines *(18,20,24*, and our unpublished data), it is quite difficult to obtain matched, repair-proficient and -deficient cell lines that contain the same *APRT* target gene deletion. The advantages of using isogenic repair gene knockout cell lines for such studies are compelling. We have used the targeted gene disruption strategy shown in **Fig. 1** to knock out the *ERCC1* gene in CHO-ATS49 cells, in order to obtain isogenic, paired *ERCC1*⁺ and *ERCC1* knockout CHO cell lines containing the same 3-bp *APRT* target gene deletion that can be used for targeted recombination experiments *(24)*.

The Xpf–Ercc1 (mammalian cell) and Rad1–Rad10 (*Saccharomyces cerevisiae*) structure-specific endonucleases perform the same function in NER; they make the 5′ structure-specific incision in the DNA-damage-containing strand during the dual-incision step of NER *(48)*. In yeast, the Rad1–Rad10 endonuclease is also required for structure-specific incision and removal of long, nonhomologous 3′-ended single-strand tails from strand-invasion or single-strand-annealing recombination intermediates during homologous recombination *(10–14)*. To investigate the role of *ERCC1* in targeted recombination, and to directly test the hypothesis that this gene is specifically required for removal of long, nonhomologous tails from strand-invasion intermediates during targeted recombination in mammalian cells, we carried out a series of *APRT* targeting experiments in isogenic *ERCC1*⁺ (ATS49tg) and *ERCC1* knockout (E1KO7-5) cell lines *(24)*, employing an "ends-in" targeted integration vector (pAG6ins0.9) in which both arms of *APRT* targeting homology are blocked by terminal nonhomologies (*see* **Fig. 2**). Although *ERCC1* gene knockout did not appear to have any statistically significant effect on the overall frequency of targeted recombination in pAG6ins0.9 *APRT* targeting experiments, analysis of APRT⁺ recombinants revealed several striking differences in recombinant class distributions obtained from *ERCC1*⁺ vs *ERCC1* knockout cells *(24)*. Analysis of 90 APRT⁺ recombinants obtained from *ERCC1*⁺ ATS49tg–pAG6ins0.9 targeting experiments demonstrated that in *ERCC1*⁺ cells, target gene conversions (*see* **Fig. 2A**; 46%), targeted integrations (*see* **Fig. 2B**; 39%), and targeting vector correction convertants (*see* **Fig. 2C**; 16%) were still obtained even when both arms of the *APRT* targeting vector were blocked by long nonhomologous tails. In contrast, analysis of 67 APRT⁺ recombinants obtained from *ERCC1* knockout E1KO7-5–pAG6ins0.9 targeting experiments revealed only a single targeted integration and no targeting vector correction conversion recombinants; 83% of the recombinants were target gene conversions, and 16% represented a novel class of aberrant targeted integration/deletion events (*see* **Fig. 2D**) that were not seen in *ERCC1*⁺ cells *(24)*. In *APRT* targeting experiments employing another "ends-in" targeted integration vector (pAG6) in which both arms of *APRT* targeting homology are blocked by very short nonhomologous tails, no significant differences in APRT⁺ recombinant class distributions for *ERCC1*⁺ vs *ERCC1* knockout cells, suggesting that these short tails can be effectively removed by *ERCC1*-independent pathways *(24)*. Results from these studies demonstrated that *ERCC1* is required for removal of long, nonhomologous tails from the 3′-OH ends of invading strands during targeted homologous recombination in mammalian cells, and suggested that removal of such end-blocking nonhomologous tails

Gene Targeting to Study Recombination 141

from recombining DNAs by Xpf–Ercc1 endonuclease may allow normal resolution of recombination intermediates that might otherwise be diverted into nonhomologous end-joining pathways leading to deleterious deletions or rearrangements.

2. Materials
2.1. Knockout and APRT Gene Targeting

1. αMEM–10% FCS medium: α-modified minimal essential medium (Irvine Scientific) supplemented with 10% fetal calf serum (FCS), penicillin (50 U/mL), and streptomycin (50 μg/mL). Penicillin and streptomycin (tissue culture grade) are available from Gibco/ Invitrogen as well as other suppliers. A formulation of this medium using dialyzed FCS (i.e., αMEM supplemented with 10% FCS that has been dialyzed against four changes of Hank's balanced salt solution [HBSS] and filter-sterilized) is referred to as αMEM–10% DFCS. Serum-free αMEM is referred to as SFMEM.
2. Puck's solution A (solution A): Physiological saline (0.8% NaCl) in 5 mM KCl, 4.2 mM NaHCO$_3$, containing 0.1% dextrose. A 10X stock solution is prepared by dissolving 80 g of NaCl, 4 g of KCl, 10 g of dextrose, and 3.5 g of NaHCO$_3$ in 1 L of H$_2$O; a 1X working solution is used after 1:10 dilution with H$_2$O and filter sterilization. Store both 10X and 1X solutions at 4°C.
3. Hank's balanced salt solution (HBSS): Physiological saline (0.8% NaCl) in 5.4 mM KCl, 0.33 mM Na$_2$HPO$_4$, 0.44 mM KH$_2$PO$_4$, 4.2 mM NaHCO$_3$, 0.5 mM MgCl$_2$, 1 mM CaCl$_2$, containing 0.1% dextrose. A 1X solution is prepared by dissolving 40 g NaCl, 2 g KCl, 0.45 g Na$_2$HPO$_4$·7H$_2$O, 0.3 g KH$_2$PO$_4$, 1.75 g NaHCO$_3$, 0.5 g MgCl$_2$·6H$_2$O, 0.7 g CaCl$_2$·2H$_2$O, and 5 g dextrose in 5 L of H$_2$O. Filter sterilize and store at 4°C.
4. Phosphate-buffered saline (PBS): Physiological saline (0.8% NaCl) in 8 mM Na$_2$HPO$_4$, 1.5 mM KH$_2$PO$_4$, 2.5 mM KCl. A 10X solution is prepared by dissolving 12 g of KCl, 12 g of KH$_2$PO$_4$, 480 g of NaCl, and 129.6 g of Na$_2$HPO$_4$·7H$_2$O in 6 L of H$_2$O. A 1X working solution is prepared by autoclaving a 1:10 dilution (in H$_2$O). Store at 4°C.
5. Trypsin–EDTA: 0.25% trypsin, 25 mM Tris·HCl, 0.5 mM EDTA in Puck's solution A. Dissolve 3 g of Trizma base (Sigma) in 460 mL of solution A and add 20 mL of 1 N HCl; dissolve 0.2 g of Na$_2$EDTA·2H$_2$O (Sigma) in 20 mL of water; dissolve 2.5 g of trypsin (Gibco-BRL) in 450 mL of solution A, while stirring. Stir all three solutions individually until homogeneous, mix the Trizma and EDTA solutions together to dissolve, then add trypsin suspension, and adjust to pH 7.4 if necessary. Adjust the final volume to 1 L with H$_2$O, filter-sterilize, and dispense in 10-mL aliquots. Store frozen (do not refreeze).
6. Tris-EDTA Buffer (TE): 10 mM Tris-HCl, 1 mM EDTA, pH 8.0. For a 10X stock solution, dissolve 12.1 g of Trizma base (Sigma) and 3.7 g of Na$_2$EDTA·2H$_2$O in 950 mL of H$_2$O; adjust to pH 8.0 with HCl and bring the final volume to 1 L; store at 4°C. For a 1X working solution, dilute 1:10 with H$_2$O and filter sterilize.

7. 10% Sodium dodecyl sulfate (10% SDS): Dissolve 100 g of SDS in 1 L (final volume) of hot water with gentle stirring; autoclave and store at room temperature.
8. 5 M Lithium acetate: Dissolve 51 g of $LiC_2H_3O_2 \cdot 2H_2O$ in 100 mL of H_2O (final volume); autoclave and store at 4°C.
9. Equilibrated phenol: Melt crystalline, H_2O-saturated redistilled phenol (stored frozen) at 37°C in water bath. Add one-half volume of 0.5 M Tris-HCl, pH 7.5, and stir at room temperature for 30–45 min. Separate and retain the lower, organic phase. Add one volume 0.1 M Tris-HCl, pH 7.5, shake in separatory funnel, separate, and retain lower phase. Repeat once. Recover the equilibrated phenol and store under a shallow layer of 0.1 M Tris-HCl, pH 7.5, in a brown glass bottle, for up to 3 wk at 4°C.
10. Crystal violet staining solution: A 10X stock solution is prepared by dissolving 20 g of crystal violet (Sigma) in 400 mL of 90% ethanol. A 1X working solution for staining is prepared by 1:10 dilution with 90% ethanol. Store at room temperature.
11. Restriction endonucleases: Available from Roche, Gibco/Invitrogen, Promega (Madison, WI), Fisher Biotech, New England Biolabs (NEB, Beverly, MA), and Stratagene, and used according to the suppliers' directions.
12. Southern blotting reagents: QuikHyb Hybridization Solution (Stratagene) is used for Southern hybridizations, using double-stranded DNA probes labeled with [α-^{32}P]dCTP by nick translation or random prime labeling.

2.2. Knockout Gene Targeting in CHO Cells

1. Cell line: CHO–ATS49tg (*see* **Note 1**).
2. *ERCC1* targeting vector: pSL1 (*see* **Note 2**).
3. High ionic strength electroporation buffer (HEP buffer): 20 mM N-2-hydroxyethylpiperizine-N'-2-ethanesulfonic acid (HEPES), 137 mM NaCl, 5 mM KCl, 0.7 mM Na_2HPO_4, 6 mM dextrose, pH 7.05 *(49)*. Just prior to electroporation, a fresh solution is prepared by dissolving 238.3 mg of HEPES, 400 mg of NaCl, 54 mg of dextrose, 18.5 mg of KCl, and 9.5 mg of $Na_2HPO_4 \cdot 7H_2O$ in 40 mL of H_2O; adjust the pH to 7.05 with 1 N NaOH. Adjust the final volume to 50 mL prior to filter sterilization.
4. G418/FIAU selection medium: 400 µg/mL of G418 (Geneticin), 0.2 µM FIAU (1-[2-deoxy-2-fluoro-β-D-arabinofuranosyl-5-iodo]-uracil) in αMEM–10% DFCS. Adjust the G418 (Gibco) to 100% activity by weight, and dissolve directly in SFMEM to 20 active mg/mL (50X), then neutralize (to color) with 1 N NaOH, filter sterilize, and store at 4°C. Prepare FIAU as a 200 µM (1000X) stock by dissolving 4.6 mg of FIAU in 0.5 mL of fresh 1 N NaOH, and then bringing the volume to 60 mL with PBS; after filter-sterilizing, aliquot, and store frozen. Prepare selection medium containing G418 alone, or both G418 and FIAU, by appropriate dilutions directly into αMEM–10% DFCS (*see* **Note 3**). FIAU was the generous gift of Bristol-Myers Squibb Pharmaceutical Research Institute (Wallingford, CT).

5. DNA lysis solution: 1% proteinase K. Dissolve proteinase K lyophilized powder (Roche) at 10 mg/mL in H_2O; stir or shake gently to avoid foaming. Prepare fresh just before use.
6. Polymerase chain reaction (PCR) primers: Forward Neo 1: 5′-CCG CTT TTC TGG ATT CAT CGA C-3′, and Forward Neo 2: 5′-GCT TCC TCG TGC TTT ACG GTA TC -3′ are nested in the pMC1neopA cassette *(50)*; Reverse *ERCC1*: 5′-TGC CAC CCC TGA CCA CAT ATA CAC-3′ was designed in a region of *ERCC1* not represented in the targeting vector, pSL1 (*see* **Note 4**). We use commercially prepared primers from Operon (Houston, TX) or Genosys (Alameda, CA).
7. PCR reagents: Reagents for PCR, including *Taq* DNA polymerase, were commerically purchased as a kit from Perkin-Elmer (GeneAmp PCR Reagent Kit, N801-0043). *Taq* Extender and TaqStart antibody were purchased from Stratagene and Clontech (Palo Alto, CA), respectively.

2.3. APRT Gene Targeting in CHO Cells

1. Cell lines: CHO–ATS49tg and E1KO7-5 (*see* **Note 5**).
2. *APRT* targeting vector: pAG6ins0.9 (*see* **Note 6**).
3. Low ionic strength electroporation buffer (LEP buffer): 272 mM sucrose, 1 mM $MgCl_2$, 7 mM $Na_2HPO_4 \cdot 7H_2O$, pH 7.1 (Bio-Rad Gene Pulser Instruction Manual, Hercules, CA). Just prior to electroporation, prepare fresh electroporation buffer by dissolving 2.33 g of sucrose and 47 mg of $Na_2HPO_4 \cdot 7H_2O$ in 20 mL of H_2O; add 25 µL of 1 M $MgCl_2$, adjust the pH to 7.1, and bring the final volume to 25 mL with H_2O prior to filter sterilization.
4. ALASA selection medium (ALASA): 25 µM alanosine, 50 µM azaserine, 100 µM adenine in αMEM–10% FCS (*see* **Note 7**). Prepare alanosine as a 12.5 mM (500X) stock by dissolving 93 mg of alanosine in 1 mL of 1 N fresh NaOH; bring to 50 mL with solution A, filter-sterilize and store frozen in small aliquots. Alanosine (NSC-529469) can be obtained from the Drug Synthesis and Chemistry Branch of the National Cancer Institute. Prepare azaserine (Sigma) as a 25 mM (500X) stock by dissolving 216.5 mg of azaserine in 1 mL of 1 N fresh NaOH; bring to 50 mL with solution A, filter-sterilize, and store frozen in small aliquots. Prepare adenine as a 5 mM (50X) stock by dissolving 42.9 mg of adenine·HCl in 50 mL of solution A. After filter-sterilizing, store at room temperature.
5. HAT selection medium (HAT): 100 µM hypoxanthine, 2 µM amethopterin, 50 µM thymidine in αMEM–10% FCS (*see* **Note 8**). Prepare hypoxanthine as a 5 mM (50X) stock by dissolving 68 mg of hypoxanthine (Sigma) in 80 mL of H_2O; add 1 mL of fresh 1 N NaOH to aid solution, filter-sterilize, and store at room temperature. Prepare thymidine as a 2.5 mM (500X) stock by dissolving 302.5 mg of thymidine (Sigma) in 50 mL of H_2O; filter-sterilize and store at 4°C. Prepare amethopterin (methotrexate) as a 1 mM (500X) stock by dissolving 11.4 mg of amethopterin (Sigma) in 25 mL of HBSS; filter-sterilize and freeze in small aliquots (light-sensitive).

3. Methods

3.1. Preparation of Targeting Vectors for Electroporation

Targeting vector DNAs are prepared for electroporation in an identical manner whether they are gene knockout vectors or *APRT* targeting vectors. For gene knockout experiments, the targeting vector is linearized at a unique restriction site such as *Sca*I in the plasmid backbone (i.e., opposite plasmid homologous *ERCC1* DNA sequence containing *NEO* and flanked by HSV-*tk* sequences; *see* **Fig. 1**). For *APRT* targeted recombination assays utilizing the pAG6ins0.9 integration vector, the vector is linearized by cutting at a unique *Sal*I restriction site within the large (949-bp) heterologous insert that disrupts the *APRT* targeting sequence homology (*see* **Fig. 2**; **Note 6**).

1. Linearize the vector by overnight restriction endonuclease digestion of plasmid DNA, using 2–3 U of restriction enzyme/µg of DNA, and employing the specific reaction buffers and conditions recommended by suppliers.
2. Treat the linearized targeting vector DNAs with fresh proteinase K (200 µg/mL) in 0.1% SDS, 1 *M* lithium acetate for 2 h at 50°C.
3. Purify the DNA by one phenol extraction (equal volume) using equilibrated phenol warmed to room temperature.
4. Recover the DNA from the separated aqueous phase by addition of two volumes of absolute ethanol, chilling on ice, and centrifuging in a microcentrifuge.
5. After rinsing with ice-cold 70% ethanol and air-drying for 20–30 min, resuspend the pellet in TE buffer overnight at 4°C to achieve a DNA concentration of 250–500 µg/mL. Store frozen. (*See* **Note 9**.)

3.2. Growth and Preparation of Cell Cultures

CHO cell cultures are grown and prepared for electroporation in an identical manner whether targeted gene knockout or *APRT* targeted recombination assays are to be performed.

1. Harvest cells from exponentially growing (approx 50–60% confluent) monolayer cultures of CHO cells grown in T-150 flasks (Corning, NY) at 37°C in a humidified atmosphere of 5% CO_2-95% air. Cell monolayers are rinsed twice with prewarmed, sterile solution A and exposed to 2 mL of trypsin–EDTA, which is aspirated and discarded after 60–90 s. Tap the flasks forcefully several times, and resuspend the detached cells in 10 mL of ice-cold αMEM–10% FCS by vigorous and repeated pipetting. After low-speed centrifugation, put the cell pellets on ice and use within 15–20 min. (*See* **Note 10**.)
2. Based on cell counts obtained using a Coulter Counter (Hialeah, FL), determine the volume of cell suspension to spin down in a 15-mL sterile conical polypropylene centrifuge tube in order to provide sufficient cells for one electroporation (1 × 10^7 cells/cuvet for knockout electroporations; 2 × 10^7 cells/cuvet for *APRT* targeted recombination experiments). (*See* **Note 11**.)

3. Using a sterile, plastic 1-mL pipet, aseptically resuspend each cell pellet in 0.9 mL of sterile, ice-cold, electroporation buffer (HEP for targeted knockout experiments; LEP for *APRT* targeted recombination experiments) containing restriction enzyme-linearized targeting vector DNA. For gene knockout experiments, a final concentration of 1×10^7 cells and 20 µg of DNA/mL of HEP buffer should be achieved; for *APRT* targeted recombination experiments, a final concentration of 2×10^7 cells and 10 µg of DNA/mL of LEP buffer should be achieved. (*See* **Note 12**.)
4. Using a sterile, plastic 1-mL pipet, immediately transfer 0.85 mL of the CHO cell/targeting vector DNA suspension in EP buffer to a prechilled, sterile electroporation cuvet (Bio-Rad, 4-mm electrode gap) and place on ice for a 10-min, pre-electroporation incubation.
5. Electroporate by delivering a single pulse of 500 V at 25 µF to the cuvet, using a Bio-Rad Gene Pulser apparatus, and immediately place the cuvet back on ice for a 10 min, post-electroporation incubation. (*See* **Note 13**.)
6. Using a sterile plastic pipet, dilute the electroporated cell suspension by pipetting the contents of the cuvet into 8.5 mL of cold αMEM–10% FCS (also add one 1-mL rinse of the cuvet with αMEM–10% FCS), yielding a final cell concentration of approx $1–2 \times 10^6$ cells/mL.
7. For gene knockout experiments, make a further dilution with ice-cold SFMEM to yield a final volume of 60 mL. Plate 0.2 mL of this suspension to each of 300 100-mm tissue-culture dishes that contain 10 mL of αMEM–10% DFCS. Incubate the dishes at 37°C for 16–24 h in a humidified atmosphere of 5% CO_2–95% air. (*See* **Note 14**.)
8. For *APRT* targeted recombination experiments, plate the resuspended cells into six 100-mm tissue-culture dishes, each containing 10 mL of αMEM–10% FCS. Incubate dishes at 37°C for approx 48 h in a humidified atmosphere of 5% CO_2–95% air before replating into selection media. (*See* **Note 15**.)

3.3. Selection and Screening of Targeted Gene Knockout Clones

1. After 24 h, add G418 to each of the 300 dishes to achieve a final concentration of 400 µg/mL (adjusted to 100% activity). After an additional 24 h of incubation, add FIAU to a final concentration of 0.2 µ*M* to all except 16 plates. Refeed dishes with αMEM–10% DFCS containing G418 and FIAU (except for 16 dishes that are refed with G418 alone) every 3–4 d until visible colonies are evident, approx 12–14 d postelectroporation.
2. Pick well-isolated G418/FIAU resistant colonies by gentle scraping with a Pipetman P-200 while pipetting up 200 µL of culture medium using sterile, widebore yellow tips. To ensure that cell clones are independent, pick only one colony per dish. For each set of six picked and resuspended colonies, inoculate two six-well dishes: 50 µL per well to a reserved master dish, and 150 µL to wells of a corresponding dish to be used for DNA isolation for screening by PCR.
3. Stain the plates after marking picked colonies with a felt pen on the bottom of the dishes; enrichment can be calculated by counting the total number of colonies

from the 16 plates that were subjected to G418 selection alone, normalizing to a per plate value, and comparing to the number of colonies per plate from the G418/FIAU selected cells.
4. After 3–5 d, cell monolayers growing in all wells of six-well DNA dishes are rinsed twice with solution A. Treat with trypsin–EDTA for 1 min. Aspirate the trypsin solution and discard. Resuspend the cells by adding αMEM–10% FCS to each well and vigorously pipetting. Pool the cell suspensions for each dish into one tube (Falcon 1063, Becton Dickinson, Franklin Lanes, NJ), harvest the cells by low-speed centrifugation, wash once with PBS, and resuspend in 50 µL of PBS. Lyse the cells by addition of 200 µL of H_2O and heating for 10 min at 93°C. To the cell lysate, add 10 µL of fresh 1% Proteinase K, incubate for 30 min at 55°C, heat for 10 additional min at 93°C to inactivate the Proteinase K, centrifuge, and aspirate the supernatant for PCR analysis.
5. The resulting lysate is used in a two-step nested PCR strategy as shown in **Fig. 1**. A 5′ primer (Forward Neo 1: 5′-CCG CTT TTC TGG ATT CAT CGA C-3′) designed to anneal to *NEO* sequence within the targeting vector, and a 3′ primer (Reverse *ERCC1*: 5′-TGC CAC CCC TGA CCA CAT ATA CAC-3′) designed using *ERCC1* sequence downstream from exon 5 and not present in the targeting vector are used with 5 µL of cell lysate from the previous step in a 50-µL total reaction volume containing 1 µL of *Taq* Extender (Statagene) and 1 µL TaqStart Antibody (Clontech). Perform 20 cycles of PCR (1 min 30 s at 95°C; 1 min at 56°C; 3 min at 73°C). Then perform a second round of PCR with 5 µL of first round amplimer using a nested *NEO* primer (Forward Neo 2: 5′-GCT TCC TCG TGC TTT ACG GTA TC -3′). After electrophoresis on 0.6% agarose gels, identify pooled colony samples exhibiting a diagnostic 1.8-kilobase (kb) targeting fragment. (*See* **Fig. 1**.)
6. To verify successful gene targeting, individually trypsinize and transfer to T-25 flasks the six cultures from each reserved master dish that correspond to the positive, pooled sample. Repeat the PCR analysis as described in **step 5** to identify which of the six clones exhibited the diagnostic recombinant fragment. Southern analysis (*see* **Subheading 3.5.**) can be used to confirm disruption of the target gene by the *NEO* sequence present in the knockout vector.

3.4. Selection of APRT Targeted Recombinant Clones

1. Approximately 48 h after electroporation, trypsinize and suspend the cells from each 100-mm dish, centrifuge, and resuspend in 10 mL of ice-cold αMEM–10% FCS. (*See* **Note 16**.)
2. Take a small aliquot from each suspended cell population to count, serially dilute, and plate into αMEM–10% FCS medium for plating efficiency (PE) determinations (six 60-mm dishes at approx 200 cells/dish). Colonies arising on PE dishes can be fixed, stained, and counted after approx 8 d of incubation.
3. Plate a small portion of the resuspended cells from each 100-mm dish into HAT selection medium (three 100-mm dishes at 150 µL cell suspension/dish) to select for GPT^+ nontargeted vector integration events and monitor electroporation effi-

ciency. Colonies arising on HAT selection dishes can be fixed, stained, and counted after approx 12 d of incubation.
4. Plate the remainder of the cells from each 100-mm dish into ALASA selection medium (six 100-mm dishes at 1.5 mL of cell suspension/dish) to select for APRT+ recombinants. Representative ALASA^r APRT+ recombinant clones can be picked for analysis after approx 14 d of incubation (*see* **Note 17**), by gently scraping off and pipetting up cells from individual colonies using a Pipetman P-200 and sterile, wide-bore yellow tips, as described in **step 2** of **Subheading 3.3.** The rest of the colonies can then be fixed, stained, and counted.

3.5. Analysis of APRT Targeted Recombinant Clones

1. Isolate genomic DNAs from each independent ALASA^r recombinant clone, as well as from each parental cell line (e.g., ATS49tg, E1K07-5) used for *APRT* targeting experiments.
2. Subject DNA samples from each recombinant clone (or parental cell line) to overnight restriction endonuclease digestion by *Mbo*II, *Hin*dIII/*Sac*I, or *Bgl*II/*Xho*I, using 2–3 U of restriction enzyme/μg of DNA, employing the specific reaction buffers/conditions recommended by suppliers.
3. Following restriction endonuclease digestion and electrophoresis (560 V-h) in 0.8% agarose gels, transfer by capillary blotting to nitrocellulose or nylon membranes. Hybridize the blot with nick-translated or random-primed ^{32}P-labeled *APRT* gene-specific probes. For *Mbo*II digests, a 1.4-kb *Eco*RI–*Xba*I *APRT* fragment probe is used; a 3.9-kb *Bam*HI fragment probe that includes the entire CHO *APRT* gene sequence is used for *Hin*dIII/*Sac*I and *Bgl*II/*Xho*I digests.
4. For Southern blotting hybridization and washing conditions, supplier's instructions for radioactive hybridization (Stratagene QuikHyb Instruction Manual, pp. 2–4) can be followed without modification.
5. The various classes of targeted *APRT* recombinants (**Fig. 2A–D**) can be readily identified on the basis of their distinctive and diagnostic restriction fragment patterns. (*See* **Note 18**.)

4. Notes

1. CHO cell lines such as CHO–ATS49tg are hemizygous for the *ERCC1* gene (*22,24,29*). We have successfully used CHO–ATS49tg, CHO-K1, and several other CHO cell lines (*22–25*, and unpublished results) for *ERCC1* gene knockout experiments as described in this chapter.
2. Construction of the *ERCC1* targeting vector pSL1 is described in **ref. 22**. In principle, a repair gene targeting vector with the replacement-type structure illustrated in **Fig. 1** could be constructed using *NEO* and *HSV-tk* markers and used to knockout a variety of NER or DSBR genes in CHO or other cultured mammalian cell lines.
3. Some positive–negative selection gene targeting protocols use FIAU in selection media containing nondialyzed serum; however, we find that FIAU counter-selection is more stringent and economical (requiring much less FIAU) if dialyzed serum is used.

4. These Forward Neo PCR primers, nested in the *NEO* casette from plasmid pMC1neopA *(50)* should be useful for any PCR screening strategy in which *NEO* is used as the gene disruption marker and the reverse PCR primer is designed from a target gene sequence outside of the homologous sequence in the targeting vector, and downstream (3′) from the exon to be disrupted.
5. CHO–ATS49tg is an APRT$^-$/HPRT$^-$ Chinese hamster cell line that was derived from CHO-AT3-2 *(15,24)*. CHO-AT3-2 cells are hemizygous for the *APRT* locus *(51)*. CHO–ATS49tg cells contain a single, mutationally altered copy of the *APRT* gene, with a 3-bp deletion in exon 5 that has eliminated an *Mbo*II restriction site *(15,52)*, providing a convenient restriction fragment length polymorphism for monitoring whether targeted recombination events have resulted in correction of the *APRT* target gene defect (*see* **refs.** *15,19,24,34,39*). E1KO7-5 is a CHO *ERCC1* knockout cell line generated by targeted disruption of the hemizygous *ERCC1* gene locus in CHO–ATS49tg cells *(24)*. We have also used several other CHO–AT3-2 or CHO–AA8-4-derived cell lines with hemizygous, mutant *APRT* alleles suitable for targeted correction for gene targeting experiments (our unpublished results).
6. The pAG6ins0.9 *APRT* targeting vector (*see* **Fig. 2**) was derived from pSV2gpt *(53)* in several steps that involved the construction of several intermediate plasmids, using conventional cloning methods *(24)*. The pAG6ins0.9 targeting vector contains 3.9 kb of *APRT* targeting homology, including a full-length Chinese hamster *APRT* gene that has been disrupted by a 949-bp heterologous insertion into the exon 3 *Xho*I site *(24)*. Linearization of this targeting vector by cleavage at a unique *Sal*I site within the large heteterologous insertion creates an "ends-in" targeted integration vector configuration in which both arms of *APRT* targeting homology (that would ordinarily flank each side of the DSB) are blocked by long terminal nonhomologies (*see* **Fig. 2**). When this *Sal*I-linearized vector is electroporated into cells, resection of the 5′-ended strands on each side of the DSB leaves long (271- or 670-nt), 3′-ended, nonhomologous tails on each arm of the targeting vector. These 3′-end-blocking, nonhomologous tails must be removed to permit target gene-templated DNA synthesis and extension of homologous ends following strand invasion, in order for targeted integration (crossover) or vector correction (conversion) recombination products to be generated from homologous recombination intermediates *(24)*. A *GPT* cassette on the pAG6ins0.9 targeting vector allows determination of the frequencies of untargeted vector integration (HAT^r, GPT$^+$ transfectants), thereby providing a reliable internal control for monitoring electroporation efficiency in *APRT* targeting experiments.
7. Alanosine and azaserine act as inhibitors of different steps in the *de novo* pathway for purine biosynthesis. APRT is a purine salvage pathway enzyme that allows cells to utilize adenine for purine biosynthesis by catalyzing the conversion of adenine to AMP. ALASA selection medium will effectively kill APRT$^-$ cells such as CHO–ATS49tg (which are unable to utilize exogenous adenine as a purine source), but will support the growth of APRT$^+$ recombinants.

8. Amethopterin (methotrexate) acts as an inhibitor of steps in the *de novo* pathways for both purine and pyrimidine biosynthesis. Both *HPRT* and its bacterial equivalent, *GPT*, are purine salvage pathway enzymes that allow cells to utilize hypoxanthine or guanine for purine biosynthesis. HAT selection medium will effectively kill HPRT⁻ cells such as CHO–ATS49tg (which are unable to utilize exogenous hypoxanthine as a purine source), but will support the growth of GPT⁺ transfectants.
9. All steps subsequent to ethanol precipitation should be performed aseptically in a tissue culture hood. DNA pellets may be left to air-dry for 20 min in a sterile hood, and then resuspended using sterile technique
10. To achieve maximal and reproducible transformation efficiency in gene targeting experiments, it is important that the cells are in exponential growth at the time cell cultures are trypsinized and harvested for electroporation.
11. Cells should be harvested such that only one cell pellet of $1-2 \times 10^7$ cells /centrifuge tube is used for one cuvet; we have experienced poorer results when we tried resuspending sufficient cells for several cuvets in a single centrifuge tube, and then distributing the suspension to multiple cuvets.
12. It is important to aspirate the supernatant fluid as completely as possible after centrifugation. Prior to resuspending cells in EP buffer, thump the conical centrifuge tube containing the cell pellet sharply several times to loosen the pellet. Then, resuspend the cells very carefully by pipeting back-and-forth slowly several times, without introducing bubbles. For *APRT* targeted recombination experiments, it is important that the cell pellet be directly suspended in LEP buffer (without a rinse step) after carefully draining off the supernatant and aspirating any remaining traces of liquid from the sides of the tube. Inclusion of an LEP buffer rinse step reduces cell survival after electroporation; this procedure is also followed for knockout gene targeting experiments using HEP buffer.
13. For knockout electroporation in HEP buffer, time constants under these conditions should be approx 0.6. For *APRT* gene targeting in LEP buffer, time constants under these conditions should be approx 5.5–6.0. The 10-min postelectroporation incubation on ice is not necessary for knockout electroporations in HEP buffer.
14. One to two cuvets represent a standard size experiment for knockout gene targeting, resulting in 300–600 100-mm dishes. In our experience, one cuvet will produce approx 180–200 G418/FIAU doubly-resistant colonies after positive–negative selection.
15. For *APRT* targeting experiments, a standard size experiment assessing the effects of one targeting vector structure on targeted homologous recombination in one cell line is three to six cuvets, resulting in 18–36 100-mm dishes prior to plating into ALASA selection. Two or more cell lines, and one or two targeting vector configurations are typically assayed at the same time in parallel experiments.
16. It is important to: (a) keep this cell suspension (and any dilutions to be used for plating) on ice to minimize cell clumping or attachment, and (b) make sure that the cells are kept well suspended throughout the plating procedure, and that the cells are uniformly distributed after plating into tissue culture dishes.

17. To ensure the independence of all APRT⁺ recombinants analyzed, only a single ALASA^r colony from each independent cell population should be picked for expansion, cryogenic preservation, and Southern blot analysis.
18. APRT⁺ recombinants that have arisen by target gene conversion (**Fig. 2A**) show loss of the 2.0-kb endogenous ATS49tg or E1KO7-5 *Mbo*II fragment and reappearance of a 1.5-kb *Mbo*II fragment characteristic of a wild-type Chinese hamster *APRT* locus *(24)*. They retain normal, unchanged (wild-type) *APRT* restriction fragment patterns for all other restriction enzymes, including *Hin*dIII (which yields a single 8.6-kb restriction fragment), and *Bgl*II/*Xho*I (which yields 2.2- and 2.4-kb fragments). APRT⁺ targeted integration recombinants (**Fig. 2B**) show diagnostic restriction fragment patterns for *Hin*dIII (13- and 4.0-kb fragments), and *Bgl*II/*Xho*I (2.2-, 3.9-, 4.6-, and 2.4-kb fragments), after hybridization with the 3.9-kb *Bam*HI fragment probe. Two subclasses of targeted integration recombinants may be distinguished after *Mbo*II digestion and hybridization with the 1.4-kb *Eco*RI–*Xba*I fragment probe: type I (*Mbo*II +/Δ) targeted integrations, in which the downstream copy of the *APRT* gene retains the original 3-bp exon 5 *Mbo*II site deletion; and type II (*Mbo*II +/+) targeted integrations, in which both copies of the *APRT* gene have a wild-type exon 5 *Mbo*II site *(24)*. Type I (*Mbo*II +/Δ) targeted integrations show both a 1.5-kb (wild-type) and 2.0-kb (mutant) *Mbo*II fragment, while type II (*Mbo*II +/+) targeted integrations show only a 1.5-kb (wild-type) *Mbo*II fragment. APRT⁺ vector correction recombinants (**Fig. 2C**) will contain both the original, endogenous, 2.0-kb (mutant) *Mbo*II fragment and a newly acquired 1.5-kb *Mbo*II fragment characteristic of a wild-type CHO *APRT* gene. After digestion with *Hin*dIII and hybridization with the 3.9-kb *Bam*HI fragment probe, vector correction recombinants show both an 8.6-kb *Hin*dIII fragment characteristic of the endogenous *APRT* locus, and a novel fragment of indeterminate size, representing the randomly integrated, wild-type (corrected vector) *APRT* sequence. After digestion with *Bgl*II/*Xho*I, vector correction recombinants will show 2.2- and 2.4-kb fragments characteristic of the endogenous *APRT* locus, plus two or more novel fragments, produced by ectopic integration of the corrected, vector-derived *APRT* sequence. Aberrant integration recombinants (**Fig. 2D**), in which a variable portion of the integrated vector sequence appears to have been deleted, are recovered only from pAG6ins0.9 targeting experiments in *ERCC1*⁻ cell lines. These recombinants can display widely differing restriction fragment patterns, depending on the size of the deleted region, but in most cases show loss of the downstream 2.4-kb *Bgl*II/*Xho*I *APRT* fragment *(24)*.

References

1. Rothstein, R. (1991) Targeting, disruption, replacement, and allele rescue: Integrative DNA transformation in yeast. *Methods Enzymol.* **194**, 281–301.
2. Capecchi, M. R. (1989) Altering the genome by homologous recombination. *Science* **244**, 1288–1292.
3. Waldman, A. S. (1992) Targeted homologous recombination in mammalian cells. *Crit. Rev. Oncol. Hematol.* **12**, 49–64.

4. Ramirez-Solis, R., Davis, A. C., and Bradley, A. (1993) Gene targeting in embryonic stem cells. *Methods Enzymol.* **225,** 855–879.
5. Capecchi, M. R. (1989) The new mouse genetics: altering the genome by gene targeting. *Trends Genet.* **5,** 70–76.
6. Wijnhoven, S. W. P., and van Steeg, H. (2003) Transgenic and knockout mice for DNA repair functions in carcinogenesis and mutagenesis. *Toxicology* **193,** 171–187.
7. Friedberg, E. C. and Meira, L. B. (2003) Database of mouse strains carrying targeted mutations in genes affecting cellular responses to DNA damage (Version 5). *DNA repair* **2,** 501–530.
8. Dhar, P. K., Sonoda, E., Fujimori, A., Yamashita, Y. M., and Takeda, S. (2001) DNA repair studies: experimental evidence in support of chicken DT40 cell line as a unique model. *J. Environ. Pathol. Toxicol. Oncol.* **20,** 273–283.
9. Winding, P. and Berchtold, M. W. (2001) The chicken B cell line DT40: a novel tool for gene disruption experiments. *J. Immunol. Methods* **249,** 1–16.
10. Schiestl, R. H. and Prakash, S. (1988) *RAD1*, an excision repair gene of *Saccharomyces cerevisiae,* is also involved in recombination. *Mol. Cell. Biol.* **8,** 3619–3626.
11. Schiestl, R. H. and Prakash, S. (1990) *RAD10* an excision repair gene of *Saccharomyces cerevisiae* is involved in the *RAD1* pathway of mitotic recombination. *Mol. Cell. Biol.* **10,** 2485–2491.
12. Fishman-Lobell, J. and Haber, J. E. (1992) Removal of nonhomologous DNA ends in double-strand break recombination: the role of the yeast ultraviolet repair gene *RAD1*. *Science* **258,** 480–484.
13. Ivanov, E. L. and Haber, J. E. (1995) *RAD1* and *RAD10*, but not other excision repair genes are required for double-strand break-induced recombination in *Saccharomyces cerevisiae*. *Mol. Cell. Biol.* **15,** 2245–2251.
14. Sugawara, N., Paques, F., Colaiacovo, M., and Haber, J. E. (1997) Role of *Saccharomyces cerevisiae* Msh2 and Msh3 repair proteins in double-strand break-induced recombination. *Proc. Natl. Acad. Sci. USA* **94,** 9214–9219.
15. Adair, G. M., Nairn, R. S., Wilson, J. H., et al. (1989) Targeted homologous recombination at the endogenous adenine phosphoribosyltransferase locus in Chinese hamster cells. *Proc. Natl. Acad. Sci. USA* **86,** 4574–4578.
16. Adair, G. M., Nairn, R. S., Wilson, J. H., Scheerer, J. B., and Brotherman, K. A. (1990) Targeted gene replacement at the endogenous *APRT* locus in CHO cells. *Somat. Cell Mol. Genet.* **16,** 437–441.
17. Porter, T., Pennington, S., Adair, G. M., Nairn, R. S., and Wilson, J. H. (1990) A novel selection system for recombinational and mutational events within an intron of a eucaryotic gene. *Nucleic Acids Res.* **18,** 5173–5180.
18. Nairn, R. S., Adair, G. M., Porter, T., et al. (1993) Targeting vector configuration and method of gene transfer influence targeted correction of the *APRT* gene in Chinese hamster ovary cells. *Somat. Cell Mol. Genet.* **19,** 363–375.
19. Scheerer, J. B. and Adair, G. M. (1994) The homology dependence of targeted recombination at the Chinese hamster *APRT* locus. *Mol. Cell. Biol.* **14,** 6663–6673.

20. Adair, G. M. and Nairn, R. S. (1995) Gene targeting, in *DNA Repair Mechanisms: Impact on Human Diseases and Cancer*, (Vos, J-M., ed.), Biomedical Landes, Austin, TX, pp. 301–328.
21. Sargent, R. G., Merrihew, R. V., Nairn, R. S., Adair, G. M., Meuth, M., and Wilson, J. H. (1996) The influence of a $(GT)_{29}$ microsatellite sequence on homologous recombination in the hamster *APRT* gene. *Nucleic Acids Res.* **24,** 746–753.
22. Rolig, R. L., Layher, S. K., Santi, B., et al. (1997) Survival, mutagenesis, and host cell reactivation in a Chinese hamster ovary cell *ERCC1* knockout mutant. *Mutagenesis* **12,** 277–283.
23. Sargent, R. G., Rolig, R. L., Kilburn, A. E., Adair, G. M., Wilson, J. H., and Nairn, R. S. (1997) Recombination-dependent deletion formation in mammalian cells deficient in the nucleotide excision repair gene *ERCC1*. *Proc. Natl. Acad. Sci. USA* **94,** 13,122–13,127.
24. Adair, G. M., Rolig, R. L., Moore-Faver, D., Zabelshansky, M., Wilson, J. H., and Nairn, R. S. (2000) Role of *ERCC1* in removal of long non-homologous tails during targeted homologous recombination. *EMBO J.* **19,** 5552–5561.
25. Sargent, R. G., Meservy, J. L., Perkins, B. D., et al. (2000) Role of the nucleotide excision repair gene *ERCC1* in formation of recombination-dependent rearrangements in mammalian cells. *Nucleic Acids Res.* **28,** 3771–3778.
26. Gottesman, M. M. (ed.) (1985) *Molecular Cell Genetics*. John Wiley, New York, NY.
27. O'Neill, J. P., Couch, D. B., Machanoff, R., San Sebastion, J. R., Brimer, P. A., and Hsie, A. W. (1977) A quantitative assay of mutation induction at the hypoxanthine-guanine phosphoribosyl transferase locus in Chinese hamster ovary cells (CHO/HGPRT system): utilization with a variety of mutagenic agents. *Mutat. Res.* **45,** 103–109.
28. Siciliano, M. J., Stallings, R. L., Humphrey, R. M., and Adair, G. M. (1986) Mutation in somatic cells as determined by electrophoretic analysis of mutagen-exposed Chinese hamster ovary cells, in *Chemical Mutagens*, Vol. 10, (de Serres, F. J., ed.), Plenum, New York, NY, pp. 509–531.
29. Thompson, L. H., Bachinski, L. L., Stallings, R. L., et al. (1989) Complementation of repair gene mutations on the hemizygous chromosome 9 in CHO: a third repair gene on human chromosome 19. *Genomics* **5,** 670–679.
30. Merrihew, R. V., Sargent, R. G., and Wilson, J. H. (1995) Efficient modification of the *APRT* gene by FLP/FRT site-specific targeting. *Somat. Cell. Mol. Genet.* **21,** 299–307.
31. te Riele, H., Maandag, E. R., and Berns, A. (1992) Highly efficient gene targeting in embryonic stem cells through homologous recombination with isogenic DNA constructs. *Proc. Natl. Acad. Sci. USA* **89,** 5128–5132.
32. Deng, C. and Capecchi, M. R. (1992) Re-examination of gene targeting frequency as a function of the extent of homology between the targeting vector and the target locus. *Mol. Cell. Biol.* **12,** 3365–3371.
33. van Duin, M., de Wit, J., Odjik, H., et al. (1986) Molecular characterization of the human excision repair gene *ERCC1*: cDNA cloning and amino acid homology with the yeast DNA repair gene *RAD10*. *Cell* **44,** 913–923.

34. Adair, G. M., Scheerer J. B., Brotherman, A., McConville, S., Wilson, J. H., and Nairn, R. S. (1998) Targeted recombination at the Chinese hamster *APRT* locus using insertion versus replacement vectors. *Somat. Cell Mol. Genet.* **24,** 91–105.
35. Hasty, P., Ramirez-Solis, R., Krumlauf, R., and Bradley, A. (1991) Introduction of a subtle mutation into the *Hox-2.6* locus in embryonic stem cells. *Nature* **350,** 243–246.
36. Valancius, V. and Smithies, O. (1991) Testing an "in-out" targeting procedure for making subtle genomic modifications in mouse embryonic stem cells. *Mol. Cell. Biol.* **11,** 1402–1408.
37. Hasty, P., Rivera-Perez, J., and Bradley, A. (1991) The length of homology required for gene targeting in embryonic stem cells. *Mol. Cell. Biol.* **11,** 5586–5591.
38. Mansour, S. L., Thomas, K. T., and Capecchi, M. R. (1988) Disruption of the proto-oncogene *int-2* in mouse embryo-derived stem cells: a general strategy for targeting mutations to nonselectable genes. *Nature* **336,** 348–352.
39. Pennington, S. L. and Wilson, J. H. (1991) Gene targeting in Chinese hamster ovary cells is conservative. *Proc. Natl. Acad. Sci. USA* **88,** 9498–9502.
40. Aratani, Y., Okazaki, R., and Koyama, H. (1992) End extension repair of introduced targeting vectors mediated by homologous recombination in mammalian cells. *Nucleic Acids Res.* **20,** 4795–4801.
41. Fujioka, K.-I., Aratani, Y., Kusano, K., and Koyama, H. (1993) Targeted recombination with single-stranded DNA vectors in mammalian cells. *Nucleic Acids Res.* **21,** 407–412.
42. Wang, Q. and Taylor, M. W. (1993) Correction of a deletion mutant by gene targeting with an adenovirus vector. *Mol. Cell. Biol.* **13,** 918–927.
43. Waldman, B. C., O'Quinn, J. R., and Waldman, A. S. (1996) Enrichment for gene targeting in mammalian cells by inhibition of poly(ADP-ribosylation). *Biochim. Biophys. Acta* **1308,** 241–250.
44. Lukacsovich, T., Waldman, B. C., and Waldman, A. S. (2001) Efficient recruitment of transfected DNA to a homologous chromosomal target in mammalian cells. *Biochim. Biophys. Acta* **1521,** 89–96.
45. Sargent, R. G., Brenneman, M., and Wilson, J. H. (1997) Repair of site-specific double-strand breaks on a mammalian chromosome by homologous and illegitimate recombination. *Mol. Cell. Biol.* **17,** 267–277.
46. Kilburn, A. E., Shea, M. J., Sargent, R. G., and Wilson, J. H. (2001) Insertion of a telomere repeat sequence into a mammalian gene causes chromosome instability. *Mol. Cell. Biol.* **21,** 126–135.
47. Meservy, J. L., Sargent, R. G., Iyer, R. R., et al. (2003) Long CTG tracts from the myotonic dystrophy gene induce deletions and rearrangements during recombination at the *APRT* locus in CHO cells. *Mol. Cell. Biol.* **23,** 3152–3162.
48. Friedberg, E. C., Walker, C., and Siede, W. (1995) *DNA Repair and Mutagenesis*. American Society for Microbiology Press, Washington, DC.
49. Chu, G., Hayakawa, H., and Berg, P. (1987) Electroporation for the efficient transfection of mammalian cells with DNA. *Nucleic Acids Res.* **15,** 1311–1326.

50. Thomas, K. R. and Capecchi, M. R. (1987) Site-directed mutagenesis by gene targeting in mouse embryo-derived stem cells. *Cell* **51,** 503–512.
51. Adair, G. M., Stallings, R. L., Nairn, R. S., and Siciliano, M. J. (1983) High-frequency structural gene deletion as the basis for functional hemizygosity of the adenine phosphoribosyltransferase locus in Chinese hamster ovary cells. *Proc. Natl. Acad. Sci. USA* **80,** 5961–5964.
52. Smith, D. G. and Adair, G. M. (1996) Characterization of an apparent hotspot for spontaneous mutation in exon 5 of the Chinese hamster *APRT* gene. *Mutat. Res.* **352,** 87–96.
53. Mulligan, R. and Berg, P. (1981) Expression of a bacterial gene in mammalian cells. *Proc. Natl. Acad. Sci. USA* **78,** 2072–2076.

13

Gene-Specific and Mitochondrial Repair of Oxidative DNA Damage

R. Michael Anson, Penelope A. Mason, and Vilhelm A. Bohr

Summary

The Southern blot gene-specific DNA damage and repair assay is a robust and flexible method for quantifying many kinds of induced damage and repair with high reproducibility. Specific nicking and loss of a restricted DNA fragment at the site of induced damage is visualized by Southern blot and quantified against a control; since the blot is gene specific, only the damage of interest is measured. Here we show how the assay may be adapted to assess mitochondrial DNA (mtDNA) damage. In the mitochondrion, 8-oxoguanine is a significant oxidative lesion; in the laboratory, photoactivated methylene blue may be used to introduce this lesion into cells. Other lesions may also be studied by using different DNA damaging agents. We find that damage induction by methylene blue is consistently far greater in the mitochondrion than the nucleus. Thus advantageously, mitochondrial 8-oxoguanine repair may be studied without mtDNA isolation or preparation, which are processes known to induce DNA damage and skew measurements. This chapter gives detailed instructions for using methylene blue and the gene-specific repair assay to accurately measure mitochondrial oxidative damage and repair rates.

Key Words: DNA damage quantification; DNA damage; DNA repair; DNA repair assay; mitochondria; mtDNA; oxidative damage; gene-specific assay; Southern blot.

1. Introduction

1.1. Mitochondrial DNA Damage Repair and Its Consequences

Reactive oxygen species (ROS, also called free radicals) are byproducts of normal aerobic respiration *(1)*. These species are also formed as a consequence of environmental insults such as pesticides, heavy metals, and ozone *(2)*. ROS, for example the superoxide ($O^\bullet$) or hydroxyl ($H^\bullet$) radical, may react with and damage DNA, causing lesions such as abasic sites and oxidized bases. These lesions are

From: *Methods in Molecular Biology: DNA Repair Protocols: Mammalian Systems, Second Edition*
Edited by: D. S. Henderson © Humana Press Inc., Totowa, NJ

often mutagenic; they allow base–base mispairing which, if not removed, can allow deleterious mutations to persist. Mitochondrial DNA (mtDNA) is found in the mitochondrial matrix; the protein complexes of the electron transport chain (ETC), which produce high levels of ROS, are also situated here.

MtDNA is subjected to unusually high levels of oxidative stress. The accumulation of oxidative lesions in the mitochondrion, leading to dysfunction or depletion, is thought to contribute largely to aging *(3,4)* and may be implicated in cancer *(5)*. Specific mtDNA mutations also cause many human disorders, usually neuromuscular and progressive, that are as yet incurable *(6)*. MtDNA repair is thus highly studied; much remains to be elucidated.

It was generally thought that mitochondria did not have the capability to repair DNA damage because of experimental observations that ultraviolet (UV)-induced pyrimidine dimers and cisplatin-induced intrastrand crosslinks accumulate in mitochondria *(7)*. Thus nucleotide excision repair (NER), the pathway that repairs these types of lesions, may well not appear in the mitochondrion. However, many base excision repair (BER) enzymes have been localized to mitochondria, including uracil DNA glycosylase (UDG), 8-oxoguanine (7-hydro-8-oxo-deoxyguanosine) DNA glycosylase (hOGG1β), and others *(8–21)*. 8-Oxoguanine is one of the most common oxidative lesions, and steady-state levels are often reported to be tenfold higher in mtDNA than nuclear, and to increase dramatically with age *(22,23)*.

BER is the main cellular mechanism for the detection and removal of alkylation and oxidative damage, and functions with good efficiencies both in the eukaryotic nucleus and the mitochondrion. Many mitochondrial repair proteins are differentially spliced isoforms of nuclear repair genes such as *UNG* and *OGG1* *(24,25)*, and show great functional similarity to their nuclear counterparts.

Another pathway possibly active in mammalian mitochondria is mismatch repair (MMR). Yeast and other lower eukaryotes possess a specific bacterial MMR homolog (MSH1) *(26)*, the deletion of which causes mtDNA rearrangement and dysfunction. While a mammalian homolog has not yet been characterized, other mismatch proteins are localized to mitochondria. Mitochondrial MutY homolog (MYH) excises the adenine from adenine:8-oxoguanine mispairs *(27,28)*, while OGG1 (also called MutM homolog) removes the oxidized base. BER and MMR often act in concert; this interplay must be fully understood so that therapeutic strategies for disease or aging based on these pathways have optimal design.

1.2. Base Excision Repair of Oxidative Damage

Mammalian BER consists of several distinct steps, conserved in both the nucleus and the mitochondrion *(29,30)*. The first step is the recognition and removal of a damaged base by DNA glycosylase to reveal an abasic site. Next comes the recognition and cleavage of the DNA backbone at this abasic site by either AP endonuclease or AP lyase. The resulting "blocked termini" are processed in one of several ways depending upon whether long or short patch BER is used. In short patch repair, dRPase processes the results of AP endonuclease cleavage, while after AP lyase cleavage, processing is via 3′ phosphodiesterase. Long patch repair allows extension of the 5′ terminus left by AP endonuclease by polymerase β, resulting in a "flap" of single-stranded DNA which is resolved using Flap endonuclease to leave a single-stranded nick. The fourth step for short patch BER is to fill the resultant gap using a polymerase (β or δ/ε), and the fifth and final step for either pathway requires ligation of the nick using either ligase III and XRCC1, or ligase I.

Thus BER may be studied via any combination of these five steps: (1) excision of the damaged base, (2) incision of the DNA backbone at the apurinic/apyrimidinic (AP) site product, (3) removal of the AP terminal fragment, (4) gap-filling synthesis, and (5) ligation of the final nick. Electromobility shift assay and DNA cleavage assays are routinely performed to visualize the earlier steps, as are polymerase incorporation and ligation assays for the later steps. MMR, which also follows a similar scheme of recognition and incision, but uses long-patch gap filling, can also be studied by these biochemical methods.

1.3. Studying mtDNA Repair Has Associated Problems

Nuclear BER is amenable to all these types of study, but analysis of mitochondrial BER is hampered by the difficulties attendant on mitochondrial isolation and purification. It is both difficult and time-consuming to purify mitochondria so that *all* extramitochondrial matter is excluded, the most approved method being density gradient purification through Percoll, and thus it is easy to dismiss mitochondrially derived data as supposed nuclear contamination. Yields from Percoll purification and other methods are low relative to total material, requiring large starting amounts; therefore organ tissues are easier to use than cell lines. When whole cells are used for experimentation, different problems are encountered. For example, the double membrane means that the mitochondrion is intractable to import studies, especially using large or charged molecules such as DNA.

Further problems arise when oxidative damage is studied in isolated mitochondria. For example, photoactivated methylene blue (MB)-induced lesions are often used to study oxidative damage repair. The primary product, 8-oxoguanine *(31–33)*, can then be quantitated before and after experimental repair. However, there is no single consensus method for inducing and measuring oxidative damage, for example, the European standards committee on oxidative DNA damage (ESCODD) has published comparison data between different methods showing great variance *(34,35)*. Likely because of this, endogenous

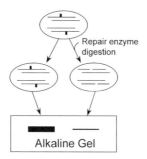

Fig. 1. The principle behind the Southern blot gene-specific damage and repair assay. A DNA repair enzyme is used as a tool to nick a restricted DNA fragment at the site of damage. In a denaturing agarose gel, the cleaved strand migrates further than the full length restriction fragment. The band is detected by Southern blot.

(and induced) levels of damage quoted in the literature for both the nucleus and mitochondria vary widely, from as little as four lesions per million bases to nearly 5000 for mitochondria. Even within a single method, variance is substantial; levels from 4 to 110 lesions per million are quoted. Further, the ratio of mitochondrial/nuclear damage is quoted as between 2-fold and 16-fold, even though mitochondria are known to efficiently repair 8-oxoguanine. For an analysis of these problems *see* **refs. *36,37*.**

1.4. Development of a Robust Assay to Visualize Oxidative Damage

Investigation into this variance has been carried out using the Southern blot gene-specific assay—a method that allows measurement of nuclear and mitochondrial damage without prior isolation *(38)* (**Fig. 1**). The data suggest that mitochondrial isolation is itself a potent inducer of mtDNA oxidative damage *(39)*; in whole cells the nuclear/mitochondrial ratio for endogenous damage is approximately equal, whereas isolated mitochondria show two to three times more lesions than the nucleus, even when antioxidants are used in the preparation.

Although the ESCODD data show that HPLC DNA purification, for example, does not induce significant extra damage when compared to nonisolating techniques such as the COMET assay *(34)*, this has limited application. Another, novel way to visualize the rate and amount of DNA lesions in vivo, without inducing artifactual damage, is to use the differential expression of DNA repair genes as biomarkers of oxidative stressors such as peroxisome proliferators *(40)*. While this method shows great promise in tracing the pathways such stressors use, it does not directly measure the extent of oxidative DNA damage.

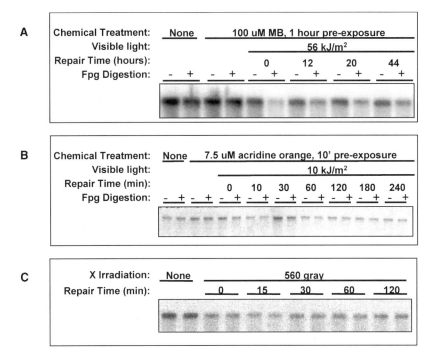

Fig. 2. Sample data obtained using the Southern blot gene-specific damage and repair assay. (**A**) Damage and repair of Fpg-sensitive base damage in mtDNA following treatment of cells with photoactivated methylene blue. (**B**) Damage and repair of Fpg-sensitive base damage in mtDNA following treatment of cells with acridine orange. (**C**) Damage and repair (measurement by comparison with DNA from untreated cells) of alkali-labile sites and strand breaks following treatment of cells with X-irradiation.

The gene-specific assay removes the need for mtDNA isolation and any associated problems of purification and induced damage without sacrificing speed or yield. This method also allows direct *in situ* correlation of nuclear and mitochondrial damage levels. Using gene-specific probes removes most of the background associated with the measurement of oxidative damage so that data show greater reproducibility compared to other detection methods (**Fig. 2**). It also means that the assay can easily be tailored specifically to particular hot spots or genes of interest. Finally it is flexible in that different types of oxidative damage induction may be tested and used for study (**Fig. 3**). In the treatment, methylene blue shows efficient and differential induction of 8-oxoguanine damage, but other agents, for example, acridine orange, have been analyzed for use. In summation, the assay has many advantages over other common ways of studying oxidative damage.

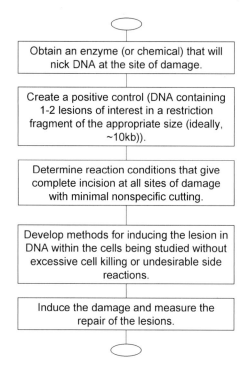

Fig. 3. A flow chart of the basic steps necessary to develop a gene-specific damage and repair assay for a new lesion.

1.5. The Southern Blot Gene-Specific Mitochondrial Repair Assay

DNA repair is a heterogeneous process. The efficiency of repair differs not only with lesion type, but also with genomic location. Protein coding regions of the genome are repaired more efficiently than noncoding regions in a process termed preferential repair. Furthermore, for many lesions, there is a hierarchy even within the coding regions. Transcription-coupled repair (TCR) allows active genes to be repaired more rapidly than inactive genes, and the transcribed strand to be repaired with greater efficiency than the nontranscribed strand *(38)*.

To measure differences in the rate of repair between different regions of the genome, and different strands within a given region, the Southern blot gene-specific DNA damage and repair assay is frequently used *(38)*. The principle behind the assay is to use a DNA repair enzyme as a tool to nick a restricted DNA fragment at the site of damage. In a denaturing agarose gel, the cleaved strand migrates further than the full-length restriction fragment (**Fig. 1**). This loss of full-length damaged DNA from the main band, when detected by Southern blot (**Fig. 2**), allows quantitation of the undamaged DNA. Film (or PhosphorImager) exposures are chosen which give a linear relationship between band intensity and the DNA content of the band. A random distribution of damage in a given restriction fragment is assumed, that allows the use of the Poisson equations in determination of the average number of adducts per strand.

The assay has been successfully applied to many different lesions, including UV damage (the major product of which is the pyrimidine dimer), 4-nitroquinoline, *N*-acetylaminofluorene, benzopyrene diol epoxide, cisplatin, various alkylating agents, and, more recently, several types of oxidative DNA damage *(41,42)*. In general, the assay allows measurement of any type of damage that can be converted to a single-stranded nick either chemically or by using a repair enzyme. For alkali-sensitive damage, aliquots of the sample may be run on neutral and alkaline gels, and the band intensities normalized to a damage-free internal standard. The intensity of the alkaline band relative to the native band may then be used to calculate the number of alkali-sensitive sites.

The rest of this chapter presents a method for applying the gene-specific repair assay to the measurement of 8-oxoguanine in mtDNA. To damage the DNA, cells are treated with photoactivated MB. This is a cell-permeant cation, and is in equilibrium in the cells with the uncharged and colorless leukomethylene blue *(43)*. When photoactivated, MB leads to DNA damage, with the formation of 8-oxoguanine the major product of the reaction *(31–33)*. When human fibroblasts are pretreated with MB and exposed to light, oxidative damage is induced in the mtDNA *(44)*. To measure the damage, the bacterial repair enzyme, formamidopyrimidine DNA glycosylase (Fpg), is used. Fpg recognizes oxidatively modified deoxyguanosine, particularly 8-oxoguanine, in double-stranded DNA, and excises it to generate a single-stranded break *(45,46)*.

A limitation in the development of gene-specific assays for oxidative damage is the induction of the lesion of interest without the induction of other damage that interferes with the measurement of repair. The protocol given here induces far more damage in mtDNA than in nuclear DNA, making it useful for repair measurements of 8-oxoguanine in mtDNA but not in nuclear DNA. In the following subheadings we describe assays to measure the repair of 8-oxoguanine in mtDNA, and this will also serve as a model in the development of new assays for measuring similar damage in nuclear DNA, or in the development of methods for measuring other lesions in both genomes. Acridine orange induces DNA damage in the nuclear DNA with reasonable efficiency *(47)*, but we find that MB is a more useful agent. Where possible, notes have been included that will assist the investigator in using this protocol as a base on which to develop assays for other lesions, using other agents and other enzymes.

The methods presented in the following are grouped into several sections. **Subheadings 3.1.** and **3.2.** describe the preparation of DNA for use as a positive control and the optimization of reaction conditions to give complete cleavage at the sites of damage while minimizing nonspecific cutting. These protocols will be of use if a new Fpg preparation of unknown specific activity or purity has been purchased, but also may serve as model protocols in the characterization of any repair enzyme for use in gene-specific repair experiments. **Subheading 3.3.** establishes the sensitivity of an untested cell line to photoactivated MB, both in terms of survival and of the levels of DNA damage that are induced. In addition, the protocols given may be used as models in testing other agents for the induction of DNA damage prior to gene-specific repair studies. **Subheadings 3.4.** and **3.5.** describe the actual repair assay in detail. Finally, **Subheading 3.6.** provides support protocols that are needed for the success of any gene-specific repair assay, regardless of the type of lesion.

2. Materials

2.1. Damaging Purified DNA In Vitro

1. Desk lamp with a 300-W tungsten bulb.
2. DNA, 200 µg/mL in dH_2O.
3. Ice.
4. Methylene blue (MB), 100 µM (Ricca Chemical Company, Arlington, TX: ε = 89125, γ_{max} = 655 nm) *(48)*.
5. Plastic tissue-culture dishes, six-well, clear (Corning cat. no. 430343 or equivalent).
6. 15-mL Polypropylene tubes.
7. TE: 10 mM Tris-HCl, pH 8.0, 1 mM EDTA.
8. 11 M Ammonium acetate.
9. Ethanol, 70% and 95% (or absolute).

2.2. Determination of Enzyme: DNA Ratios and Digestion Times

1. 10X Alkaline loading buffer: 25% Ficoll, 10 mM EDTA, 0.025% (w/v) bromocresol purple, 0.5 M NaOH (NaOH is not added until the day of use).
2. Damaged DNA in 1X reaction buffer: 9.5 µg of DNA in a 285-µL final volume.
3. Eppendorf tubes, 0.5-mL capacity.
4. Fpg stock. Fpg in sterile Fpg storage buffer: 50 mM N-2-hydroxyethylpiperazine-N'-2-ethanesulfonic acid (HEPES)-KOH, pH 7.5, 100 mM NaCl, 1 mM EDTA, 50% glycerol. Fpg is stable for at least a year at –20°C in this buffer. It is available from Trevigen, Inc., Gaithersburg, MD. (Inquire about pricing for large-quantity purchases.)
5. 5X Reaction buffer (per milliliter; prepare just before use):
   a. 500 µL 10X Fpg reaction buffer: 500 mM Tris-HCl, pH 7.5–8.0, 500 mM KCl, 10 mM EDTA: Sterile aliquots may be kept frozen for 2–3 yr.

b. 50 µL of 50 mg/mL bovine serum albumin (BSA; Roche).
c. 450 µL of dH_2O.
6. 1X Reaction buffer: Prepare by 1:5 dilution of 5X reaction buffer.
7. Undamaged DNA: 9.5 µg of DNA in a 285-µL final volume.

2.3. Determining Conditions for Damage Induction in Untested Cell Lines

1. Tissue culture plates, 96-well (Costar 3596 or equivalent; Costar Corporation, Cambridge, MA).
2. Dulbecco's phosphate-buffered saline (DPBS) (with calcium and magnesium) (Gibco/Invitrogen).
3. DPBS (without calcium and magnesium) (Gibco).
4. DPBS (without calcium and magnesium) (Gibco) + 1% D-glucose. Filter-sterilize using a 0.22-µm filter.
5. MB (see **Subheading 2.1., item 4**), 200 µM in DPBS (with calcium and magnesium) containing 1% D-glucose: filter-sterilize using a 0.22-µm filter.
6. Ethidium bromide, 20 µg/mL.
7. Fluorescence-capable plate reader, imaging system, or microscope.

2.4. Damage Induction and Repair Incubation

1. Cells in 15-cm tissue-culture plates (Falcon 3025 or equivalent, Becton Dickinson Labware, Lincoln Park, NJ).
2. DPBS (with calcium and magnesium).
3. DPBS (without calcium and magnesium).
4. MB in DPBS (with calcium and magnesium) containing 1% D-glucose. (Prepare at concentration established in **Subheading 3.3.**) Filter-sterilize using a 0.22-µm filter.
5. Repair media (normal growth medium): 5'-bromo-2'-deoxyuridine (Sigma, St. Louis, MO: γ_{max} = 278 nm, ε = 8820) and 5'-fluoro-2'-deoxyuridine (Sigma: γ_{max} = 268 nm, ε = 7570) may be added to final concentrations of 10 and 1 µM, respectively (see **Note 1**). Sterile 1000X stock solutions of these reagents (in water) may be stored frozen for extended periods of time. Spectrophotometrically determine exact concentrations of each separately prior to mixing.

2.5. Cell Lysis and DNA Isolation

1. 10X proteinase K (Gibco): 5 mg of proteinase K/mL of lysis buffer. Make immediately prior to use.
2. Cell scraper.
3. DNase-free RNase, 100 µg/mL.
4. Ice slurry.
5. Lysis buffer: 0.5 M Tris-HCl, pH 8.0, 20 mM EDTA; 10 mM NaCl, 1% sodium dodecyl sulfate (SDS) Stable for months at room temperature.
6. Saturated NaCl (approx 6 M).

2.6. Restriction Digestion

1. Minigel apparatus and supplies.
2. Restriction enzyme and 10X restriction buffer.
3. Sample DNA.

2.7. Purification of Parental DNA (CsCl Gradient and Fluorescent Localization)

1. Centricon 30 filtration tubes (Amicon, Inc., Beverly, MA).
2. Cesium chloride.
3. Coffee grinder (optional).
4. DNA standards (1, 10, 20, 40, and 80 ng/μL) (*see* **Note 2**).
5. Ethidium bromide, 40 μg/mL.
6. Metal spatula (for stirring salt in ultracentrifuge tubes).
7. Mineral oil.
8. Restricted DNA.
9. Small beaker and foil.
10. Syringe and 18-gage needles.
11. Ultra-Clear 16 × 76 mm ultracentrifuge tubes with lids (Beckman, Palo Alto, CA; *see* **Note 3**).

2.8. Gel Electrophoresis

1. Agarose, molecular biology grade.
2. Alkaline running buffer: 1 mM EDTA, 30 mM NaOH.
3. Gel apparatus with buffer recirculation system.
4. Lambda *Hin*dIII-digested marker DNA.
5. 500 mM Na-EDTA, pH 8.0.
6. 10 M Sodium hydroxide.

2.9. Southern Transfer

1. Blotting paper.
2. Ethidium bromide, 10 mg/mL.
3. Hybrisol (Oncor, Rockville, MD) (*see* **Note 4**).
4. 0.25 M Hydrochloric acid.
5. Neutral gel wash: 0.5 M Tris-HCl, pH 7.5, 1.5 M NaCl.
6. pH paper.
7. 1 M Sodium hydroxide.
8. 2X SSPE: 2.2 mM EDTA, 0.02 M sodium phosphate, pH 7.0, 0.36 M NaCl.
9. Transfer apparatus.
10. Ultraviolet source, handheld, short range.

2.10. Hybridization

1. Blotting paper.
2. Hybrisol (Oncor, Rockville, MD) (*see* **Note 4**).

3. Labeled ribonucleotide triphosphate.
4. PES: 40 mM sodium phosphate, pH 7.2, 1% SDS, 1 mM EDTA.
5. PSE: 250 mM sodium phosphate, pH 7.2, 2% SDS, 1 mM EDTA.
6. Riboprobe template.
7. SP6/T7 Transcription Kit (Roche).
8. 2X SSPE: *see* **Subheading 2.9., item 8**.

3. Methods
3.1. Damaging Purified DNA In Vitro

1. Isolate and restrict the DNA as described in **Subheadings 3.6.1.** to **3.6.3.** (*see* **Note 5**). Resuspend it in dH$_2$O at 200 µg/mL (*see* **Note 6**). Be certain to retain an equal amount for use as an undamaged (negative) control.
2. Prepare one foil-covered 15-mL polypropylene tube for each 200 µg of DNA to be treated. To each, add 400 µL of 11 M ammonium acetate.
3. Place a six-well dish on ice, and adjust a lamp so that a 300-W tungsten bulb is 18 cm from the dish (*see* **Note 7**).
4. Dilute 100 µM MB 1:25 with dH$_2$O. Use a foil-covered tube. (Make a volume equal to the volume of your DNA solution.)
5. **Turn out the white lights**, and then mix the DNA and MB solutions 1:1 (*see* **Note 8**).
6. Aliquot the MB–DNA mixture into a six-well dish, 2 mL/well.
7. Turn on the lamp for 5 min. (This will give one to three lesions/10 kb. *See* **Note 9**.) The lid may be on or off of the dish as long as it is clear in the visible spectrum.
8. Transfer the DNA from the six-well dish to the prepared tubes containing ammonium acetate solution, one well/tube. Ethanol-precipitate the DNA. Continue to work in dim blue light until the pellet has been washed twice with 70% ethanol. Do not allow the pellet to dry: centrifuge briefly a final time to bring all drops of ethanol to the bottom of the tube, and use a pipet to remove the last bit of ethanol.
9. Resuspend the DNA in 500 µL of TE/tube, quantitate, and dilute to 200 µg/mL.

3.2. Determination of Enzyme: DNA Ratios and Reaction Times (see **Note 10**)

1. Prepare Fpg by serial dilution into 1X reaction buffer; 45 µL is needed of each of the following concentrations:
   a. 10 ng/µL
   b. 1 ng/µL
   c. 0.1 ng/µL
   d. 0.01 ng/µL
2. Aliquot the damaged DNA, 15 µL/tube, into 18 of 36 Eppendorf tubes. Aliquot the undamaged DNA into the remaining 18 tubes.
3. To one of the Eppendorf tubes containing damaged DNA, add 5 µL of Fpg dilution a, repipeting to mix. Place in a 37°C water bath, and set a timer for 80 min

(*see* **Note 11**). At 1-min intervals, repeat for Fpg dilution b, then for c, and so forth, and repeating for all four dilutions using undamaged DNA.
4. Repeat **step 3**, setting a timer for 40 min. Repeat for 20 min and for 10 min.
5. Stop the reactions at the appropriate times by the addition of 2.2 µL of 10X alkaline loading buffer with finger-vortexing to mix. Add 5 µL of 1X reaction buffer and 2.2 µL of 10X alkaline loading buffer to the remaining tubes of DNA, which will not be treated with Fpg and will serve as the 0 time point. Incubate all tubes at 37°C for 15 min to fully denature the DNA.
6. Centrifuge samples briefly to bring condensation to the bottom of the tube.
7. Run an alkaline gel and transfer for use in a Southern blot as described in **Subheadings 3.6.4.** and **3.6.5.** Probe for mitochondrial sequences (*see* **Subheading 3.6.6.** and **Note 12**).
8. Quantitate band volumes for the fragment of interest (in this case the mitochondrial genome) in the Southern blot, and calculate the average number of enzyme incisions per band, as described in **Subheading 3.5.**
9. Mark as unusable any points for which the band intensity is too weak to accurately quantitate.
10. a. Note the highest enzyme:DNA ratio at which little or no cutting was seen in the control (undamaged) DNA after 20 min. This is the highest enzyme: DNA ratio that will be of use.
    b. At each enzyme concentration and time point, subtract the number of incisions observed in control (undamaged) DNA from the number seen for damaged (MB-treated) DNA. Plot "specific incisions" as a function of enzyme concentration for the 40-min time point (*see* **Note 13**). Note the lowest ratio that gives the number of incisions corresponding to the number of lesions (that is, that gives complete cutting). It is unlikely that a lower ratio will be useful, but *see* **Note 11**.
11. Plot the number of incisions as a function of reaction time for the lowest enzyme:DNA ratio determined to be effective (in **step 10**, above). Note the time at which the reaction reached completion. The reaction can occur quickly, and may be complete at the earliest time assayed.
12. Select the enzyme/DNA ratio and time to use in subsequent experiments based on the above data. Bear in mind that longer reaction times allow easier staggering of reactions (starting tubes at, for example, 1-min intervals) and that it is best to allow a safety margin both in enzyme concentration and in reaction times so that minor pipeting errors, errors in DNA quantitation, or fluctuations in pipeting times will not affect the outcome of the reaction. (That is, it will still go to completion, and nonspecific cutting will be absent.)

3.3. Determining Conditions for Damage Induction in Untested Cell Lines (see **Note 14**)

Plate cells in four 96-well dishes. The objective is to have the cells reach approx 80% confluence approx 24 h after plating. (For human fibroblasts, plate at 5000 cells/well.) Leave column 12 as a blank (no cells).

3.3.1. Effect of MB Concentration on Cell Survival

1. Once the cells have reached 80% confluence, rinse two of the dishes twice with DPBS (+Ca, +Mg) and then change the medium, replacing it with 100 µL of DPBS (+Ca, +Mg) containing 1% D-glucose. (Rapid changing can be accomplished by inverting the dish onto three folded paper towels that have been sprayed lightly with 70% alcohol to destroy any microbials that might be on the top layer. The towels absorb the old medium. A multichannel pipet is then used to replace the medium.)
2. **Turn off the white lights: all subsequent steps must be performed under dim blue light** (*see* **Note 8**).
3. Add 100 µL of the 200 µM MB to the first column. Repipet to mix, then transfer 100 µL to the next column. Repeat the serial dilution until column 10 is reached.
4. Return the cells to the 37°C incubator for 1 h.
5. After pretreatment is complete, rinse each plate twice with DPBS (–Ca, –Mg), leaving the second wash on the cells. Expose one dish to 4 min of light (*see* **Notes 7** and **15**), and then discard the final wash on both dishes.
6. Fix the cells in methanol or Histochoice Tissue Fixative for 10 min, and then allow to dry for several hours. Stain with 20 µg/mL of ethidium bromide (*see* **Note 16**), discard the excess, and read on a fluorescence plate reader, a fluorescence imaging system, or by counting several fields in a fluorescence microscope.
7. Subtract any background fluorescence (the column without cells) and express the fluorescence in each column as a percentage of the fluorescence seen in the MB-untreated column (i.e., column 11).
8. Two curves are obtained corresponding to cell attachment, with and without damage, immediately after treatment as a function of MB concentration.
9. Some cell lines are more sensitive than others to some types of damage. Note the concentration of MB that gives 50% survival after light exposure (as measured by attachment) for each plate.

3.3.2. Effect of Preexposure to MB

1. Dilute an appropriate amount of MB using DPBS (+Ca, +Mg) + 1% D-glucose. The final concentration should be the one just determined experimentally.
2. **Turn off the white lights: all subsequent steps must be performed under dim blue light** (*see* **Note 8**).
3. For the remaining two 96-well dishes, remove the medium from two columns, wash twice with DPBS (+Ca, +Mg), and replace with the MB. Set a timer for 2 h, and return cells to the 37°C incubator. Every 30 min, repeat for two more columns of cells. (Three columns serve as controls, and the fourth is the cell-free blank.)
4. After pretreatment is complete, rinse each plate twice with DPBS (–Ca, –Mg), leaving the second wash on the cells. Expose one dish to 4 min of light. Discard the final wash.
5. Fix the cells in methanol or Histochoice tissue fixative for 10 min, and then allow to dry for several hours. Stain with 20 µg/mL of ethidium bromide (*see* **Note 16**),

discard the excess, and read on a fluorescence plate reader, a fluorescence imaging system, or by counting several fields in a fluorescence scope.
6. Subtract any background fluorescence (the column without cells) and express the fluorescence in each column as a percentage of the fluorescence seen in the MB-untreated column.
7. Two curves are obtained corresponding to cell attachment, with and without damage, immediately after treatment as a function of preexposure time.
8. Note the amount of variability with time. Ideally, minor variations in pretreatment time will not have an effect on the cellular response. If it does, however, this must be kept in mind so that pretreatment times are extremely precise, or else other pretreatment times must be tested.

3.3.3. Determining the Level of Initial Damage (see **Note 14**)

Plan an experimental timetable that will allow you to stagger the treatment of individual plates in a convenient manner as described in the following protocol, with duplicate plates receiving each of the following light exposures (in minutes): 0, 2, 4, and 6 (for a total of eight plates). In addition, three plates should serve as a full control (treated with neither MB nor light).

1. Plate cells in 15 (or more, if a margin for error is desired) 15-cm tissue-culture plates (*see* **Note 17**). Allow to grow until approx 80% confluent.
2. For each plate, wash the cells twice with DPBS (+Ca, +Mg).
3. Add the MB solution (at the concentration determined in **Subheading 3.3.1.**). All cell treatments involving MB are performed under a dim blue safelight (*see* **Note 8**).
4. Return the cells to the incubator (for the amount of time established in **Subheading 3.3.2.**).
5. After the preincubation with MB is complete, wash the cells twice with DPBS (–Ca, –Mg). Leave the second wash on the cells.
6. Expose the cells to the appropriate amount of light (0, 2, 4, or 6 min).
7. Proceed to isolate the DNA and analyze the damage as described in **Subheading 3.5.**, combining the lysates from each time point. **Protect the lysates from white light!** The desired level of damage is 1–1.5 lesions/fragment (*see* **Notes 18** and **19**).

3.4. Damage Induction and Repair (see **Note 20**)

Plan an experimental timetable that will allow you to stagger the treatment of individual plates in a convenient manner. Each plate will have to be treated as described in the following protocol, with three or more plates used for each repair point. A duplicate determination (six plates in total) of the initial damage level is advised. Repair times can be varied: typical for 8-oxoguanine might be 0.5, 1, 2, 4, 8, and 24 h. In addition, three plates should serve as a negative control (and not be treated with MB or light), and three as a MB control (and not be treated with light). The latter serves to guard against accidental light exposure during DNA isolation.

1. Plate cells in 15-cm tissue-culture plates. Allow the cells to grow until approx 80% confluent.
2. Wash cells in one plate twice with DPBS (+Ca, +Mg).
3. Add the MB solution.
4. Return the cells to the incubator (for the amount of time established in **Subheading 3.3.2.**).
5. After the preincubation with MB is complete, wash the cells twice with DPBS (–Ca, –Mg). Leave the second wash on the cells.
6. Expose the cells to the amount of light determined in **Subheading 3.3.3.**, to induce the correct number of lesions.
7. Remove the DPBS:
    a. For the control cells and "0 min repair" time point: proceed to isolate the DNA and analyze the damage as described below, combining the lysates from each time point. **Protect the lysates from white light!**
    b. For the cells that will be allowed to repair the damage: add 25 mL of repair medium, and return the cells to the incubator. At the appropriate time, lyse as in (a).
8. Isolate and restrict the DNA as described in **Subheading 3.6.** (*see* **Note 1**).
9. Prepare Fpg by dilution into 1X reaction buffer. The final concentration should be based on the enzyme:DNA ratios determined in **Subheading 3.2.** to be optimal for the Fpg preparation being used. 5 µL of each sample is required for analysis. (Remember to allow for the positive control, which is the purified DNA that was prepared in **Subheading 3.1.**)
10. Prepare 2.2 µg of DNA in 33 µL of 1X reaction buffer for each sample (*see* **Note 21**).
11. Aliquot 15 µL into each of two 500-µL Eppendorf tubes (*see* **Note 22**).
12. To one tube from each pair, add 5 µL of the Fpg dilution, repipeting to mix. To the other, add 5 µL of 1X reaction buffer. Place in a 37°C water bath, and set a timer for the reaction time determined earlier to be optimal in **Subheading 3.3.2.**
13. Repeat **step 12** for each sample, staggering the additions by at least 1 min so that reaction times are constant for all samples.
14. Stop the reactions at the appropriate times by the addition of 2.2 µL of 10X alkaline loading buffer with finger-vortexing to mix. Incubate all tubes at 37°C for 15 min to fully denature the DNA. At this time, an aliquot of *Hin*dIII-digested λ DNA molecular weight marker (approx 2–5 µg for easy visualization on an alkaline gel and after Southern transfer) should be prepared and denatured in the same way.
15. Centrifuge samples briefly to bring condensation to the bottom of the tube.
16. Run an alkaline gel as described in **Subheading 3.6.4.**
17. Transfer the gel for use in a Southern blot as described in **Subheading 3.6.5.**
18. Probe for mitochondrial sequences as described in **Subheading 3.6.6.**

3.5. Data Analysis
3.5.1. Signal Measurement

Quantify band volumes for the fragment of interest (in this case the mitochondrial genome) in the Southern blot (*see* **Note 23**), and calculate the average number of enzyme incisions in each sample using the Poisson distribution as follows:

$$\text{Incisions} = -\ln(BIE/BIC)$$

where BIC refers to the band intensity in a "minus enzyme" lane, and BIE refers to the band intensity in an enzyme-treated lane (*see* **Note 24**). Mark as unusable any points for which BIE is too weak to accurately quantify.

3.5.2. Data and Error Analysis for an Experimental Series

1. If more than one gel was run from a single biological experiment, obtain a mean value for each time point, and treat this as a single determination for purposes of error analysis.
2. Calculate the mean and standard error for the number of lesions at each time point based on several biological experiments.
3. Comparison of two sets of repair curves: There are several methods for analyzing data resulting from the serial measurement of a process such as DNA repair. The most commonly used is to simply treat each time point as a separate measure, and compare the two curves one point at a time using a Student's t-test. However, it is difficult to interpret the multiple (and far from independent) P values obtained. A much more acceptable approach is to evaluate the data using a multifactor analysis of variance (ANOVA) (where the factors are the times allowed for repair, the individual experiments, and the comparison of biological interest).

   Another, simpler alternative is to compare summary measures. Two that are especially relevant are the initial rate of repair and the maximum repair level attained. The initial (absolute) rate of repair may be estimated as follows:

   $$(D_0 - D_{t1})/t_1$$

   where D_0 = initial damage, D_{t1} = the damage at the first repair point, and t_1 = time elapsed between damage and first repair measurement.
4. a. When the results are presented graphically, repair is commonly expressed as a percentage rather than a raw lesion frequency. Percentage repair is calculated as follows:

   $$([(D_0 - D_t)/D_0]) \, 100$$

   where D_0 = initial damage and D_t = the damage at time t.

   b. It is simplest to show the error in each time point by reporting lesions/10 kb ± SEM in the text or a table prior to conversion to a percentage, since the number expressed as a percentage includes not only error in the level of damage measured for each time point, but also the error in the initial time point and covariance owing to the influence of the initial level of damage on subsequent levels within a single experiment.

3.6. Support Protocols

3.6.1. Cell Lysis and DNA Isolation

1. After cells have been washed with DPBS, remove the DPBS and add 2 mL of lysis buffer/15 cm dish. Wait at least 5 min. At this stage, dishes may be kept for several hours at room temperature or at 37°C.

Repair of Oxidative DNA Damage

2. Use a cell scraper to assist in pouring the lysate into 15- or 50-mL conical polypropylene tubes. Protect from light by covering the tubes with aluminum foil if the DNA has been exposed to MB. When all the time points have been collected, add 0.1 volume of 10X proteinase K and incubate overnight at 37°C.
3. Add one-fourth volume of saturated NaCl solution to the warm lysate. Mix thoroughly and gently; heat briefly to 55°C if necessary to dissolve precipitate. Thorough mixing is necessary to prevent the formation of large SDS–protein complexes that will trap high molecular weight DNA.
4. Cool, with swirling, by immersing the tube in ice slurry. As the SDS precipitates, so does the protein to which it is bound.
5. Centrifuge for 30 min at 500*g* or greater.
6. Pour the supernatant (which contains DNA) into a fresh conical tube. Ethanol-precipitate the DNA (no extra salt is needed). Centrifuge briefly a final time to bring all drops of ethanol to the bottom of the tube, and use a pipet to remove the last bit of ethanol. Resuspend the DNA in 0.1 mL of TE for each 1×10^6 cells.
7. Treat with 100 µg/mL of RNase A for 3 h at 37°C.
8. Ethanol-precipitate the DNA, again resuspending in 0.1 mL TE for each 1×10^6 cells.
9. Quantitate the DNA (*see* **Note 17**).

3.6.2. Restriction Digestion

1. Select an appropriate restriction enzyme. (*Pvu*II, for example, is useful to linearize human mtDNA.)
2. Restrict an aliquot of the DNA using approx 10 U/µg of DNA for 3 h at 37°C. Save approx 400 ng of unrestricted DNA to test for DNA integrity on a neutral agarose minigel. DNA integrity reflects cell death, DNA damage, and potential problems with the DNA isolation. The DNA concentration should be approx 200 µg/mL or less during the restriction, in the buffer recommended by the manufacturer of the restriction enzyme (*see* **Note 25**).
3. If the DNA is not fully restricted prior to proceeding with the experiment, the results will not be useable. It is worthwhile at this stage, therefore, to check the restriction using a small, native 0.5% agarose gel. At this time, also run 200 ng/lane of the unrestricted DNA. A tight band should be observed if the DNA has not degraded. The restricted DNA should be of lower molecular weight than the unrestricted DNA. If necessary, dilute the DNA further and repeat the restriction.

3.6.3. Purification of Parental DNA:
CsCl Gradient and Fluorescent Localization (see **Notes 1** and **3**)

1. Weigh 6 g of CsCl into each tube. (Use the coffee grinder if the CsCl has clumped: this will speed **step 3**.)
2. Place the tube in a small beaker containing aluminum foil to hold the tube in an upright position. Tare the tube holder and the tube containing the CsCl on a balance.
3. Add the restricted sample to the centrifuge tube, then add TE until the total liquid added weighs 4.70 g. Stir with a metal spatula until the CsCl is in solution.

4. Tare the empty tube holder.
5. Place the tube containing the sample DNA and CsCl in the holder and a top next to it. Add mineral oil to reach about 4 mm from the top. Note the weight. (It will be approx 22–22.3 g.)
6. Place the cap assembly on the tube, making sure that it snaps into the bottom position. Tighten the hex nut.
7. Repeat **steps 5** and **6** with the other tubes, but add oil by weight so that all of the tubes are within 0.01 g of one another. Pair the tubes so that they are as closely matched by weight as possible. It may be necessary to add a tiny amount of oil with a syringe at this time to one or two of the tubes.
8. Place a screw into the top of each cap assembly.
9. Centrifuge at 124,000g in a Beckman Ti-50 rotor (37,000 rpm) at 25°C for 36 h. Make sure that the deceleration is gradual (approx 6 min from 250g [approx 500 rpm] to 0).
10. Fasten the tube to a stand. Use a syringe filled with air to puncture the top of the tube, being careful not to disturb the gradient. Place fraction tubes under the gradient (1.5-mL Eppendorf tubes are most convenient). Use a small-gage needle to puncture the bottom of the tube, and collect approx 400-µL fractions by dripping the gradient from the bottom of the tube. Repeat until all samples have been fractionated.
11. Quantitate the DNA in each fraction using fluorescence: mix 2.5 µL of each fraction with 2.5 µL of 40 µg/mL ethidium bromide (*see* **Note 16**) on plastic wrap. At the same time, use 1, 10, 20, 40, and 80 ng of DNA (*see* **Note 2**) to create a standard curve. Photograph the fluorescence on a UV transilluminator. Use densitometry to quantify the dots on the negative (*see* **Note 26**).
12. Combine the appropriate fractions (those in the peak nearest the top of the centrifuge tube) and remove CsCl by using Centricon 30 tubes (or equivalent) (*see* **Note 27**).

3.6.4. Gel Electrophoresis

1. For a 20-kb gene of interest, use a 0.6% (w/v) gel, and increase it by approx 0.2% for each 5-kb decrease in fragment length. (Thus, for the linearized mammalian mitochondrial genome, a 0.75% gel is used.) Prepare the alkaline agarose gel by mixing the agarose, water and EDTA (final concentration, 1 mM). Dissolve the agarose in a microwave, then equilibrate it in a 55°C water bath for approx 1 h. Add 10 M NaOH to a 30 mM final concentration (*see* **Note 28**), and then pour it immediately into the gel tray, preferably in a cold room. Check that no bubbles have formed on the gel comb (*see* **Note 29**).
2. Place the gel in the bed, fill the bed with running buffer, and prepare the system for buffer recirculation (*see* **Note 30**).
3. Prior to running the gel, prepare the required amount of 10X alkaline loading buffer. In addition, prepare enough 1X loading buffer to allow the loading of 20 µL/well.
4. Preload the gel using the 1X alkaline loading buffer. Let it sit in the wells for approx 1 min, and inspect for leaks. Note which wells are unusable, if any. Prerun the gel at approx 1.5 V/cm for 1 h to clear the wells.

5. Load the samples, being extremely careful to load quantitatively. Gel-loading pipet tips can be very useful for this purpose.
6. Perform elecrophoresis overnight (assuming a 25 cm gel) at 1.5 V/cm with buffer recirculation.

3.6.5. Southern Transfer

1. Following electrophoresis, wash the gel at room temperature for 45 min in neutral gel wash.
2. Stain for 45 min in 1 µg/mL of ethidium bromide (*see* **Note 16**), then destain in dH$_2$O for 30 min and photograph the gel.
3. Transfer the gel to 0.25 N HCl and leave for 30 min.
4. Transfer to 1 N NaOH and leave for 45 min. The gel is now ready for blotting.
5. Rapid transfer can be achieved with a Posiblot apparatus (Stratagene, La Jolla, CA), following the manufacturer's directions. Traditional overnight methods are also acceptable.
6. Following transfer, rinse the membrane several times with 2X SSPE. Using pH paper, check the pH of the SSPE wash; the procedure is complete when it remains at pH 7.0.
7. Pat the membrane dry between two pieces of blotting paper, then bake at 80°C under vacuum for 2 h. This allows visualization of the DNA on the blot by UV shadowing using a handheld short-range UV source. Use a No. 2 pencil to label the membrane and to note the position of the molecular weight markers.
8. Prehybridize overnight or longer in Hybrisol (*see* **Note 4**).

3.6.6. Hybridization

1. Prepare a riboprobe against the sequence of interest (*see* **Notes 31** and **32**) using a kit and following the manufacturer's instructions.
2. Hybridize overnight in Hybrisol at 45°C (*see* **Note 4**).
3. After hybridization, wash twice with 2X SSPE at room temperature to remove unbound probe.
4. Wash twice with PSE at 65°C (or lower, depending on the probe) and then once with PES at 65°C (or lower, depending on the probe) to remove nonspecifically bound probe.
5. Use blotting paper to lightly blot the membrane dry, and then place in plastic wrap (*see* **Note 33**) and expose to X-ray film (or place in a PhosphorImager cassette) (see **Note 23**).

4. Notes

1. CsCl centrifugation to separate parental and daughter DNA may be necessary if cell replication, or mitochondrial replication in the absence of cell replication, is able to occur during the repair period. (In at least some cell lines, MB prevents both.) It requires that 5'-bromo-2'-deoxyuridine and 5'-fluoro-2'-deoxyuridine be included in the repair medium, so that parental and daughter DNA differ in density. It should also be noted that variations in GC/AT content may cause some sequences (e.g., the nuclear ribosomal sequences) to move away from the bulk of

the DNA in the gradient. The mobility of specific sequences can be monitored by the use of dot or slot blots. (Also, since mtDNA can replicate independently of the nuclear DNA, dot or slot blots are necessary if visualization of parental and daughter mtDNA peaks is desired.)
2. DNA standards can be prepared from any high molecular weight DNA by serial dilution of DNA that has been extensively purified and quantitated by UV absorption.
3. This protocol is written for a Beckman ultracentrifuge, and may require modification if a different ultracentrifuge is used.
4. Other hybridization solutions will also work.
5. a. If a new lesion or new enzyme is being characterized, it is useful to work with plasmids at this stage, rather than total mammalian DNA. Transitions from supercoiled to nicked circle to linear can easily be monitored in neutral gels with ethidium fluorescence, rather than the more laborious Southern blot methods required for specific sequences in total DNA. This also avoids concerns about the stability of the lesion in the alkaline conditions used to denature DNA for the Southern blot method.
   b. If a new lesion is being characterized, its stability under the conditions used to denature the DNA for the overnight gel must be ascertained (unless this is known from the literature). Identical aliquots of damaged plasmid are loaded onto neutral and denaturing gels both with and without prior treatment with the repair enzyme. The percentage remaining supercoiled is determined and compared between the two gels. Values should be determined in duplicate or triplicate for precision. (Since ethidium bromide does not efficiently bind denatured DNA, the use of a Southern blot is advised.) Nonalkaline-based methods for denaturing DNA are also available (glyoxal). Here too, however, a test of lesion stability would be required.
6. The protocol is written for 400 µg, but more or less DNA can be treated depending on the number of experiments planned. Five to ten micrograms will be used on each repair gel as a positive control, and an additional approx 20 µg will be used to characterize a new preparation of enzyme.
7. The light exposure, if measured, should be 50–70 kJ/m^2. (Measurement may be made with an IL1400A Radiometer/Photometer [International Light, Inc., Newburyport, MA]; owing to the intensity of the light, light must be measured through a screen that reduces the intensity by a known amount.)
8. Work with MB should be done in a dim blue light. Safelights for photosensitive agents can generally by made with inexpensive plastic available from a local industrial plastic supplier. Obtain small samples of approximately the same color as the reagent, and check the spectrum on a scanning spectrometer. Choose the one that most closely matches the reagent to construct the safelight.
9. The goal is to obtain two or three lesions/10 kb of double-stranded DNA (dsDNA). If there is an independent method to measure this level, it is recommended that it be used. Otherwise, the enzyme itself must be used, which may require one to two iterations. That is, a rough estimate of optimal enzyme condi-

tions is used to quantitate damage in a DNA sample. This is used to select the sample with two or three lesions/10 kb. This sample is then used to further refine the optimal enzyme conditions.

10. This is suggested for each new batch of enzyme. At the very least, a new batch of enzyme should be directly compared to an old batch using both undamaged and damaged DNA. It is critical to test each new preparation, and to work in conditions where cutting is complete and nonspecific cutting is not extensive.

11. Eighty minutes is arbitrarily the longest time point. It is possible that a longer incubation, perhaps even overnight, could be used with lower enzyme concentrations to conserve enzyme.

12. It is useful to optimize the assay conditions by probing for mtDNA for two reasons: first, the genome is completely sequenced and second, it is multicopy and so sensitivity is increased (and less DNA is needed per lane, thereby saving on enzyme costs). Remember that assay sensitivity is optimal between one and two lesions/ fragment, and so if another gene is to be assayed, the restriction of the mitochondrial genome and initial damage levels should be adjusted to approximate the desired conditions. Endogenous damage has not been seen in the mtDNA of cultured cells (*44,49*), and is thus not a confounding factor.

13. A clear plateau should be seen. If it is not seen, the reaction is not going to completion. The number of incisions seen at the plateau corresponds to the actual number of lesions induced in the damaged DNA.

14. Three points should be considered:
    a. Each time a new cell type is to be treated, it is advisable to test its sensitivity to the agent being used. In addition, the cells should be as close as possible in terms of growth state and culture conditions to the conditions that will be used in the actual experiment.
    b. Differences in initial damage levels are a confounding variable when two treatment populations are to be compared. If, for example, the time required for a given level of repair is twice as great in one population as in another, this could be explained by a twofold difference in the number of lesions removed per minute, or by a twofold difference in the initial level of damage. It is therefore important to establish conditions that yield similar initial lesion frequencies if two populations are to be compared.
    c. Differences in initial lesion frequency may be unavoidable in a single treatment population. For example, there are 2.38 times more guanines in the heavy mitochondrial strand than in the light strand. Randomly induced damage to guanine will, therefore, result in more initial damage in the heavy strand than in the light strand. In this case, it is biologically relevant to treat the two strands as a single system, ignoring the difference in initial lesion frequency. If the probability of any lesion being repaired is equal, and not a function of strand placement, the fraction of initial lesions should be the same in each strand at any given time in the repair process. If, on the other hand, it is not random, the fraction remaining should depend on which strand is being examined.

15. The light source for all experiments is, ideally, a 300-W tungsten bulb situated beneath the plate providing 18.5 mW/cm^2 of visible light. A filter may be necessary to provide even intensity across the dish. To prevent heating, tissue culture plates should be separated from the light source by a ventilated chamber. Exposure from above is also possible, using a conventional desk lamp: in this case, an orbital shaker at low speed can be used to move the plate in such a way that light exposure is not higher in the center of the plate than at the periphery, and a small fan can be used to blow air across the plate to prevent heat buildup (keeping the dish closed to avoid contamination).
16. **Ethidium bromide is a carcinogen**. Handle with care and dispose of waste in an appropriate manner.
17. The expected yield is approx 7–10 μg/1 × 10^6 cells. DNA can be quantitated using UV spectroscopy, or, to conserve DNA, fluorescent quantitation is also effective using a kit such as the Picogreen dsDNA Quantitation Kit (Molecular Probes, Eugene, OR). (The cell-culture conditions suggested in these protocols should yield enough DNA for analysis of single-copy genes. If only multicopy genes such as in mtDNA are to be assayed, smaller plates and fewer cells may be used.)
18. Cell death prior to the measurement of initial damage does not affect the repair determination. Cell death subsequent to the initial measurement becomes a problem only if the dying cells had a higher level of initial damage than the surviving cells, which would lead to an overestimation of the extent of repair. Nevertheless, it is best to avoid cell death if possible, and if DNA damage is sufficient (one lesion/restriction fragment of interest) at levels that allow all cells to survive, that is the dose that should be used. If, on the other hand, a higher dose must be used, one option is to limit repair measurements to earlier times, before death or detachment occur.
19. A final series of survival studies, done at the selected MB concentration, pre-exposure time, and light exposure, is advisable, this time also measuring attachment during the repair period. Ideally, little or no cell death will occur after damage, during the repair period. If it does, it is usually assumed that cell death occurred randomly, rather than that a subset of cells received a greater amount of initial damage, unless there is reason to think otherwise.
20. a. When two cell lines are compared, ideally they will have completed a similar percentage of their *in vitro* lifespan. This can sometimes be arranged for mortal cell lines simply by growing the "younger" line an extra passage or two. For this reason, for mortal cell lines the replicative status of the cells should be assayed by determining the "30-h labeling index" prior to the repair experiment. This index is the fraction of cells that can incorporate labeled nucleotides into their DNA during a 30-h incubation. This may be done either nonisotopically, using a kit such as the BrdU Labeling and Detection Kit (Boehringer-Mannheim), or by tritium incorporation. For data analysis, the log of the percentage of nondividing cells is proportional to the completed lifespan *(50,51)*.

b. No two cell lines, even sister clones, have exactly the same maximum population doubling level (PDL). PDL refers to the population doubling level of cells in culture. It is a more objective measure of cell growth than passage number, since the latter does not give any information on the split (e.g., 1:2 or 1:4) that was used at each passage. In this context, it is worth noting that immortalized cell lines may show genetic drift, so that after many cell divisions they may not be identical to the parent line or to cells of earlier passage. Calculate the PDL of the culture. The formula for the change in PDL from one passage to the next is log (final cell number/initial cell number) divided by log2. If the cell line being tested has been transfected, calculate the posttransfection PDL.
21. The DNA concentration should be increased 5- to 10-fold if single-copy genes are to be studied, rather than the multicopy mitochondrial genome.
22. It is critical for the success of the assay that the two tubes contain exactly the same amount of DNA.
23. These protocols absolutely require a method for quantitating the relative signal from the bands. A system such as the PhosphorImager (Molecular Dynamics, Inc., Sunnyvale, CA) is ideal due to its extended linear range. However, photographic film and densitometry may also be used if care is taken to remain within the linear range of the film.
24. The sensitivity of the assay is limited at both extremely high lesion frequencies (which cause weak, poorly quantifiable bands in the "minus enzyme" lanes) and at extremely low lesion frequencies (in which case the "minus" and "plus" enzyme lanes both give approximately the same signal, and their ratio is thus greatly altered by even slight differences in loading or transfer).
25. DNA that is not fully dissolved will not restrict.
26. If exact quantitation is not desired, and only the location of the parental and daughter peaks is sought, then densitometry is not necessary. The peaks will be clearly visible.
27. CsCl may also be removed by dialysis if preferred.
28. The agarose will caramelize (turn brown) if it is too hot when the NaOH is added.
29. Since bubble formation is a problem at the cooler gel-pouring temperatures used for alkaline agarose gels, a gravy pitcher, purchased from a local grocery store, can be used to prevent bubbles from reaching the gel tray.
30. Buffer recirculation is necessary with alkaline gels. This can be accomplished in two ways: over the gel (in which case, run the gel long enough for the DNA to enter the agarose prior to beginning the recirculation), or under it. For the latter, have the gel tray sitting on a spacer, and cover the top of the gel with glass (or heavy plastic such as Plexiglas). The latter method has two advantages: diffusion of smaller DNA fragments and marker dyes is minimized, and the recirculation can be begun as soon as the gel is loaded.
31. The use of riboprobes is recommended for two reasons: strand-specific probing is possible, and stripping the membrane without any residual probe remaining is easily accomplished. (A room temperature wash in 0.4 M NaOH for 1 h, fol-

lowed by one or two neutralizing washes containing 0.5% SDS, is usually sufficient.)
32. The riboprobe template plasmids used by our laboratory were prepared by cloning mitochondrial sequences amplified by PCR into the PCR II vector (Invitrogen, Carlsbad, CA) between the T7 and SP6 promoter sequences. The sequences were from 652 to 3226 and from 5905 to 7433 of the mitochondrial genome (numbering based on the Anderson sequence *[52]*).
33. The plastic must be wrinkle free, front and back, for accurate quantitation. Minor variations in the distance between the blot and the film (or PhosphorImager screen, Molecular Dynamics, Inc. Sunnyvale, CA) will alter the signal intensity significantly. Clear plastic page protectors are not subject to wrinkling and can be substituted for plastic wrap.

References

1. Brown, M. D. and Wallace, D. C. (1994) Molecular basis of mitochondrial DNA disease. *J. Bioenerg. Biomembr.* **26**, 273–289.
2. Kumagai, Y., Koide, S., Taguchi, K., et al. (2002) Oxidation of proximal protein sulfhydryls by phenanthraquinone, a component of diesel exhaust particles. *Chem. Res. Toxicol.* **15**, 483–489.
3. Wallace, D. C. (2001) A mitochondrial paradigm for degenerative diseases and ageing. *Novartis Found. Symp.* **235**, 247–263.
4. Souza-Pinto, N. C. and Bohr, V. A. (2002) The mitochondrial theory of aging: involvement of mitochondrial DNA damage and repair. *Int. Rev. Neurobiol.* **53**, 519–534.
5. Hileman, E. A., Achanta, G., and Huang, P. (2001) Superoxide dismutase: an emerging target for cancer therapeutics. *Expert. Opin. Ther. Targets.* **5**, 697–710.
6. Wallace, D. C. (1999) Mitochondrial diseases in man and mouse. *Science* **283**, 1482–1488.
7. Clayton, D. A., Doda, J. N., and Friedberg, E. C. (1975) Absence of a pyrimidine dimer repair mechanism for mitochondrial DNA in mouse and human cells. *Basic Life Sci.* **5B**, 589–591.
8. Anderson, C. T. and Friedberg, E. C. (1980) The presence of nuclear and mitochondrial uracil-DNA glycosylase in extracts of human KB cells. *Nucleic Acids Res.* **8**, 875–888.
9. Croteau, D. L., ap Rhys, C. M., Hudson, E. K., Dianov, G. L., Hansford, R. G., and Bohr, V. A. (1997) An oxidative damage-specific endonuclease from rat liver mitochondria. *J. Biol. Chem.* **272**, 27,338–27,344.
10. Thyagarajan, B., Padua, R. A., and Campbell, C. (1996) Mammalian mitochondria possess homologous DNA recombination activity. *J. Biol. Chem.* **271**, 7,536–27,543.
11. Driggers, W. J., Grishko, V. I., LeDoux, S. P., and Wilson, G. L. (1996) Defective repair of oxidative damage in the mitochondrial DNA of a xeroderma pigmentosum group A cell line. *Cancer Res.* **56**, 1262–1266.

12. Shen, C. C., Wertelecki, W., Driggers, W. J., LeDoux, S. P., and Wilson, G. L. (1995) Repair of mitochondrial DNA damage induced by bleomycin in human cells. *Mutat. Res.* **337,** 19–23.
13. Driggers, W. J., LeDoux, S. P., and Wilson, G. L. (1993) Repair of oxidative damage within the mitochondrial DNA of RINr 38 cells. *J. Biol. Chem.* **268,** 22,042–22,045.
14. LeDoux, S. P., Wilson, G. L., Beecham, E. J., Stevnsner, T., Wassermann, K., and Bohr, V. A. (1992) Repair of mitochondrial DNA after various types of DNA damage in Chinese hamster ovary cells. *Carcinogenesis* **13,** 1967–1973.
15. Snyderwine, E. G. and Bohr, V. A. (1992) Gene- and strand-specific damage and repair in Chinese hamster ovary cells treated with 4-nitroquinoline 1-oxide. *Cancer Res.* **52,** 4183–4189.
16. Pettepher, C. C., LeDoux, S. P., Bohr, V. A., and Wilson, G. L. (1991) Repair of alkali-labile sites within the mitochondrial DNA of RINr 38 cells after exposure to the nitrosourea streptozotocin. *J. Biol. Chem.* **266,** 3113–3117.
17. Tomkinson, A. E., Bonk, R. T., Kim, J., Bartfeld, N., and Linn, S. (1990) Mammalian mitochondrial endonuclease activities specific for ultraviolet-irradiated DNA. *Nucleic Acids Res.* **18,** 929–935.
18. Domena, J. D., Timmer, R. T., Dicharry, S. A., and Mosbaugh, D. W. (1988) Purification and properties of mitochondrial uracil-DNA glycosylase from rat liver. *Biochemistry* **27,** 6742–6751.
19. Tomkinson, A. E., Bonk, R. T., and Linn, S. (1988) Mitochondrial endonuclease activities specific for apurinic/apyrimidinic sites in DNA from mouse cells. *J. Biol. Chem.* **263,** 12,532–12,537.
20. Satoh, M. S., Huh, N., Rajewsky, M. F., and Kuroki, T. (1988) Enzymatic removal of O6-ethylguanine from mitochondrial DNA in rat tissues exposed to *N*-ethyl-*N*-nitrosourea in vivo. *J. Biol. Chem.* **263,** 6854–6856.
21. Myers, K. A., Saffhill, R., and O'Connor, P. J. (1988) Repair of alkylated purines in the hepatic DNA of mitochondria and nuclei in the rat. *Carcinogenesis* **9,** 285–292.
22. Richter, C. (1995) Oxidative damage to mitochondrial DNA and its relationship to ageing. *Int. J. Biochem. Cell Biol.* **27,** 647–653.
23. Richter, C. (1992) Reactive oxygen and DNA damage in mitochondria. *Mutat. Res.* **275,** 249–255.
24. Nishioka, K., Ohtsubo, T., Oda, H., et al. (1999) Expression and differential intracellular localization of two major forms of human 8-oxoguanine DNA glycosylase encoded by alternatively spliced OGG1 mRNAs. *Mol. Biol. Cell* **10,** 1637–1652.
25. Nilsen, H., Otterlei, M., Haug, T., et al. (1997) Nuclear and mitochondrial uracil-DNA glycosylases are generated by alternative splicing and transcription from different positions in the UNG gene. *Nucleic Acids Res.* **25,** 750–755.
26. Dzierzbicki, P., Koprowski, P., Fikus, M. U., Malc, E., and Ciesla, Z. (2004) Repair of oxidative damage in mitochondrial DNA of *Saccharomyces cerevisiae*: involvement of the MSH1-dependent pathway. *DNA Repair* **3,** 403–411.
27. Ohtsubo, T., Nishioka, K., Imaiso, Y., et al. (2000) Identification of human MutY homolog (hMYH) as a repair enzyme for 2-hydroxyadenine in DNA and detec-

tion of multiple forms of hMYH located in nuclei and mitochondria. *Nucleic Acids Res.* **28,** 1355–1364.
28. Shinmura, K., Yamaguchi, S., Saitoh, T., et al. (2000) Adenine excisional repair function of MYH protein on the adenine:8-hydroxyguanine base pair in double-stranded DNA. *Nucleic Acids Res.* **28,** 4912–4918.
29. LeDoux, S. P. and Wilson, G. L. (2001) Base excision repair of mitochondrial DNA damage in mammalian cells. *Prog. Nucleic Acid Res. Mol. Biol.* **68,** 273–284.
30. Bohr, V. A., Stevnsner, T., and Souza-Pinto, N. C. (2002) Mitochondrial DNA repair of oxidative damage in mammalian cells. *Gene* **286,** 127–134.
31. Boiteux, S., Gajewski, E., Laval, J., and Dizdaroglu, M. (1992) Substrate specificity of the Escherichia coli Fpg protein (formamidopyrimidine-DNA glycosylase): excision of purine lesions in DNA produced by ionizing radiation or photosensitization. *Biochemistry* **31,** 106–110.
32. Schneider, J. E., Price, S., Maidt, L., Gutteridge, J. M., and Floyd, R. A. (1990) Methylene blue plus light mediates 8-hydroxy 2′-deoxyguanosine formation in DNA preferentially over strand breakage. *Nucleic Acids Res.* **18,** 631–635.
33. Ravanat, J. L. and Cadet, J. (1995) Reaction of singlet oxygen with 2′-deoxyguanosine and DNA. Isolation and characterization of the main oxidation products. *Chem. Res. Toxicol.* **8,** 379–388.
34. ESCODD (2003) Measurement of DNA oxidation in human cells by chromatographic and enzymic methods. *Free Radic. Biol. Med.* **34,** 1089–1099.
35. ESCODD (2002) Comparative analysis of baseline 8-oxo-7,8-dihydroguanine in mammalian cell DNA, by different methods in different laboratories: an approach to consensus. *Carcinogenesis* **23,** 2129–2133.
36. Beckman, K. B. and Ames, B. N. (1999) Endogenous oxidative damage of mtDNA. *Mutat. Res.* **424,** 51–58.
37. Beckman, K. B. and Ames, B. N. (1996) Detection and quantification of oxidative adducts of mitochondrial DNA. *Methods Enzymol.* **264,** 442–453.
38. Bohr, V. A. (1991) Gene specific DNA repair. *Carcinogenesis* **12,** 1983–1992.
39. Anson, R. M., Hudson, E., and Bohr, V. A. (2000) Mitochondrial endogenous oxidative damage has been overestimated. *FASEB J.* **14,** 355–360.
40. Rusyn, I., Asakura, S., Pachkowski, B., et al. (2004) Expression of base excision DNA repair genes is a sensitive biomarker for in vivo detection of chemical-induced chronic oxidative stress: identification of the molecular source of radicals responsible for DNA damage by peroxisome proliferators. *Cancer Res.* **64,** 1050–1057.
41. Bohr, V. A. and Anson, R. M. (1995) DNA damage, mutation and fine structure DNA repair in aging. *Mutat. Res.* **338,** 25–34.
42. Bohr, V. A. (1994) Gene-specific damage and repair of DNA adducts and cross-links, in *IARC Sci.*, Publ. No. 125 (Hemminki, K., Dipple, A., Shuker, D. E. G., Kadlubar, F. F. Segerback, D., and Bartsch, H., eds.) International Agency for Research on Cancer, Lyon, France, pp. 361–369.
43. Sass, M. D., Caruso, C. J., and Axelrod, D. R. (1967) Accumulation of methylene blue by metabolizing erythrocytes. *J. Lab. Clin. Med.* **69,** 447–455.

44. Anson, R. M., Croteau, D. L., Stierum, R. H., Filburn, C., Parsell, R., and Bohr, V. A. (1998) Homogenous repair of singlet oxygen-induced DNA damage in differentially transcribed regions and strands of human mitochondrial DNA. *Nucleic Acids Res.* **26,** 662–668.
45. Tchou, J., Bodepudi, V., Shibutani, S., et al. (1994) Substrate specificity of Fpg protein. Recognition and cleavage of oxidatively damaged DNA. *J. Biol. Chem.* **269,** 15,318–15,324.
46. Tchou, J., Kasai, H., Shibutani, S., Chung, M. H., Laval, J., Grollman, A. P., and Nishimura, S. (1991) 8-oxoguanine (8-hydroxyguanine) DNA glycosylase and its substrate specificity. *Proc. Natl. Acad. Sci. USA* **88,** 4690–4694.
47. Taffe, B. G., Larminat, F., Laval, J., Croteau, D. L., Anson, R. M., and Bohr, V. A. (1996) Gene-specific nuclear and mitochondrial repair of formamidopyrimidine DNA glycosylase-sensitive sites in Chinese hamster ovary cells. *Mutat. Res.* **364,** 183–192.
48. Robinson, J. W. (ed.) (1974) *CRC Handbook of Spectroscopy*, vol. II. CRC, Cleveland, OH.
49. Higuchi, Y. and Linn, S. (1995) Purification of all forms of HeLa cell mitochondrial DNA and assessment of damage to it caused by hydrogen peroxide treatment of mitochondria or cells. *J. Biol. Chem.* **270,** 7950–7956.
50. Cristofalo, V. J. (1976) Thymidine labelling index as a criterion of aging in vitro. *Gerontology* **22,** 9–27.
51. Cristofalo, V. J. and Sharf, B. B. (1973) Cellular senescence and DNA synthesis. Thymidine incorporation as a measure of population age in human diploid cells. *Exp. Cell Res.* **76,** 419–427.
52. Anderson, S., Bankier, A. T., Barrell, B. G., et al. (1981) Sequence and organization of the human mitochondrial genome. *Nature* **290,** 457–465.

14

Quantitative PCR-Based Measurement of Nuclear and Mitochondrial DNA Damage and Repair in Mammalian Cells

Janine H. Santos, Joel N. Meyer,
Bhaskar S. Mandavilli, and Bennett Van Houten

Summary

In this chapter, we describe a gene-specific quantitative polymerase chain reaction (QPCR)-based assay for the measurement of DNA damage, using amplification of long DNA targets. This assay has been extensively used to measure the integrity of both nuclear and mitochondrial genomes exposed to different genotoxins, and has proved particularly valuable in identifying reactive oxygen species-mediated mitochondrial DNA (mtDNA) damage. QPCR can be used to quantify the formation of DNA damage, as well as the kinetics of damage removal. One of the main strengths of the assay is that it permits monitoring the integrity of mtDNA directly from total cellular DNA without the need for isolating mitochondria, or a separate step of mtDNA purification. Here we discuss advantages and limitations of using QPCR to assay DNA damage in mammalian cells. In addition, we give a detailed protocol for the QPCR assay that helps facilitate its successful deployment in any molecular biology laboratory.

Key Words: Mammalian cells; DNA quantitative; damage; repair; QPCR.

1. Introduction
1.1. Principle of the Assay

The quantitative polymerase chain reaction (QPCR) assay of DNA damage is based on the principle that many kinds of DNA lesions can slow down or block the progression of DNA polymerase *(1)*. Therefore, if equal amounts of DNA from differently treated samples are QPCR amplified under identical conditions, DNA with fewer lesions will amplify to a greater extent than more

From: *Methods in Molecular Biology: DNA Repair Protocols: Mammalian Systems, Second Edition*
Edited by: D. S. Henderson © Humana Press Inc., Totowa, NJ

damaged DNA *(2,3)*. For example, DNA from a biological sample exposed to ultraviolet (UV) radiation will be amplified less than the DNA from a corresponding control sample *(4)*. Damage can be expressed in terms of lesions per kilobase mathematically (*see* **Subheading 3.4.4.**), by assuming a Poisson distribution of lesions. In addition, DNA repair kinetics can be followed by measuring restoration of amplification of the target DNA over time, after the removal of the DNA damaging agent. QPCR can be performed using genomic DNA from cultured cells or extracted DNA from tissue obtained from treated animals (such as rat or mouse).

1.2. Advantages of the Assay

Strengths of QPCR include its sensitivity, the requirement for only nanogram amounts of total (genomic) DNA, its applicability to measurement of gene-specific DNA damage and repair, and the fact that it can be used to directly compare damage to nuclear DNA (nDNA) and to mitochondrial DNA (mtDNA) from the same sample.

Gene-specific QPCR is highly sensitive because of the use of "long" PCR methodology that permits the quantitative amplification of fragments of genomic DNA between 10 and 25 kb in length *(5,6)*. As a result, very low levels of lesions (approx 1 per 10^5 kb) can be detected, permitting the study of DNA damage and repair at levels of lesions that are biologically relevant.

Because this is a PCR-based assay, it is possible to use a few nanograms of total genomic DNA, which allows analysis of a much wider range of biological samples than is feasible with other methods (such as Southern blots or high-performance liquid chromagraphy [HPLC] with electrochemical detection), that require 10–50 µg of total cellular DNA.

Any gene (or region of DNA) that can be specifically PCR-amplified can be studied using QPCR. Thus, it is possible to compare the rate of damage and/or repair in regions that are hypothesized to be more quickly repaired than others. For example, using this method, it was demonstrated that normal human fibroblasts showed higher rates of repair in the actively transcribed hypoxanthine guanine phosphoribosyl transferase gene than in the nontranscribed β-globin gene *(4)*. This study also demonstrated that repair deficiencies in cells from patients with xeroderma pigmentosum could be clearly detected with this assay.

Finally, the use of genomic DNA, which includes nuclear and mitochondrial genomes, allows direct comparison of the degree of damage and/or repair in nDNA vs mtDNA in the same biological sample. In fact, QPCR has been used successfully to quantify damage and repair in nDNA and mtDNA in a wide variety of cells and tissues after exposure to many types of genotoxicants *(7–25)*.

The importance of damage to the nuclear genome is widely recognized. We discuss below the consequences of damage to the mitochondrial genome.

1.3. Mitochondrial DNA Damage

The mammalian mitochondrial genome is a circular molecule present in multiple (often 2–10) copies in each mitochondrion, with hundreds to thousands of mitochondria per cell (more mitochondria are generally present in cells with high energy requirements). The human mtDNA encodes 2 rRNAs, 22 tRNAs, and 13 polypeptides, all of which are involved in oxidative phosphorylation through the electron transport chain (ETC). While the important role of nDNA damage in human pathological conditions such as cancers is well known, increasing attention is being paid to the association of mtDNA damage with various human diseases *(26,27)*. Some of these include neurodegenerative disorders such as Alzheimer's, Parkinson's and Huntington's disease *(28–30)*; hereditary diseases such as Leber hereditary optic neuropathy and Kearns-Sayre syndrome *(31)*; cancer *(32)*; and aging *(33–35)*. The MITOMAP web-site (http://www.mitomap.org) provides additional information and links related to mtDNA mutations and deletions and the pathological conditions associated with them.

Substantial evidence suggests that mtDNA may be more vulnerable than nDNA to certain kinds of damage, in particular reactive oxygen species (ROS)-mediated lesions *(7,14,19)*. Several reasons may underlie this observation, including the immediate proximity of mtDNA to the ETC in the inner mitochondrial membrane, which is the main source of endogenous ROS production. ROS are generated at substantial rates under normal circumstances by the ETC; it is estimated that as much as 1–5% of the oxygen consumed in vitro by mitochondrial preparations is released as superoxide and hydrogen peroxide (H_2O_2) *(36,37)*, although the in vivo level of production is probably lower *(38)*. The rate of ROS generation by the ETC can be increased by exposure to some xenobiotics (e.g., certain redox-cycling compounds and ETC inhibitors; **refs. 35,39**), and lipophilic xenobiotics tend to accumulate in the mitochondrial membranes (reviewed in **ref. 40**). Redox-active metals, such as iron and copper that can participate in Fenton chemistry, are in close proximity to, or can directly bind to mtDNA *(41,42)*. In addition, mtDNA lacks many of the protective protein structures associated with nDNA, and it is believed that repair of mtDNA lesions occurs only via base excision repair *(35,43,44)*. In fact, while oxidative damage can be repaired in the mtDNA, bulky DNA adducts, which are removed by the nucleotide excision repair machinery, are not removed from the mitochondrial genome (reviewed in **ref. 40**).

Regardless of the reasons, in the past decades various studies using different methodologies have identified higher rates of damage in the mtDNA than in the nDNA of the same biological sample, most notably ROS-mediated damage *(7,14,45–48)*. Early studies often relied on DNA extraction techniques that caused extensive DNA oxidation, resulting in reports of artifactually high levels of adducts *(49)*. Moreover, in some cases the higher levels of damage observed in mtDNA may have been due to the additional handling necessary to first isolate mitochondria from whole-cell (or tissue) extracts in order to obtain nDNA-free mtDNA *(50)*. Thus, the ability of the long PCR assay to measure mtDNA damage without manipulation of mitochondria, and compare the mtDNA to nDNA damage in the same samples, is of great utility.

1.4. Limitations of the Long PCR Assay

Two limitations are associated with QPCR. First, DNA lesions that do not significantly stall progression of DNA polymerase, such as 8-hydroxydeoxyguanosine (8-OHdG) will not be detected with high efficiency. However, agents that cause oxidative stress, such as H_2O_2 or ionizing radiation, are very unlikely to produce only one type of lesion. In fact, it is estimated that only 10% of H_2O_2-induced damage is 8-OHdG *(51)*. Second, although we can identify the presence of damage on the DNA template, the specific nature of the lesion cannot be inferred by QPCR alone. Both of these limitations are shared by some of the other methods available. An additional hypothetical concern is that, in terms of the nuclear genome, typically only one or a few genes or regions are amplified. Thus, if the nDNA damage induced by a given agent were highly region specific, such as for the *p53* gene *(52)*, the results of this assay could possibly be skewed (indicating no or low DNA damage if the target regions were principally away from the fragments amplified, or too many lesions if the fragments amplified were preferentially damaged). It is expected, however, that for most DNA damaging agents, lesions will be introduced randomly along the genome. This concern is not relevant for the mtDNA since the vast majority of the mitochondrial genome can be amplified using QPCR.

2. Materials

2.1. DNA Sample Extraction

High molecular weight DNA is essential to efficiently amplify long genomic targets. We have found that the DNA purified using the QIAGEN Genomic Tip and Genomic DNA Buffer Set Kit (Qiagen, cat. nos. 10323 and 19060, respectively) is of high quality and quite reproducible from sample to sample. Finally the purified DNA is very stable, yielding comparable amplification over long periods of storage.

2.2. DNA Quantitation

1. PicoGreen dye (Molecular Probes, cat. no. P-7581). We perform quantitation of the purified genomic DNA, as well as of PCR products, fluorimetrically using

the PicoGreen dsDNA quantitation reagent. The free dye has very low fluorescence but exhibits a >1000-fold increase in fluorescence signal on binding to dsDNA. The assay displays a linear correlation between dsDNA quantity and fluorescence over a wide range of concentrations, and is extremely sensitive (limit of detection is approx 25 pg/mL). For detailed information about PicoGreen see the manufacturer's web-site. Upon first use, we store 50-µL aliquots of the PicoGreen reagent at –20°C. Vials are thawed only immediately prior to use.

2. 20X TE buffer: 200 mM Tris-HCl, 20 mM EDTA, pH 7.5, which is diluted to 1X and is stored at room temperature.
3. Lambda (λ)/HindIII DNA (Gibco; cat. no. 15612-013) to generate a standard curve. Both the diluted (λ) DNA standard and the diluted samples are kept at 4°C. For long-term storage, we routinely store the concentrated samples at –20°C and try to avoid many cycles of freeze–thaw.
4. Fluorescence reader capable of measuring fluorescence with 485 nm excitation and 535 nm emission; a 96-well plate reader format is most convenient (e.g., FL600 Microplate Fluorescence, from Bio-Tek).

2.3. PCR Reagents

1. GeneAmp XL PCR Kit (Applied Biosystems) for long QPCR. The kit includes $rTth$ DNA polymerase XL (400 U; 2 U/µL), 3.3X XL PCR buffer, and 25 mM Mg(OAc)$_2$. These reagents are stored at –20°C, with buffer and Mg(OAc)$_2$ vials in current use kept at 4°C.
2. Bovine serum albumin (BSA). Because of the long run time of our PCR programs, we add BSA (100 ng/µL final concentration) to the PCR mix to increase the stability of the polymerase *(6)*.
3. Deoxyribonucleoside triphosphates (dNTPs), purchased separately from Amersham Biosciences (cat. no. 27-2035-01). Prepare a solution of 10 mM total dNTPs (2.5 mM of each nucleotide) and store in 100-µL aliquots at –80°C to minimize degradation. The dNTPs are thawed immediately prior to use and are not reused.
4. Primer stocks are maintained at –20°C and aliquots of the working concentration (10 µM) are kept at 4°C. We routinely dilute the primers in sterile deionized water. We have purchased oligonucleotides from several vendors and find that the primers work well with no purification beyond standard desalting.

2.4. Dedicated Equipment and Workstations

In addition to high-quality reagents, the most important factor for the success of QPCR is the diligent avoidance of sample cross-contamination with PCR product. We use sterile technique for all steps. The constant use of disposable gloves, when handling samples and reagents, is essential to avoid the introduction of nucleases, foreign DNA, or other contaminants that can cause degradation of the template or inhibition of the polymerase during cycling. In addition, we have found that it is extremely important to have distinct, dedicated workstations for different steps of the procedure, preferably in physically separate laboratories (*see also* **Note 1**). We also suggest that micropipets, racks, tubes, tips, and other materials used for QPCR be exclusively used for the assay. We set up PCRs in a hood that is sterilized with UV light (that will also extensively damage any potentially contaminating product-carryover DNA) immediately before each use.

3. Methods

3.1. DNA Extraction

DNA template integrity is essential for the reliable amplification of long PCR targets *(53)*. Although various kits are commercially available for DNA isolations, procedures that involve phenol extraction should be avoided due to potential introduction of artifactual DNA oxidation. The DNA extraction kit from QIAGEN gives rise to templates of relatively high molecular weight and highly reproducible yield. The protocol for DNA isolation is followed as suggested by the manufacturer.

Note that the tissue protocol is used irrespective of whether tissue or cells are being studied, since the protocol for DNA extraction of cultured cells involves isolation of nuclei and hence loss of mtDNA.

Samples that cannot be processed immediately after experiments should be stored at –80°C until DNA is extracted. (*See* **Note 2**.)

3.2. Quantitation of DNA Template

The success of QPCR is absolutely dependent on the accurate quantitation of the DNA present in the samples *(6)*. We have adopted PicoGreen as means to quantify DNA. The DNA concentration of the samples is calculated based on a DNA standard curve, plotting the fluorescence values on a Microsoft Excel spreadsheet (*see* **Fig. 1**; template available on request).

We perform quantitation in a minimum of two different steps, called pre- and final quantitation. The first gives a rough estimate of the initial amount of DNA in each sample. At the end of this first step, the amount of DNA necessary to make a 10 ng/µL solution of DNA is calculated. The final quantitation uses this latter solution to calculate the exact amount of DNA needed to dilute samples to 3 ng/µL, which is the amount of template routinely used for QPCR in our laboratory. Our protocol for quantitation is as follows:

1. Dilute λ/*Hin*dIII DNA (in 1X TE buffer) yielding different concentrations to generate a standard curve (e.g., from 1.25 to 20 ng/µL of DNA).
2. Add 90 µL of 1X TE buffer to each well that will be used (for standards and samples).
3. Add 10 µL of each λ DNA standard per well (producing a curve of 0–200 ng DNA/well), at least in duplicate (*see* **Note 3**).
4. For prequantitation, pipet 2 µL of the sample DNA in duplicate.

Pre Quant standard	reading	read1	read 2	mean
200	8477	8945	8281	8613
100	4475.5	4572	4651	4611.5
50	2278	2411	2417	2414
25	1181	1267	1367	1317
12.5	578.5	684	745	714.5
0	0	138	134	136

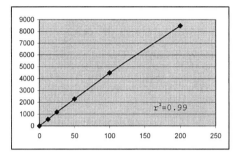

DNA samples	read 1	read 2	mean	net	conc (ng/ul)	DNA(ul) 10 ng/ul	TE(ul) 10 ng/ul
1	8255	9897	9076	8940	104.4	11.5	108.5
2	610	1514	1062	926	9.8	122.6	-2.6
3	3195	3046	3120.5	2984.5	34.1	35.2	84.8
4	8164	8725	8444.5	8308.5	97.0	12.4	107.6

Fig. 1. Spreadsheet used for DNA quantitation (top). Example depicts fluorescence values obtained during the first step of quantitation, prequantitation, and the graph shows values obtained for the standard curve. Above the graph are all calculations related to the DNA standard curve. First and second columns represent, respectively, the concentrations of DNA used as standards and the final fluorescence readings (where background fluorescence was subtracted) obtained for each concentration. Third and fourth columns show the raw fluorescence readings. These values were averaged (last column). Below the graph is an example of values obtained for an experimental set of DNA samples. First and second columns show raw fluorescence readings for each sample; third column is the mean of those values. "Net" represents the average of the readings with background fluorescence subtracted. DNA concentration is calculated based on the slope of the standard curve. The last two columns show, respectively, the amount of DNA and of TE buffer necessary to dilute the sample DNA to 10 ng/μL.

5. Prepare a solution containing the PicoGreen reagent (5 μL of reagent per milliliter of 1X TE). Mix this solution and add 100 μL into each well containing the DNA samples.
6. Incubate for 10 min at room temperature in the dark (the plate can be covered with foil paper).
7. Read the fluorescence. For the FL600 Microplate Fluorescence Reader, we use the following parameters: excitation and emission wavelengths 485 nm and 530

nm, respectively; sensitivity limit 75 and shaking of the plate set at level 3 for 20 s. (*See* **Note 4**.)

3.3. Quantitation of PCR Products

The PicoGreen reagent has proven efficient for quantitation not only of DNA template but also of PCR products. In fact, the accuracy of the data obtained with this assay is comparable to or can exceed the reproducibility that is accomplished with ^{32}P radiolabeled nucleotides (Chen, Y. and Van Houten, B., unpublished observation) followed by subsequent agarose gel electrophoresis. Analysis of PCR products is performed similarly to the DNA quantitation (*see* **Subheading 3.2.**), using 10 µL of the PCR products, and subtracting the fluorescence of a PCR run without template. For data analysis *see* **Subheading 3.4.4.**

IMPORTANT: When first developing the assay in your laboratory it is essential to assess the PCR products by agarose gel electrophoresis to verify the size of the product, and to ensure that no other spurious products are generated.

3.4. QPCR

3.4.1. Primer Selection

Appropriate primer selection is highly important, and is empirically based. In general, the oligonucleotides should be 20–24 bases in length with a G + C content of approx 50% and a T_m of approx 68°C. The selected primers should be evaluated for secondary structures using appropriate software since the formation of artifacts such as primer–dimers can compete with the QPCR *(6)*. In addition, the production of one unique band should be verified by gel electrophoresis prior to further use. **Table 1** shows the sequences of the oligonucleotides currently in use in our laboratory to amplify human, mouse and rat target genes. (*See also* **Note 5**.)

3.4.2. PCR

Once the primers are selected, finding the optimal reaction conditions is the next step. Different target genes and different primers usually require distinct conditions. Our laboratory has established optimal concentrations of reagents to amplify specific genes of our interest.

1. Using the GeneAmp XL PCR kit, the PCRs are prepared as follows:
    a. 15 ng of DNA (total).
    b. 1X buffer.
    c. 100 ng/µL final concentration of BSA.
    d. 200 µM final concentration of dNTPs (*see* **Note 6**).
    e. 20 pmol of each primer.
    f. 1.3 mM final concentration of Mg^{++}.
    g. Water to a total volume of 45 µL.

Table 1
Gene Targets and Primer Pairs for QPCR

Human primers

17.7-kb fragment from the 5′ flaking region near the β-globin gene, accession number, J00179

44329	5′-TTG AGA CGC ATG AGA CGT GCA G-3′	Sense
62007	5′-GCA CTG GCT TAG GAG TTG GAC T-3′	Antisense
48510	5′-CGA GTA AGA GAC CAT TGT GGC AG-3′	Sense

Note: Using 48510 and 62007 we get a robust 13.5-kb fragment which is of adequate length to give suficent sensitivity.

16.2-kb mitochondria fragment, accession number, J01415

15149	5′-TGA GGC CAA ATA TCA TTC TGA GGG GC-3′	Sense
14841	5′-TTT CAT CAT GCG GAG ATG TTG GAT GG-3′	Antisense
14620	5′-CCC CAC AAA CCC CAT TAC TAA ACC CA-3′	Sense
5999	5′-TCT AAG CCT CCT TAT TCG AGC CGA-3′	Sense

Note: Using primers 14841 and 5999 we get a fragment of 8.9-kb

10.4-kb fragment encompassing exons 2–5 of the hprt gene, accession number, J00205

14577	5′-TGG GAT TAC ACG TGT GAA CCA ACC-3′	Sense
24997	5′-GCT CTA CCC TCT CCT CTA CCG TCC-3′	Antisense

12.2-kb region of the DNA polymerase gene β, accession number, L11607

2372	5′-CAT GTC ACC ACT GGA CTC TGA AC-3′	Sense
3927	5′-CCT GGA GTA GGA ACA AA ATT GCT-3′	Antisense

Mouse primers

8.7-kb fragment from the β-globin gene, accession number, X14061

21582	5′-TTG AGA CTG TGA TTG GCA ATG CCT-3′	Sense
30345	5′-CCT TTA ATG CCC ATC CCG GAC T-3′	Antisense

6.5-kb region of the DNA polymerase gene β, accession number, AA79582

MBFor1	5′-TAT CTC TCT TCC TCT TCA CTT CTC CCC TGG-3′	Sense
MBEX1B	5′-CGT GAT GCC GCC GTT GAG GGT CTC CTG-3′	Antisense

10-kb mitochondria fragment

2372	5′-GCC AGC CTG ACC CAT AGC CAT ATT AT-3′	Sense
13337	5′-GAG AGA TTT TAT GGG TGT ATT GCG G-3′	Antisense

117-bp mitochondria fragment

13597	5′-CCC AGC TAC TAC CAT CAT TCA AGT-3′	Sense
13688	5′-GAT GGT TTG GGA GAT TGG TTG ATG-3′	Antisense

Rat primers

12.5-kb fragment from the clusterin (TRPM-2) gene, accession number, M64733

5781	5′-AGA CGG GTG AGA CAG CTG CAC CTT TTC-3′	Sense
18314	5′-CGA GAG CAT CAA GTG CAG GCA TTA GAG-3′	Antisense

3.4-kb mitochondria fragment

13559	5′-AAA ATC CCC GCA AAC AAT GAC CAC CC-3′	Sense
10633	5′-GGC AAT TAA GAG TGG GAT GGA GCC AA-3′	Antisense

235-bp mitochondria fragment

14678	5′-CCT CCC ATT CAT TAT CGC CGC CCT TGC-3′	Sense
14885	5′-GTC TGG GTC TCC TAG TAG GTC TGG GAA-3′	Antisense

2. Begin the PCR by a "hot start." Bring the reaction mixture to 75°C prior to addition of enzyme (1 U/reaction; dilute 0.5 µL of the polymerase in 4.5 µL of sterile water; *see* **Note 7**) and subsequent cycling.

Primers and magnesium concentrations may need to be optimized for different genes (*see* **Note 8**). In addition, add the reaction components to the PCR tube in a consistent order. The DNA template should be added first, followed by the PCR mix and finally the enzyme (as a hot start). In our laboratory reactions are set up at room temperature. Include a control that contains no genomic DNA—any signal produced in this sample would be indicative of a carryover problem. This is a serious problem and can only be cured by strict adherence to the conditions described above and starting with all new reagents.

3.4.3. Cycle Number and Thermal Parameters

The usefulness of the QPCR assay for the detection of DNA damage requires that amplification yields be directly proportional to the starting amount of template. These conditions must be met by keeping the PCR in the exponential phase. The first step towards this criterion is to perform cycle tests to determine quantitative conditions for the gene of interest *(53)*. This can be accomplished using a nondamaged sample and a "50% control" containing half of the amount of the nondamaged template (1.5 ng/µL of DNA). This control should give a 50% reduction of the amplification signal (*see* **Note 9**). Thus, a cycle test should identify a range of cycles over which the product amplification is exponential and 50% controls are very close to 50%. Once the optimal number of cycles is identified, always run this 50% control as a quality control.

Another concern when performing QPCR is finding the optimal thermal conditions for amplification of your target gene. As mentioned before, QPCR in our laboratory is routinely performed using hot start, which produces cleaner PCR products because it prevents nonspecific annealing of primers to each other, as well as to template, before enzyme addition. Keep in mind that the melting temperature of the primers and the annealing temperature used in the PCR determine how stably and specifically the primers hybridize to the DNA template. Thus, it is important to check this parameter with suitable software beforehand, and annealing temperatures must be experimentally optimized. **Table 2** shows the most favorable conditions for human and rodent amplifications currently used in our laboratory.

3.4.4. Data Analysis

Analysis of data obtained by the PicoGreen protocol described above is done using a Microsoft Excel spreadsheet (*see* example in **Fig. 2**). The fluorescence readings (of the duplicate samples) are averaged and the blank value (from no-DNA control) is subtracted. These values are used to calculate the "relative amplification," which refers to the comparison between amplification

Table 2
PCR Conditions for Human and Rodent Targets

Target		Primer set	Mg^{++} conc. (mM)	T_m (°C)	Cycle number
Human	Large mito	15149/14841	1.3	64	26
	Large mito	5999/14841	1.3	64	19
	Small mito	14620/14841	1.3	60	19
	hprt	14577/24997	1.3	64	29
	β-Globin	48510/62007	1.3	64	27
	β-Pol	3927/2372	1.3	64	26
Mouse	Large mito	3278/13337	1.0	65	20
	Small mito	13688/13597	1.1	60	18
	β-Globin	21582/30345	1.1	65	25
	β-Pol	MBFor1/MBEX1B	1.1	65	25
Rat	Large mito	10633/13559	1.1	65	18
	Small mito	14678/14885	1.1	60	18
	Clusterin	5781/18314	1.1	65	28

of treated-samples with nontreated (or undamaged) control. This is accomplished simply by dividing the respective fluorescence values. These results are then used to determine the lesion frequency per fragment at a particular dose, such that lesions/strand (average for both strands) at dose, $D = -\ln A_D/A_C$. This equation is based on the "zero class" of a Poisson expression. Note that a Poisson distribution requires an assumption that DNA lesions are randomly distributed (for details *see* **ref. 6**).

3.4.5. Normalization to mtDNA Copy Number

It is known that DNA content can vary in mitochondria from different cells or tissues, depending, for instance, on energy requirements. Thus, in samples from distinct areas of a specific organ, one could expect discrepancies in the ability to amplify the mtDNA based not on different levels of lesions within the sample, but simply from fluctuation in the number of copies of the mitochondrial genome present. Therefore, to normalize for mitochondrial copy number, we routinely amplify an additional short fragment (no longer than 300 base pairs) of the mitochondrial gene under study. The idea is that the amplification of the short fragment reflects only undamaged DNA due to the low probability of introducing lesions in small segments. The results obtained with the short sequence are used to monitor the copy number of the mitochondrial genome and, more importantly, to normalize the data obtained with the large (7 to 15-kb) fragment.

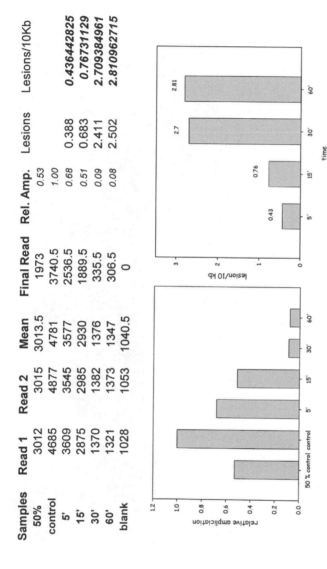

Samples	Read 1	Read 2	Mean	Final Read	Rel. Amp.	Lesions	Lesions/10Kb
50%	3012	3015	3013.5	1973	0.53		
control	4685	4877	4781	3740.5	1.00		
5'	3609	3545	3577	2536.5	0.68	0.388	0.436442825
15'	2875	2985	2930	1889.5	0.51	0.683	0.76731129
30'	1370	1382	1376	335.5	0.09	2.411	2.709384961
60'	1321	1373	1347	306.5	0.08	2.502	2.810962715
blank	1028	1053	1040.5	0			

Fig. 2. Representation of the raw fluorescence values obtained after PCR amplification of the mitochondrial genome of mammalian fibroblasts exposed to 200 μM hydrogen peroxide for the indicated times. Column one, sample identification; columns two and three, raw fluorescence readings for each sample; column four, average of values from first two columns; these values are then background corrected (column 5). Relative amplification (column 6) is calculated comparing the values of the treated samples with undamaged control and is plotted in the left graph. Lesion frequency (column 7) is obtained based in the values plotted on column 6 and are expressed as lesions per 10 kb of the mitochondrial genome (column 8 and right graph).

4. Notes

1. It is extremely important not to open the PCR tubes after the last cycle in the same laboratory where the reactions were setup. Small DNA quantities can volatilize and contaminate other reactions, particularly if the tube is still hot, and completed reactions contain very high quantities of PCR products. The inclusion of a blank sample (where no DNA is added) helps to ensure that no contamination has occurred with spurious DNA or PCR products. This sample should give no DNA band, if checked on a gel, and minimal fluorescence (as gauged by PicoGreen staining).
2. When extracting DNA, vortex-mix the samples well prior to lysis and again before adding them to the columns. This vortex-mixing does not affect the subsequent amplification of the DNA.
3. Load samples as well as standard DNA in duplicate and average the fluorescence reading of the two wells. This helps increase the accuracy of the readings and, thus, of the estimated DNA concentration.
4. Make sure the samples are well homogenized (by vortex-mixing, for example) prior to quantitation. If samples are still highly concentrated after the first dilution (i.e., well above 10 ng/μL), we recommend an additional round of quantitation. This ensures accuracy of the concentration of the final 3 ng/μL solution.
5. Primers: since the same batch of primers when used over long period of time (several months) can give rise to lower amplification, it is advisable to make new dilutions from time to time. Always protect primer stocks from unnecessary temperature fluctuation and contamination. If frozen, primer stocks should be completely thawed prior to use.
6. dNTPs: higher misincorporation frequency for the enzyme and reduction in effective magnesium concentration can occur if dNTPs exceed 200 μM.
7. *rTth* Polymerase: increasing amounts of the thermostable polymerase beyond 2.5 U per reaction can increase the production of nonspecific amplification products.
8. The optimal concentration of magnesium must be determined for each set of primers and template. The *rTth* polymerase is extremely sensitive to magnesium; we advise that amplification of the fragment of interest be evaluated using varying quantities of Mg^{++}, starting from 0.9 mM and increasing by 0.1 mM increments.
9. During each set of amplifications we routinely amplify a control sample in which only 50% of the template is added to the QPCR. Depending on the DNA quality and the products being amplified, relative amplification ranging from 40 to 60% is considered acceptable. Any experiments that are outside this range are not satisfactory and the entire set of reactions is discarded. It may be necessary to reoptimize the PCR by varying the number of cycles to establish a linear response to increasing template concentrations from 1.25 ng to 30 ng.

References

1. Ponti, M., Forrow, S. M., Souhami, R. L., D'Incalci, M., and Hartley, J. A. (1991) Measurement of the sequence specificity of covalent DNA modification by antineoplastic agents using *Taq* DNA polymerase. *Nucleic Acids Res.* **19,** 2929–2933.

2. Jennerwein, M. M. and Eastman, A. (1991) A polymerase chain reaction-based method to detect cisplatin adducts in specific genes. *Nucleic Acids Res.* **19,** 6209–6214.
3. Kalinowski, D., Illenye, S., and Van Houten, B. (1992) Analysis of DNA damage and repair in murine leukemia L1210 cells using a quantitative polymerase chain reaction assay. *Nucleic Acids Res.* **20,** 3485–3494.
4. Van Houten, B., Cheng, S., and Chen, Y. (2000) Measuring DNA damage and repair in human genes using quantitative amplification of long targets from nanogram quantities of DNA. *Mutat. Res.* **460,** 81–94.
5. Van Houten, B., Chen, Y., Nicklas, J.A., Rainville, I.R., and O'Neill, J.P. (1998) Development of long PCR techniques to analyze deletion mutations of the human hprt gene. Mutat. Res. **403,** 171–175.
6. Ayala-Torres, S., Chen, Y., Svoboda, T., Rosenblatt, J., and Van Houten, B. (2000) Analysis of gene-specific DNA damage and repair using quantitative PCR. *Methods* **22,** 135–147.
7. Yakes, F. M. and Van Houten, B. (1997) Mitochondrial DNA damage is more extensive and persists longer than nuclear DNA damage in human cells following oxidative stress. *Proc. Natl. Acad. Sci. USA* **94,** 514–519.
8. Mandavilli, B. S., Ali, S. F., and Van Houten, B. (2000) DNA damage in brain mitochondria caused by aging and MPTP treatment. *Brain Res.* **885,** 45–52.
9. Moon S. K., Thompson L. J., Madamanchi, N., et al. (2001) Aging, oxidative stress, and proliferative capacity in cultured mouse aortic smooth muscle cells. *Am. J. Physiol. Heart Circ. Physiol.* **280,** 2779–2788.
10. Denissenko, M. F., Cahill, J., Koudriakova, T. B., Gerber, N., and Pfeifer, G. P. (1999) Quantitation and mapping of aflatoxin B1-induced DNA damage in genomic DNA using aflatoxin B1-8,9-epoxide and microsomal activation systems. *Mutat. Res.* **425,** 205–211.
11. Ballinger, S. W., Patterson, C., Knight-Lozano, C. A., et al. (2002) Mitochondrial integrity and function in atherogenesis. *Circulation* **106,** 544–549.
12. Jin, G. F., Hurst, J. S., and Godley, B. F. (2001) Rod outer segments mediate mitochondrial DNA damage and apoptosis in human retinal pigment epithelium. *Curr. Eye Res.* **23,** 11–19.
13. Sawyer, D. E., Mercer, B. G., Wiklendt, A. M., and Aitken, R. J. (2003) Quantitative analysis of gene-specific DNA damage in human spermatozoa. *Mutat. Res.* **529,** 21–34.
14. Santos, J. H., Hunakova, L., Chen, Y., Bortner, C., and Van Houten, B. (2003) Cell sorting experiments link persistent mitochondrial DNA damage with loss of mitochondrial membrane potential and apoptotic cell death. *J. Biol. Chem.* **278,** 1728–1734.
15. Yanez, J. A., Teng, X. W., Roupe, K. A., Fariss, M. W., and Davies, N. M. (2003) Chemotherapy induced gastrointestinal toxicity in rats: involvement of mitochondrial DNA, gastrointestinal permeability and cyclooxygenase-2. *J. Pharmacol. Pharmaceut. Sci.* **6,** 308–314.

16. O'Brien, T., Xu, J., and Patierno S. R. (2001) Effects of glutathione on chromium-induced crosslinking and DNA polymerase arrest. *Mol. Cell Biochem.* **222**, 173–182.
17. Chandrasekhar, D. and Van Houten, B. (1994) High resolution mapping of UV-induced photoproducts in the *E. coli lacI* gene: inefficient repair in the nontranscribed strand correlates with high mutation frequency. *J. Mol. Biol.* **238**, 319–332.
18. Yakes, F. M., Chen, Y., and Van Houten, B. (1996) PCR-based assays for the detection and quantitation of DNA damage and repair, in *Technologies for Detection of DNA Damage and Mutations*, (Pfeifer, G. P., ed.). Plenum Press, New York, NY, pp. 171–184.
19. Salazar, J. J. and Van Houten, B. (1997) Preferential mitochondrial DNA injury caused by glucose oxidase as a steady generator for hydrogen peroxide in human fibroblasts. *Mutat Res.* **385**, 139–149.
20. Chen, K. H., Srivastava, D. K., Yakes, F. M., et al. (1998) Up-regulation of base excision repair correlates with enhanced protection against a DNA damaging agent in mouse cell lines. *Nucleic Acids Res.* **26**, 2001–2007.
21. Horton, J. K., Roy, G., Piper, J. T., et al. (1999) Characterization of a chlorambucil-resistant human ovarian carcinoma cell line overexpressing glutathione s-transferase μ. *Biochem. Pharmacol.* **58**, 693–702.
22. Deng, G., Su, J. H., Ivins, K. J., Van Houten, B., and Cottman, C. (1999) Bcl-2 facilitates recovery from DNA damage after oxidative stress. *Experimental Neurol.* **159**, 309–318.
23. Ballinger, S. W., Patterson, C., Yan, C. N., et al. (2000) Hydrogen peroxide- and peroxynitrite-induced mitochondrial DNA damage and dysfunction in vascular endothelial and smooth muscle cells. *Circ. Res.* **786**, 960–966.
24. Chandrasekhar, D. and Van Houten, B. (2000) In vivo formation and repair of cyclobutane pyrimidine dimers and 6-4 photoproducts measured at the gene and nucleotide level in *E. coli*. *Mutat. Res.* **450**, 19–40.
25. Sobol, R. W., Watson, D. E., Nakamura, J., et al. (2002) Mutator phenotype associated with a gene-environment interaction: effect of base excision repair deficiency and methylation-induced genotoxic stress. *Proc. Natl. Acad. Sci. USA* **99**, 6860–6865.
26. Wallace, D. C. (1999) Mitochondrial diseases in man and mouse. *Science* **283**, 1482–1488.
27. DiMauro, S. and Schon, E. A. (2001) Mitochondrial DNA mutations in human disease. *Am. J. Med. Genet.* **106**, 18–26.
28. Wallace, D. C., Shoffner, J. M., Trounce, I., et al. (1995) Mitochondrial DNA mutations in human degenerative diseases and aging. *Biochim. Biophys. Acta* **1271**, 141–151.
29. Bowling, A. C. and Beal, M. F. (1995) Bioenergetic and oxidative stress in neurodegenerative diseases. *Life Sci.* **56**, 1151–1171.
30. Schapira, A. H. V. (1998) Mitochondrial dysfunction in neurodegenerative disorders. *Biochim. Biophys. Acta* **1366**, 225–233.

31. Wallace, D. C. (1994) Mitochondrial DNA mutations in diseases of energy metabolism. *J. Bioenerg. Biomembr.* **26**, 241–250.
32. Penta, J. S., Johnson, F. H., Wachsman, J. T., and Copeland, W. C. (2001) Mitochondrial DNA in human malignancy. *Mutat. Res.* **488**, 119–133.
33. Hudson, E. X., Hogue, B. A., Souza-Pinto, N. C., et al. (1998) Age-associated change in mitochondrial DNA damage. *Free Radic. Res.* **29**, 573–579.
34. Cadenas E. and Davies, K. J. (2000) Mitochondrial free radical generation, oxidative stress, and aging. *Free Radic. Biol. Med.* **29**, 222–230.
35. Mandavilli, B. S., Santos, J. H., and Van Houten, B. (2002) Mitochondrial DNA repair and aging. *Mutat. Res.* **509**, 127–151.
36. Boveris, A. and Cadenas, E. (1982) Superoxide and hydrogen peroxide in mitochondria, in *Free Radicals in Biology*, (Pryor, W. A., ed.). Academic Press, San Diego, CA, pp. 65–90.
37. Turrens, J. F. and Boveris, A. (1980) Generation of superoxide anion by the NADH dehydrogenase of bovine heart mitochondria. *Biochem. J.* **191**, 421–427.
38. Beckman, K. B. and Ames, B. N. (1999) Endogenous oxidative damage of mtDNA. *Mutat. Res.* **424**, 51–58.
39. Kowaltowski, A. J. and Vercesi, A. E. (1999) Mitochondrial damage induced by conditions of oxidative stress. *Free Radic. Biol. Med.* **26**, 463–471.
40. Sawyer, D. E. and Van Houten, B. (1999) Repair of DNA damage in mitochondria. *Mutat. Res.* **434**, 161–176.
41. Massa, E. M. and Giulivi, C. (1993) Alkoxyl and methyl radical formation during cleavage of tert-butyl hydroperoxide by a mitochondrial membrane-bound, redox active copper pool: an EPR study. *Free Radic. Biol. Med.* **14**, 559–565.
42. Walter, P. B., Beckman, K. B., and Ames, B.N. (1999) The role of iron and mitochondria in aging, in *Understanding the Process of Aging: The Roles of Mitochondria, Free Radicals, and Antioxidants* (Cadenas, E., and Packers, L., eds.). Marcel Dekker, New York, NY, pp. 203–227.
43. Croteau, D. L., Stierum, R. H., and Bohr, V. A. (1999) Mitochondrial DNA repair pathways. *Mutat. Res.* **434**, 137–148.
44. Bohr, V. A. (2002) Repair of oxidative DNA damage in nuclear and mitochondrial DNA, and some changes with aging in mammalian cells. *Free Radic. Biol. Med.* **32**, 804–812.
45. Zastawny, T. H., Dabrowska, M., Jaskolski, T., et al. (1998) Comparison of oxidative base damage in mitochondrial and nuclear DNA. *Free Radic. Biol. Med.* **24**, 722–725.
46, Richter, C., Park, J. W., and Ames, B. N. (1998) Normal oxidative damage to mitochondrial and nuclear DNA is extensive. *Proc. Natl. Acad. Sci. USA* **85**, 6465–6467.
47. Mecocci, P., MacGarvey, U., Kaufman, A.E., et al. (1993) Oxidative damage to mitochondrial DNA shows marked age-dependent increases in human brain. *Ann. Neurol.* **34**, 609–616.
48. Mecocci, P., MacGarvey, U., and Beal, M. F. (1994) Oxidative damage to mitochondrial DNA is increased in Alzheimer's disease. *Ann. Neurol.* **36**, 747–751.

49. Helbock, H. J., Beckman, K. B., Shigenaga, M. K., et al. (1998) DNA oxidation matters: the HPLC-electrochemical detection assay of 8-oxo-deoxyguanosine and 8-oxoguanine. *Proc. Natl. Acad. Sci. USA* **95,** 288–293.
50. Anson, R. M., Hudson, E., and Bohr, V. A. (2000) Mitochondrial endogenous oxidative damage has been overestimated. *FASEB J.* **14,** 355–360.
51. Termini, J. (2000) Hydroperoxide-induced DNA damage and mutations. *Mutat. Res.* **450,** 107–124.
52. Quan, T. and States, J. C. (1996) Preferential DNA damage in the p53 gene by benzo[a]pyrene metabolites in cytochrome P4501A1-expressing xeroderma pigmentosum group A cells. *Mol. Carcinogen.* **16,** 32–43.
53. Cheng, S., Chen, Y., Monforte, J. A., Higuchi, R., and Van Houten, B. (1995) Template integrity is essential for PCR amplification of 20- to 30-kb sequences from genomic DNA. *PCR Methods Appl.* **4,** 294–298.

15

Measuring the Formation and Repair of DNA Damage by Ligation-Mediated PCR

Gerd P. Pfeifer

Summary

There is a need to analyze the formation of DNA lesions in specific sequence contexts. The formation and repair of DNA damage at specific locations in the genome is modulated by the DNA sequence, by DNA methylation patterns, by the transcriptional status of the locus, and by chromatin proteins associated with the DNA. The only method currently available to allow a precise sequence mapping of DNA lesions in mammalian cells is the ligation-mediated polymerase chain reaction (LM-PCR). I describe the technical details of LM-PCR as exemplified by the mapping of DNA damage products in ultraviolet (UV) light-irradiated cells.

Key Words: DNA adducts; DNA repair; ligation-mediated PCR; (6–4) photoproduct; pyrimidine dimer; ultraviolet (UV) light.

1. Introduction

Several types of DNA lesions are formed on irradiation of cells with ultraviolet (UV) light *(1,2)*. The two most frequent ones are the cyclobutane pyrimidine dimers (CPDs) and the pyrimidine (6–4) pyrimidone photoproducts [(6–4) photoproducts; (6–4) PPs]. CPDs are formed between the 5,6 bonds of two adjacent pyrimidines. The (6–4) PPs are charcterized by covalent bonds between positions 6 and 4 of two adjacent pyrimidines and arise through a rearrangement mechanism. CPDs are several times more frequent than (6–4) PPs *(3)*. Both photoproducts are mutagenic, but it is believed that the CPD is the more mutagenic lesion in mammalian cells *(4)*. CPDs persist much longer in mammalian DNA than (6–4) PPs owing to a significantly faster repair of (6–4)PPs *(5,6)*. Perhaps because of the inefficient recognition of CPDs by the general nucleotide excision repair pathway, cells have developed other means to cope with this lesion. CPDs are subject to a specialized transcription-coupled repair pathway *(7,8)*, which removes these lesions selectively from the template strand of genes transcribed by RNA polymerase II.

Nucleotide excision repair plays an important role in preventing UV-induced mutagenesis and carcinogenesis. Several human genetic disorders are characterized by a defect in DNA repair. Cells from patients suffering from xeroderma pigmentosum (XP) are hypersensitive to UV light *(9)*. XP is a genetic disease characterized by seven different functional complementation groups. The incidence of skin cancer in certain XP patients is increased by several thousand-fold relative to the normal population *(10)* and this probably is a consequence of a severe deficiency in repair of UV photolesions.

In our previous work, we developed a technique, based on the ligation-mediated polymerase chain reaction (LM-PCR), which can be used to analyze the distribution and repair of UV photoproducts along specific human genes at the DNA sequence level *(11–22)*. LM-PCR methods for the detection of (6–4) photoproducts *(13)* and cyclobutane pyrimidine dimers *(14)* are available. LM-PCR provides a sufficient level of sensitivity even when rather low UV doses (10–20 J/m^2 of UV-C) are used for irradiation, and the repair of CPDs can be measured reliably at these doses *(11,12,15,18–22)*. Since (6–4) photoproducts are much less frequent than CPDs and the detection method produces a higher background in nonirradiated DNA, the repair of this lesion has not yet been analyzed successfully by LM-PCR.

The ability of LM-PCR to detect DNA adducts depends on the specific conversion of the adducts into strand breaks with a 5′-phosphate group. The (6–4) photoproducts and their Dewar isomers can be converted by heating UV-irradiated DNA in piperidine *(23)*. CPDs can be mapped at the DNA sequence level by cleavage with specific enzymes such as T4 endonuclease V *(24,25)*. T4 endonuclease V cleaves the glycosidic bond of the 5′ base in a pyrimidine dimer and also cleaves the sugar phosphate backbone between the two dimerized pyrimidines. The digestion products still contain a dimerized pyrimidine base at the cleavage site. We found that these fragments could be amplified efficiently by LM-PCR only after photoreversal of the cyclobutane ring with *E. coli* photolyase to result in a normal base on a 5′ terminal sugar-phosphate *(14)*.

The LM-PCR technique is based on the ligation of an oligonucleotide linker onto the 5′ end of each DNA molecule that was created by the strand cleavage reactions. This ligation provides a common sequence on all 5′ ends allowing exponential PCR to be used for signal amplification. Thus by taking advantage of the specificity and sensitivity of PCR, one needs only a microgram of mammalian DNA per lane to obtain good quality DNA sequence ladders. The general LM-PCR procedure is outlined in **Fig. 1**. The first step of the procedure is

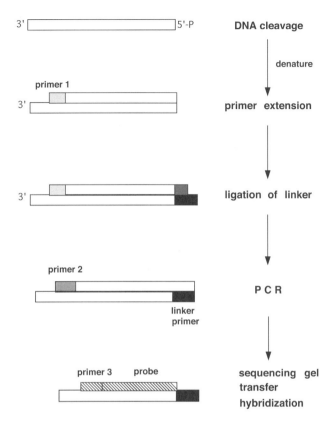

Fig. 1. Outline of the ligation-mediated PCR procedure. The steps include cleavage and denaturation of genomic DNA, annealing and extension of primer 1, ligation of the linker, PCR amplification of gene-specific fragments with primer 2 and the linker-primer, and detection of the sequence ladder by gel electrophoresis, electroblotting and hybridization with a single-stranded probe made with primer 3.

cleavage of DNA, generating molecules with a 5′-phosphate group by converting UV photolesions into strand breaks. Then, primer extension of a gene-specific oligonucleotide (primer 1) generates molecules that have a blunt end on one side. Linkers are ligated to these blunt ends, and then an exponential PCR amplification of the linker-ligated fragments is done using the longer oligonucleotide of the linker (linker-primer) and a second gene-specific primer (primer 2). After 18–20 PCR amplification cycles, the DNA fragments are separated on sequencing gels, electroblotted onto nylon membranes and hybridized with a gene-specific probe to visualize the sequence ladders. The arrangement of primers in a typical LM-PCR primer set is illustrated in **Fig. 2**.

Fig. 2. Arrangement of primers in an LM-PCR primer set to analyze UV photoproducts on a specific DNA strand (shown as a single strand). Primer 1 is used for linear primer extension before ligation, primer 2 is used for PCR, and primer 3 is used to make a single-stranded hybridization probe from a cloned template.

LM-PCR can be used for sequence mapping of DNA adducts whenever it is possible to convert the adducted nucleotides into strand breaks with remaining 5′ phosphate groups. Many DNA glycosylases produce abasic sites that can be treated with alkali to result in a strand break of this type. Other base excision repair enzymes that have β-δ-lyase activity, such as the *E. coli* Fpg or Nth proteins, produce appropriate strand breaks directly. DNA glycosylases have been used to map the formation and repair of oxidative DNA damage products and alkylated bases *(26–30)*. Bulky DNA adducts can be cleaved with the *E. coli* UvrABC complex, which makes incisions in most cases four nucleotides 3′ to the adduct. These 3′ incisions can be detected by LM-PCR and this approach has been used for mapping of adducts produced by polycyclic aromatic hydrocarbons or other chemicals *(31–35)*. In this chapter, I have focused on the mapping of UV damage and provide detailed protocols for analysis of UV photoproducts and their repair rates by ligation-mediated PCR. The protocols can easily be adapted for mapping of other types of DNA damage and its repair.

2. Materials

2.1. Irradiation of Cells

1. UV light source: Light sources emitting 254 nm light are available in most laboratories as germicidal lamps or UV crosslinking devices. UV-B irradiation can be performed with UV-B lamps such as a Philips TL 20W/12RS lamp.
2. UVX radiometer (Ultraviolet Products, San Gabriel, CA).

2.2. DNA Isolation

1. Buffer A: 0.3 M sucrose, 60 mM potassium chloride, 15 mM sodium chloride, 60 mM Tris-HCl, pH 8.0, 0.5 mM spermidine, 0.15 mM spermine, 2 mM EDTA.
2. Nonidet P-40 (NP-40).
3. Buffer B: 150 mM NaCl, 5 mM EDTA, pH 8.0.
4. Buffer C: 20 mM Tris-HCl, pH 8.0, 20 mM NaCl, 20 mM EDTA, 1% sodium dodecyl sulfate (SDS).
5. Proteinase K.

6. DNase-free RNase A.
7. Phenol, equilibrated with 0.1 M Tris-HCl, pH 8.
8. Chloroform.
9. Ethanol.
10. 3 M Sodium acetate, pH 5.2.
11. TE buffer: 10 mM Tris-HCl, pH 7.6, 1 mM EDTA.

2.3. Cleavage of DNA at Sites of UV Photodamage

1. Piperidine (Fluka), 1 M, freshly prepared.
2. 10X T4 endonuclease V buffer: 500 mM Tris-HCl, pH 7.6, 500 mM NaCl, 10 mM EDTA, 10 mM dithiothreitol (DTT), 1 mg/mL of bovine serum albumin (BSA).
3. T4 endonuclease V. This enzyme is commercially available from Epicentre Technologies (Madison, WI), or from Texagen (Plano, TX).
4. *E. coli* photolyase. This enzyme was kindly provided by Dr. A. Sancar (University of North Carolina at Chapel Hill). It is also available from Trevigen (Gaithersburg, MD).
5. Two 360-nm black lights (Sylvania 15W F15T8).
6. TE buffer: 10 mM Tris-HCl, pH 7.6, 1 mM EDTA.

2.4. Estimation of Cleavage Frequency by Alkaline Agarose Gels

1. Agarose.
2. 50 mM NaCl, 4 mM EDTA.
3. Running buffer: 30 mM NaOH, 2 mM EDTA.
4. Loading dye: 50% glycerol, 1 M NaOH, 0.05% bromocresol green.
5. 0.1 M Tris-HCl, pH 7.5.
6. Ethidium bromide (1 µg/mL).

2.5. Ligation-Mediated PCR

1. Oligonucleotide primers for primer extension. The primers used as primer 1 (Sequenase primers) are 15- to 20-mers with a calculated T_m of 48–56°C (*see* **Note 1**). Primers are prepared as stock solutions of 50 pmol/µL in water or TE buffer and are kept at –20°C.
2. 5X Sequenase buffer: 250 mM NaCl, 200 mM Tris-HCl, pH 7.7.
3. Mg-DTT-dNTP mix: 20 mM MgCl$_2$, 20 mM DTT, 0.25 mM of each dNTP.
4. Sequenase 2.0 (USB), 13 U/µL.
5. 300 mM Tris-HCl, pH 7.7.
6. 2 M Tris-HCl, pH 7.7.
7. Linker. The double-stranded linker is prepared in 250 mM Tris-HCl, pH 7.7, by annealing a 25-mer (5′-GCGGTGACCCGGGAGATCTGAATTC, 20 pmol/µL) to an 11-mer (5′-GAATTCAGATC, 20 pmol/µL) by heating to 95°C for 3 min and gradually cooling to 4°C over a time period of 3 h. Linkers can be stored at –20°C for at least 3 mo. They are thawed and kept on ice.
8. Ligation mix: 13.33 mM MgCl$_2$, 30 mM DTT, 1.66 mM ATP, 83 µg/mL of BSA, 3 U/reaction T4 DNA ligase (Promega), and 100 pmol of linker/reaction (= 5 µL of linker).

9. *E. coli* tRNA.
10. 2X *Taq* polymerase mix: 20 m*M* Tris-HCl, pH 8.9, 80 m*M* NaCl, 0.02% gelatin, 4 m*M* MgCl$_2$, and deoxyribonucleoside triphosphates (dNTPs) at 0.4 m*M* each.
11. Oligonucleotide primers for PCR. The primers used in the amplification step (primer 2) are 20- to 30-mers with a calculated T_m between 60 and 68°C (*see* **Note 2**). Ten picomoles of the gene-specific primer (primer 2) and 10 pmol of the 25-mer linker-primer (5'-GCGGTGACCCGGGAGATCTGAATTC) are used per reaction along with 3 U of *Taq* polymerase, and these components can be included in the 2X Taq polymerase mix.
12. *Taq* polymerase.
13. Mineral oil.
14. 400 m*M* EDTA, pH 7.7.

2.6. Sequencing Gel Analysis of Reaction Products

1. Formamide loading buffer: 94% formamide, 2 m*M* EDTA, pH 7.7, 0.05% xylene cyanol, 0.05% bromophenol blue.
2. 1 *M* TBE: 1 *M* Tris-base, 0.83 *M* boric acid, 10 m*M* EDTA, approx pH 8.3.
3. Whatman 3MM and Whatman 17 paper (Clifton, NJ).
4. Gene Screen nylon membranes (New England Nuclear).
5. Electroblotting apparatus (Owl Scientific) and high-amperage power supply.
6. An appropriate plasmid or PCR product containing the sequences of interest.
7. Oligonucleotide primer to make the hybridization probe. This primer is used together with the cloned template and *Taq* polymerase to make single-stranded hybridization probes (*see* **Note 3**).
8. [^{32}P]dCTP (3000 Ci/mmol).
9. 7.5 *M* Ammonium acetate.
10. Hybridization buffer: 0.25 *M* sodium phosphate, pH 7.2, 1 m*M* EDTA, 7% SDS, 1% BSA.
11. Washing buffer: 20 m*M* sodium phosphate, pH 7.2, 1 m*M* EDTA, 1% SDS.
12. Kodak XAR-5 film.

2.7. Data Analysis

Molecular Dynamics phosphorimager (Sunnyvale, CA) and ImageQuant™ software.

3. Methods

3.1. Irradiation of Cells

Approximately 1–5 × 10^6 cells are typically used for irradiation. Cells that grow as monolayers in Petri dishes, such as fibroblasts or keratinocytes, are irradiated with a germicidal (UV-C) lamp after removal of the medium and washing with phosphate-buffered saline (PBS). It is also possible to use a UV-B irradiation source (*see* **Note 4**). UV doses are measured with a UV radiometer. Typical UV doses for DNA repair assays of CPDs are 10–20 J/m^2 of 254 nm light (*see* **Note 5**).

3.2. DNA Isolation

1. Lyse the cells after UV irradiation by adding to the plate 10 mL of buffer A containing 0.5% NP-40. This step will release nuclei and removes most of the cytoplasmic RNA. Transfer the suspension to a 50-mL tube. Incubate on ice for 5 min.
2. Centrifuge at 1000g for 5 min at 4°C.
3. Wash the nuclear pellet once with 15 mL of buffer A.
4. Resuspend nuclei thoroughly in 2–5 mL of buffer B, add one volume of buffer C, containing 600 µg/mL of proteinase K (added just before use). Incubate for 1 h at 37°C.
5. Add DNase-free RNase A to a final concentration of 100 µg/mL. Incubate for 30 min at 37°C (*see* **Note 6**).
6. Extract with one volume of buffer-saturated phenol. Then, extract with 0.5 volumes of phenol and 0.5 volumes of chloroform. Repeat this step until the aqueous phase is clear and no interface remains. Finally, extract with 1 volume of chloroform.
7. Add 0.1 volumes of 3 M sodium acetate, pH 5.2, and precipitate the DNA with 2.5 volumes of ethanol at room temperature.
8. Centrifuge at 2000g for 1 min (*see* **Note 7**). Wash the pellet with 75% ethanol and air-dry briefly.
9. Dissolve the DNA in TE buffer to a concentration of approx 0.2 µg/µL. Keep at 4°C overnight. The DNA should be well dissolved before T4 endonuclease V cleavage.

3.3. Cleavage of DNA at Sites of UV Photodamage

3.3.1. (6–4) Photoproducts

To obtain DNA fragments with a 5′ phosphate group at the positions of (6–4) photoproducts, DNA is heated in 1 M piperidine. This will destruct the photolesion and create strand breaks with 5′ phosphate groups since the sugar residue at the 3′-base of the (6–4) photoproduct is destroyed by β-elimination.

1. Dissolve 10–50 µg of UV-irradiated DNA in 100 µL of 1 M piperidine.
2. Heat the DNA at 90°C for 30 min in a heat block (use lid locks to prevent tubes from popping). Cool samples briefly on ice after heating.
3. Add 10 µL of 3 M sodium acetate, pH 5.2, and 2.5 volumes of ethanol. Place on dry ice for 20 min.
4. Centrifuge at 14,000 rpm in an Eppendorf centrifuge for 15 min.
5. Wash twice with 1 mL of 75% ethanol.
6. Remove traces of remaining piperidine by drying the sample overnight in a vacuum concentrator. Dissolve DNA in TE buffer to a concentration of approx 0.5 to 1 µg/µL.
7. Determine the frequency of (6–4) photoproducts by separating 1 µg of the DNA on a 1.5% alkaline agarose gel along with appropriate size markers.

3.3.2. Cyclobutane Pyrimidine Dimers

DNA is first incubated with T4 endonuclease V and then with *E. coli* photolyase to create fragments with 5′-phosphate groups and ligatable ends.

1. The UV-irradiated DNA (about 10 μg in 50 μL) is mixed with 10 μL of 10X T4 endonuclease V buffer and a saturating amount of T4 endonuclease V in a final volume of 100 μL. Saturating amounts of T4 endonuclease V activity can be determined by incubating UV-irradiated (20 J/m^2) genomic DNA with various enzyme dilutions and separating the cleavage products on alkaline agarose gels (*see* **step 8** below). Incubate at 37°C for 1 h.
2. Add dithiothreitol to a final concentration of 10 m*M*. Add 5 μg of *E. coli* photolyase under yellow light.
3. Irradiate the samples in 1.5-mL tubes from two 360-nm black lights (Sylvania 15W F15T8) filtered through 0.5-cm thick window glass for 1 h at room temperature at a distance of 3 cm.
4. Extract once with phenol–chloroform.
5. Precipitate the DNA by adding one-tenth volume of 3 *M* sodium acetate, pH 5.2, and 2.5 volumes of ethanol. Leave on dry ice for 20 min. Centrifuge the samples for 10 min at 14,000*g* at 4°C.
6. Wash the pellets with 1 mL of 75% ethanol and air-dry.
7. Dissolve the DNA in TE buffer to a concentration of about 0.5–1 μg/μL.
8. Determine the frequency of CPDs by running 1 μg of the samples on a 1.5% alkaline agarose gel.

3.4. Estimation of Cleavage Frequency by Alkaline Agarose Gels

The approximate size of the fragments obtained after cleavage of UV-irradiated DNA is determined on an alkaline 1.5% agarose gel.

1. Prepare a 1.5% alkaline agarose gel by suspending agarose in 50 m*M* NaCl, 4 m*M* EDTA, and microwaving. Pour the gel.
2. After the gel solidifies, soak it in running buffer for at least 2 h.
3. Dilute the DNA sample with one volume of loading dye. Incubate for 15 min at room temperature. Load the samples.
4. Run the gel at 40 V for 3–4 h.
5. Neutralize the gel by soaking for 60 min in 500 mL of 0.1 *M* Tris-HCl, pH 7.5.
6. Stain with ethidium bromide (1 μg/mL) for 30 min.
7. Destain in water for 30 min.

3.5. Ligation-Mediated PCR

1. Mix in a siliconized 1.5-mL tube: 1–2 μg of cleaved DNA (*see* **Note 8**), 0.6 pmol of primer 1, and 3 μL of 5X Sequenase buffer in a final volume of 15 μL.
2. Incubate at 95°C for 3 min, then at 45°C for 30 min.
3. Cool on ice, centrifuge 5 s.
4. Add 7.5 μL of cold, freshly prepared Mg–DTT–dNTP mix.
5. Add 1.5 μL of Sequenase, diluted 1:4 in cold 10 m*M* Tris-HCl, pH 7.7.

6. Incubate at 48°C, 15 min, then cool on ice.
7. Add 6 µL 300 m*M* Tris-HCl, pH 7.7.
8. Incubate at 67°C, 15 min (heat inactivation of Sequenase).
9. Cool on ice, centrifuge 5 s.
10. Add 45 µL of freshly prepared ligation mix.
11. Incubate overnight at 18°C.
12. Incubate 10 min at 70°C (heat inactivation of ligase).
13. Add 8.4 µL of 3 *M* sodium acetate, pH 5.2, 10 µg of *E. coli* tRNA, and 220 µL of ethanol.
14. Place the samples on dry ice for 20 min.
15. Centrifuge for 15 min at 4°C in an Eppendorf centrifuge.
16. Wash the pellets with 950 µL of 75% ethanol.
17. Remove ethanol residues in a SpeedVac or by air-drying.
18. Dissolve the pellets in 50 µL of H_2O and transfer to 0.5-mL siliconized tubes.
19. Add 50 µL of freshly prepared 2X *Taq* polymerase mix containing the primers and the enzyme and mix by pipetting.
20. Cover the samples with 50 µL of mineral oil and centrifuge briefly.
21. Cycle 18–20 times at 95°C, 1 min, 60–66°C, 2 min, and 76°C, 3 min. The temperature during the annealing step is at the calculated T_m of the gene-specific primer.
22. To completely extend all DNA fragments and add an extra nucleotide through *Taq* polymerase's terminal transferase activity, an additional *Taq* polymerase step is performed (*see* **Note 9**). One unit of fresh *Taq* polymerase per sample is added together with 10 µL of reaction buffer. Incubate 10 min at 74°C.
23. Stop reaction by adding sodium acetate to 300 m*M*, EDTA to 10 m*M*, and add 10 µg of tRNA.
24. Extract with 70 µL of phenol and 120 µL of chloroform (premixed).
25. Add 2.5 volumes of ethanol and place on dry ice for 20 min.
26. Centrifuge samples for 15 min in an Eppendorf centrifuge at 4°C.
27. Wash the pellets in 1 mL of 75% ethanol.
28. Dry the pellets in a vacuum concentrator.

3.6. Sequencing Gel Analysis of Reaction Products

1. Dissolve the pellets in 1.5 µL of water and add 3 µL of formamide loading buffer.
2. Heat the samples to 95°C for 2 min prior to loading. Loading is performed with a very thin flat tip. Load only one half of the samples or less. The gel is 0.4 mm thick and 60 cm long, consisting of 8% polyacrylamide (ratio acrylamide to *bis*-acrylamide = 29:1) and 7 *M* urea in 0.1 *M* TBE. To allow identification of the sequence position of the UV-specific bands, include Maxam–Gilbert sequencing standards prepared from genomic DNA as previously described *(36)*.
3. Run the gel until the xylene cyanol marker reaches the bottom. Fragments below the xylene cyanol dye do not hybridize significantly.
4. After the run, transfer the gel (i.e., the bottom 40 cm of it) to Whatman 3MM paper and cover with Saran Wrap®.

5. Electroblotting of the gel piece can be performed with a transfer box available from Owl Scientific (*see* **Note 10**). Pile three layers of Whatman 17 paper, 43 × 19 cm, presoaked in 90 m*M* TBE, onto the lower electrode. Squeeze the paper with a roller to remove air bubbles between the paper layers. Place the gel piece covered with Saran Wrap onto the paper and remove the air bubbles between the gel and the paper by wiping over the Saran Wrap with a soft tissue. Remove the Saran wrap and cover the gel with a GeneScreen nylon membrane cut somewhat larger than the gel and presoaked in 90 m*M* TBE. Put three layers of presoaked Whatman 17 paper onto the nylon membrane carefully removing trapped air with a roller. Place the upper electrode onto the paper. Perform the electroblotting procedure at a current of 1.6 A. After 30 min, remove the nylon membrane and mark the DNA side. A high-amperage power supply is required for this transfer.
6. After electroblotting, dry the membrane briefly at room temperature. Then crosslink the DNA by UV irradiation. UV irradiation is performed in a commercially available UV crosslinker.
7. The hybridization is performed in rotating 250-mL plastic or glass cylinders in a hybridization oven. Soak the nylon membranes briefly in 90 m*M* TBE. Roll them into the cylinders by unspooling from a thick glass rod so that the membranes stick completely to the walls of the cylinders without air pockets. The prehybridization is done with 15 mL of hybridization buffer for 10 min. For hybridization, dilute the labeled probe into 7 mL of hybridization buffer. Prehybridization as well as hybridization are performed at 62°C.
8. To prepare labeled single-stranded probes, 200–300 nucleotides in length, use repeated primer extension by *Taq* polymerase with a single primer (primer 3) on a double-stranded template DNA *(17)*. This can be either plasmid DNA restriction-cut approx 200–300 nucleotides 3′ to the binding site of primer 3 or a PCR product containing the target area of interest. To prepare the single-stranded probe, mix 50 ng of the respective restriction-cut plasmid DNA (or 10 ng of the gel-purified PCR product) with primer 3 (20 pmol), 100 µCi of [^{32}P]dCTP, 10 µ*M* of the other three dNTPs, 10 m*M* Tris-HCl, pH 8.9, 40 m*M* NaCl, 0.01% gelatin, 2 m*M* MgCl$_2$, and 3 U of *Taq* polymerase in a volume of 100 µL. Perform 35 cycles at 95°C (1 min), 60–66°C (1 min), and 75°C (2 min). Recover the probe by phenol–chloroform extraction, addition of ammonium acetate to a concentration of 0.7 *M*, ethanol precipitation at room temperature, and centrifugation.
9. After hybridization, wash each nylon membrane with 2 L of washing buffer at 60°C. Perform several washing steps in a dish at room temperature with prewarmed buffer. After washing, dry the membranes briefly at room temperature, wrap in Saran Wrap, and expose to Kodak XAR-5 films or phosphoimager. If the procedure has been done without error, a result can be seen after 0.5–8 h of exposure with two intensifying screens at –80°C.

3.7. Data Analysis

Data analysis (*see* **Note 11**) is routinely performed by PhosphorImager analysis using a Molecular Dynamics scanner or equivalent equipment. For quantitation of repair rates, nylon membranes are exposed to the PhosphorImager and radioactivity is determined in all CPD-specific bands of the sequencing gel that show a consistent and measurable signal above background. Background values (from the control lanes without UV irradiation) are subtracted. A repair curve can be established for each CPD position that gives a sufficient signal above background. The time at which 50% of the initial damage is removed can then be determined from this curve (*see* **refs.** *11,19* for examples].

4. Notes

1. Calculation of the T_m is done with the Oligo™ computer program (Molecular Biology Insights Inc., Cascade, CO). If a specific target area is to be analyzed (e.g., a defined sequence position), primer 1 should be located approx 100 nucleotides upstream of this target.
2. Primer 2 is designed to extend 3' to primer 1. Primer 2 can overlap several bases with primer 1, but we have also had good results with a second primer that overlapped only one or two bases with the first.
3. The primer that is used to make the single-stranded probe (primer 3) should be on the same strand just 3' to the amplification primer (primer 2) and should have a T_m of 60–68°C (*see* **Fig. 2**). It should not overlap more than 8–10 bases with primer 2.
4. UV-B light emitted from sunlamps sufficiently penetrates the plastic material of Petri dishes and can be administered from the bottom of the dish without the need to remove the cell culture medium before irradiation. Polystyrene plastic dishes have a lower wavelength cutoff of 295–300 nm. When UV-B is used, a dose of about 500–1000 J/m² (measured with a 310 nm sensor) produces equivalent frequencies of CPD lesions as 10–20 J/m² of UV-C.
5. At these UV doses, the average photoproduct frequencies are one CPD every 5–10 kb and one (6–4) photoproduct every 20–40 kb.
6. Although RNase is probably active only for a very short time in the proteinase K solution, this step seems to aid in removal of traces of RNA.
7. If the initial number of cells was very low, the DNA may need to be pelleted at 10,000*g* for 10 min.
8. DNA concentration measurement before LM-PCR is critical for DNA repair assays. To avoid wide variations in DNA concentration, the same number of cells is used as starting material for irradiation at each time point, and care is taken that no material is lost during the DNA isolation procedure. DNA concentrations are measured by A_{260} optical density reading. It is important that the DNA is completely in solution before these measurements are made.
9. If this step is omitted, double bands may occur.
10. The advantages of the hybridization approach over the end-labeling technique *(37)* have been discussed previously *(36)*.
11. LM-PCR can be used for quantitation of UV damage frequencies at different nucleotide positions. There are some limitations, however, owing to variations in

ligation and PCR amplification of the different fragments. Fortunately, these variations are minor at sequences containing stretches of pyrimidines, which are the primary targets for formation of UV photoproducts. DNA repair experiments measure relative lesion frequencies over a time course at a defined sequence position. Therefore, amplification bias does not play a role in measuring repair rates for individual positions.

Acknowledgments

We thank A. Sancar for kindly providing *E. coli* photolyase. This work has been supported by a grant from the National Institute of Environmental Health Sciences (ES06070) to G.P.P.

References

1. Tornaletti, S. and Pfeifer, G. P. (1996) UV damage and repair mechanisms in mammalian cells. *BioEssays* **18**, 221–228.
2. Pfeifer, G. P. (1997) Formation and processing of UV photoproducts: effects of DNA sequence and chromatin environment. *Photochem. Photobiol.* **65**, 270–283.
3. Yoon, J.-H., Lee, C.-S., O'Connor, T., Yasui, A., and Pfeifer, G. P. (2000) The DNA damage spectrum produced by simulated sunlight. *J. Mol. Biol.* **299**, 681–693. [Erratum: Yoon et al. (2000) *J. Mol. Biol.* **302**, 1019–1020.]
4. You, Y. H., Lee, D. H., Yoon, J. H., Nakajima, S., Yasui, A., and Pfeifer, G. P. (2001) Cyclobutane pyrimidine dimers are responsible for the vast majority of mutations induced by UVB irradiation in mammalian cells. *J. Biol. Chem.* **276**, 44,688–44,694.
5. Mitchell, D. L. and Nairn, R. S. (1989) The biology of the (6–4) photoproduct. *Photochem. Photobiol.* **49**, 805–819.
6. Mitchell, D. L., Brash, D. E., and Nairn, R. S. (1990) Rapid repair kinetics of pyrimidine (6–4) pyrimidone photoproducts in human cells are due to excision repair rather than conformational change. *Nucleic Acids Res.* **18**, 963–971.
7. Mellon, I., Spivak, G., and Hanawalt, P. C. (1987) Selective removal of transcription blocking DNA damage from the transcribed strand of the mammalian DHFR gene. *Cell* 51, 241–249.
8. Bohr, V. A., Smith, C. A., Okumoto, D. S., and Hanawalt, P. C. (1985) DNA repair in an active gene: removal of pyrimidine dimers from the DHFR gene of CHO cells is much more efficient than in the genome overall. *Cell* **40**, 359–369.
9. Cleaver, J. E. (1968) Defective repair replication of DNA in xeroderma pigmentosum. *Nature* **218**, 652–656.
10. Hanawalt, P. C. and Sarasin, A. (1986) Cancer-prone hereditary diseases with DNA processing abnormalities. *Trends Genet.* **2**, 124–129.
11. Dammann, R. and Pfeifer, G. P. (1997) Lack of gene- and strand-specific DNA repair in RNA polymerase III transcribed human *tRNA* genes. *Mol. Cell. Biol.* **17**, 219–229.
12. Gao, S., Drouin, R., and Holmquist, G. P. (1994) DNA repair rates mapped along the human *PGK*-1 gene at nucleotide resolution. *Science* **263**, 1438–1440.

13. Pfeifer, G. P., Drouin, R., Riggs, A. D., and Holmquist, G. P. (1991) In vivo mapping of a DNA adduct at nucleotide resolution: detection of pyrimidine (6–4) pyrimidone photoproducts by ligation-mediated polymerase chain reaction. *Proc. Natl. Acad. Sci. USA* **88,** 1374–1378.
14. Pfeifer, G. P., Drouin, R., Riggs, A. D., and Holmquist, G. P. (1992) Binding of transcription factors creates hot spots for UV photoproducts in vivo. *Mol. Cell. Biol.* **12,** 1798–1804.
15. Tornaletti, S. and Pfeifer, G. P. (1994) Slow repair of pyrimidine dimers at p53 mutation hotspots in skin cancer. *Science* **263,** 1436–1438.
16. Tornaletti, S. and Pfeifer, G. P. (1996) Ligation-mediated PCR for analysis of UV damage, in *Technologies for Detection of DNA Damage and Mutations*, (Pfeifer, G. P., ed.). Plenum, New York, NY, pp. 199–209.
17. Törmänen, V. T. and Pfeifer, G. P. (1992) Mapping of UV photoproducts within *ras* protooncogenes in UV-irradiated cells: correlation with mutations in human skin cancer. *Oncogene* **7,** 1729–1736.
18. Tu, Y., Tornaletti, S., and Pfeifer, G. P. (1996) DNA repair domains within a human gene: selective repair of sequences near the transcription start site. *EMBO J.* **15,** 675–683.
19. Tommasi, S., Oxyzoglou, A. B., and Pfeifer, G. P. (2000) Cell cycle-independent removal of UV-induced pyrimidine dimers from the promoter and the transcription initiation domain of the human *CDC2* gene. *Nucleic Acids Res.* **28,** 3991–3998.
20. Hu, W., Feng, Z., Chasin, L. A., and Tang, M. S. (2002) Transcription-coupled and transcription-independent repair of cyclobutane pyrimidine dimers in the dihydrofolate reductase gene. *J. Biol. Chem.* **277,** 38,305–38,310.
21. Zhu, Q., Wani, M. A., El-Mahdy, M., and Wani, A. A. (2000) Decreased DNA repair efficiency by loss or disruption of p53 function preferentially affects removal of cyclobutane pyrimidine dimers from non-transcribed strand and slow repair sites in transcribed strand. *J. Biol. Chem.* **275,** 11,492–11,497.
22. Tu, Y., Bates, S., and Pfeifer, G. P. (1997) Sequence-specific and domain-specific DNA repair in xeroderma pigmentosum and Cockayne syndrome cells. *J. Biol. Chem.* **272,** 20,747–20,755.
23. Lippke, J. A., Gordon, L. K., Brash, D. E., and Haseltine, W. A. (1981) Distribution of UV light-induced damage in a defined sequence of human DNA: detection of alkaline-sensitive lesions at pyrimidine nucleoside-cytidine sequences. *Proc. Natl. Acad. Sci. USA* **78,** 3388–3392.
24. Gordon, L. K. and Haseltine, W. A. (1980) Comparison of the cleavage or pyrimidine dimers by the bacteriophage T4 and *Micrococcus luteus* UV-specific endonucleases. *J. Biol. Chem.* **255,** 12,047–12,050.
25. Radany, E. H. and Friedberg, E. C. (1980) A pyrimidine dimer-DNA glycosylase activity associated with the v gene product of bacterophage T4. *Nature* **286,** 182–185.
26. Ye, N., Holmquist, G. P., and O'Connor, T. R. (1998) Heterogeneous repair of *N*-methylpurines at the nucleotide level in normal human cells. *J. Mol. Biol.* **284,** 269–285.

27. Rodriguez, H., Drouin, R., Holmquist, G. P., et al. (1995) Mapping of copper/hydrogen peroxide-induced DNA damage at nucleotide resolution in human genomic DNA by ligation-mediated polymerase chain reaction. *J. Biol. Chem.* **270,** 17633–17640.
28. Cloutier, J. F., Drouin, R., Weinfeld, M., O'Connor, T. R., and Castonguay, A. (2001) Characterization and mapping of DNA damage induced by reactive metabolites of 4-(methylnitrosamino)-1-(3-pyridyl)-1-butanone (NNK) at nucleotide resolution in human genomic DNA. *J. Mol. Biol.* **313,** 539–557.
29. Cloutier, J. F., Castonguay, A., O'Connor, T. R., and Drouin, R. (2001) Alkylating agent and chromatin structure determine sequence context-dependent formation of alkylpurines. *J. Mol. Biol.* **306,** 169–188.
30. Akman, S. A., O'Connor, T. R., and Rodriguez, H. (2000) Mapping oxidative DNA damage and mechanisms of repair. *Ann. NY Acad. Sci.* **899,** 88–102.
31. Denissenko, M. F., Pao, A., Tang, M., and Pfeifer, G. P. (1996) Preferential formation of benzo[a]pyrene adducts at lung cancer mutational hotspots in p53. *Science* **274,** 430–432.
32. Denissenko, M. F., Koudriakova, T. B., Smith, L., O'Connor, T. R., Riggs, A. D., and Pfeifer, G. P. (1998) The p53 codon 249 mutational hotspot in hepatocellular carcinoma is not related to selective formation or persistence of aflatoxin B1 adducts. *Oncogene* **17,** 3007–3014.
33. Denissenko, M. F., Pao, A., Pfeifer, G. P., and Tang, M. (1998) Slow repair of bulky DNA adducts along the nontranscribed strand of the human *p53* gene may explain the strand bias of transversion mutations in cancers. *Oncogene* **16,** 1241–1247.
34. Smith, L. E., Denissenko, M. F., Bennett, W. P., Li, H., Amin, S., Tang, M., and Pfeifer, G. P. (2000) Targeting of lung cancer mutational hotspots by polycyclic aromatic hydrocarbons. *J. Natl. Cancer Inst.* **92,** 803–811.
35. Hu, W., Feng, Z., and Tang, M. S. (2003) Preferential carcinogen-DNA adduct formation at codons 12 and 14 in the human *K-ras* gene and their possible mechanisms. *Biochemistry* **42,** 10,012–10,023.
36. Pfeifer, G. P. and Riggs, A. D. (1993) Genomic sequencing. *Methods Mol. Biol.* **23,** 169–181.
37. Mueller, P. R. and Wold, B. (1989) In vivo footprinting of a muscle specific enhancer by ligation mediated PCR. *Science* **246,** 780–786.

16

Immunochemical Detection of UV-Induced DNA Damage and Repair

Marcus S. Cooke and Alistair Robson

> There nearly always is method in madness. It's what drives men mad, being methodical.
> — G. K. Chesterton

Summary
 Because of a substantial rise in the incidence of skin cancer in the United Kingdom and elsewhere a greater awareness of the role of sun-induced cutaneous genetic damage has developed. This, in turn, has increased interest in the cellular mechanisms responsible for tumorigenesis, and the need to develop experimental methodologies to investigate these mechanisms. DNA represents a most important cellular target for ultraviolet radiation (UVR), leading to the formation of various DNA damage products. A number of these products, such as the cyclobutane pyrimidine dimer, have been implicated in the pathogenesis of various UVR-related conditions. In this chapter we detail a number of methods for assessing UVR-induced DNA damage using two antisera which recognize cyclobutane thymine dimers (T–T). Immuno-approaches have a number of benefits over chromatographic techniques, and have been applied herein to quantitatively and qualitatively assess the presence of T–T in cultured keratinocytes, human skin, and urine, providing information about lesion induction and repair.

Key Words: Antibodies; DNA damage; immunoassay; repair; skin; ultraviolet radiation; urine.

1. Introduction

Protection by the Earth's atmosphere results in only a proportion of the ultraviolet radiation (UVR; UV-C: 100–280 nm; UV-B: 280–315 nm; UV-A: 315–400 nm), emitted from the sun, actually reaching the Earth's surface. Of the total UVR, UV-C is entirely excluded, the remainder comprising 5% UV-B, and 95% UV-A. This UVR has been speculated as being an etiological factor, if not the primary cause, of numerous dermatological disorders, which include

From: *Methods in Molecular Biology: DNA Repair Protocols: Mammalian Systems, Second Edition*
Edited by: D. S. Henderson © Humana Press Inc., Totowa, NJ

skin aging, solar keratosis, malignant melanoma, basal cell carcinoma and squamous cell carcinoma, as well as a variety of ill-understood photo-related inflammatory dermatoses. Exposure to UVR may lead to mutation *(1)*, carcinogenesis *(2)*, or cytotoxicity *(3)*, via the induction of DNA damage. The interaction between UVR and DNA may produce a number of oxidative *(4,5)* and non-oxidative lesions *(6)*, and alterations in the Earth's stratospheric ozone are likely to lead to variation in the frequency and type of such damage *(7)*, making the ability to monitor such damage essential. A major, non-oxidative product of UVR exposure of DNA is the cyclobutane pyrimidine dimer (P–P), formed between adjacent pyrimidine bases, consisting of thymine-thymine (T–T) dimers as major products with smaller amounts of thymine–cytosine (T–C), less still, cytosine–thymine (C–T) and least of all, cytosine–cytosine (C–C).

Methods for assessing UVR-induced dimer damage to DNA include high performance liquid chromatography (HPLC) *(8)*, ^{32}P-postlabeling *(9)*, and alkaline gel electrophoresis *(10)*. Essentially no mass spectrometric methods existed until the development of a gas chromatography–mass spectrometric technique *(11)*, a precedent which allowed the establishment of other MS approaches, such as liquid chromatography with tandem mass spectrometry *(12)*. Initial capital outlay for equipment and the time-consuming sample workup are problems largely avoided by immunochemical techniques, in addition to the benefit of lesion localization within a sample. However, the above techniques, and in particular the mass spectrometric approaches, will continue to be used to correlate results obtained from novel assays. Immunoassay has been a very successful approach to quantitation of UVR-induced DNA lesions within DNA after extraction, in cells *(13)* and in tissues *(14)*. Using UV-C-irradiated DNA, or poly-thymidilic acid, we have previously generated two polyclonal antisera, both of which recognize T–T, denoted Ab529 and 479, respectively *(15,16)*. The methods detailed here describe the immunochemical detection of UV-induced DNA damage, and its repair.

2. Materials

2.1. Antibody Purification on Protein A Columns

1. 1-mL Plastic syringe (Becton Dickinson Labware, Plymouth, UK)
2. Protein A insolublized on Sepharose CL-4B (Sigma, Sigma-Aldrich Chemical Company, Poole, UK).
3. 1 M Tris-HCl, pH 8.0; 100 mM Tris-HCl, pH 8.0; 10 mM Tris-HCl, pH 8.0 (All from Sigma).
4. 100 mM glycine, pH 3.0 (Sigma).

2.2. UV Irradiation of DNA

1. Calf thymus DNA was from Calbiochem (Nottingham, UK) and only batches with an A_{260}/A_{280} ratio greater than or equal to 1.8 were used.

2. UV-C lamp (254 nm ±1 nm, Knight Optical Technologies, Leatherhead, Surrey, UK). (*See* **Note 1**.) UV-B (λ_{max} 302 nm, range of 272–352 nm; <2 % UV-C, Model-UVM, Chromatovue lamp from Knight Optical Technologies. (*See* **Note 2**.) Doses of UV-C and UV-B were calculated using an optical radiometer (MP100) in conjunction with the appropriate UV-C (MC125) and UV-B (MP131) sensor (Knight Optical Technologies).

2.3. Keratinocyte Culture

1. Human SV40 immortalized RHT keratinocytes (passage no. 101) were a kind gift from Professor Irene Leigh, Department of Dermatology, London Hospital Medical College.
2. RM+ medium (Life Technologies Ltd, Paisley, Scotland): consists of Dulbecco's minimum essential medium (DMEM) containing L-glutamax, supplemented with antibiotics and mitogens: Penicillin G–streptomycin (penicillin 100 U/mL, streptomycin 100 µg/mL; Flow Laboratories, Irvine, UK) and mitogens: 0.41 µg/mL of hydrocortisone, $1 \times 10^{-10} M$ choleratoxin, 51 µg/mL of transferrin, $2 \times 10^{-11} M$ lyothyronine, $1.8 \times 10^{-4} M$ adenine, 51 µg/mL of insulin, 10 ng/mL of epidermal growth factor.
3. Ham's nutrient mixture F12 (Ham's F12; Imperial Laboratories Ltd., Andover, UK) at a ratio of 3:1 with heat-inactivated fetal calf serum (10 % v/v in PBS; Sigma).
4. Hank's balanced salt solution (HBSS) was from Gibco (Paisley, UK).
5. Cell culture plastics were from Becton Dickinson Labware, apart from the Nunc Lab-Tek eight-well chamber slides (Life Technologies Ltd., Paisley, Scotland).
6. Trypan blue: 0.4% w/v in 0.8% w/v sodium chloride, 0.06% dipotassium hydrogen phosphate (purchased ready prepared from Sigma).

2.4. Qualitative Detection of UV-B-Induced Damage in Keratinocytes

1. Methanol (Fisher Scientific, Loughborough, UK).
2. Ethanol (Fisher Scientific, Loughborough, UK).
3. Normal goat serum (NGS; DAKO Ltd, High Wycombe, UK)
3. 0.01 M Phosphate buffered saline (PBS), pH 7.4; prepared from the tablet form, Sigma.
4. FITC-labeled, goat antirabbit immunoglobulins (DAKO Ltd).
5. Vectashield (Vector Laboratories, Peterborough, UK)

2.5. Quantitative Detection of DNA Damage From UV-Irradiated Keratinocytes

1. Nunc Delta Petri dishes (Life Technologies Ltd.)
2. Tris–EDTA buffer: 10 mM Tris-HCl, 1 mM EDTA, pH 7.5 (Sigma).
3. Buffer 1: 5 mM trisodium citrate, 20 mM NaCl, pH 6.5 (Sigma).
4. Buffer 2: 20 mM Tris-HCl, 20 mM EDTA, 1.5% v/v sarkosyl, pH 8.5 (Sigma).
5. Buffer 3: 10 mM Tris-HCl, 1 mM EDTA, pH 7.5 (Sigma).

6. Pronase E (Sigma).
7. 7.5 M Ammonium acetate (Sigma), made up in ultrapure (UP, 18 MΩ) water.
8. To improve binding of DNA to the enzyme-linked immunosorbent assay (ELISA) plate, they are first coated with poly-L-lysine (Sigma), diluted 25 µg/mL. Coat ELISA plates (96-well, Immuno plate Maxisorp, Nunc) with 50 µL/well, overnight at 4°C in a humidified chamber (see **Note 3**).
9. Dried skimmed milk (Tesco Stores Ltd., Cheshunt, UK).
10. Tween-20 (Sigma).
11. Peroxidase-labeled goat antirabbit immunoglobulin G (DAKO Ltd.).
12. Perborate capsules and orthophenylenediamine tablets (Sigma).
13. 2 M Sulfuric acid, diluted in ultrapure water (Fisher Scientific Ltd., Loughborough, UK).

2.6. Immunohistochemical Detection of T–T Induction and Repair in Skin Biopsies From UV-B Irradiated Human Subjects

1. Xylene (Fisher Scientific Ltd.).
2. 99% and 95% (v/v) ethanol (Fisher Scientific Ltd.), diluted in distilled water.
3. Biotinylated goat antirabbit immunoglobulin, and peroxidase-labeled streptavidin–biotin complex (DAKO Ltd.).
4. Hydrogen peroxide (3% v/v in distilled water; Sigma).
5. Diaminobenzidine (Sigma).
6. 0.1% Mayer's hematoxylin (purchased ready prepared from Sigma).
7. XAM (BDH, Poole, Dorset).
8. Sections of biopsies, taken from UV-B–irradiated volunteers, were obtained and processed by Dr. Alistair Robson (formerly from the Department of Pathology, University of Leicester). The biopsies originated from the Department of Dermatology, University Hospital of Wales, Cardiff. Full details of the irradiations, biopsy procedure, fixation etc. are described in Anstey et al. *(17)*. Briefly, twice the minimal erythema dose (see **Note 4**) of UV-B (300 ± 10 nm), from a 100-W xenon arc lamp coupled to an Oriel Corporation grating monochromator, was delivered to the buttock skin of six human volunteers. Post-irradiation, 4-mm punch biopsies were taken 5, 8, 24, and 48 h later, along with adjacent, unirradiated, control skin. The samples were then formalin fixed and mounted in paraffin wax.

3. Methods

The methods described in the following outline (1) preparation of antibodies/antisera prior to use; (2) UV irradiation of DNA; (3) culture of keratinocytes; (4) qualitative detection of UV-induced DNA damage in cells; (5) quantitative detection of UV-induced DNA damage in cells; (6) immunohistochemical detection of UV-induced DNA damage and repair; and (7) ELISA detection of urinary T–T, a putative product of DNA repair.

3.1. Antibody Purification on Protein A Columns

3.1.1. Preparation of the Protein A Column

Purification of IgG fraction Ab529 was undertaken prior to use. Prepare the Protein A column as follows:

1. Using the plunger, place a small amount of glass wool at the base of a 1-mL syringe.
2. Add ultrapure water to 143 mg of Protein A–Sepharose CL-4B until it swells, forming a slurry.
3. Using a transfer pipet, carefully load the syringe with the slurry, avoiding air bubbles. This should half fill the column, leaving room for the antiserum/washes. Some liquid, containing Sepharose, will pass through the column, collect this and add it to the top of the column.
4. Once the slurry has become a gel, wash the column with 1.5 mL of ultrapure water.

3.1.2. Purification of the Antiserum

1. Adjust the pH of the crude antibody preparation (serum, tissue culture supernatant, or ascites) to 8.0 by adding 1/10 volume of 1 M Tris-HCl (pH 8.0).
2. Pass the antibody solution down the column. These columns bind 10–20 mg of antibody per milliliter of wet beads (one molecule of Protein A binds two molecules of antibody). Serum contains approx 10 mg/mL of total IgG (therefore can only use approx 1 mL of neat, undiluted antiserum). It is a wise precaution to collect all fractions that pass through the column, starting with the addition of the antiserum.
3. Wash the column with 10 column volumes (10 mL) of 100 mM Tris-HCl, pH 8.0.
4. Wash the column with 10 column volumes (10 mL) of 10 mM Tris-HCl, pH 8.0.
5. Elute the IgG from the column with 100 mM Glycine, pH 3.0. Add this buffer stepwise, approx 500 µL per sample. Collect the eluate in 1.5-mL Eppendorf tubes containing 50 µL of 1 M Tris-HCl, pH 8.0. Mix each tube gently to bring the pH back to neutral. Avoid bubbling and frothing as this denatures the proteins.
6. Identify the Ig-containing fractions by absorbance at 280 nm (1 OD unit = approx 0.8 mg of protein/mL)

3.2. UV Irradiation of DNA

For purposes of antiserum characterization *(16)*, determination of an assay's limit of detection *(18)*, or as an assay's positive control, UV irradiation of pre-extracted DNA is sometimes required. Make up solutions of DNA in ultrapure water (18 MΩ), determining the DNA concentration spectrophotometrically at 260 nm (A_{260} of 1 = 50 µg DNA/mL). Irradiate a 1-mL aliquot (1 mg/mL), having determined the output of the chosen UV lamp (*see* **Notes 5** and **6**).

3.3. Keratinocyte Culture

Grow the human, immortalized RHT keratinocytes to confluence in RM+, routinely passaging them every 3–4 d as required. Maintain the cells in 250-mL tissue culture flasks under a humidified atmosphere of 95% air, 5% CO_2 at 37°C. Determine cell number by the trypan blue assay, as follows. Mix a 20-µL volume of cell suspension with an equal volume of trypan blue. Apply the resultant mix to a hemocytometer. The cells excluding the dye are deemed to be viable and counted under a ×40 microscope objective. Total cell number is calculated thus: viable cells per mL = viable cells counted × 10^4 × dilution factor.

3.4. Qualitative Detection of UV-B-Induced Damage in Keratinocytes

This approach provides localization of the DNA damage within UV-irradiated cells, and requires few cells for analysis.

3.4.1. Culturing of Cells on Chamber Slides

Grow the human, immortalized RHT keratinocytes until 50% confluence, at which point seed into eight-well chamber slides, and leave overnight until they reach 60–80% confluence.

3.4.2. UV Irradiation of Cells

1. Remove the lids and the medium from the chamber slides, wash with HBSS, and place on ice for the duration of the irradiation.
2. For the purposes of T–T induction, irradiate the cells with 0, 0.135, 0.27, and 1.08 kJ/m² using a UV-B lamp *(13)*. Sham irradiate control wells on the same slide, by covering them with aluminum foil. Otherwise treatment is exactly the same.
3. Immediately following treatment, fix the cells by the addition of 1:1 methanol–acetone and leave at 4°C for at least 10 min. Following fixation, allow the cells to air-dry, and store at 4°C until analysis.

3.4.3. Immunocytochemical Staining

1. Rehydrate the cells by the introduction of 0.01 *M* PBS, for 10 min.
2. Block nonspecific binding by preincubation of the chamber slides with normal goat serum (2 % v/v in 0.01 *M* PBS).
3. Detect T–T, following UV-B irradiation, by incubating the cells with Ab529 diluted 1:5000 in normal goat serum (2% v/v in 0.01 *M* PBS), for 1 h at 37°C.
4. Carefully wash the slides with 0.01 *M* PBS.
5. Dilute the fluorescein isothiocyanate (FITC)-labeled goat antirabbit IgG secondary antibody to 1:200 in normal goat serum (2% v/v in 0.01 *M* PBS), to localize binding of the primary antiserum.
6. Carefully wash all chambers slides with 0.01 *M* PBS, prior to a final wash in deionized water, and then allow the slides to air-dry.

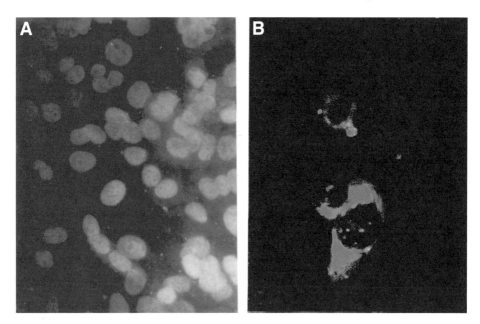

Fig. 1. (**A**) RHT keratinocytes irradiated with 1.08 kJ/cm^2 UV-B, and processed as described in **Subheading 3.4.** prior to indirect immunofluorescent staining with an anti-thymine dimer antiserum (Ab529), utilizing an FITC-labeled secondary (×400 magnification). Intense staining was noted using fluorescence microscopy, appearing localized in the nucleus. (**B**) Unirradiated RHT keratinocytes processed as described for **A** (×400 magnification). No staining was apparent. Weak background cytoplasmic autofluorescence is visible.

7. Once dry, remove the chambers and then mount the slide in Vectashield. The FITC is light sensitive; therefore, wrap the slides in aluminum foil and store at 4°C

Figure 1 represents an example of positive and negative controls, reproduced from Cooke et al. *(13)*, with permission (Web-site: http://www.tandf.co.uk).

3.5. Quantitative Detection of DNA Damage From UV-Irradiated Keratinocytes

Compared to immunocytochemical analysis, greater numbers of cells are required for ELISA of the extracted DNA from UV-irradiated keratinocytes.

3.5.1. Cell Culture and UV Irradiation

1. Culture the cells, as described **above**, until eight 250-mL flasks are confluent. Subsequently, seed the cells into eight Nunc Delta Petri dishes and grow to near confluence, with each dish representing a single UV-B dose.

2. Remove the medium and wash the cells with HBSS. Place the Petri dishes on ice and irradiate (0–0.8 kJ/m^2 UV-B) *(13)*.
3. Following irradiation, remove the cells from the Petri dish by a brief incubation with Tris–EDTA buffer (TE). As the cells begin to lift, neutralize the TE with an equal volume of RM+, and pellet at 400g for 10 min at 4°C. Discard the supernatants and extract the DNA using Pronase E extraction (*see* below).

3.5.2. Pronase E DNA Extraction (see **Note 7**)

1. Resuspend the cells in 5–10-mL of PBS and centrifuge at 700g for 15 min.
2. Resuspend the cells in 1.75 mL of ice-cold buffer 1.
3. Add 2 mL of buffer 2 and mix vigorously. (*See* **Note 8**.)
4. Add 2 mg of Pronase E in 0.5 mL of buffer 1, and incubate overnight. Subsequently add 2 mL of buffer 3, along with 0.5 mL of 7.5 M ammonium acetate, and mix by inversion after each addition.
5. Finally precipitate the DNA by the addition of 18 mL of ice-cold ethanol. Spool the DNA onto a "pipet hook" (*see* **Note 9**) and wash in 70% ethanol, and then in 100% ethanol for 10 min each.
6. Remove residual ethanol by a stream of nitrogen.

3.5.3. Direct ELISA of Extracted DNA From UV-Irradiated Cells

This methodology was originally described in Cooke et al. *(13)*.

1. Redissolve the dried DNA pellets in 1 mL of ultrapure water and determine the DNA content by absorbance at 260 nm.
2. To bind double-stranded, UV-irradiated DNA to the 96-well ELISA plate, dilute the DNA to a final concentration of 50 µg/mL with ultrapure water (18 MΩ). (*See* **Note 10**.)
3. Add 50 µL per well, in triplicate for each treatment/dose. Do not bind DNA in the outermost edges of the ELISA plate (*see* **Note 11**). These wells may be used for all the subsequent antibody steps to evaluate "background," nonspecific binding of the antibodies used.
4. Incubate the plate in a humidified environment at 37°C for 1 h (*see* **Note 12**).
5. After incubation, wash the plate three times with 0.01 M PBS, pH 7.4, preferably with an automated platewasher (e.g., Anthos 2001, Anthos Labtec Instr.).
6. Block the free sites by incubation with 150 µL/well of 4% (w/v) dried skimmed milk in PBS (4% milk/PBS) for 1 h at 37°C in a humidified environment. (*See* **Note 13**.)
7. Wash the plate with PBS three times.
8. Dilute the test antiserum (e.g., IgG fraction of Ab529; **ref.** *18*) to its working concentration (1/5000 in 4% milk/PBS) and add to the plate (50 µL/well). Incubate the plate for 1 h, as described above in **step 4**.
9. Following incubation, wash the plate three times with PBS containing 0.05% (v/v) Tween-20.
10. The secondary (detecting) antibody can then be added. To detect the IgG fraction of Ab529, use a peroxidase-labeled, goat anti-rabbit IgG. Dilute the immunoglo-

bulin 1:2000 in milk–PBS, and apply to the entire plate (50 µL/well). Incubate the plate as described in **step 4**.
11. After incubation, wash the plate with PBS–Tween-20 solution, as in **step 9**.
12. Prepare the substrate solution, immediately before use. Add the contents of one perborate capsule to 100 mL of deionized water. To 20 mL of this add one 10-mg orthophenylenediamine tablet, and dissolve by mixing.
13. Add 50 µL/well of the substrate solution to the plate, and incubate for 15–30 min at room temperature, in the dark, in order for the color to develop (*see* **Note 14**).
14. Stop the reaction by the addition of 2 M H_2SO_4 (25 µL/well).
15. Determine the resulting absorbance by reading at 492 nm using a plate reader. The final data should account for any background (*see* **Note 15**), nonspecific binding of the antibodies used, that is, subtract a mean background value from the mean of all determinations.

3.6. Immunohistochemical Detection of T–T Induction and Repair in Skin Biopsies of UV-B-Irradiated Human Subjects

3.6.1. Pretreatment of Tissue Sections

Dewax formalin-fixed, paraffin-embedded sections by immersing in xylene for 5 min. Then rehydrate the sections by immersion in 99% and 95% ethanol solutions (v/v in distilled water), prior to rinsing in distilled water and equilibrating in PBS *(18)*.

3.6.2. Immunohistochemical Detection of Damage

1. Drain the slides, and excess buffer blotted from around the sections.
2. Block the sections by covering with 20% normal goat serum (v/v in PBS), and incubate in a covered chamber for 10 min at room temperature (*see* **Note 16**).
3. Without rinsing, drain the slides, and remove excess serum from around the sections by blotting.
4. Add the primary antiserum (Ab529; diluted 1:2000 in 2% normal goat serum, v/v in PBS).
5. Incubate the slides in a covered chamber for 30 min at room temperature.
6. Following two consecutive 5-min washes in PBS, drain the slides of excess buffer, and cover with biotinylated goat antirabbit immunoglobulin diluted 1:1000 in 2% normal goat serum in PBS (v/v).
7. Incubate the slides for 30 min in a humidified chamber.
8. Give the slides two 5-min washes in PBS (fresh PBS for each wash).
9. Cover the slides with the peroxidase-labeled streptavidin–biotin complex (following the manufacturer's instructions) and incubate for 30 min.
10. Again, give the slides two 5-min washes in PBS and then one 5-min wash in distilled water.
11. To demonstrate peroxidase activity, drain and blot around the sections, prior to covering them with freshly prepared substrate (0.5% w/v diaminobenzidine in 9.5 mL of PBS, activated by 3% hydrogen peroxide; approx 0.5 mL/slide).

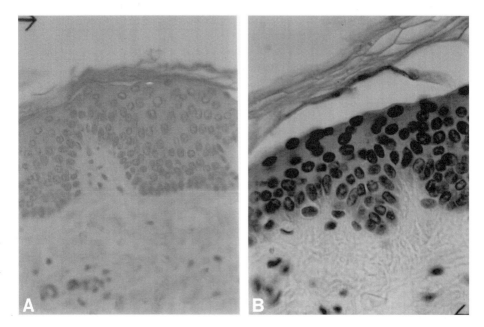

Fig. 2. Immunohistochemical localization of T–T, using Ab529, in sections from human skin biopsies: **(A)** control, unirradiated, and **(B)** UV-B-irradiated (2 MED) taken 2 h post-irradiation (×400 magnification). The (blue) nuclear counterstain is hematoxylin, T–T identified by Ab529 were localized using a peroxidase-labeled secondary antibody. (Reprinted from Cooke et al. *(18)*, *Journal of Immunological Methods*, with permission from Elsevier.)

12. Incubate 5–15 min in a covered chamber.
13. Then carefully wash the slides under running tap water for 5 min.
14. Counterstain with Mayer's hematoxylin: Place the slides in the solution for approximately ten seconds, then wash again in running tap water prior to dehydration (by immersion in a series of 95 and 99% ethanol solutions, then finally xylene) and mounting in resinous mountant (e.g., XAM).

Figure 2 represents an example of immunohistochemical detection of T–T in section of skin taken from UV-B–irradiated volunteers.

3.7. Competitive ELISA for the Detection of Urinary T–T

Damaged DNA products are eliminated by a variety of repair enzymes and may be detected as nucleoside derivatives. The level of these products depends on the equilibrium between the rates of damage and repair, reflected by the amount of lesion excreted into urine. The development of assays to measure such lesions in urine may offer a noninvasive means by which biomonitoring of genotoxic exposure, and an individual's ability to process DNA lesions could be determined, along with presenting a means to assess therapy-related risk *(16)*. The procedure is based on a competitive ELISA, using the IgG fraction of antiserum 479 *(16)*.

3.7.1. Urine Collection

Collect urine samples in 20-mL plastic universal tubes, without any additives, and store at –80°C until analysis. Following thawing and centrifugation (300g for 10 min), process the supernatants in the competitive ELISA according to **Subheading 3.7.2.**

3.7.2. Analysis of Urine

The method is based upon that previously described *(16,19)*, with modifications for the target lesion and antiserum used. Also, *see* **Notes 10–15**, for direct ELISA.

1. Prepare poly-L-lysine–coated plates, as described above (*see* **Subheading 2.5., item 8**).
2. Coat the 96-well ELISA plates with single-stranded (*see* **Note 10**) UV-C-irradiated DNA (50 kJ/cm^2; *see* **Notes 5** and **6**), 50 µg/mL in PBS, 50 µL/well, by incubation at 37°C in a humidified chamber for 1 h.
3. Wash the plate three times with PBS.
4. Block free sites on the plate by incubation of a 4% milk–PBS solution, in all wells, for 1 h at 37°C in a humidified chamber.
5. Wash the plate three times with PBS.
6. Apply 50 µL of (undiluted) urine to appropriate wells (usually in at least duplicate). Dilute the purified IgG fraction of antiserum 479 (primary antiserum) in 4% milk–PBS to twice its working concentration (i.e., 3.75 µg/mL) and apply 50 µL/well to columns 2–12. Columns 1 (no primary Ab) and 12 (no competitor) are used as controls, therefore include 50 µL of PBS in place of primary antibody or competitor. Incubate for 1 h at 37°C.
7. Wash the plate three times with PBS containing 0.05% (v/v) Tween-20.
8. Add the goat antirabbit IgG specific peroxidase-labeled secondary antibody, diluted 1:2000 in 4% milk–PBS (w/v), 100 µL/well and incubate for 1 h at 37°C.
9. Wash three times with PBS–Tween.
10. Prepare the substrate solution immediately before use. Add the contents of one perborate capsule to 100 mL of deionized water. To 20 mL of this add one 10-mg orthophenylenediamine tablet, dissolve by mixing. This substrate solution (100 µL) is added to each well and incubated in the dark, at room temperature for up to 30 min. Stop the reaction with 100 µL of 2 M H$_2$SO$_4$/well.
11. Read the resulting absorbances at 492 nm using an ELISA plate reader.
12. Urinary creatinine measurements. In our laboratory this task is undertaken via collaboration with the Department of Chemical Pathology, Leicester Royal Infir-

mary, Leicester, UK. Assays exist for determining urinary creatinine, for example, from Sigma (555-A), based on the method of Slot *(20)*.
13. Results from the ELISA are expressed as percentage inhibition, according to the following equation:

$$\frac{\text{absorb. of replicates} - \text{absorb. no primary Ab}}{\text{absorb. no competitor} - \text{absorb. no primary Ab}}$$

14. The results are further corrected for urine concentration, by dividing by the urinary creatinine value.

4. Notes

1. UVR wavelengths shorter than approx 290 nm do not reach the Earth's surface *(21)* owing to absorption by stratospheric ozone; nevertheless, use of a UV-C source is appropriate in a test system because of its effectiveness at inducing DNA damage.
2. The UV-B source does not represent the spectral characteristics of solar irradiance (the latter composed of 4.1% UV-B, and 95.9% UV-A) *(21)*. However, it is suitable for a broad study of wavelength-dependent induction of DNA damage.
3. This may be done well in advance, the plates sealed with Nescofilm, or masking tape, and stored at 4°C until use.
4. MED, minimal erythemal dose, the dose of UVR required to induce erythema, which persists for 24 h.
5. Not only will use of a radiometer allow accurate determination of UV lamp output, but also obviate the need for having the sample at any fixed distance. The distance between the sample and lamp may be altered to make timing of exposures straightforward.
6. The radiometers described in this chapter describe the output of the UV lamps in $\mu W/cm^2$ or mW/cm^2. To give a particular dose, an exposure time will need to be calculated. Knowing that watts (W) are equivalent to joules/s, this can be substituted into the lamp output (giving $\mu J/s/cm^2$ or $mJ/s/cm^2$, respectively). Rearrangement of these equations will allow an exposure time (sec) to be calculated for a required exposure dose (J/cm^2).
   Note: $1 \text{ J/cm}^2 = 10 \text{ kJ/m}^2$; $1 \text{ J/m}^2 = 0.1 \text{ mJ/cm}^2 = 0.001 \text{ kJ/m}^2$.
7. For the measurement of oxidatively damaged DNA, the method employed to extract DNA appears to be crucial if adventitious damage is to be avoided. As the lesion of interest, T–T, is not a product of free radical or reactive oxygen species attack of DNA, the issue of artifactual oxidation is not raised.
8. For the analysis of DNA lesions other than T–T, RNA is removed at **step 4**, as follows: add 250 µg of RNase A in RNase buffer A (50 mM Tris-HCl, 10 mM EDTA, 10 mM NaCl, pH 6.0) and incubate at 37°C for 1 h. However, for analysis of T–T, this step may be omitted.
9. Pipet hooks are made by briefly heating the narrow end of a glass pipet in a flame until it can be manipulated into a small hook.

10. Greater sensitivity may be achieved if the DNA of interest is rendered single-stranded, as this often improves antibody access to the antigen. This may be achieved by boiling the DNA solution in a screw-top Eppendorf tube for 10 min, followed by rapid cooling on ice, prior to binding to the ELISA plate, as described in **Subheading 3.5.3., steps 1** and **2**. While this is appropriate for nonoxidative lesions, such as those of interest in this chapter, the formation of artifactual oxidative damage is a risk accompanying such processing.
11. Do not bind DNA in the outermost edges of the ELISA plate, as there are suggestions that so-called "edge effects" may arise in the outer wells, as they are not fully surrounded by wells. This improves consistency in results across the plate.
12. A humidified chamber may simply consist of an airtight box, large enough for a number of plates, containing wetted paper towels.
13. Pipetting of multiple wells (e.g., addition of primary or secondary antibody) can be achieved simultaneously by use of a multichannel pipet. This also aids reproducibility across the ELISA plate.
14. To test that the peroxidase label on secondary antibody is functional, retain a little of this solution, and add to any leftover substrate solution. A color should rapidly form, and aids troubleshooting, if required.
15. Typical background absorbances are 0.001–0.01 absorbance units.
16. While purpose-built humidified chambers may be purchased, they may simply comprise an airtight box, containing two 10-mL plastic pipets and wetted tissue. The two pipets support the slides, and keep them raised off the wetted tissue.

Acknowledgments

We thank past and present members of our laboratories for discussions, along with financial support from The UK Food Standards Agency, Arthritis Research Campaign, and Leicester Dermatology Research Fund.

References

1. Peak, M. J., Peak, J. G., Moehring, M. P., and Webb, R. B. (1984) Ultraviolet action spectra for DNA dimer induction, lethality, and mutagenesis in *Escherichia coli* with emphasis on the UVB region. *Photochem. Photobiol.* **40,** 613–620.
2. Ananthaswamy, H. N. and Pierceall, W. E. (1990) Molecular mechanisms of ultraviolet radiation carcinogenesis. *Photochem. Photobiol.* **52,** 1119–1136.
3. Clingen, P. H., Arlett, C. F., Cole, J., et al. (1995) Correlation of UVC and UVB cytotoxicity with the induction of specific photoproducts in T-lymphocytes and fibroblasts from normal human donors. *Photochem. Photobiol.* **61,** 163–170.
4. Mitchell, D. L., Jen, J., and Cleaver, J. E. (1991) Relative induction of cyclobutane dimers and cytosine photohydrates in DNA irradiated in vitro and in vivo with ultraviolet-C and ultraviolet-B light. *Photochem. Photobiol.* **54,** 741–746.
5. Doetsch, P. W., Zasatawny, T. H., Martin, A. M., and Dizdaroglu, M. (1995) Monomeric base damage products from adenine, guanine, and thymine induced by exposure of DNA to ultraviolet radiation. *Biochemistry* **34,** 737–742.
6. Cadet, J., Anselmino, C., Douki, T., and Voituriez, L. (1992) Photochemistry of nucleic acids in cells. *J. Photochem. Photobiol. B* **15,** 277–298.

7. Freeman, S. E., Hacham, H., Gange, R. W., Maytum, D. J., Sutherland, J. C., and Sutherland, B. M. (1989) Wavelength dependence of pyrimidine dimer formation in DNA of human skin irradiated in situ with ultraviolet light. *Proc. Natl. Acad. Sci. USA* **86**, 5605–5609.
8. Cadet, J., Gentner, N. E., Rozga, B., and Paterson, M. C. (1983) Rapid quantitation of ultraviolet-induced thymine-containing dimers in human cell DNA by reversed-phase high-performance liquid chromatography. *J. Chromatogr.* **280**, 99–108.
9. Bykov, V. J., Kumar, R., Forsti, A., and Hemminki, K. (1995) Analysis of UV-induced DNA photoproducts by 32P-postlabelling. *Carcinogenesis* **16**, 113–118.
10. Freeman, S. E., Blackett, A. D., Monteleone, D. C., Setlow, R. B., Sutherland, B. M., and Sutherland, J. C. (1986) Quantitation of radiation-, chemical-, or enzyme-induced single strand breaks in nonradioactive DNA by alkaline gel electrophoresis: application to pyrimidine dimers. *Anal. Biochem.* **158**, 119–129.
11. Podmore, I. D., Cooke, M. S., Herbert, K. E., and Lunec, J. (1996) Quantitative determination of cyclobutane thymine dimers in DNA by stable isotope-dilution mass spectrometry. *Photochem. Photobiol.* **64**, 310–315.
12. Douki, T., Court, M., Sauvaigo, S., Odin, F., and Cadet, J. (2000) Formation of the main UV-induced thymine dimeric lesions within isolated and cellular DNA as measured by high performance liquid chromatography-tandem mass spectrometry. *J. Biol. Chem.* **275**, 11,678–11,685.
13. Cooke, M. S., Mistry, N., Ladapo, A., Herbert, K. E., and Lunec, J. (2000) Immunochemical quantitation of UV-induced oxidative and dimeric DNA damage to human keratinocytes. *Free Radic. Res.* **33**, 369–381.
14. Potten, C. S., Chadwick, C. A., Cohen, A. J., et al. (1993) DNA damage in UV-irradiated human skin in vivo: automated direct measurement by image analysis (thymine dimers) compared with indirect measurement (unscheduled DNA synthesis) and protection by 5-methoxypsoralen. *Int. J. Radiat. Biol.* **63**, 313–324.
15. Herbert, K. E., Mistry, N., Griffiths, H. R., and Lunec, J. (1994) Immunochemical detection of sequence-specific modifications to DNA induced by UV light. *Carcinogenesis* **15**, 2517–2521.
16. Ahmad, J., Cooke, M. S., Hussieni, A., et al. (1999) Urinary thymine dimers and 8-oxo-2′-deoxyguanosine in psoriasis. *FEBS Lett.* **460**, 549–553.
17. Anstey, A., Marks, R., Long, C., et al. (1996) In vivo photoinduction of metallothionein in human skin by ultraviolet irradiation. *J. Pathol.* **178**, 84–88.
18. Cooke, M. S., Podmore, I. D., Mistry, N., et al. (2003) Immunochemical detection of UV-induced DNA damage and repair. *J. Immunol. Methods* **280**, 125–133.
19. Cooke, M. S., Mistry, N., Wood, C., Herbert, K. E. and Lunec, J. (1997) Immunogenicity of DNA damaged by reactive oxygen species—implications for anti-DNA antibodies in lupus. *Free Radic. Biol. Med.* **22**, 151–159.
20. Slot, C. (1965) Plasma creatinine determination. A new and specific Jaffe reaction method. *Scand. J. Clin. Lab. Invest.* **17**, 381–387.
21. Gasparro, F. P. and Brown, D. B. (2000) Photobiology 102: UV sources and dosimetry—the proper use and measurement of "photons as a reagent." *J. Invest. Dermatol.* **114**, 613–615.

17

A Dot-Blot Immunoassay for Measuring Repair of Ultraviolet Photoproducts

Shirley McCready

Summary

The method described here makes use of a polyclonal antiserum to measure repair of the principal photoproducts induced in DNA by short-wave ultraviolet light (UV-C)—pyrimidine-pyrimidone 6–4 photoproducts ([6–4]PPs) and cyclobutane pyrimidine dimers (CPDs). DNA extracted from irradiated cells is applied to a nitrocellulose dot-blot and quantified using an enzyme-conjugated secondary antibody and a color assay. Though the polyclonal antiserum contains antibodies to both cyclobutane pyrimidine dimers and (6–4) photoproducts, repair of these can be measured separately by differential destruction of one or other photoproduct. The method is useful for measuring repair in total genomic DNA. It is more sensitive than most other methods and is sufficiently sensitive to measure repair of damage induced by doses of 10 J/m^2 of UV-C in DNA from mammalian cells.

Key Words: Cyclobutane pyrimidine dimers; dot blot; immunoassay; 6–4 photoproducts; UV damage.

1. Introduction

The dot-blot method described here can be used to measure repair of the principal photoproducts induced in DNA by short-wave ultraviolet light (UV-C)—pyrimidine-pyrimidone 6–4 photoproducts ([6–4]PPs) and cyclobutane pyrimidine dimers (CPDs). The method is used to measure the overall rate of repair in total genomic DNA. One of the advantages it has over other methods is its sensitivity—the assay is sufficiently sensitive to measure repair of damage induced by doses of 10 J/m^2 of UV-C with ease and could be used for lower doses. It is also very versatile and has been successfully used to measure repair in human

From: *Methods in Molecular Biology: DNA Repair Protocols: Mammalian Systems, Second Edition*
Edited by: D. S. Henderson © Humana Press Inc., Totowa, NJ

cells and in yeasts as well as a variety of other organisms (**refs. *1–3***; and McCready, unpublished). In addition, the DNA does not have to be especially intact for this assay, unlike polymerase chain reaction (PCR) methods and alkaline gel methods that rely on high and uniform integrity of the extracted DNA.

To use the method, it is necessary to raise polyclonal antiserum to UV-irradiated DNA. The antiserum must be characterized for its ability to recognize damage that can be photoreactivated by *Escherichia coli* photolyase (CPDs) and damage that is not photoreactivated (predominantly [6–4]PPs *[4,5]*). The antiserum, containing activities against CPDs and (6–4)PPs, can be used to measure total lesions. Alternatively, it can be used to measure each type of photoproduct individually by destroying one or the other lesion in the DNA before carrying out the assay. (6–4)PPs can be destroyed by treating DNA samples with hot alkali before applying DNA to the blotting membrane. CPDs can be destroyed in DNA after it has been applied to the blot by treating the entire blot with *E. coli* photolyase and visible light.

Cells are irradiated and samples are harvested immediately and after suitable incubation periods. DNA can be extracted from the cells by a variety of procedures—commercially available kits, or by phenol or phenol–chloroform extraction. It is of crucial importance to equalize the amounts of DNA in samples from the different time points, and this is best done by running aliquots on an agarose gel and estimating relative amounts by densitometry. Concentrations must be adjusted and checked on gels for as many times as necessary until the DNA concentrations are uniform. Each DNA sample is then divided into two, and one half is treated with hot alkali to destroy (6–4)PPs. Dilution series of the samples are then applied to dot blots. For measuring repair of (6–4)PPs, one blot is exposed to a crude preparation of photolyase and illuminated with visible light to destroy CPDs. The blots are then exposed to polyclonal antiserum, then to a biotinylated secondary antibody, and then to an alkaline phosphatase-conjugated avidin. Nitroblue tetrazolium is used as substrate so that a blue color stains the DNA dots containing UV lesions. Over a certain range, the amount of blue color is proportional to the amount of damage. Blots contain their own in-built calibration curves, namely, the dilution series of the time-zero samples. The amount of damage remaining in post-incubation samples is quantitated by densitometry and reference to the time-zero dilution series.

The method was originally developed for measuring repair rates in yeast and has been described in detail previously *(6)*. Exactly the same method can be used for measuring repair in human or other mammalian cells as well as in plants, bacteria and archaea. The only difference is in the details of the repair experiments and the method for DNA extraction. The method below is for HeLa cells grown in suspension or as monolayers in Petri dishes.

2. Materials

All buffers and bacterial growth media are sterilized by autoclaving.

2.1. Production of Polyclonal Antiserum

1. Isotonic saline: 0.15 M NaCl, pH 7.0.
2. High molecular weight calf thymus DNA: dissolve at 1 mg/mL in isotonic saline.
3. Methylated bovine serum albumin (MBSA) (Sigma): dissolve at 2 mg/mL in sterile water and add an equal volume of 2X isotonic saline (final concentration 1 mg/mL in isotonic saline).
4. Poly(dA) poly(dT) (Sigma).

2.2. Preparation of Crude Photolyase

1. Luria broth with tetracycline (20 µg/mL final concentration).
2. Isopropylthio-β-D-galactopyranoside (IPTG): 0.2 g/mL in water (840 mM).
3. Lysis buffer: 50 mM Tris-HCl, pH 7.4, 1 mM EDTA, 100 mM NaCl, 10 mM β-mercaptoethanol, autoclaved before adding β-mercaptoethanol.
4. Storage buffer: 50 mM Tris-HCl, pH 7.4, 1 mM EDTA, 10 mM dithiothreitol (DTT), 50% glycerol, autoclaved before adding DTT.
5. *E. coli* strain PMS 969 [PHR1] (Kindly provided by Prof. Aziz Sancar; **ref. 7**).

2.3. Repair Experiments and DNA Isolation

1. Minimal essential medium supplemented (MEM) with 10% fetal calf serum.
2. Earle's balanced salts without phenol red (Invitrogen).

2.4. Preparation and Processing of Dot Blots

1. Control DNA containing thymine dimers: Irradiate herring sperm DNA (0.1 mg/mL) in 10 nM acetophenone in an open Petri dish with midwave ultraviolet light (UV-B) (e.g., using a Westinghouse FS20 sunlamp). Under these conditions the only detectable photoproducts are thymine dimers *(8)*.
2. 1 N NaOH. Always make fresh.
3. Neutralizing solution: 3 M potassium acetate in 5 M acetic acid.
4. Nitrocellulose membrane (e.g., Schleicher and Schuell, Germany) (*see* **Note 1**).
5. 1 M Ammonium acetate.
6. 5X Saline sodium citrate (SSC): 0.75 M sodium chloride, 0.075 M sodium citrate, pH 7.0.
7. 1% Gelatin: warm in water to dissolve.
8. Carrier DNA: preparation of unirradiated DNA, prepared in the same way as the experimental samples. Dissolve at approx 1 mg/mL.
9. Phosphate-buffered saline (PBS): 20 mM sodium phosphate, 150 mM NaCl.
10. PBNT: PBS containing 0.5% normal goat serum, 0.5% bovine serum albumin (BSA), 0.05% Tween-20.
11. PBX: PBS containing 0.1% Tween-20.

12. Biotinylated antirabbit antiserum and alkaline phosphatase-conjugated Extr-Avidin (ExtrAvidin Alkaline Phosphatase staining kit, Sigma EXTRA-3A) (*see* **Note 2**).
13. Tris-buffered saline (TBS): 50 mM Tris-HCl, pH 7.4, 150 mM NaCl.
14. Alkaline phosphatase buffer: 100 mM NaCl, 5 mM MgCl$_2$, 100 mM Tris base (should be pH 9.5 without needing adjustment).
15. Alkaline phosphatase substrate: nitroblue tetrazolium, 5-bromo-4-chloro-3-indolyl phosphate (Invitrogen) (*see* **Note 2**).
16. PBS–EDTA: PBS containing 0.75% EDTA.
17. Any image analysis system suitable for quantitative analysis of color intensity on dot blots. (e.g., Bio-Rad Molecular Imager).

3. Methods
3.1. Preparation and Characterization of the Polyclonal Antiserum

The antiserum is raised in rabbits, following the protocol described by Mitchell and Clarkson (*4*).

1. Phenol-extract and ethanol-precipitate calf thymus DNA. Dissolve in isotonic saline at a concentration of 1 mg/mL.
2. Irradiate the DNA with UV-C in an open Petri dish on ice, giving a total dose of 100 kJ/m^2.
3. Readjust the concentration of the DNA to 0.4 mg/mL by isotonic saline as appropriate.
4. Prepare 1 mL of immunogen by mixing 0.5 mL of irradiated, heat-denatured DNA with 0.5 mL of MBSA. Mix well and filter-sterilize.
5. For the first injection, emulsify 1 mL of immunogen with 1 mL of complete Freund's adjuvant. Give four subsequent injections every 2 wk using incomplete adjuvant. Two weeks after the last injection, administer a booster of 200 µg of poly(dA) · poly(dT) DNA irradiated with a dose of 250 kJ/m^2. Preimmune serum and test bleeds taken after each injection must be checked for activity. Harvest the antiserum 2 wk after the booster. The exact details of this protocol must be approved and possibly modified according to local rules for animal handling, for example, with regard to use of adjuvants.
6. Test bleeds: Prepare test strips by applying a dilution series of denatured herring-sperm DNA, which has been irradiated with UV-C at 50 J/m^2 to dot blots, in the same way as for the repair assay (*see* **Subheading 3.5.**). Process the test strips in exactly the same way as for the repair assay (*see* **Subheading 3.6.**). The activity of the antiserum against total lesions and nondimer photoproducts should be monitored (**Fig. 1**).

3.2. Preparation of Crude Photolyase

This method is based on the first part of the purification procedure for photolyase described by Sancar et al. (*7*).

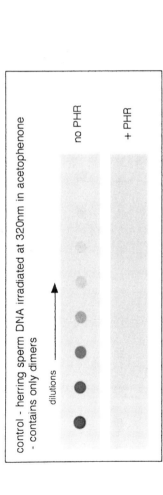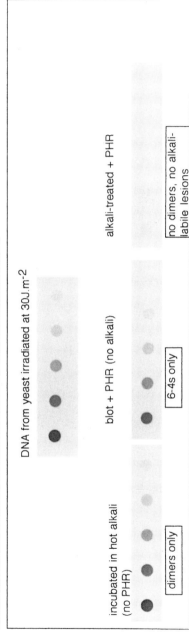

Fig. 1. Strip tests for polyclonal antiserum. The control DNA (top panel) contains only CPDs, which are completely removed by incubating the blot in photolyase (PHR) under visible light illumination (photoreactivation). Yeast DNA incubated in hot alkali (lower left) contains only CPDs, which are completely removed if the blot is treated with photolyase (lower right). Photolyase treatment alone removes CPDs (lower middle) and leaves alkali-labile sites, which are principally or entirely (6–4)PPs.

1. Grow *E. coli* [PHR1] in 1 L of Luria broth containing tetracycline (25 mg/L) to $OD_{600} = 1.0–1.1$. Add IPTG to 0.5 mM. Grow for a further 12 h.
2. Harvest the cells by centrifugation and wash in lysis buffer.
3. Resuspend in 20 mL of ice-cold lysis buffer. Divide into three, and sonicate (four 30-s pulses on ice). Keep the lysate cool.
4. Centrifuge at 31,000g (16,000 rpm) in a Sorvall SS-34 rotor at 4°C for 20 min.
5. Centrifuge the supernatant in an ultracentrifuge at 120,000g (35,000 rpm) in a Beckman Ti50 rotor at 4°C for 1 h.
6. To 20 mL of supernatant, add 8.6 g of ammonium sulfate, slowly, over a 1-h period, keeping on ice and swirling to dissolve well.
7. Centrifuge the yellow precipitate, in a sterile Corex tube, at 8000g (8000 rpm) in an SS-34 rotor for 30 min at 4°C.
8. Dissolve the precipitate in 5 mL of ice-cold storage buffer. Add 100-µL aliquots to precooled 0.5-mL microcentrifuge tubes and store at –70°C.
9. The photolyase preparation should be tested for photoreactivating activity on test strips (**Fig. 1**).

3.3. Repair Experiment

All operations should be carried out in a hot room at 37°C and all media and buffers should be pre-warmed.

1. For suspension cultures, centrifuge the cells and resuspend in Earle's balanced salts (Invitrogen) without phenol red at 2×10^5 cells/mL. Irradiate (or mock-irradiate) the cell suspension in plastic Petri dishes using a dose of 10 J/m^2 (with the lid off —UV does not penetrate plastic) with gentle agitation. You will need 10 mL of cell suspension for each time point. After irradiation, centrifuge the cells and resuspend in normal medium and place in the dark for further growth.
2. For cells grown in monolayers, seed the cells at 10^6 cells per 10-cm plate and allow to grow for 2 d. Prior to irradiation, pour off the medium and replace with Earle's balanced salts without phenol red. Irradiate the plates with the lid off using a dose of 10 J/m^2. One 10-cm Petri dish should be irradiated for each time point. After irradiation, pour off the balanced salts solution and replace with 10 mL of MEM containing 10% fetal calf serum (FCS).
3. Immediately after irradiation, harvest cells from one dish onto ice; wash in ice-cold PBS and immediately extract DNA. This will serve as the time-zero sample.
4. The remainder of the dishes should be incubated for times up to 24 h before DNA extraction.

3.4. DNA Extraction

There are many commercially available kits for genomic DNA preparation and any of these is suitable. Alternatively, DNA can be extracted by phenol–chloroform extraction as described in Sambrook et al. (**9**).

1. Isolated DNA should be salt-washed in 70% ethanol and finally redissolved at a concentration of approx 100–200 µg/mL in sterile distilled water.

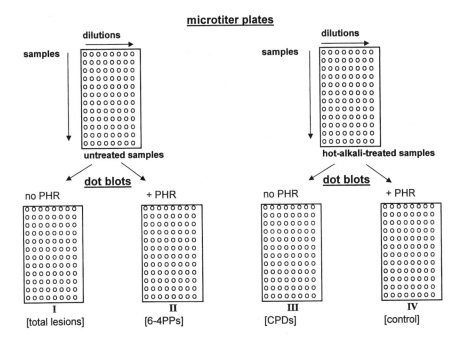

Fig. 2. Layout of microtiter plates and dot blots for a repair experiment on human cells. Doubling dilutions of the DNA samples are set up in microtiter plates and transferred to dot blots in the array illustrated.

2. Run 5-μL aliquots on agarose gels, and stain with ethidium bromide. Scan the gel and compare concentrations by densitometry. Adjust the concentrations, and run aliquots again on gels. Repeat until all the samples have identical DNA concentrations (*see* **Note 3**).

3.5. Preparation of Dot Blots

A suitable layout of the dot blots is shown in **Fig. 2**. Four blots are prepared—two identical blots for the hot-alkali-treated samples and two for the untreated samples. One of the latter is destined for photolyase treatment to measure (6–4)PPs. The other is used to measure total lesions. One of the blots with alkali-treated samples is used to measure CPDs. The other is treated with photolyase and serves as a control—both lesions should have been destroyed on this blot.

1. Divide 400 μL of each DNA sample into two 200-μL aliquots. To one, add 22 μL of freshly made 1 N NaOH.
2. Incubate at 90°C for 30 min, and cool on ice for 5 min. Then add 110 μL of neutralizing solution and 70 μL of water.

3. Treat the second 200-µL aliquot the same way, but omit the 90°C incubation.
4. Transfer 100-µL aliquots of all samples into siliconized microtiter plates, and set up a twofold dilution series in 1 *M* ammonium acetate in a 96-well microtiter plate as indicated in **Fig. 2**.
5. Transfer samples onto nitrocellulose filters, 100 µL per dot, using a vacuum dot-blotting apparatus. Rinse the filters in 1 *M* ammonium acetate and then in 5X SSC, dry, and bake at 80°C.

3.6. Developing the Dot Blots and Quantitating DNA Damage

The method is derived from that described by Wani et at. *(10)*.

1. Incubate the blots overnight in a 1% gelatin solution at 37°C.
2. Incubate the blots destined for measurement of (6–4)PPs in 20 mL of 50 m*M* Tris-HCl, pH 7.6, containing 100 µL of crude photoreactivating enzyme. Incubate the blots in individual plastic boxes for 5 min in the dark followed by 1 h under two 60-W desk lamps, using a piece of plate glass to cut out wavelengths below 320 nm.
3. Rinse all blots in PBS.
4. Incubate all blots at 37°C for 1 h in 20 mL of PBNT containing 1 mL of denatured crude unirradiated carrier DNA (to bind any nonspecific antibody) and 1 µL/mL (i.e., 1:1000) anti-UV-DNA polyclonal antiserum.
5. Wash the blots four times in PBX.
6. Incubate the blots for 1 h at 37°C in PBNT containing 1:1000 biotinylated antirabbit antiserum.
7. Wash the blots three times in PBX followed by two washes in TBS.
8. Incubate for 1 h at 37°C in 20 mL of TBS containing 1:1000 alkaline phosphatase-conjugated ExtrAvidin.
9. Wash the blots thoroughly in several changes of TBS, and then incubate, in the dark, in 15 mL of substrate solution for 5–10 min. Watch the reaction and stop before the background begins to go blue, by adding 25 mL of PBS-EDTA. Rinse the blots in water (*see* **Note 4**). Examples of processed dot blots are shown in **Fig. 3**.
10. Dry the blots and scan using a scanning densitometer with an image analysis facility. Measure the intensity of the blue color in the dots and set up a calibration curve for each set of samples using the serial dilutions of the time-zero sample as standards. Calculate the lesions remaining in the samples from each of the time points as a percentage of the lesions in the time-zero sample, as in **Fig. 4**. (*See* **Note 5**.)

4. Notes

1. Several types of membrane have been tried for this method. Nitrocellulose gives the lowest background and cleanest results. Nylon gives very high background and is not suitable.
2. Several different enzyme-linked assays and different substrates were used when setting up this assay. The one described here gave a low background and good sensitivity.

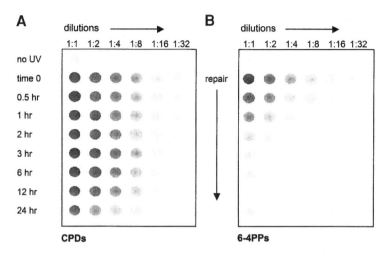

Fig. 3. Blots showing repair of cyclobutane pyrimidine dimers and (6–4) photoproducts in HeLa cells after a dose of 10 J/m^2. (**A**) Samples have been treated with hot alkali prior to loading onto the blot to destroy (6–4) photoproducts. So only CPDs remain. (**B**) Samples have not been treated with hot alkali. Instead the entire blot has been treated with photolyase under visible light illumination to photoreactivate CPDs. Only (6–4)PPs remain.

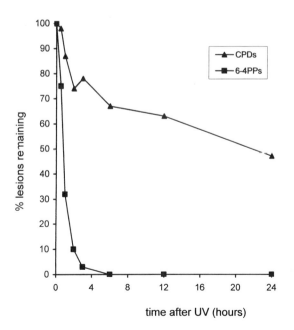

Fig. 4. Repair of CPDs and (6–4)PPs in HeLa cells after a UV dose of 10 J/m^2, calculated from scans of the blots shown in **Fig. 3**.

3. It is crucial to equalize the DNA in the samples from the various time points. This cannot be done accurately with a spectrophotometer and is difficult to do accurately even with a fluorimeter. The gel method described is the only one we have found to be adequate.
4. When incubating with the substrate, it is essential to keep the solution dark, to agitate the solution, to keep the blot well covered, and to stop the reaction before the background begins to go blue.
5. Although the method is only semiquantitative, it gives very reproducible results provided care is taken to choose dilutions where the intensity of the blue color is not near saturation, that is, choose the linear part of the calibration curve.

References

1. McCready, S. J. and Cox, B. S. (1993) The repair of 6–4 photoproducts in *Saccharomyces cerevisiae*. *Mutat. Res.* **293,** 233–240.
2. McCready, S. J., Carr, A. M., and Lehmann, A. R. (1993) The repair of cyclobutane pyrimidine dimers and 6–4 photoproducts in *Schizosaccharomyces pombe*. *Mol. Microbiol.* **10,** 885–890.
3. McCready, S. J. (1996) Induction and repair of UV photoproducts in the salt-tolerant archaebacteria, *Halobacterium cutirubrum, Halobacterium halobium* and *Haloferax volcanii. Mutat. Res.* **364,** 25–32.
4. Mitchell, D. L. and Clarkson, J. M. (1981) The development of a radioimmunoassay for the detection of photoproducts in mammalian cell DNA. *Biochem. Biophys. Acta* **655,** 54–60.
5. Mitchell, D. L. and Nairn, R. S. (1989) The biology of the (6–4) photoproduct. *Photochem. Photobiol.* **49,** 805–820.
6. McCready, S. J. (1999) A dot blot immunoassay for UV photoproducts. *Methods Mol. Biol.* **113,** 147–156.
7. Sancar, A., Smith, F. W., and Sancar, G. B. (1984) Purification of *Escherichia coli* DNA photolyase. *J. Biol. Chem.* **259,** 6028–6032.
8. Lamola, A. A. (1969) Specific formation of thymine dimers in DNA. *Photochem. Photobiol.* **9,** 291–294.
9. Sambrook, J., Frisch, E. F., and Maniatis, T. (1989) *Molecular Cloning, a Laboratory Manual.* Cold Spring Harbor Laboratory Press, Plainview, NY.
10. Wani, A. A., d' Ambrosio, S. M., and Nasir, A. K. (1987) Quantitation of pyrimidine dimers by immunoslot blot following sublethal UV-irradiation of human cells. *Photochem. Photobiol.* **46,** 477–482.

18

Quantification of Photoproducts in Mammalian Cell DNA Using Radioimmunoassay

David L. Mitchell

Summary

Over the past 20 yr, the use of polyclonal and monoclonal antibodies to quantify damage in DNA has burgeoned. Immunoassays offer distinct advantages over other analytical procedures currently used to measure DNA damage including adaptability, sensitivity and selectivity. This combination of attributes allows for the development of powerful analytical techniques to visualize and quantify specific types of DNA damage in cells and organisms exposed to subtoxic levels of xenobiotics with distinct advantages over the other procedures in the analysis of DNA damage in human and environmental samples. Radioimmunoassay (RIA) is readily applied to a variety of biological materials and has typically been used to measure DNA damage in cell and organ cultures, tissue sections and biopsies, buccal cells, bone marrow aspirates, peripheral blood lymphocytes, and urine. Here we describe the use of a very sensitive RIA for the specific quantitation of cyclobutane dimers and (6–4) photoproducts in DNA extracted from mammalian cells and tissues.

Key Words: Cyclobutane dimer; (6–4) photoproduct; polyclonal antibody; radioimmunoassay; ultraviolet light.

1. Introduction

Radioimmunoassay (RIA) is a competitive binding assay between an unlabeled and a radiolabeled antigen for binding to antibody raised against that antigen. Yalow and Berson *(1)* received a Nobel Prize in Medicine for its development. For detailed theory and troubleshooting of RIA refer to Harlow and Lane *(2)* or Chard *(3)*. We have adapted this technique to the measurement of specific DNA photoproducts in the DNA of ultraviolet (UV)-irradiated cells *(4–6)*. The following description is given for quantification of cyclobutane pyrimidine dimers (CPDs) and pyrimidine(6–4)pyrimidinone

From: *Methods in Molecular Biology: DNA Repair Protocols: Mammalian Systems, Second Edition*
Edited by: D. S. Henderson © Humana Press Inc., Totowa, NJ

photodimers ([6–4]PDs) in DNA using RIA. For convenience, the radiolabeled antigen is referred to as the "probe" and the unlabeled competitor as "sample" or "standard." The amount of radiolabeled antigen bound to antibody is determined by separating the antigen–antibody complex from free antigen by secondary antibody by immunoprecipitation (**Fig. 1**). The amount of radioactivity in the antigen–antibody complex in the presence of known amounts of competitor (i.e., standards) is used to quantify the amount of unknown sample present in the reaction. The sensitivity of the RIA is determined by the affinity of the antibody and specific activity of the radiolabeled antigen. Using high-affinity antibody and probe labeled to a high specific activity, the reaction can be limited to such an extent that extremely low levels of damage in sample DNA can be quantified. This particular procedure has resulted from 20–25 yr of research and has proven to be a reliable and facile technique for measuring DNA damage and repair end points. That is not to say that variations on this basic procedure will not be as productive or useful in DNA damage and repair studies.

2. Materials

1. H_2O (high-performance liquid chromatography [HPLC] or Millipore-filtered).
2. Salmon testes (or calf thymus) DNA (Sigma).
3. Acetone.
4. Methylated bovine serum albumin (Sigma A1009) (*see* **Note 1**).
5. DNA nick-translation kit (Roche cat. no. 976 776).
6. ^{32}P-Labeled deoxynucleotide triphosphates (dNTPs).
7. TE buffer: 10 mM Tris-HCl, pH 8.0, 1 mM EDTA.
8. ^{14}C-labeled thymidine deoxyribonucleoside ([^{14}C]TdR).
9. Lysis buffer A: 10 mM Tris, pH 8.0, 1 mM EDTA, 0.5% sodium dodecyl sulfate [SDS], and 0.3 mg/mL of Proteinase K [Roche cat. no. 745-723] (*see* **Note 2**).
10. Lysis buffer B: 10 mM Tris-HCl, pH 8.0, 1 mM EDTA, 0.5% SDS, 100 µg/mL of DNase-free RNase A (Roche cat. no. 109-169) (*see* **Note 2**).
11. Chloroform:isoamyl alcohol (24:1 v/v).
12. Tris-saturated phenol (pH > 7.6) (Roche cat. no. 100-997).
13. 10X TES: 100 mM Tris-HCl, pH 8.0, 10 mM EDTA, 1.5 M NaCl.
14. Gelatin (type B: bovine skin) (Sigma cat. no. G-9382).
15. RIA buffer: 1X TE + 0.15% gelatin (*see* **Note 3**).
16. Normal rabbit serum (Calbiochem cat. no. 566442); stored frozen in 200-µL aliquots (*see* **Note 4**).
17. Goat antirabbit IgG [Calbiochem cat. no. 539844 or cat. no. 539845 (bulk)]; stored frozen in 0.5-mL aliquots (*see* **Note 5**).
18. Tissue Solubilizer (NCS-II from Amersham; cat. no. NNCS.502) supplemented with 10% (v/v) H_2O.
19. Scintillation cocktail (e.g., ScintiSafe from Fisher) containing 1 mL/L of acetic acid (*see* **Note 6**).

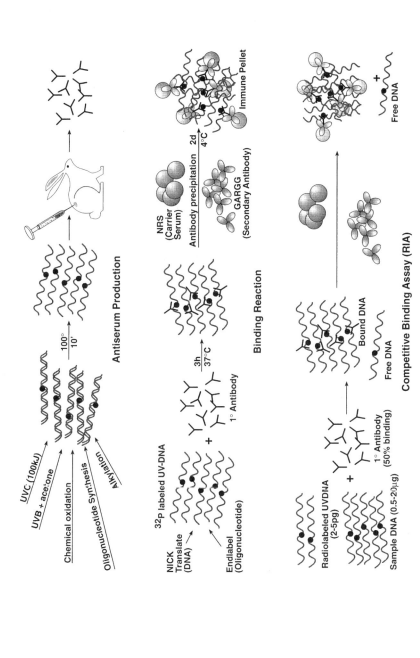

Fig. 1. Diagram of RIA protocol. (Top) Antibodies are raised against a specific type of DNA lesion, e.g., CPD. A variety of methods and treatments can be used to damage or modify bases in DNA for use as an immunogen. (Middle) Binding activity of antiserum is characterized. (Bottom) RIA is used to measure lesion levels in sample DNA.

3. Methods

3.1. Preparation of Immunogen

1. Dilute commercial DNA (e.g., salmon testes or calf thymus from Sigma) to 1 mg/mL in 10 mL of sterile H_2O (as determined by optical density at 260 nm) (*see* **Note 7**).
2. UV-irradiate the dilute double-stranded (or single-stranded) DNA using one of the following protocols:
   a. Prepare the immunogen for anti-CPD sera by irradiating DNA diluted in 10% acetone (**ref. 7**) with approx 75 kJ/m^2 of UV-B light in a glass 100-mm plate (*see* **Note 8**).
   b. Prepare the immunogen for anti-(6–4)PD sera by irradiating DNA with 60 kJ/m^2 UV-C light (*see* **Note 9**).
3. Heat denature the UV-irradiated DNA at 100°C for 10 min.
4. Electrostatically couple the UV-irradiated DNA (heat denatured or not; *see* **Note 9**) to methylated bovine serum albumin (MBSA) (*see* **Note 10**) *(8)*.

3.2. Immunization Schedule

1. Initially inject four New Zealand White female rabbits (*see* **Note 11**) subcutaneously at 10 sites (100 µL each) with 0.5 mL of immunogen mixed with an equal volume of Freund's Complete Adjuvant (final concentration of 1.0 mg/mL of UV-DNA).
2. Subsequently, inject rabbits at 2-wk intervals except that Freund's Incomplete, rather than Complete, Adjuvant is mixed with 0.5 mL of immunogen giving a final concentration of 0.4 mg/mL of UV-DNA.
3. At 10–12 d following the second injection, draw 1 mL of serum and evaluate the binding affinity using immunoprecipitation (*see* **Subheading 3.3.**).
4. Continue immunization at 2-wk intervals until sufficient binding activity is attained, at which time draw antiserum (60–80 mL) from the animal using heart puncture (*see* **Note 12**).
5. Dispense antiserum into 1-mL aliquots and store at –20°C (*see* **Note 13**).

3.3. Determination of Antiserum Binding Using Immunoprecipitation (Fig. 1)

1. Nick-translate DNA (0.1 µg) with [^{32}P]dCTP or [^{32}P]TTP to give a specific activity of at least $5 \times 10^8 – 10^9$ cpm/µg (*see* **Note 14**). A typical reaction includes:
   a. 2 µL of 10X buffer (from kit).
   b. 2 µL of dATP (for poly[dA]:poly[dT]); or 2 µL each dATP and dGTP (for DNA).
   c. 0.5 µL of poly(dA)–poly(dT) or *Clostridium perfringens* DNA (diluted to 20 µg/100 µL).
   d. 12.5 µL [^{32}P]TTP (10 mCi/mL, NEN or Amersham).
   e. 3–4 µL of enzyme mix (from kit).

f. Incubate for 30–45 min at 15°C.
g. Separate radiolabeled ligand from free dNTPs using a nick column (Amersham Biosciences) equilibrated and eluted with 1X TE buffer.
2. Irradiate ^{32}P-labeled probe with 30 kJ/m^2 UV-C light (*see* **Notes 9** and **15**).
3. Restore the volume (due to evaporation) with H$_2$O and dilute 2500- to 5000-fold in RIA buffer (yielding 2.5–5.0 pg of probe in 50 µL of buffer) (*see* **Note 16**).
4. Add 1 mL of RIA buffer to duplicate 12-mm disposable culture tubes (*see* **Note 17**).
5. Add 50 µL of antiserum diluted in RIA buffer at half-log increments from 1:1000 to 1:1,000,000 (dilution prior to dispensing). Dispense duplicate tubes without antiserum to determine background.
6. Add 50 µL of diluted ^{32}P-labeled probe (from **step 3**) and vortex-mix well.
7. Incubate 3–4 h with gentle rotation (optional) in a 37°C dry incubator.
8. Separately, add 50 µL of normal rabbit serum diluted 1:40 in RIA buffer and 50 µL of goat antirabbit IgG diluted 1:20 and vortex-mix well.
9. Incubate at 4°C for 2 d until the immune pellet (translucence) develops.
10. Centrifuge the tubes at approx 2500*g* (approx 3500–4000 rpm) for 30–45 min.
11. Decant the supernatant, invert the tubes onto absorbent paper in a test tube rack, and drain for 5–10 min (*see* **Note 18**).
12. Swab the lip of the test tube with a cotton-tipped applicator wrapped in tissue (to remove any accumulated liquid).
13. Add 100 µL of NCS Tissue Solubilizer supplemented with 10% H$_2$O and incubate at 37°C (or room temperature) with rotation until the immune pellet is completely dissolved (*see* **Note 19**).
14. Add 2 mL of scintillation cocktail (e.g., ScintiSafe from Fisher) supplemented with 1 mL/L of acetic acid and vortex-mix.
15. Either place the tube directly in a 20-mL scintillation vial or decant the sample into the scintillation vial and wash twice with 4 mL of additional scintillation cocktail.
16. Count ^{32}P using liquid scintillation counter.

3.4. Treatment and Isolation of Cultured Mammalian Cell DNA

1. Plate 2.5–3 × 10^6 cells in 7 × 100 mm plates (duplicate or triplicate plates can be used) with media (α-MEM) containing 0.005–0.01 mCi/mL of [^{14}C]TdR 2 d prior to irradiation (*see* **Note 20**).
2. For a DNA repair experiment, irradiate all but one plate (-UV control) with 10–20 J/m^2 UV-C light or UV-B equivalent (*see* **Note 21**). The perimeter of the plates should be swabbed with a cotton-tipped applicator to remove cells that would otherwise be shielded from the irradiation. Pour off the media and wash the plates once with 1X TES (or PBS).
3. Harvest one irradiated plate (with duplicate) at the time of irradiation by scraping with a rubber policeman into a 15-mL polypropylene centrifuge tube (trypsinization can be used to lift adherent cells from the plate). Additional plates should be harvested at, for example, 1.5, 3, 6, 24, and 48 h post-irradiation for repair studies.

4. Centrifuge at approx 150g (500–1000 rpm) for 5 min to pellet the cells; decant the buffer.
5. Add 4 mL of lysis buffer A or B, mix vigorously, and incubate overnight at 37°C or 2–3 h at 60°C.
6. Extract with 4 mL of Sevag (chloroform–isoamyl alcohol, 24:1 v/v) and transfer the aqueous phase to a 30-mL Corex tube.
7. Add 0.4 volume (1.6 mL) of 5 M sodium acetate (*see* **Note 22**) and 2.5 volumes (14 mL) of ice-cold absolute ethanol. Place in a freezer overnight.
8. Centrifuge the sample at 12,000g in an SS-34 rotor (10,000 rpm) at 0°C for 20 min to pellet DNA. Decant the supernatant away from the side containing the pellet.
9. Wash the pellet with 5–10 mL of 70% ice-cold ethanol.
10. Invert and partially dry the pellet for 30–60 min at 23°C and resuspend in 1.5 mL of sterile H_2O or TE buffer. Allow several hours with periodic vortex-mixing for the pellet to completely resuspend. Drying the pellet too long will make resuspension difficult.
11. Determine DNA concentration using absorption or spectrofluorometry (*see* **Note 20**).
12. After heat denaturation count 20–50 µL to determine the level of [^{14}C]DNA (if cells were prelabeled).
13. Place in refrigerator at 4°C for short-term or in –20°C freezer for long-term storage.

3.5. Competitive Binding Assay (RIA)

The RIA is simply the basic immunoprecipitation reaction outlined in **Subheading 3.3.** into which a standard or sample DNA has been added to compete with the radiolabeled probe for antibody binding (**Fig. 1**). Hence, the procedure is exactly the same as that used for immunoprecipitation with the following additions/modifications:

1. A single dilution of antiserum is used. This dilution is determined from immunoprecipitation analyses of binding activity (**Subheading 3.3.**) and should yield 30–60% of the radiolabeled probe in the immune pellet.
2. For quantification of CPDs or (6–4)PDs a dose response of heat-denatured UV-irradiated salmon testes DNA is used to generate a standard curve. We routinely use doses of 3, 10, 30, 100, and 300 J/m^2 as our standard curve and assay the same amount of standard as sample DNA (*see* **Note 23**). When relative, rather than exact, amounts of CPDs or (6–4)PDs are adequate for experimental purposes (as in repair kinetics experiments) the sample harvested at the time of irradiation is titrated in half-log increments to determine the optimal amount required for assay.
3. Unlabeled competitor mammalian DNA, radioactive ligand, and diluted antibody are incubated together for 3 h at 37°C with gentle rotation (optional). It is prudent to perform a preliminary titration of sample DNA to determine the amount

required for adequate inhibition in the RIA. The total volume and concentration of sample DNA added can vary within certain limitations (*see* **Note 24**).

3.6. Data Analysis

1. A sample Excel spreadsheet for quantification of CPDs or (6–4)PDs is shown in **Fig. 2**. Formulae for quantifying CPDs are shown in **Fig. 2A**. An identical spreadsheet can be used to quantify (6–4)PPs.
2. A sample Excel spreadsheet for quantification of *relative* photoproducts (PDs) remaining at specific times post UV-irradiation (e.g., in a DNA repair experiment) is shown in **Fig. 2B**.

4. Notes

1. Methylated BSA can be frozen and thawed *ad infinitum*. MBSA is added dropwise with a Pasteur pipet (approx 50 µL/drop) until the UV-DNA is significantly opaque (i.e., until further addition of MBSA does not change the cloudiness of the solution).
2. Lysis buffer A is used for "crude" extractions in which the DNA has been prelabeled with [^{14}C]TdR and the RIA is set up to determine *relative* amounts of photoproduct remaining in a DNA repair experiment (as shown in **Fig. 2B**). In such an experiment actual quantification of damage is not required since the 0 h sample is titrated to serve as a standard curve. In this case it is necessary only to analyze equivalent amounts of ^{14}C. Lysis buffer B is used when numbers of photoproducts per megabase of DNA are required. In this case the concentration of the DNA sample is critical and care must be taken to assure accurate quantification. A more standard DNA isolation procedure is called for which includes lysis in the presence of RNase, followed by Proteinase K digestion, organic extractions with equal volumes of phenol, phenol–Sevag (1:1), and Sevag, and precipitation with 2 volumes of ethanol in the presence of 0.4 volume 5 *M* ammonium acetate. After precipitation and washing (with 70% ethanol) duplex DNA is quantified using absorbance at 260 nm (assuming the $A_{260/280}$ is >1.7) or spectrofluorometry using a DNA-specific dye (e.g., Hoescht or DAPI). From **Fig. 2A,B** it is evident that equivalent amounts of standard and sample are required to determine photoproduct concentrations.
3. RIA buffer consists of 1X TES to which 0.15% gelatin (w/v) (Sigma) has been added to reduce nonspecific binding. The gelatin is heated into solution using a hotplate magnetic stirrer (not microwave) and heated to precisely 39–40°C. Overheating (by as much as 1°C) will result in prohibitive background! The cause of this is unknown. This is the first step that should be checked when troubleshooting poor or erratic binding and high background.
4. Normal rabbit serum (NRS) from Calbiochem has been titrated and we have found that a 1:40 dilution is optimal for immune pellet formation. Obviously other sources are readily available; however, we suggest that titrations be performed using the described immunoprecipitation protocol to determine optimal dilution.

A

Tube #	Sample	UV Dose	ul/(x)ug	ug/ml	cpm	cpm-bkg	%inhibition		CPD/mb			
1	-Ab					=AVERAGE(F2,F3)						
=A2+1	+Ab					=F4-$G$3	=AVERAGE(G4,G5)					
=A4+1	standard		3		x	=F5-$G$3						
			3			=F6-$G$3	=(1-($G6/$H$5))*100		=LOG(C6^8.1)	=RSQ(H1:H5,I1:I5)	Corr.Coef.
=A6+1	standard		10	=(x/E6)*1000		=F7-$G$3	=(1-($G7/$H$5))*100		=LOG(C7^8.1)		
			10		x	=F8-$G$3	=(1-($G8/$H$5))*100		=LOG(C8^8.1)	=SLOPE(H1:H5,I1:I5)	Slope
=A8+1	standard		30	=(x/E8)*1000		=F9-$G$3	=(1-($G9/$H$5))*100		=LOG(C9^8.1)		
			30		x	=F10-$G$3	=(1-($G10/$H$5))*100		=LOG(C10^8.1)	=INTERCEPT(H1:H5,I1:I5)	Y-intercept
=A10+1	standard		100	=(x/E10)*1000		=F11-$G$3	=(1-($G11/$H$5))*100		=LOG(C11^8.1)		
			100		x	=F12-$G$3	=(1-($G12/$H$5))*100		=LOG(C12^8.1)		
=A12+1	sample	x		=(x/E12)*1000		=F13-$G$3	=(1-($G13/$H$5))*100		=LOG(C13^8.1)		
				=(x/E14)*1000	x	=F14-$G$3	=(1-($G14/$H$5))*100	=10^(($H14-$K$10)/$K$8)			
=A14+1	sample	y				=F15-$G$3	=(1-($G15/$H$5))*100	=10^(($H15-$K$10)/$K$8)			
				=(x/E16)*1000	x	=F16-$G$3	=(1-($G16/$H$5))*100	=10^(($H16-$K$10)/$K$8)			
						=F17-$G$3	=(1-($G17/$H$5))*100	=10^(($H17-$K$10)/$K$8)			

B

Tube #					cpm	cpm-bkg	%inhibition					
1	-Ab					=AVERAGE(G2,G3)						
=A2+1	+Ab					=G4-$H$3	=AVERAGE(H4,H5)					
	Cell line	Dose	Time (h)	ul/(x)cpm	cpm/20 ul	=G5-$H$3		% PD remaining				
=A4+1	-UV		0h	(y)ul		=G7-$H$3	=AVERAGE(H7,H8)					
=A7+1	+UV		0h	=E13*0.1		=G8-$H$3						
						=G9-$H$3	=(1-($H9/#REF!))*100	10	=LOG(J9)	=RSQ(I9:I14,K9:K14)	Corr.Coef.
=A9+1	+UV		0h	=E13*0.3		=G10-$H$3	=(1-($H10/#REF!))*100	10	=LOG(J10)		
						=G11-$H$3	=(1-($H11/#REF!))*100	30	=LOG(J11)	=SLOPE(I9:I14,K9:K14)	Slope
=A11+1	+UV		0h	(y)ul		=G12-$H$3	=(1-($H12/#REF!))*100	30	=LOG(J12)		
						=G13-$H$3	=(1-($H13/#REF!))*100	100	=LOG(J13)	=INTERCEPT(I9:I14,K9:K14)	Y-intercept
=A13+1	+UV	1.5		(y)ul		=G14-$H$3	=(1-($H14/#REF!))*100	100	=LOG(J14)		
						=G15-$H$3	=(1-($H15/#REF!))*100		=10^(($I15-$L$13)/$L$11)		
=A15+1	+UV	3		(y)ul		=G16-$H$3	=(1-($H16/#REF!))*100		=10^(($I16-$L$13)/$L$11)		
						=G17-$H$3	=(1-($H17/#REF!))*100		=10^(($I17-$L$13)/$L$11)		
=A17+1	+UV	6		(y)ul		=G18-$H$3	=(1-($H18/#REF!))*100		=10^(($I18-$L$13)/$L$11)		
						=G19-$H$3	=(1-($H19/#REF!))*100		=10^(($I19-$L$13)/$L$11)		
=A19+1	+UV	24		(y)ul		=G20-$H$3	=(1-($H20/#REF!))*100		=10^(($I20-$L$13)/$L$11)		
						=G21-$H$3	=(1-($H21/#REF!))*100		=10^(($I21-$L$13)/$L$11)		
						=G22-$H$3	=(1-($H22/#REF!))*100		=10^(($I22-$L$13)/$L$11)		

5. Goat antirabbit IgG can be purchased from Calbiochem in bulk or in smaller aliquots. The bulk product requires a greater concentration than the individual 5-mL aliquots and we suggest, as previously, that the optimal dilution be determined using the immunoprecipitation protocol.
6. Tissue solubilizer is very basic and will chemoluminesce in scintillation cocktail resulting in very high counts. This effect will naturally dissipate within 12–18 h but can be immediately quenched by the addition of acetic acid in the cocktail.
7. Commercial DNA does not usually require repurification. However, the purity should be checked using the $A_{260/280}$ with values >1.7 acceptable.
8. The UV-B source consists of four Westinghouse FS20 sunlamps filtered through cellulose acetate (Kodacel from Kodak) with a wavelength cutoff of 290 nm *(9)*. Dosimetry is determined with an appropriate photometer/radiometer (e.g., IL1400 photometer coupled to a SCS 280 probe). At a distance of approx 10 cm the fluence rate is approx 5 J/m²/s, hence, exposure times of approx 4 h are required for adequate CPD induction. DNA diluted in photosensitizing solvent should not be irradiated in plastic. The DNA is extensively dialyzed post-irradiation to remove any acetone.
9. The UV-C source consists of a bank of 5 Philips Sterilamp G8T5 bulbs emitting predominantly 254 nm light. At a distance of approx 20 cm the fluence rate is approx 14 J/m²/s and at this fluence rate the average duration of exposure is approx 2 h. This high UV-C dose results in extensive fragmentation and denaturation of the DNA, precluding heat denaturation prior to MBSA coupling.
10. Methylated BSA is added dropwise (approx 50 µL each drop) until the UV-irradiated DNA solution turns cloudy.
11. We have found that individual rabbits have very different immune responses *(5)*, hence we recommend that at least four animals be used for raising anti-UV DNA antibodies until an antiserum with high sensitivity is obtained (e.g., 50% inhibition in the RIA using 1 µg of denatured DNA containing approx 10 CPDs or (6–4)PDs per megabase).
12. An alternate strategy has yielded high titer antisera. In brief, if an adequate titer (and sensitivity) is not reached after several biweekly injections, a 4–6-wk hiatus followed by a single injection with 100 µg/100 µL UV-DNA may yielded a more robust immune response.
13. Repeated freezing and thawing of antiserum is to be strictly avoided since this can severely reduce binding activity.
14. Both CPD and (6–4)PD frequencies are greatest in nucleic acid substrates containing a high A:T content. Hence, optimal substrates for the radiolabeled probe include *C. perfringens* DNA (70% A:T) as well as the homopolymer poly(dA)–poly(dT). The higher the specific activity the less probe can be used (*see* **Note 16**).

Fig. 2. *(Opposite page)* Microsoft Excel spreadsheets showing calculations used in RIA experiments. **(A)** Formulae used for quantifying CPDs. **(B)** Formulae for quantifying relative levels of photoproducts (PD).

Hence, DNA or polynucleotide can be labeled with >1 nucleoside triphosphate to increase specific activity.

15. Facile irradiation of small volumes of DNA can be achieved using a 25-mm plate or 24-well culture plate in which a depression has been made in a parafilm covering. By drilling a small hole in the bottom of the well air is released to prevent puckering.
16. The amount of probe added to the RIA determines its sensitivity. It is essential to use 10 pg or less and have enough cpm in the assay to yield useful binding (and inhibition) data. Hence, if a 1:5000 dilution of probe leaves <500 cpm in 50 µL, a greater concentration must be used. Good assay conditions should be limited to at least 500 cpm/50 µL (added to reaction) at a probe dilution not to exceed 1:1250.
17. We use 12-mm culture tubes (from Fisher or Baxter) that have colored labels. This helps separate the components of the RIA (i.e., binding conditions, standard curve and sample groups) for more facile visual recognition and less error.
18. Pellets can be inverted for 15–30 min to allow drainage but care should be taken that pellets do not slide down the face of the tube. A second wash is optional and may reduce background.
19. It is extremely important that the immune pellet be *completely* solubilized but not allowed to dry. Partial solubility will result in reduced and variable counts in duplicate samples.
20. Prelabeling with [^{14}C]TdR is optional. Radiolabeled sample DNA allows for the facile dilution to equivalent amounts of DNA for assay (assuming the specific activities of the samples are the same). For repair experiments, using equivalent ^{14}C counts ensures that only DNA at the time of irradiation (excluding any nascent DNA) is assayed, thus avoiding any artifactual repair due to cell proliferation and DNA dilution. Unlabeled DNA can be used for DNA damage measurements as long as the concentration of DNA is confidently known and any "proliferation effects" are accounted for. We measure DNA concentrations after sample denaturation and immediately prior to assay using an Oligreen fluorescent reagent (Molecular Probes, Eugene, OR). The amount of [^{14}C]TdR added to the cells and the duration of the prelabeling depends on the doubling time of the particular cells being studied. Our experience with transformed human fibroblasts and Chinese hamster ovary cells (with doubling times of 24 h) has shown that 2-d incubation with label results in specific activities of 1000–3000 cpm/µg of DNA.
21. We have found that unfiltered FS20 sunlamps induce 1/10 the amount of CPDs in DNA as UV-C irradiation. FS20 sunlamps filtered through cellulose acetate induce 1/100 the amount of damage as that produced by UV-C irradiation.
22. 5 *M* Sodium acetate should be filter-sterilized and stored at 4°C. This buffer should be checked prior to use for growth of contaminating organisms (cloudiness).
23. The damage frequencies used in the standard curves (**Fig. 2A,B**) are determined from independent, nonimmunological measures of CPDs and (6–4)PDs. For our

light source the rate of CPD induction was determined using a supercoiled plasmid assay and T4 endonuclease V; (6–4)PD induction was determined using the same assay and photoinduced alkali-labile site (PALS) analysis *(10)*.
24. Sample volumes <100 µL do not significantly affect the reaction conditions (e.g., total binding). Sample volumes >100 µL can be used; however, the total reaction volume should be increased accordingly (i.e., doubled). We have found that sample amounts up to 20 µg yield reasonable data.

References

1. Yalow, R. S. and Berson, S. A. (1959) Assay of plasma insulin in human subjects by immunological methods. *Nature* **184,** 1648–1649.
2. Harlow, E. and Lane, D. (1988) *Antibodies: A Laboratory Manual,* Cold Spring Harbor Laboratory Press, Plainview, NY, pp. 553–612.
3. Chard, T. (1990) *An Introduction to Radioimmunoassay and Related Techniques* (4th Revised Ed.), *Laboratory Techniques in Biochemistry and Molecular Biology,* Vol. 6, Part II (Burdon, R. H. and van Knippenberg, P. H., eds.), Elsevier, Amsterdam, The Netherlands.
4. Mitchell, D. L. and Clarkson, J. M. (1981) The development of a radioimmunoassay for the detection of photoproducts in mammalian cell DNA. *Biochim. Biophys. Acta* **655,** 54–60.
5. Mitchell, D. L. (1996) Radioimmunoassay of DNA damaged by ultraviolet light, in *Technologies for Detection of DNA Damage and Mutations* (Pfeifer, G. P., ed.), Plenum, New York, NY, pp. 73–85.
6. Jeffrey, W. H. and Mitchell, D. L. (2001) Measurement of UVB induced DNA damage in marine planktonic communities, in *Methods in Marine Microbiology* (Paul, J., ed.), Academic Press, New York, NY, pp. 469–488.
7. Lamola, A. A. and Yamane, T. (1967) Sensitized photodimerization of thymine in DNA. *Proc. Natl. Acad. Sci. USA* **58,** 443–446.
8. Plescia, O. J., Braun, W., and Palczuk, N. C. (1964) Production of antibodies to denatured deoxyribonucleic acid (DNA). *Proc. Natl. Acad. Sci. USA* **52,** 279–285.
9. Rosenstein, B. S. (1984) Photoreactivation of ICR 2A frog cells exposed to solar UV wavelengths. *Photochem. Photobiol.* **40,** 207–213.
10. Mitchell, D. L., Brash, D. E., and Nairn, R. S. (1990) Rapid repair of pyrimidine (6–4)pyrimidone photoproducts in human cells does not result from change in epitope conformation. *Nucleic Acids Res.* **18,** 963–971.

19

DNA Damage Quantitation by Alkaline Gel Electrophoresis

Betsy M. Sutherland, Paula V. Bennett, and John C. Sutherland

Summary

Quantifying DNA lesions provides a powerful way to assess the level of endogenous damage or the damage level induced by radiation, chemical or other agents, as well as the ability of cells to repair such damages. Quantitative gel electrophoresis of experimental DNAs along with DNA length standards, imaging the resulting dispersed DNA and calculating the population average length allows accurate measurement of lesion frequencies. Number average length analysis provides high sensitivity and does not require any specific distribution of lesions within the DNA molecules. These methods are readily applicable to strand breaks and ultraviolet radiation induced pyrimidine dimers, but can also be used—with appropriate modifications—for ionizing radiation-induced lesions such as oxidized bases and abasic sites.

Key Words: agarose gel; alkaline electrophoresis; DNA damage; DNA length standards; gel electrophoresis; pyrimidine dimers; strand breaks.

1. Introduction

Physical and chemical agents in the environment, those used in clinical applications, or those encountered during recreational exposures to sunlight induce damages in DNA. Understanding the biological impact of these agents requires quantitation of the levels of such damages in laboratory test systems as well as in field or clinical samples. Alkaline gel electrophoresis provides a sensitive (down to a few lesions per 5 megabases), rapid method of direct quantitation of a wide variety of DNA damages in nanogram quantities of nonradioactive DNAs from laboratory, field, or clinical specimens, including higher plants and animals. This method stems from velocity sedimentation studies of DNA populations, and from the simple methods of agarose gel electrophoresis. Our laboratories have developed quantitative agarose gel methods, analytical descriptions of DNA migration during electrophoresis on agarose gels *(1–6)*, and electronic imaging for accurate determinations of DNA mass *(7–9)*. Although all these components improve sensitivity and throughput of large numbers of samples *(7,8,10)*, a simple version using only standard molecular biology equipment allows routine analysis of DNA damages at moderate frequencies. We present here a description of the methods, as well as a brief description of the underlying principles, required for a simplified approach to quantitation of DNA damages by alkaline gel electrophoresis.

2. Materials

All solutions for DNA isolation, cleavage, and gel electrophoresis are sterilized by appropriate means. Gels are handled using powder-free gloved hands.

2.1. DNAs

1. Molecular length standards: DNA length standards should span the lengths of experimental DNAs. Static field gel electrophoresis resolves only molecules less than approx 50 kilobases (kb) *(11)*; thus standards should include DNAs ≤50 kb. Commercially available standards include λ DNA (48.5 kb), and *Hin*dIII digest of λ (23.1, 9.4, 6.6, 4.4, 2.3, 2, and 0.56 kb). Because DNA conformation affects mobility on electrophoretic gels, neither circular nor supercoiled DNAs should be used as molecular length standards for linear DNAs. For both λ DNA and *Hin*dIII digest of λ, aliquot into single-use portions and store at –20°C.
2. Tris–EDTA buffer (TE): 10 mM Tris, pH 8.0, 1 mM EDTA.
3. Agarose for sample embedding (SeaPlaque or InCert agarose, FMC, Rockland ME).
4. Proteinase K, 10 mg/mL stock in 10 mM Tris-HCl, pH 7.5. Prepare Proteinase K solutions at 1 mg/mL in TE, and in 10 mM Tris, pH 7.5, 1 mM CaCl$_2$, then predigest solutions for 1 h at 37°C. Check for endonuclease activity (integrity of supercoiled DNA); incubate supercoiled DNA with Proteinase K solutions at 37°C for 1 h and overnight in both buffers. If satisfactory, purchase large quantities of that lot. Prepare stock, 10 mg/mL in L buffer with 1% sarcosyl (for cells) or 2% sarcosyl (for tissues). (*See* **Subheading 2.3., item 7.**)
5. Phenylmethylsulfonyl fluoride (PMSF) (40 µg/mL in isopropanol; store at –20°C.)
6. Lesion-specific endonuclease (store enzymes in 40% glycerol at –20°C.) Pyrimidine dimer-specific endonucleases include the *Micrococcus luteus* UV endonuclease and bacteriophage T4 endonuclease V (commercially available from Epicentre, Madison, WI). Preparations must be checked for nonspecific nucleases (cleavage of supercoiled DNA without cyclobutyl pyrimidine dimers [CPDs]), as well as activity (CPD sites incised/volume/time in standard conditions, for example, 4×10^{15} CPD incised /µL/h), or specific activity (CPD incised/*protein*/time). Activities reported as "micrograms of irradiated DNA cleaved/unit protein/unit time" are not useful, since the level of dimers in "irradiated" DNA depends on the UV wavelength, exposure, and DNA base composition.

a. Endonuclease for CPDs: *Micrococcus luteus* UV endonuclease or T4 endonuclease V.
b. Endonuclease buffer: 30 mM Tris-HCl, pH 7.6, 1 mM EDTA, 40 mM NaCl.
c. Endonuclease buffer containing 0.1 mM dithiothreitol (DTT) and 0.1 mg/mL of purified bovine serum albumin (BSA) (New England Biolabs, Beverly, MA). Add DTT and BSA just before use.

7. Ethidium bromide: Prepare a 10 mg/mL ethidium bromide solution using double distilled water. Stir the solution using an electric stirring motor and stirbar until the ethidium is well dissolved. Filter the solution through a 0.2-µm filter. Subdivide the stock solution into portions appropriate to approx 1 wk's use. Keep one tube (capped and wrapped with foil) at room temperature; store stock at –20°C. **Caution:** ethidium bromide is a mutagen. Investigators should wear gloves, and handle the solution as a potential hazard. Ethidium is also light sensitive, keep stock in subdued light.

2.2. Agarose Gels

All solutions should be sterilized by appropriate means.

1. Bio-Rad Mini-Sub Cell, tray for 6.5 cm × 10 cm gel, or as appropriate for gel system used.
2. LE agarose (FMC).
3. Deionized, double-distilled water.
4. 5 M NaCl.
5. 0.1 M EDTA, pH 8.
6. Time tape (TimeMed Labeling Systems, Inc., Burr Ridge, IL).
7. Plastic ruler or spacer (0.02 in. thick).
8. Alkaline electrophoresis solution: 30 mM NaOH, 2 mM EDTA *(11)*.
9. Leveling plate and small spirit level.
10. Dust cover (plastic shoebox).
11. 70% Ethanol.
12. Lint-free tissue.
13. Microwave oven.
14. Plexiglas gel tray, cleaned thoroughly with hot water and detergent (ascertained not to produce fluorescent residues) immediately after last use.
15. 15-Well comb for tray.
16. Alkaline stop mix: one part alkaline dye mix (0.25% bromocresol green in 0.25 N NaOH, 50% glycerol): one part 6 N NaOH.
17. Disposable bacteriological loop (1 µL, USA Scientific, Ocala, FL).

2.3. Demonstration UV Experiment

1. Low-pressure mercury lamp; emits principally 254 nm UV.
2. Meter for 254-nm UV. Commercial UV meters have filters transmitting limited wavebands, with the meter output weighted to specific spectral distributions. Other spectral distributions will not be measured accurately, and radiation of wavelengths not transmitted by the filter will not be recorded. Thus the output of

a "UV-A" lamp reported by a UV-A meter may give an accurate measure of UV-A radiation, but this measurement will not reflect any UV-B also emitted from the lamp. UV-B radiation can be orders of magnitude more effective in inducing biological damage than UV-A *(12,13)*.
3. Human cultured cells.
4. Phosphate-buffered saline (PBS): 170 mM NaCl, 3.4 mM KCl, 10.1 mM Na$_2$HPO$_4$, 1.8 mM KH$_2$PO$_4$.
5. Plumb line.
6. Trypsin solution: 0.05% trypsin (w/v) in Hank's buffered saline plus 0.1% EDTA.
7. L buffer: 20 mM NaCl, 0.1 M EDTA, 10 mM Tris-HCl, pH 8.3.
8. L buffer containing 0.2% *n*-lauroyl sarcosine (Sigma, St. Louis, MO).
9. Red bulbs for room illumination (GE, 25 W red Party Bulb) *(14)*.

2.4. Electrophoresis

1. Bio-Rad MiniSub Cell electrophoresis apparatus or equivalent. It is important to measure the voltage across a gel. To do this, drill two small holes (each large enough for insertion of a volt meter probe) a known distance (e.g., 10 cm) apart, in the top cover of the gel apparatus above the two ends of the gel. Place gel of standard size and composition, and electrophoresis solution of standard composition and volume in the apparatus as usual. Insert the probes into the gel, and begin electrophoresis. Read voltage on the voltmeter; knowing the distance between the two probes, calculate the voltage/centimeter of *gel*. (Note that voltages measured between electrodes of the gel apparatus vary with the individual apparatus, and thus cannot be applied to a different apparatus, whereas voltages per distance of the gel can be.)
2. Power supply, Hoefer PS250/2.5 Amp or equivalent.
3. Pump for buffer recirculation.
4. Chiller, Lauda WKL 230 or equivalent.
5. Cooling bath for immersion of electrophoresis apparatus.

2.5. Gel Processing

1. Stainless steel or glass pan.
2. Deionized, distilled water.
3. 1 M Tris-HCl, pH 8 (stock).
4. 10 mg/mL of Ethidium bromide stock.
5. Vinyl, powder-free gloves.
6. Suction apparatus with water trap.
7. Gel platform rocker, variable speed (Bellco, Vineland, NJ).

2.6. DNA Visualization and Quantitation

1. UV transilluminator.
2. Polaroid camera system and Polaroid type 55 P/N film.
3. Densitometer.
4. Step density wedge.

3. Methods
3.1. Evaluation of DNA Length Standards

All DNA length standards should be checked for integrity on alkaline agarose gels. Commercial DNAs are usually evaluated on neutral gels; such gels do not reveal single strand breaks that will interfere with the use of the DNA as a length standard on alkaline gels.

1. Electrophorese higher molecular length standard DNAs on a static field, alkaline, 0.4% agarose gel (along with other DNAs of previously verified size); neutralize the gel, stain with ethidium, destain, and photograph. The DNA should appear as a single band, with little evidence of heterodispersity from single strand breaks.
2. Evaluate restriction digests for integrity (as above) and for complete digestion on a static field *neutral* gel: the number and sizes of bands should correspond to those expected. Incomplete digests contain partial digestion products, which may be confusing if their lengths are assigned incorrectly. If photographic conditions provide a linear response to DNA mass, the mass of DNA in each band should be directly proportional to its length.

3.2. Experimental DNAs

DNA damages are induced by many chemical and physical agents in the environment, in everyday life, in the workplace and during recreational activity. Analysis of the level of damage, cellular metabolism of those damages, and the levels and kinds of residual, unrepaired damages that can lead to mutations is essential for understanding the consequences of exposure to such agents. The exact experiment that is carried out will depend on the question being asked: What is the level of damage induced by a certain concentration of chemical or dose of radiation? How efficiently does cell type A remove those damages relative to cell type B? and so on.

DNA damages in most linear DNAs can be measured, for example, viral DNAs *(15,16)*, bacterial DNAs, simple eukaryotes, higher plants *(10,13,17–21)* and higher animals, including human tissues *(6,12,22–26)*. For each species, the isolation procedure must be verified to yield DNA of suitable size and amenity to enzyme digestion. In the gel method, sensitivity (lower limit of lesion frequency measurable) depends directly on the DNA size, and thus the larger the experimental DNA, the greater the sensitivity of lesion measurement. For lesions other than frank strand breaks, cleavage by a lesion-recognizing enzyme is required for lesion quantitation; sample DNAs must be free from contaminants that interfere with enzyme cleavage at lesion sites or produce extraneous cleavages at nonlesion sites.

3.3. Preparation of Agarose Gels

Gels for quantitation must provide both a resolving medium to separate DNAs according to size and an optical medium for accurate measurement of

DNA mass (low background fluorescence, no extraneous particles, especially those fluorescing at the wavelengths emitted by the DNA-binding fluorophore).

1. Rinse the leveling plate with distilled water, then with 70% ethanol, and dry with lint-free tissue.
2. Wipe the gel tray and comb with ethanol using lint-free tissue.
3. Place Time medical labeling tape neatly on the open ends of the gel tray, press the tape to seal; the tape under the tray must be flat and even.
4. Adjust the comb to the proper height for the gel tray.
5. Store the tray and comb under a clean dust cover.
6. Place 50 mL of H_2O in an approx 250-mL bottle, place approx 100 mL of H_2O into a second 250-mL bottle.
7. Add 0.4 g of LE agarose to the 50 mL of H_2O in the bottle. **Do not cap bottles (HAZARD!)**
8. Microwave the bottles on high (650 W oven) for 8 min; watch to prevent liquid overflow or excess evaporation. Add additional warm water to the agarose solution if necessary. (Some specialty agaroses should be autoclaved for 20 min on liquid cycle.)
9. Pour warmed water into a clean, dust-free, sterile graduated cylinder.
10. Add approx 20 mL of warm water to the agarose solution, swirl; add 1 mL of 5 M NaCl and 0.1 mL of 0.1 M EDTA (per 100 mL final volume), swirl to mix.
11. Discard the water from the warmed cylinder.
12. Pour the agarose solution into the warmed cylinder; bring to 100 mL with heated water. Pour the agarose back into the (empty) warm bottle and swirl to mix. Inspect the agarose solution for incomplete dissolution of agarose particles, or dust, fibers, or other particles.
13. Agarose solution may be capped and placed in a 55°C bath no more than 2 h; discard if the solution becomes inhomogeneous.
14. Using the warmed (or rewarmed, if necessary) cylinder, measure the required volume of agarose (35 mL per 6.5 cm × 10 cm gel). With the gel tray on a leveling plate, remove the comb from the tray. Pour the agarose slowly into the gel tray. Reset the comb exactly perpendicular to the long axis of the gel tray.
15. Replace the dust cover over the gel and allow to set approx 1 h (0.4% gel, room temperature).
16. Pour cold electrophoresis solution over the gel; pick the comb up on a slant to remove one edge, then the rest of the comb.
17. Cover the gel with electrophoresis solution (prevents well collapse, equilibrates gel).
18. Transfer the gel to the apparatus (preleveled and checked for solution recirculation) containing chilled electrophoresis solution; equilibrate approx 1 h by recirculating the electrophoresis solution. Set the apparatus on black paper to aid visualizing the wells.

3.4. Example Protocol for DNA Damage Analysis

It is beyond the scope of this chapter to discuss planning and execution of all such experiments; we will use as an example the quantitation of CPD induction in cultured human cells by increasing UV exposures. (*See* **Notes 1–3**.)

3.4.1. UV Irradiation and Room Lighting

A low-pressure mercury lamp provides DNA-damaging 254 nm (UV-C) for samples with little shielding (e.g., monolayer of cultured human cells).

Caution: UV-C is an eye hazard; wear UV-opaque glasses with side shields!

1. Turn the lamp on approx 15 min before use; after warm up, wrap the end approx 3 in. of bulb with foil.
2. Take care that the sides of the dish do not shade cells at edge. Use plumb line to locate the position for cell irradiation exactly under the bulb.
3. Remove the medium from the cells; rinse two or three times with ice-cold PBS. Keep the cells cold to minimize repair. Irradiate suspension cells in PBS at low optical density at 254 nm (not in a narrow tube from above, which may suffer from inaccurate dosimetry).
4. To prevent photorepair, use red lamps (GE Party Bulb, 25 W red) for room lighting.

3.4.2. Preparation of Agarose Buttons

1. Melt FMC Sea Plaque or InCert agarose (2% in TE) and place at 45°C.
2. Immediately after UV irradiation, suspend the cells in PBS (10^6 human cells/mL, approx 10^4 cells/10 µL).
3. Mix 1 mL of cells at 2×10^6 cells/mL with 1 mL of agarose.
4. Pipet 10 µL aliquots of suspension into "buttons" onto a Petri dish on ice, and let solidify.
5. Immerse the buttons immediately in Proteinase K solution, transfer to multiwell dish or 35-mm suspension culture dish, seal with Parafilm, and incubate at 37°C.

3.4.3. Proteinase K Digestion

1. Replace the Proteinase K solution daily for 4 d.
2. Check for complete removal of proteins by electrophoresing DNA on 0.4% alkaline agarose gels (rinse the buttons with TE, and denature; see **Subheading 3.4.4, steps 9–11**). If DNA remains at the well–gel interface, digestion is incomplete; after adequate removal of cellular proteins, DNA samples electrophorese readily into an alkaline gel.
3. Treat samples showing incomplete digestion with Proteinase K as above.
4. Rinse buttons twice with ice-cold TE, twice with 10 mM Tris-HCl, pH 7.6, 1 mM EDTA, 40 µg/mL of PMSF at 45°C for 1 h, then rinse with TE.
5. Store the buttons at 4°C in L buffer containing 2% sarcosyl.

3.4.4. Endonuclease Digestion

1. Wash the buttons in 5 volumes of ice-cold TE and soak in TE (twice, 5 volumes, 20 min each).
2. Transfer to the buffer appropriate for the lesion-specific endonuclease (twice, 1 h), then to endonuclease buffer containing 0.1 mM DTT and 0.1 mg/mL of purified BSA.

3. Use companion buttons (replicate buttons from each experimental sample) for each dimer determination.
4. Calculate the quantity of UV endonuclease for the "+ endonuclease" sample from the endonuclease activity (see **Subheading 2.1., item 6**, UV Endonuclease), the quantity of DNA per button, and the maximum expected CPD level. The validity of the assays depends on cleavage at all lesion sites; sufficient endonuclease must be used to give complete cleavage. (Check by incubating replicate buttons containing DNA with the highest damage levels—as well as undamaged DNA to check nonspecific cleavage—with increasing quantities of endonuclease. Determine quantity of endonuclease for complete cleavage, add excess enzyme to each "+ endonuclease" sample.)
5. Incubate the samples on ice for at least 60 min.
6. In the "+ endonuclease" sample, replace buffer by buffer plus endonuclease.
7. Add buffer without endonuclease to the "–endonuclease" sample.
8. Incubate the samples on ice for 30 min, then transfer to 37°C, incubate 60 min.
9. Rinse the buttons with TE.
10. Add 10 µL of alkaline stop solution; incubate at room temperature for 30 min.
11. Rinse the buttons with alkaline electrophoresis solution just prior to loading onto the gel.

3.4.5. Preparation of Molecular Length Standard DNAs for Electrophoresis

1. Dilute molecular length standard DNAs into TE at $\leq$ 80 ng/µL *(27)*.
2. Add alkaline stop solution (2 µL/10 µL of DNA solution or button).
3. Incubate the length standards under the same conditions as the experimental DNAs.

3.4.6. Sample Loading

Buttons are loaded into wells with the gel on a counter rather than in the apparatus.

1. Remove the gel and tray from the apparatus; place on a clean, lint-free tissue. Protect the gel surface by covering it with plastic wrap or film.
2. For a 15-well gel, use lanes 1, 8, and 15 for the molecular length standards (*see* **Fig. 1**), leaving 12 lanes for six sample pairs. The "+" and "–" endonuclease samples of each pair are placed in adjacent lanes; to avoid bias in analysis, code the experimental sample PAIRS; place members of different pairs at coded locations on the gel.
3. Place the tubes containing the samples close to the gel.
4. Pick up individual button from the solution using a plastic disposable loop.
5. Deposit each button in a well (containing alkaline electrophoresis solution); it should slip readily into the well. Generally buttons are not sealed into the wells; however, approx 5–15 µL of 0.4% agarose may be micropipetted into each well so that button does not become displaced.

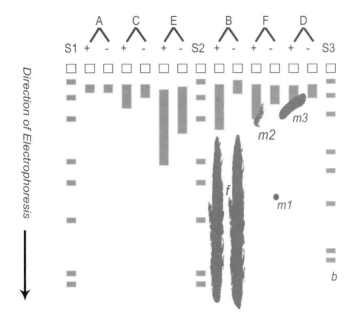

Fig. 1. Schematic diagram of an alkaline electrophoretic gel for DNA damage quantitation. Molecular length standard DNAs (S1, S2, and S3) are shown in lanes 1, 8, and 15. In the experiment shown, 6 experimental sample pairs (A, B, C, D, E, F) are included on the gel. The "+ endonuclease" and "– endonuclease" members of each sample pair are placed in adjacent lanes, but (to avoid bias in analysis) the pairs are not necessarily arranged in experimental order. The italic labels refer to specific experimental problems frequently encountered (*see* **Table 1**).

6. Load the molecular length standard DNAs formed into buttons along with the experimental buttons. If the standards are in solution, they should be loaded *after* the gel is replaced in the electrophoresis apparatus. Replicate length standards are in lanes 1, 8, and 15.
7. After the gel tray and all samples are inserted, check the apparatus with a spirit level, and level if necessary before electrophoresis is begun.

3.4.7. Electrophoresis

DNA migration, and thus resolution of DNA molecules, on electrophoretic gels can be affected by temperature. To achieve temperature uniformity, immerse the gel apparatus in a cooling, recirculating bath filled so that the cooling solution reaches the level of the gel within the apparatus. Or, set the apparatus in a pan of crushed ice water, taking care that ice does not fall into the electrophoresis solution; replenish the ice periodically. Level the gel in the cooling apparatus.

Table 1
Trouble-Shooting Quantitative Agarose Gels

Problem	Possible cause(s)	Solution(s)
A. No DNA visible	1. Sample not loaded. 2. Insufficient DNA loaded. 3. Nuclease degradation of DNA. 4. Ethidium bromide photobleached. 5. Electrophoresis polarity reversed.	1. Load sample. 2. Load more DNA. 3. Discard degraded DNA. 4. Use fresh ethidium. 5. Reverse polarity.
B. Gel lanes crooked	1. Gel not level during pouring. 2. Gel rig not leveled. 3. Thermal currents over rig.	1. Use leveling plate. 2. Use spirit level. 3. Place box over rig.
C. DNA "smiles"	1. Wells collapsed. 2. Wells dried out.	1,2. Remove comb, add buffer to wells and over gel.
D. DNA migration depends on amount of DNA.	1. DNA too concentrated.	1. Dilute DNA samples.
E. DNA lanes slant in photograph.	1. Comb crooked when gel poured. 2. Gel photographed at slant.	1. Align comb precisely. 2. Check that marker lanes are exactly parallel and straight.
F. "Fuzzy" cloud of ethidium-stained material near lane bottom.	1. RNA from sample. 2. RNA from endonuclease.	1. RNase sample. 2. RNase endonuclease.
G. "Unirradiated" sample cleaved by endonuclease.	1. Sample actually was irradiated. 2. Endonuclease contains nonspecific cleaving activity. 3. Nonsterile buffer, tube, tip.	1. Check the sample history. 2. Use a better endonuclease. 3. Use freshly sterilized buffer, and so forth.

H. "Minus endo" sample degraded.	1. Poor extraction method or technique. 2. Nonsterile buffer, tube, tip.	1. Evaluate method. 2. Use freshly sterilized buffer, and so forth.
I. DNA length standards contain extra bands.	1. Incomplete restriction digest.	1. Carry out new digestion; check completeness of digestion.
J. DNA length standards missing bands.	1. Wrong DNA or restriction enzyme. 2. Smaller bands electrophoresed off end of gel.	1. Check DNA and enzyme. 2. Use shorter electrophoresis time or lower voltage.
K. High background fluorescence on gel.	1. Too much ethidium in staining solution. 2. Bacterial contamination in agarose solution. 3. Agarose contains DNA contaminant. 4. Agarose prepared from solution with bacterial/fungal/viral contaminant.	1. Check ethidium stain. 2. Make fresh agarose. 3. Use high-quality agarose. 4. Discard solutions; use freshly prepared.
L. Gel will not set.	1. Wrong agarose used. 2. Dry agarose stored in moist conditions; has adsorbed water from atmosphere. 3. Agarose incompletely melted.	1. Use agarose intended for gel electrophoresis. 2. Store agarose powder in presence of desiccant. 3. Melt agarose thoroughly.
M. Fluorescent particles on gel: 1. Specks	1. Dust in agarose solution or in gel. 2. Dust on gel.	1. Use filtered solution. 2. Cover gel.
2. Strands	1. Lint in agarose solution.	1. Dry glassware on lint-free wipe. 2. Wipe gel apparatus, trays with lint-free wipe.
3. Globs	1. Ethidium aggregates on gel.	1. Filter ethidium stock. 2. Discard working ethidium solution, use fresh.

(**Caution: potential safety hazard!** Take care that leads from the power supply do not become submerged in the cooling water!)

The Bio-Rad MiniSub Cell apparatus allows for recirculation of electrophoretic solution using a simple pump. Set the apparatus up for recirculation.

1. Fill the apparatus with 250 mL of prechilled alkaline electrophoresis solution.
2. BEFORE inserting the gel, check that the solution circulates and tubing apparatus does not leak.
3. Begin electrophoresis (approx 1.5 V/cm; the value depends on DNA size) for 30 min without recirculation of electrophoresis solution.
4. Start recirculation of the electrophoresis solution, and continue throughout the electrophoresis.
5. Use a timed, voltage-controlled power supply to electrophorese for the correct period.
6. After electrophoresis, remove the gel and tray from the apparatus, and process the gel (*see* below).
7. Immediately after electrophoresis, remove the electrophoresis solution from the apparatus (alkaline solution is corrosive to electrodes). Discard used alkaline electrophoresis solution.
8. Rinse the apparatus and tubing thoroughly and invert on lint-free tissue in a dust-free location to dry.

3.4.8. Neutralization of the Alkaline Gel

1. After electrophoresis, remove the gel and tray from the apparatus (alkaline solution makes gels slick, so take care the gel does not slide out of the apparatus onto the floor!). Wear powder-free vinyl gloves to protect hands, and to protect the gel from fingers. 0.4% gels are fragile, handle carefully.
2. Rinse the gel surface (while the gel is in the gel tray) in a gentle stream of distilled water.
3. Transfer the gel gently to a pan.
4. Add water carefully to the pan, at a position away from the gel.
5. Rock the pan gently, and remove the water using a suction device (holding the suction device away from the gel).
6. From stock 1 M Tris, pH 8, make 500 mL of neutralizing solution, 0.1 M Tris, pH 8 in ddH$_2$O.
7. Pour 250 mL into the pan, away from the gel; place the pan on gel rocker for approx 20 min.
8. Remove the neutralizing solution carefully.
9. Add 250 mL of fresh 0.1 M Tris, neutralize the gel for at least 40 min. For high molecular length DNAs, which diffuse slowly, the gel can be neutralized overnight. For complex DNAs, complete renaturation (i.e., restoration of the original double-stranded conformation) is not usual; more likely is formation of double-stranded hairpins, which still retain partially single-stranded character.

3.4.9. DNA Staining

1. Prepare stain (250 mL, 1 µg/mL ethidium bromide in ddH$_2$O) in a clean, dust-free cylinder.
2. Remove the final neutralizing solution, and then pour ethidium solution into the pan, well away from the gel. Do not pipet stock ethidium solution just above the gel surface, as this produces uneven gel staining.
3. Stain the gel for 15 min.
4. Remove the ethidium solution.
5. Rinse the gel gently with double-distilled water.
6. Fill the gel pan (approx 2/3 full) with water, and destain the gel for at least two changes, 15 min each.
7. Additional destaining time (overnight for high molecular length DNAs) and fresh water reduces nonspecific ethidium background. Gels may be destained at 4°C; however, bubbles appearing in the gel during warming will interfere with DNA quantitation.

3.4.10. DNA Visualization and Quantitation

For number average length calculations, we need to know the position of DNA molecules on the gel AND quantity of DNA at each position. We need know only relative—not absolute—masses of DNA molecules of different sizes in different lanes. Thus, with uniform gel, ethidium background, and transilluminator, and DNA staining uniform across the gel (dependent only on DNA mass), we need a recording system giving a signal proportional to DNA mass. Photographic film is widely used for recording fluorescence from DNA, but its response to fluorescence is linear over a very limited range, determined by DNA concentrations, gel conditions and photographic conditions (film type, temperature of storage and use, exposure, processing). *See* **refs. *28,29*** for a discussion.

To determine the linear range for specific experimental conditions, prepare a standard alkaline agarose gel, and electrophorese increasing DNA masses (a few to several hundred nanograms per lane) in different gel lanes. Electrophorese and process the gel as usual; photograph the gel, scan the DNA lanes recorded on the film with a densitometer, and determine the relationship of quantity of DNA to densitometric response ("area" of each band). Plot DNA quantity vs "area" of that band, noting the threshold, linear response range and saturation. In all damage determinations, use DNA concentrations within the linear range.

3.4.11. Photography of Ethidium Fluorescence on Electrophoretic Gels

1. Place the neutralized, stained, and destained gel on the transilluminator. If the transilluminator is uneven (shows "stripes" corresponding to lamps), orient the gel so that illumination down a lane is constant.

2. Photograph the gel with film generating a negative. Do not attempt to obtain quantitative data from a positive print, as its darkening (measured in reflectance) does not reflect reliably the fluorescence to which it was exposed.
3. Process the film according to standardized conditions (*see* calibration, previously mentioned).
4. Dry in a dust-free environment. Streaking or fingerprints on the negative interfere with accurate DNA mass quantitation.

3.4.12. Densitometry

Test the densitometer's linearity of response to film darkening:

1. Align the gel precisely on the transilluminator. Since DNA migration is a function of its molecular length, the film must be aligned precisely so that an x position on the densitometer trace uniformly represents DNA migration in all lanes. Align same-sized molecular length standards in different gel lanes at the same x migration position on the densitometer trace.
2. Obtain traces (intensity of fluorescence as a function of migration position on the gel) for each molecular length and experimental sample lane. For densitometers with computer output, data may be stored and the quantitative values used for further manipulations. However, analog outputs (traces of DNA mass as a function of lane migration) can also be used.

3.5. Theory of Analysis

Suppose that an initial DNA population contains N_0 molecules, and k strand breaks are introduced directly (e.g., by X-rays) or by lesion-specific endonucleases. Each strand break increases the total number of DNA molecules by one, resulting in a final population of $N_+ = N_0 + k$ DNA molecules. To determine the number of strand breaks we count the number of DNA molecules before and after introduction of the breaks, that is, $k = N_+ - N_0$. Although this theory is simple, there are problems with implementation. First, we must count DNA molecules; accuracy in this simple counting approach would require samples of exactly the same size, which is never easy. Normalizing by the total mass of DNA avoids both problems.

3.5.1. "Normalizing" Removes the Need for Samples of Equal DNA Mass

Rather than determining the number of molecules, we determine the number of molecules per unit mass of DNA. This ratio is not changed by variations in the sample size if the sampled material is homogeneous. We could express the DNA mass in a variety of units. The most useful are the total number of individual bases or base pairs. (We use bases if we are measuring lesions affecting one DNA strand, for example, single-strand breaks, and base pairs for damages involving both strands such as double-strand breaks. In all that follows, "or base pairs" is implied whenever we give DNA masses in "bases"). We can imagine assigning an index number, i, to each DNA molecule, and determining its length in bases. If L_i represents the length (mass) of that molecule and if there are N DNA molecules in the sample, then the total mass of DNA is $\Sigma_i L_i$ where i goes from 1 to N. Our measure of strand breaks is

$$\frac{N}{\Sigma_i L_i}$$

The units are "molecules per base," but we usually express DNA mass in some multiple of bases, hence giving normalized values of, for example, molecules per megabases. The reciprocal of the molecules per base is the average number of bases per molecule. Formally, this is called the number average length of the population, $\bar{L}$. From the definitions given above,

$$\bar{L} = \frac{\Sigma_i L_i}{N}$$

which is just the normal definition for the average size of a population. Inducing breaks increases the number of molecules and decreases their average length. Our measure of the breaks produced by a given treatment is the number of breaks per unit length of DNA, that is, the frequency of strand breaks, ϕ, which is expressed in terms of number average lengths by Eq. 1, where the subscripts 0 and + indicate initial and final (untreated and treated) populations, respectively.

$$\phi = \left(\bar{L}_+\right)^{-1} - \left(\bar{L}_0\right)^{-1} \tag{1}$$

3.5.2. Determining the Number of DNA Molecules Per Unit DNA Length by Gel Electrophoresis

Fluorescence from ethidium bromide is directly proportional to the mass of each molecule. That is, $f_i = k\, L_i$, where f_i is the fluorescence from molecule i, and k is a constant of proportionality that depends on many experimental factors. Mass normalization eliminates the need to determine the value of k, as long as it is the same for all DNA molecules in a sample (i.e., lane of a gel). Instead of determining $\bar{L}$ by summing over i, suppose we separate the DNA molecules as a function of length, for example, by gel electrophoresis. If n_L is the number of molecules in a sample of length L, then the total fluorescence due to all the molecules of this length, f_L, is given by $f_L = k\, n_L\, L$. The total number of DNA molecules in the sample is $\Sigma_L n_L$ and the total number of bases in the same sample is $\Sigma_L (n_L L)$, where the sums extend from 1 to the number of bases in the largest DNA molecule. In terms of these sums,

$$\bar{L} = \frac{\Sigma_L n_L L}{\Sigma_L n_L}$$

We can replace n_L by $f_L/(kL)$ and the product $n_L L$ by f_L/k. Thus, if k is the same for all molecules in the sample, the average length of the DNA in the sample is given by Eq. 2.

$$\bar{L} = \frac{\Sigma_L f_L}{\Sigma_L \dfrac{f_L}{L}} \qquad (2)$$

This equation for $\bar{L}$ indicates a sum over all values of L, the length of the DNA molecules, but we obtain the distribution of the DNA as a function of, for example, the distance of travel during electrophoresis. While DNA lengths can only be discrete (integer) values, the distances moved by molecules of different lengths are continuous values (real numbers). Thus, the sums in the expression for $\bar{L}$ are replaced by integrals as shown in Eq. 3.

$$\bar{L} = \frac{\int f(x)\,dx}{\int \dfrac{f(x)}{L(x)}dx} \qquad (3)$$

where $f(x)\,dx$ is the intensity of fluorescence from a region of width dx at location x, while $L(x)$ is the length of the DNA molecules at this position, and x can be thought of either as the migration distance, or more generally a "separation coordinate." The limits of integration must span the values of x for which there is measurable DNA. $L(x)$ is called the dispersion function of the separation system and is treated here as a continuous function of x. The actual values of x never appear explicitly in the equation for $\bar{L}$, only the values of f and L associated with given values of x. Thus, we can express x in any convenient units. For digital data "pixels" are a good choice. Because pixels divide the data into discrete intervals, the integrals in the equation for $\bar{L}$ revert (approximately) to sums. While it is convenient to think of x as the migration distance, it is actually just a particular position on the gel along the direction of electrophoresis. Therefore, we can choose any origin for the x-axis, not just the lower edge of the loading wells.

We can either determine the dispersion function empirically, or obtain analytical functions that describe it. Although an empirical dispersion function could be used, analytical dispersion functions facilitate calculation of $\bar{L}$ directly from Eq. 3. For both static field and unidirectional pulsed field (*30*) gel electrophoresis, the dispersion function is reasonably approximated by a logistic function of the form shown in Eq. 4.

$$L(x) = L_m \left(\frac{x_0 - x}{x - x_\infty}\right)^q \qquad (4)$$

where x_0 and x_∞ are the locations on the gel of (hypothetical) molecules of "zero" and "infinite" length, respectively, and L_m is the length of the molecules that migrated to a position exactly halfway between x_0 and x_∞. [That is, $L_m = L(x_m)$, where $x_m = (x_0 + x_\infty)/2$]. The exponent q is usually in the range from 0.8 to 1.2, that is, close to unity. The evolution of dispersion functions leading to Eq. 4, is detailed elsewhere *(31)*, as is the relationship between these parameters and the conditions under which the gel was made and run *(32)*. Once the values of x_0, x_∞, L_m, and q are known for a particular gel, we can compute $L(x)$ for every value of x between x_0 and x_∞. The four parameters can be determined by nonlinear fitting to a data set containing the distances of migration of a set of DNA molecules of known length, which must be included in the same gel as the experimental samples being studied.

3.5.3. Alternate Determination of Number Average Length: Median Length

The presence of $L(x)$ in the denominator of the equation for $\bar{L}$ can produce experimental difficulty when the length approaches very small values. So long as most of the DNA is large, the fluorescence (or other label) will not give a significant signal for positions on the gel corresponding to small DNA molecules, and the integration can be truncated before x gets too close to x_0 [which causes $L(x)$ to approach 0]. For DNA with many strand breaks, there may be significant signal for values of x near x_0. For such cases, we can obtain approximate values of the number average length of the population from either the length average length or the median length *(1)*. Median length, L_{med} is the length of the DNA molecules that migrate to the position x_{med}, the value of x that divides the mass of DNA exactly in half. Formally, we can define the median distance of migration of a DNA sample from Eqs. 5 and 6. Note that L_{med} and x_{med} are not related to the variables L_m and x_m that were introduced in the discussion of dispersion functions.

$$\int_{x_\infty}^{x_{med}} f(x)dx = \frac{1}{2}\int_{x_\infty}^{x_0} f(x)dx \quad (5)$$

$$L_{med} = L(x_{med}) \quad (6)$$

3.5.4. Relationship of Median Lengths to Number Average Length

There are two special cases where there are known relationships between $\bar{L}$ and L_{med}. For a population of molecules all of which are exactly the same size, $\bar{L} = L_{med}$. If the population contains molecules of more than one length, L_{med} will be greater than $\bar{L}$, because larger molecules are weighted more heavily. The second special case is where each molecule in an initial homogeneous population has been broken randomly several times, as, for example, during extraction. A population of DNA molecules from higher organisms (where the initial length is the length of the chromosomes) that has been reduced in length sufficiently that the resulting distribution can be separated in a static field gel should fit this requirement quite well. Under these conditions, the number average length of the population is given by Eq. 7 *(33)*.

$$\bar{L} = 0.6\ L_{med} \tag{7}$$

Thus, the error associated with estimating the number average length of a population using the median length is never worse than a factor of 5/3, and in the common situation of DNA broken extensively during extraction, should be much better.

3.6. Obtaining Median Lengths and Calculating Lesion Frequencies

This discussion presumes access to molecular biology equipment, but not specialized equipment for high-sensitivity, high-throughput DNA lesion quantitation (alkaline pulsed field gels, quantitative electronic imaging, computerized analysis). In this simple approach DNA median molecular lengths are calculated *(15)* and from them, number average molecular lengths *(33)*.

3.6.1. Determination of DNA Dispersion Function

1. Compare lane traces of molecular length standards. The peak positions of the DNAs of the same molecular length should exactly coincide. If so, one lane of standards establishes a DNA dispersion function for the entire gel. If the traces do not coincide, standard lanes near individual experimental samples should be used to calculate separate DNA dispersion in different gel areas.
2. Determine x, y coordinates of each DNA length standard. (x corresponds to the migration position of the peak of a DNA band; y is the molecular length of that DNA in base pairs.)
3. Plot these points on linear–linear scales.
4. Fit a curve through the data points. This DNA dispersion function relates size of DNA molecules to migration position on this gel. Since migration position is affected by exact electrophoresis conditions, DNA dispersion curves must be determined for each gel (or gel region, *see* **Subheading 3.5.2.**).

3.6.2. Determination of Median and Number Average Molecular Lengths

The median molecular length is the molecular length in the middle of the DNA mass, that is, the molecular size of which one-half the DNA molecular mass is larger, and one-half is smaller. The manual method described in the following indicates the calculation; it could also be done by a computer "area" computation.

DNA Damage Quantitation

1. Photocopy the DNA lane (photocopier paper is quite uniform).
2. Handle photocopies with powder-free gloves to ensure that neither oils nor moisture from hands, nor powder from the gloves interferes with measurement.
3. Cut out the trace of an experimental DNA lane carefully with scissors.
4. Determine the weight (W) of the trace using an analytical balance; calculate $W/2$.
5. Estimate x position (x_1) corresponding to middle of DNA mass; cut trace vertically at x_1.
6. Weigh one of the resulting half-traces, yielding w_1.
7. If $w_1 = W/2$, refer x_1 to the dispersion plot, and determine the corresponding molecular length L_{med}, the median molecular length of that DNA population.
8. If $w_1 \neq W/2$, gradually slice the larger half vertically until its weight equals $W/2$.
9. Locate the x position of this slice giving one-half the weight in that portion of the lane trace on the dispersion curve; the corresponding length value is the median molecular length, L_{med}.
10. Calculate the number average molecular length, $\bar{L}$ from Eq. 7.

3.6.3. Computation of DNA Lesion Frequency

Calculate the frequency of DNA lesions according to Eq. 1, where $\bar{L}_+$ is the number average length of the treated sample, and $\bar{L}_0$ is the number average length of the untreated sample. For samples in which DNA lesions were revealed by lesion-specific agent cleavage, "treated" refers to samples treated with that agent, while "untreated" refers to the companion part of the sample not treated with the agent. This approach provides high sensitivity, as the experimental DNA is extracted, then split into samples for agent-specific cleavage. It also allows determination of levels of background lesions. For strand breaks induced directly by radiation, chemicals, and so forth, the "treated" sample is the one exposed to the radiation or chemical, and the "untreated" sample is the unexposed one. This determination is more difficult, as DNAs in samples to be compared are extracted independently. Reproducible isolation procedures are essential for accurate calculation of directly induced strand breaks. (*See* **Note 4**.)

4. Notes

1. Quantifying modified bases affecting one strand: Lesions other than cyclobutyl pyrimidine dimers can readily be quantified by a similar approach, using enzymes that recognize those lesions. Many lesion-recognizing enzymes are now commercially available, including *E. coli* Nth protein (principal substrate, oxidized pyrimidines), and *E. coli* Fpg protein (principal substrate, oxidized purines). In evaluating the activity and specificity of these preparations, it is important to compare the substrate and reaction conditions used by the manufacturer with those used by the researchers who characterized the substrates of these enzymes. Oxidized purines and oxidized pyrimidines can be quantified using alkaline denaturing gels as discussed for pyrimidine dimers *(34)*.

2. Quantifying abasic sites affecting one strand: Unlike the base modifications discussed in **Note 1**, abasic sites are converted to strand breaks by alkaline conditions, and thus cannot be quantified on alkaline gels. However, by use of denaturing, nonalkaline separation media, for example, glyoxal-containing gels *(35)*, abasic sites can also be measured by the same analytical approach. Glyoxal gels require substantially more DNA than do alkaline gels.
3. Quantifying lesions affecting both strands: Many genotoxic agents (e.g., ionizing radiation) also produce damages affecting both DNA strands. The best known of these is the double-strand break (DSB), which can be considered two closely spaced single strand breaks on opposing strands. The DSB is but one of many complex damages containing two or more lesions; others include oxidized purine clustered damages (or clusters), oxidized pyrimidine clusters, and abasic clusters. These complex damages are formed in DNA irradiated in solution and in irradiated cells *(16)*. These complex damages can be recognized by lesion-specific enzymes, and quantified using nondenaturing gels *(5,16)*. By use of specific pulsed field electrophoresis regimes, such damages can be measured with high sensitivity, for example, at a few per 10^9 base pairs *(4)*.
4. High-sensitivity measurements: The methods described above (static field electrophoresis, photographic recording of DNA mass, computation of median molecular length) will give quite adequate measurement of DNA damages down to approx 2/Mb. We can compare that value to a relevant biological dose: the D_{37} for 254 nm exposure of mammalian cells is approx 7 J/m^2, and 1 J/m^2 of 254 nm radiation induces about 6.5 CPD per million bases. Thus the D_{37} induces approx 45 CPD/Mb, indicating that the gel method can readily measure responses within the range of high cell survival.

For higher sensitivity measurement, three major changes are required: first, higher molecular length DNAs are needed; for methods of obtaining high molecular length DNA from various higher organisms, the reader is referred to references *(6,20,36,37)*. Second, these large DNAs must be separated, readily carried out by pulsed field electrophoresis *(30,38–42)*. Third, a method of quantifying DNA with a linear response and large dynamic range *(7,43)* allows more accurate measurement of DNA mass, especially at the leading edge of the DNA peak, corresponding to the smaller molecules in the population. Fourth, computerized calculation of the number average molecular length, rather than through its estimation through calculation of the median molecular length, allows much higher sensitivity of lesion measurement.

Acknowledgments

Research supported by grants from the Low Dose Program of the Office of Biological and Environmental Research of the U. S. Department of Energy, the U. S. National Aeronautics and Space Administration, Office of Biological and Physical Research, the National Space Biomedical Research Institute, and the National Institutes of Health (CA86897) to B. M. S., and from the National Institutes of Health (EB002121) to J. C. S.

References

1. Freeman, S. E., Blackett, A. D., Monteleone, D. C., Setlow, R. B., Sutherland, B. M., and Sutherland, J. C. (1986) Quantitation of radiation-, chemical-, or enzyme-induced single strand breaks in nonradioactive DNA by alkaline gel electrophoresis: application to pyrimidine dimers. *Anal. Biochem.* **158,** 119–129.
2. Sutherland, J. C., Bergman, A. M., Chen, C. -Z., Monteleone, D. C., Trunk, J., and Sutherland, B. M. (1988) Measurement of DNA damage using gel electrophoresis and electronic imaging, in *Electrophoresis 88* (Schafer-Nielsen, C., ed.). VCH Verlagsgesellschaft, Weinheim, Germany, pp. 485–499.
3. Sutherland, J. C., Monteleone, D. C., Trunk, J. G., Bennett, P. V., and Sutherland, B.M. (2001) Quantifying DNA damage by gel electrophoresis, electronic imaging and number average length analysis. *Electrophoresis* **22,** 843–854.
4. Sutherland, B. M., Bennett, P. V., Georgakilas, A. G., and Sutherland, J.C. (2003) Evaluation of number average length analysis in quantifying double strand breaks in genomic DNAs. *Biochemistry* **42,** 3375–3384.
5. Sutherland, B. M., Bennett, P. V., and Sutherland, J. C. (1996) Double strand breaks induced by low doses of gamma rays or heavy ions: quantitation in nonradioactive human DNA. *Anal. Biochem.* **239,** 53–60.
6. Bennett, P. V., Gange, R. W., Hacham, H., et al. (1996) Isolation of high molecular length DNA from human skin. *BioTechniques* **21,** 458–463.
7. Sutherland, J. C., Lin, B., Monteleone, D. C., Mugavero, J., Sutherland, B. M., and Trunk, J. (1987) Electronic imaging system for direct and rapid quantitation of fluorescence from electrophoretic gels: application to ethidium bromide-stained DNA. *Anal. Biochem.* **163,** 446–457.
8. Sutherland, J. C. (1990) Electronic imaging systems for quantitative electrophoresis of DNA, in *Non-Invasive Techniques in Biology and Medicine* (Freeman, S. E., Fukishima, E., and Green, E. R., eds.). San Francisco Press, San Francisco, CA, pp. 125–134.
9. Sutherland, J. C. (1993) Electronic imaging of electrophoretic gels and blots, in *Advances in Electrophoresis*, Vol. 6 (Chrambach, A., Dunn, M. J. and Radola, B. J., eds.). VCH Publishers, New York, NY, and Weinheim, Germany, pp. 1–42.
10. Sutherland, J. C., Monteleone, D. C., and Sutherland, B. M. (1997) Computer network for data acquisition, storage and analysis. *J. Photochem. Photobiol. B* **40,** 14–22.
11. McDonell, M., Simon, M. N., and Studier, F. W. (1977) Analysis of restriction fragments of T7 DNA and determination of molecular weights of electrophoresis of neutral and alkaline gels. *J. Mol. Biol.* **110,** 119–143.
12. Freeman, S. E., Hacham, H., Gange, R. W., Maytum, D., Sutherland, J. C., and Sutherland, B. M. (1989) Wavelength dependence of pyrimidine dimer formation in DNA of human skin irradiated in situ. *Proc. Natl. Acad. Sci. USA* **86,** 5605–5609.
13. Quaite, F. E., Sutherland, B. M., and Sutherland, J. C. (1992) Action spectrum for DNA damage in alfalfa lowers predicted impact of ozone depletion. *Nature* **358,** 576–578.

14. Sutherland, J. C. and Sutherland, B. M. (1975) Human photoreactivating enzyme: action spectrum and safelight conditions. *Biophysical J.* **15,** 435–440.
15. Sutherland, B. M. and Shih, A. G. (1983) Quantitation of pyrimidine dimer content of nonradioactive deoxyribonucleic acid by electrophoresis in alkaline agarose gels. *Biochemistry* **22,** 745–749.
16. Sutherland, B. M., Bennett, P. V., Sidorkina, O., and Laval, J. (2000) DNA damage clusters induced by ionizing radiation in isolated DNA and in human cells. *Proc. Natl. Acad. Sci. USA* **97,** 103–108.
17. Sutherland, B. M., Quaite, F. E., and Sutherland, J. C. (1994) DNA damage action spectroscopy and DNA repair in intact organisms: alfalfa seedlings, in *Stratospheris Ozone Depletion/UV-B Radiation in the Biosphere*, Vol. 18 (Biggs, R. H. and Joyner, M. E. B., eds.). NATO ASI Series, Springer-Verlag, Berlin, Germany, pp. 97–106.
18. Hidema, J., Kumagai, T., Sutherland, J. C., and Sutherland, B. M. (1996) Ultraviolet B-sensitive rice cultivar deficient in cyclobutyl pyrimidine dimer repair. *Plant Physiol.* **113,** 39–44.
19. Quaite, F. E., Sutherland, B. M., and Sutherland, J. C. (1992) Quantitation of pyrimidine dimers in DNA from UVB-irradiated alfalfa (*Medicago sativa L.*) seedlings. *Appl. Theor. Electrophoresis* **2,** 171–175.
20. Quaite, F. E., Sutherland, J. C., and Sutherland, B. M. (1994) Isolation of high-molecular-weight plant DNA for DNA damage quantitation: relative effects of solar 297 nm UVB and 365 nm radiation. *Plant Mol. Biol.* **24,** 475–483.
21. Quaite, F. E., Takayanagi, S., Ruffini, J., Sutherland, J. C., and Sutherland, B. M. (1994) DNA damage levels determine cyclobutyl pyrimidine dimer repair mechanisms in alfalfa seedlings. *Plant Cell* **6,** 1635–1641.
22. Freeman, S. E., Gange, R. W., Matzinger, E. A., and Sutherland, B. M. (1986) Higher pyrimidine dimer yields in skin of normal humans with higher UVB sensitivity. *J. Invest. Dermatol.* **86,** 34–36.
23. Freeman, S. E., Gange, R. W., Sutherland, J. C., and Sutherland, B. M. (1987) Pyrimidine dimer formation in human skin. *Photochem. Photobiol.* **46,** 207–212.
24. Freeman, S. E., Gange, R. W., Sutherland, J. C., Matzinger, E. A., and Sutherland, B. M. (1987) Production of pyrimidine dimers in DNA of human skin exposed in situ to UVA radiation. *J. Invest. Dermatol.* **88,** 430–433.
25. Hacham, H., Freeman, S. E., Gange, R. W., Maytum, D. J., Sutherland, J. C., and Sutherland, B. M. (1990) Does exposure of human skin *in situ* to 385 or 405 nm UV induce pyrimidine dimers in DNA? *Photochem. Photobiol.* **52,** 893–896.
26. Sutherland, B. M. and Bennett, P. V. (1995) Human white blood cells contain cyclobutyl pyrimidine dimer photolyase. *Proc. Natl. Acad. Sci. USA* **92,** 9732–9736.
27. Doggett, N. A., Smith, C. L., and Cantor, C. R. (1992) The effect of DNA concentration on mobility in pulsed field gel electrophoresis. *Nucleic Acids Res.* **20,** 859–864.
28. Ribeiro, E. A. and Sutherland, J. C. (1991) Quantitative gel electrophoresis of DNA: resolution of overlapping bands of restriction endonuclease digests. *Anal. Biochem.* **194,** 174–184.

29. Ribeiro, E., Larcom, L. L., and Miller, D. P. (1989) Quantitative fluorescence of DNA intercalated ethidium bromide on agarose gels. *Anal. Biochem.* **181,** 197–208.
30. Sutherland, J. C., Monteleone, D. C., Mugavero, J. H., and Trunk, J. (1987) Unidirectional pulsed-field electrophoresis of single- and double-stranded DNA in agarose gels: analytical expression relating mobility and molecular length and their application in the measurement of strand breaks. *Anal. Biochem.* **162,** 511–520.
31. Sutherland, J. C., Reynolds, K. J., and Fisk, D. J. (1996) Dispersion functions and factors that determine resolution for DNA sequencing by gel electrophoresis. *Proc. Soc. Photo Opti. Instr. Eng.* **2680,** 326–340.
32. Sutherland, J. C. (1997) Linking electrophoretic resolution with experimental conditions. *Proc. Soc. Photo Opti. Instr. Eng.* **2985,** 47–60.
33. Veatch, W. and Okada, S. (1969) Radiation-induced breaks of DNA in cultured mammalian cells. *Biophys. J.* **9,** 330–346.
34. Sutherland, B. M., Bennett, P. V., Sidorkina, O., and Laval, J. (2000) Clustered damages and total lesions induced in DNA by ionizing radiation: oxidized bases and strand breaks. *Biochemistry* **39,** 8026–8031.
35. Drouin, R., Rodriguez, H., Gao, S. W., et al. (1996) Cupric ion/ascorbate/hydrogen peroxide-induced DNA damage: DNA-bound copper ion primarily induces base modifications. *Free Radic. Biol. Med.* **21,** 261–273.
36. Bennett, P. V. and Sutherland, B. M. (1993) Quantitative detection of single-copy genes in nanogram samples of human genomic DNA. *BioTechniques* **15,** 520–525.
37. Bennett, P. V., Hada, M., Hidema, J., et al. (2001) Isolation of high molecular length DNA: alfalfa, pea, rice, sorghum, soybean and spinach. *Crop Sci.* **41,** 167–172.
38. Chu, G., Vollrath, D., and Davis, R. W. (1986) Separation of large DNA molecules by contour-clamped homogeneous electric fields. *Science* **234,** 1582–1585.
39. Gardiner, K., Laas, W., and Patterson, D. (1986) Fractionation of large mammalian DNA restriction fragments using vertical pulsed-field gradient gel electrophoresis. *Somat. Cell Mol. Genet.* **12,** 185–195.
40. Serwer, P. (1987) Gel electrophoresis with discontinuous rotation of the gel: An alternative to gel electrophoresis with changing direction of the electrical field. *Electrophoresis* **8,** 301–304.
41. Sutherland, J. C., Emrick, A. B., and Trunk, J. (1989) Separation of chromosomal length DNA molecules: pneumatic apparatus for rotating gels during electrophoresis. *Electrophoresis* **10,** 315–317.
42. Gardiner, K., Laas, W., and Patterson, D. (1986) Fractionation of large mammalian DNA restriction fragments using vertical pulsed-field gradient gel electrophoresis. *Som. Cell Mol. Genet.* **12,** 185–195.
43. Sutherland, J. C., Sutherland, B. M., Emrick, A., et al. (1991) Quantitative electronic imaging of gel fluorescence with charged coupled device cameras: applications in molecular biology. *BioTechniques* **10,** 492–497.

20

The Comet Assay

*A Sensitive Genotoxicity Test
for the Detection of DNA Damage and Repair*

Günter Speit and Andreas Hartmann

Summary

The comet assay (single-cell gel electrophoresis) is a simple and sensitive method for studying DNA damage and repair. In this microgel electrophoresis technique, a small number of cells suspended in a thin agarose gel on a microscope slide is lysed, electrophoresed, and stained with a fluorescent DNA-binding dye. Cells with increased DNA damage display increased migration of chromosomal DNA from the nucleus toward the anode, which resembles the shape of a comet. The assay has manifold applications in fundamental research for DNA damage and repair, in genotoxicity testing of novel chemicals and pharmaceuticals, environmental biomonitoring, and human population monitoring. This chapter describes a standard protocol of the alkaline comet assay and points to some useful modifications.

Key Words: Alkaline comet assay; alkali-labile sites; biomonitoring; crosslinks; DNA strand breaks; excision repair; genotoxicity testing; single-cell gel electrophoresis.

1. Introduction

The comet assay (single-cell gel electrophoresis) is a useful technique for studying DNA damage and repair with manifold applications. In this microgel electrophoresis technique, a small number of cells suspended in a thin agarose gel on a microscope slide is lysed, electrophoresed, and stained with a fluorescent DNA-binding dye. Cells with increased DNA damage display increased migration of chromosomal DNA on electrophoresis from the nucleus toward the anode, which resembles the shape of a comet (**Fig. 1**). In its alkaline version, which is mainly used, DNA strand breaks and alkali-labile sites become apparent, and the extent of DNA migration correlates with the amount of DNA

From: *Methods in Molecular Biology: DNA Repair Protocols: Mammalian Systems, Second Edition*
Edited by: D. S. Henderson © Humana Press Inc., Totowa, NJ

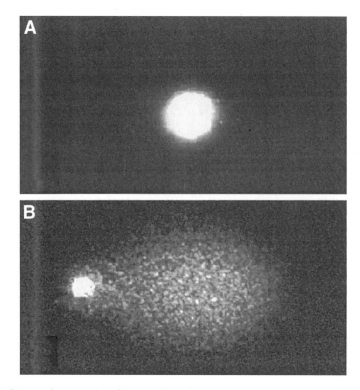

Fig. 1. Photomicrographs of human lymphocytes in the comet assay. (**A**) Untreated cell (control). (**B**) Cell exhibiting increased DNA migration after mutagen treatment.

damage in the cell. The comet assay combines the simplicity of biochemical techniques for detecting DNA single strand breaks and/or alkali-labile sites with the single-cell approach typical of cytogenetic assays. The advantages of the comet assay include its simple and rapid performance, its sensitivity for detecting DNA damage, the analysis of data at the level of the individual cell, the use of extremely small cell samples, and the usability of virtually any eukaryote cell population. Apart from image analysis, which greatly facilitates and enhances the possibilities of comet measurements, the cost of performing the assay is extremely low. The comet assay has already been used in many studies to assess DNA damage and repair induced by various agents in a variety of cells in vitro and in vivo *(1,2)*. The test has widespread applications in genotoxicity testing in vitro and in vivo *(3,4)*, DNA damage and repair studies *(1,5)*, environmental biomonitoring *(6,7)* and human population monitoring *(8)*.

The alkaline version (pH >13) of the comet assay introduced by Singh and co-workers *(9)* detects a broad spectrum of DNA lesions, that is, DNA single-

and double-strand breaks and alkali-labile sites. Modified versions of the assay introduced by Olive and co-workers *(10)* involved lysis in alkaline buffer followed by electrophoresis at either neutral or mild alkaline (pH 12.1) conditions to detect DNA double-strand breaks or single-strand breaks, respectively *(2)*. However, because the majority of genotoxic agents induce many more single-strand breaks and alkali-labile sites than double-strand breaks, the alkaline version (pH >13) of the comet assay has been identified as having the highest sensitivity for detecting induced DNA damage and has been recommended for genotoxicity testing *(3)*. Important improvements of the test procedure were introduced by Klaude and co-workers in 1996 *(11)*. The use of agarose-precoated slides in combination with drying of gels and fixation of the comets led to a further simplification and a much better handling of the test. The comet assay is especially suited for studies involving high numbers of samples because it can be performed in a high-throughput fashion and analysis of slides can be automated *(12–15)*.

1.1. Detection of DNA Damage

A broad spectrum of DNA-damaging agents increases DNA migration in the comet assay, such as ionizing radiation, hydrogen peroxide and other radical-forming chemicals, alkylating agents, polycyclic aromatic hydrocarbons (PAHs) and other adduct-forming chemicals, radiomimetic chemicals, various metals or UV-irradiation *(1)*. In principle, the alkaline version of the comet assay detects all kinds of directly induced DNA single-strand breaks and any lesion that can be transformed into a single-strand break under alkaline conditions (i.e., alkali-labile sites). Crosslinks (DNA–DNA or DNA–protein), as induced by nitrogen mustard, cisplatin, cyclophosphamide, or formaldehyde, may cause problems in the standard protocol. The induction of crosslinks reduces the ability of the DNA to migrate in the agarose gel by stabilizing chromosomal DNA *(16,17)*. Crosslinks can be detected by adjusting the duration of unwinding and/or electrophoresis to such an extent that control cells exhibit significant DNA migration. A lower extent in DNA migration in treated samples compared to controls would then indicate an induction of crosslinks *(18)*. Another possibility is to induce DNA migration with a second agent (e.g., ionizing radiation, methyl methanesulfonate [MMS]) and to determine the reduced migration in the presence of the crosslinking agent *(16,17)*. Posttreatment of samples with Proteinase K allows one to distinguish between DNA–DNA and DNA–protein crosslinks *(17)*.

In addition to directly induced strand breakage, processes that introduce single-strand nicks in DNA, such as incision during excision repair processes, are also detectable. In some cases (e.g., UV, PAHs) the contribution of excision repair to the induced DNA effects in the comet assay seems to be of major importance *(19)*. Some specific classes of DNA base damage can be detected with the comet assay in conjunction with lesion-specific endonucleases. These enzymes, applied to the slides for a short time after lysis, nick DNA at sites of specific base alterations and the resulting single-strand breaks can be quantified in the comet assay. Using this modification of the comet assay, oxidized DNA bases have been detected with high sensitivity with the help of endonuclease III, formamidopyrimidine-DNA-glycosylase (FPG) or cell extracts in in vitro tests and samples from human studies *(20–22)*.

1.2. Measuring DNA Repair

A widely used approach for determining DNA repair is to monitor time-dependent removal of lesions (i.e., the decrease in DNA migration) after treatment with a DNA-damaging agent. The comet assay has been successfully used to follow the rejoining of strand breaks induced by ionizing radiation or reactive oxygen species *(23,24)* as well as the repair of various kinds of DNA damage induced by chemical mutagens *(25,26)*. A useful extension of repair studies includes the additional use of lesion-specific enzymes *(20)* or cell extracts *(24)*. Thereby, the repair of specific types of DNA lesions can be followed and, because of its high sensitivity, this approach enables the analysis of very low ("physiological") levels of DNA damage *(27)*. A common alternative approach is the use of repair inhibitors or repair-deficient cells. Incubation of cells with inhibitors of DNA- (repair-) synthesis leads to an accumulation of incomplete repair sites as DNA breaks *(19,28)*. Mutant cell lines either with a specific defect in a repair pathway (e.g., xeroderma pigmentosum) or with a hypersensitivity toward specific DNA damaging agents (e.g., various mutant rodent cell lines) are well suited to elucidate DNA repair pathways and the biological consequences of disturbed DNA repair or to evaluate the repair competence of cells *(19,29–31)*. While the standard version of the comet assay provides information on DNA damage and repair in the whole genome of a cell, the introduction of a combination of the comet assay with fluorescence in situ hybridization (FISH) in addition allows one to measure DNA damage and repair in specific genomic regions *(32,33)*.

The purpose of this protocol is to provide information on the application of the alkaline comet assay for the investigation of DNA damage and repair in mammalian cells in vitro. For establishing the method, we recommend starting with experiments using blood samples and the induction of DNA damage by a standard mutagen (e.g., MMS). The method described here is based on a protocol established by R. Tice according to the original work of Singh et al. *(9)* and includes the modifications introduced by Klaude and co-workers *(11)*. An outline of the protocol is diagrammed in **Fig. 2**.

The Comet Assay

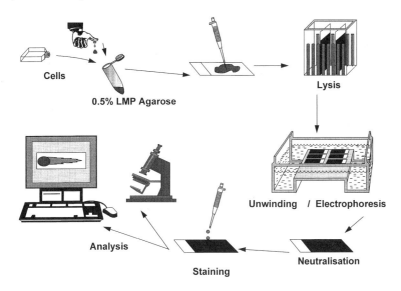

Fig. 2. Scheme for the performance of the comet assay.

2. Materials

1. Microscope slides (with frosted end).
2. Cover slips (24 × 60 mm).
3. Normal melting-point agarose.
4. Low-melting-point (LMP) agarose.
5. Horizontal gel electrophoresis unit.
6. Fluorescence microscope equipped with an excitation filter of 515–560 nm and a barrier filter of 590 nm.
7. Phosphate-buffered saline (PBS) (without Ca^{2+} and Mg^{2+}).
8. Lysing solution (1L): 2.5 M NaCl, 100 mM EDTA, 10 mM Tris (set pH to 10.0 with approx 7 g of solid NaOH). Store at room temperature. Final lysing solution (100 mL, made fresh): Add 1 mL of Triton X-100 and 10 mL of dimethyl sulfoxide (DMSO) to 89 mL of lysing solution, and then refrigerate (4°C) for 60 min before use.
9. Electrophoresis buffer: 300 mM NaOH, 1 mM EDTA. Prepare from stock solutions of 10 N NaOH (200 g/500 mL of distilled H_2O), 200 mM EDTA (14.89 g/200 mL of dH_2O, pH 10.0). Store at room temperature. For 1X buffer, mix 45 mL of NaOH, 7.5 mL of EDTA, and add water to 1500 mL (total volume needed depends on gel box capacity). Mix well. Make fresh before each run.
10. Neutralization buffer: 0.4 M Tris-HCl, pH 7.5. Store at room temperature.
11. Ethidium bromide staining solution: 10X stock: 200 µg/mL. Store at room temperature. For 1X stock (20 µg/mL), mix 1 mL with 9 mL of dH_2O and filter. **Caution:** Ethidium bromide is a mutagen. Handle with care.

3. Methods (see Notes 1–3)

3.1. Preparation of Slides

1. Clean slides with ethanol before use. Wear gloves.
2. Scratch slides with a diamond pen, drawing a line width-wise approx 5 mm from the end of the slide to improve the adhesion of the agarose.
3. For the bottom layer, prepare 1.5% normal melting agarose (300 mg in 20 mL of PBS) and boil until the agarose is completely melted. Dip the slides briefly into hot (>60°C) agarose. The agarose should reach to and cover half of the frosted part of the slide to ensure that the agarose will stick properly. Wipe off the agarose from the bottom side of the slide and lay the slide horizontally. This step has to be performed quickly to ensure a good distribution of agarose. Dry the slides overnight at room temperature. Slides can be stored for several weeks.
4. Prepare 0.5% LMP agarose (100 mg in 20 mL of PBS). Microwave or heat until near boiling and the agarose dissolves. Place the LMP agarose in a 37°C water bath to cool.
5. Add 120 µL of LMP agarose (37°C) mixed with 5000–50,000 cells (see **Subheading 3.2.**) in approx 5–10 µL (do not use more than 10 µL). Add a cover slip, and place the slide in a refrigerator for approx 2 min (until the agarose layer hardens). Using approx 10,000 cells results in approx 1 cell per microscope field (×250 magnification). From this step until the end of electrophoresis, direct light irradiation should be avoided to prevent additional DNA damage.
6. Gently slip off the cover slip and slowly lower the slide into cold, freshly made lysing solution. Protect from light, and place at 4°C for a minimum of 1 h. Slides may be stored for extended periods of time in cold lysing solution (but generally not longer than 4 wk). If precipitation of the lysing solution is observed, slides should be rinsed carefully with distilled water before electrophoresis.

3.2. Preparation of Cells (see Notes 4 and 5)

1. Whole blood: Mix approx 5 µL of whole blood with 120 µL of LMP agarose, and layer onto the slide.
2. Isolated lymphocytes: Add 4 mL of whole blood to a tube with 4 mL of prewarmed (37°C) Ficoll. Centrifuge for 25 min at approx 320g. Carefully remove the lymphocytes and resuspend them in 8 mL of RPMI 1640 medium. Centrifuge again for 10 min at approx 180g. Remove the supernatant and repeat the washing step. Incubate the cells for 30 min at 37°C. Centrifuge for 10 min at approx 130g, discard the supernatant and resuspend the pellet in 375 µL of RPMI 1640 medium. Count the cells and adjust to 1500 cells/µL. Mix 10 µL of the suspension with 120 µL of LMP agarose and layer onto the slide.
3. Cell cultures:
   a. Monolayer cultures: Gently trypsinize the cells (for approx 2 min with 0.15% trypsin, stop by adding serum or complete cell culture medium) to yield

approx 1×10^6 cells/mL. Add 10 µL of cell suspension to 120 µL of LMP agarose, and layer onto the slide.

b. Suspension cultures: Add approx 15,000 cells in 10 µL (or smaller volume) to 120 µL of LMP agarose and layer onto the slide.

3.3. Electrophoresis and Staining (see Notes 6–9)

1. After at least 1 h at 4°C, gently remove the slides from the lysing solution.
2. Place the slides in the gel box near the anode (+) end, positioning them as close together as possible.
3. Fill the buffer reservoirs with electrophoresis buffer (4°C) until the slides are completely covered (avoid bubbles over the agarose). Perform the electrophoresis in an ice bath (4°C).
4. Let the slides sit in the alkaline buffer for 20–60 min to allow unwinding of the DNA and the expression of alkali-labile damage. For most experiments with cultured cells, 20 min are recommended.
5. Turn on the power supply to 25 V (approx 0.8–1.5 V/cm, depending on gel box size) and adjust the current to 300 mA by slowly raising or lowering the buffer level. Depending on the purpose of the study and on the extent of migration in the control samples, allow the electrophoresis to run for 20–40 min. For most experiments, 20 min is recommended.
6. Turn off the power. Gently lift the slides from the buffer and place on a staining tray. Coat the slides with drops of neutralization buffer, and let sit for at least 5 min. Repeat two more times.
7. Drain the slides, rinse carefully with distilled water, and let them dry (inclined) at room temperature. (If kept clean and dry, slides can be stored for months before staining.) To stain, rinse the slides briefly in distilled water, add 30 µL of 1X ethidium bromide staining solution, and cover with a cover slip. Antifade can be used to prevent slides from drying or fading out if necessary, that is, when automated analysis is used *(13)*.

Slides should be stained one by one and evaluated immediately. It is possible to rinse stained (evaluated) slides in distilled water, remove the cover slip, let the slides dry and stain them at a later time point for reevaluation.

3.4. Evaluation of DNA Effects (see Note 10)

For visualization of DNA damage, observations are made of ethidium bromide-stained DNA at ×250 (or ×400) magnification using a fluorescence microscope. Generally, 50 randomly selected cells per sample are analyzed. In principle, evaluation can be done in four different ways:

1. Image analysis systems are used to quantitate DNA damage. Parameters such as percentage DNA in the tail, tail moment, tail length are commonly used. It is important to note that the same parameters (e.g., tail moment) may be calculated differently among image analysis systems. For the purpose of interlaboratory comparison of DNA damage parameters, "percentage DNA in the tail" is probably the most suited.

2. Cells are scored visually according to tail size into five classes (from undamaged, 0, to maximally damaged, 4). Thus the total score for 50 comets can range from 0 (all undamaged) to 200 (all maximally damaged) *(23)*.
3. The percentage of cells with tail vs those without is determined.
4. Cells are analyzed using a calibrated scale in the ocular lens of the microscope. For each cell, the image length (diameter of the nucleus plus migrated DNA) is measured in microns, and the mean is calculated.

For the statistical analysis of comet assay data, a variety of parametric and nonparametric statistical methods are used. The most appropriate means of statistical analysis depends on the kind of study and has to take into account the various sources of assay variability. For a powerful statistical analysis of in vitro test data, appropriate replication and repeat experiments have to be performed *(3,34,35)*. For example, the median DNA migration of 50 cells per sample and the mean of two or three samples per data point may be determined. Also, the mean from repeat experiments can be determined. The use of the median should be preferred over the average because a normal size distribution is usually not observed. Analyses are based mainly on changes in group mean response but attention should also be paid to the distribution among cells, which often provides additional important information. Recommendations for appropriate statistical analyses of comet assay data have been published *(34,35)*.

4. Notes

1. Many technical variables have been used including the concentration and amount of LMP agarose, the composition of the lysing solution and the lysis time, the alkaline unwinding, the electrophoresis buffer and electrophoretic conditions, DNA-specific dyes for staining, etc. (for details, see **ref. 1**). Some of these variables may affect the sensitivity of the test. To allow for a comparison obtained in different laboratories and for a critical evaluation of data, it is absolutely necessary to clearly describe the technical details of the method employed.
2. Although the protocol described here detects a broad spectrum of DNA-damaging agents with high sensitivity, modifications have been suggested that further increase the sensitivity and may be advantageous for certain applications *(36,37)*. These modifications include the addition of radical scavengers to the electrophoresis buffer (to reduce damage during prolonged electrophoresis), the addition of Proteinase K to the lysis solution (to remove residual proteins that might inhibit DNA migration) and the use of the DNA dye YOYO-1 (to increase the sensitivity for the detection of migrated DNA).
3. The simplicity of the comet assay combined with the need for only low numbers of cells per sample enables the conduct of in vitro studies with high efficiency. Therefore, the comet assay can be used in a high-throughput fashion *(13,15)*. Furthermore, the introduction of automated image analysis systems for comet assay slides can further speed up test performance *(12,14)*.

4. Many other cell types have been used and it is an advantage of the comet assay that virtually any eukaryote cell population is amenable to analysis. The comet assay is particularly suited for the investigation of organ- or tissue-specific genotoxic effects in vivo *(3,4)*, the only requirement being the preparation of an intact single-cell suspension.
5. For the demonstration of a positive effect, mix 200 µL of heparinized whole blood with 50 µL of a $2.5 \times 10^{-4}\ M$ MMS solution (final concentration: $5 \times 10^{-5}\ M$), incubate for 1 h at 37°C and then use 10 µL for the test.
6. For each cell type the method should be adjusted empirically to obtain valid and reproducible results. It is important to define the optimal time for alkaline treatment and electrophoresis. It is recommended that the conditions must be such that the DNA from the control cells exhibit, on average, some migration. This effect ensures sensitivity and enables an evaluation of intralaboratory experiment-to-experiment variability *(3)*.
7. The temperature during alkaline treatment and electrophoresis significantly influences the amount of DNA migration *(38)*. It is necessary to establish stable and reproducible conditions and it may be useful to place the gel electrophoresis unit in a jar filled with ice or in a cold room.
8. If specific types of base damage are to be determined by using lesion-specific endonucleases or cell extracts, the standard protocol has to be modified in the following way: After at least 1 h at 4°C, gently remove slides from the lysing solution and wash three times in enzyme buffer. Drain the slides and cover with 200 µL of either buffer or enzyme in buffer. Seal with a cover slip and incubate for 30 min at 37°C. Remove the cover slip, rinse the slides with PBS and place them on the electrophoretic box *(20–22,24)*.
9. For in vitro tests, cells are usually incubated with the test substances for a defined period of time, then mixed with LMP agarose and added to the slide. A modified protocol that may be performed in combination with the standard comet assay suggests treatment of samples after lysis. Under these conditions, the lysed cells are no longer held under the regulation of any metabolic pathway or membrane barrier *(39)*.
10. It is strongly recommended to include some measure of cytotoxicity in any study, as increased DNA migration may also occur as a result of nongenotoxic cell killing. However, such an effect may depend on the cell type used. While no increased DNA migration had been observed in human leukocytes *(40)* or cell lines such as V79 *(40,41)* and L5178Y *(15)*, TK-6 cells showed increased DNA migration after treatment with nongenotoxic cytotoxins when viability in treated cultures fell below 75% *(42)*. Therefore, acute cytotoxic effects should be determined by trypan blue exclusion measurements or flurochrome-mediated viability tests. Furthermore, individual dead or dying cells may be identified by their characteristic microscopic image, that is, necrotic or apoptotic cells result in comets with small or non-existent head and large, diffuse tails *(43)*. These cells are commonly called "hedgehogs," "ghost cells," "clouds," or "nondetectable cell nuclei (NDCN)." Such cells have been detected after treatment with cytotoxic, non-

genotoxic agents *(41,42,44)*. However, since these microscopic images are also seen after treatment with high doses of radiation or high concentrations of strong mutagens, such comets are not uniquely diagnostic for apoptosis/necrosis *(45,46)*. For the evaluation of genotoxic effects, it is recommended to record these cells but to exclude them from image analysis under the principle that they represent dead cells.

References

1. Tice, R. R. (1995) The single cell gel/comet assay: A microgel electrophoretic technique for the detection of DNA damage and repair in individual cells, in *Environmental Mutagenesis* (Phillips, D. H. and Venitt, S., eds.), BIOS Scientific Publishers, Oxford, UK, pp. 315–339.
2. Fairbairn, D. W., Olive, P. L., and O'Neill, K. L. (1995) The comet assay: a comprehensive review. *Mutat. Res.* **339,** 37–59.
3. Tice, R. R., Agurell, E., Anderson, D., et al. (2000) The single cell gel /comet assay: Guidelines for in vitro and in vivo genetic toxicology testing. *Environ. Mol. Mutagen.* **35,** 206–221.
4. Hartmann, A., Agurell, E., Beevers, C., et al. (2003) Recommendations for conducting the in vivo alkaline Comet assay. *Mutagenesis* **18,** 45–51.
5. Collins, A. R. (2004) The comet assay for DNA damage and repair: principles, applications, and limitations. *Mol. Biotechnol.* **26,** 249–261.
6. Lee R. F. and Steinert, S. (2003) Use of the single cell gel electrophoresis/comet assay for detecting DNA damage in aquatic (marine and freshwater) animals. *Mutat. Res.* **544,** 43–64.
7. Cotelle, S. and Férard, F. (1999) Comet assay in genetic ecotoxicology: a review. *Environ. Mol. Mutagen.* **34,** 246–255.
8. Moller, P., Knudsen, L. E., Loft, S., and Wallin, H. (2000). The comet assay as a rapid test in biomonitoring occupational exposure to DNA-damaging agents and effect of confounding factors. *Cancer Epidemiol. Biomarkers Prev.* **9,** 1005–1015.
9. Singh, N. P., McCoy, M. T., Tice, R. R., and Schneider, E. L. (1988) A simple technique for quantification of low levels of DNA damage in individual cells. *Exp. Cell Res.* **175,** 184–191.
10. Olive, P. L. (1989) Cell proliferation as a requirement for development of contact effect in Chinese hamster V79 spheroids. *Radiat. Res.* **117,** 79–92.
11. Klaude, M., Erikson, S., Nygren, J., and Ahnström, G. (1996) The comet assay: mechanisms and technical considerations. *Mutat. Res.* **363,** 89–96.
12. Boecker, W., Rolf, W., Bauch, T., Muller, W. U., and Streffer, C. (1999). Automated comet assay analysis. *Cytometry* **35,** 134–144.
13. McNamee, J. P., McLean, J. R., Ferrarotto, C. L., and Bellier, P. V. (2000) Comet assay: rapid processing of multiple samples. *Mutat Res.* **466,** 63–69.
14. Frieauff, W., Hartmann, A., and Suter, W. (2001) Automatic analysis of slides processed in the comet assay. *Mutagenesis* **16,** 133–137.
15. Kiskinis, E., Suter, W., and Hartmann, A. (2002) High-throughput comet assay using 96-well plates. *Mutagenesis* **17,** 37–43.

16. Pfuhler, S. and Wolf, H. U. (1996) Detection of DNA-crosslinking agents with the alkaline comet assay. *Environ. Mol. Mutagen.* **27,** 196–201.
17. Merk, O. and Speit, G. (1999) Detection of crosslinks with the comet assay in relationship to genotoxicity and cytotoxicity. *Environ. Mol. Mutagen.* **33,** 167–172.
18. Fuscoe, J. C., Afshari, A. J., George, M. H., et al. (1996) In vivo genotoxicity of dichloroacetic acid: evaluation with the mouse peripheral blood micronucleus assay and the single cell gel assay. *Environ. Mol. Mutagen.* **27,** 1–9.
19. Speit, G. and Hartmann, A. (1995) The contribution of excision repair to the DNA-effects seen in the alkaline single cell gel test (comet assay). *Mutagenesis* **10,** 555–559.
20. Collins, A. R., Duthie, S. J., and Dobson, V.L. (1993) Direct enzymic detection of endogenous oxidative base damage in human lymphocyte DNA. *Carcinogenesis* **14,** 1733–1735.
21. Dennog, C., Hartmann, A., Frey, G., and Speit, G. (1996) Detection of DNA damage after hyperbaric oxygen (HBO) therapy. *Mutagenesis* **11,** 605–609.
22. Speit, G., Schütz, P. Bonzheim, I., Trenz, K., and Hoffmann, H. (2004) Sensitivity of the FPG protein towards alkylation damage in the comet assay. *Toxicol. Lett.* **146,** 151–158.
23. Collins, A. R., Ai-guo, A., and Duthie, S. J.(1995): The kinetics of repair of oxidative DNA damage (strand breaks and oxidised pyrimidines) in human cells. *Mutat. Res.* **336,** 69–77.
24. Collins, A. R., Dusinska, M., Horvathova, E., Munro, E., Savio, M., and Stetina, R. (2001). Inter-individual differences in repair of DNA base oxidation, measured in vitro with the comet assay. *Mutagenesis* **16,** 297–301.
25. Hartmann, A. and Speit, G. (1996) The effect of arsenic and cadmium on the persistence of mutagen-induced DNA lesions in human cells. *Environ. Mol. Mutagen.* **27,** 98–104.
26. Hartmann, A. and Speit, G. (1995) Genotoxic effects of chemicals in the single cell gel (SCG) test with human blood cells in relation to the induction of sister chromatid exchanges (SCE). *Mutat. Res.* **346,** 49–56.
27. Collins, A. R., Harrington, V., Drew, J., and Melvin, R. (2003) Nutritional modulation of DNA repair in a human intervention study. *Carcinogenesis* **24,** 511–515.
28. Gedik, C. M., Ewen, S. W. B., and Collins, A. R. (1992) Single-cell gel electrophoresis applied to the analysis of UV-C damage and its repair in human cells. *Int. J. Radiat. Biol.* **62,** 313–320.
29. Green, M. H. L., Lowe, J. E., Harcourt, S. A., et al. (1992) UV-C sensitivity of unstimulated and stimulated human lymphocytes from normal and xeroderma pigmentosum donors in the comet assay: a potential diagnostic technique. *Mutat. Res.* **273,** 137–144.
30. Helbig, R. and Speit, G. (1997) DNA effects in repair-deficient V79 Chinese hamster cells studied with the comet assay. *Mutat. Res.* **377,** 279–286.
31. Tebbs, R. S., Flannery, M. L., Meneses, J. J., et al. (1999) Requirement for the Xrcc1 DNA Base Excision Repair Gene During Early Mouse Development. *Dev. Biol.* **208,** 513–529.

32. Rapp, A., Bock, C., Dittmar, H., and Greulich, K. O. (2000) UV-A breakage sensitivity of human chromosomes as measured by COMET-FISH depends on gene densitiy and not on the chromosome size. *Photochem. Photobiol. B.* **56,** 109–117.
33. McKenna, D. J., Gallus, M., McKeown, S. R., Downes, C. S., and McKelvey-Martin, V. J. (2003) Modification of the alkaline Comet assay to allow simultaneous evaluation of mitomycin C-induced DNA cross-link damage and repair of specific DNA sequences in RT4 cells. *DNA Repair* **2,** 879–890.
34. Lovell, D. P., Thomas, G., and Dubow, R. (1999) Issues related to the experimental design and subsequent statistical analysis of in vivo and in vitro comet studies. *Teratog. Carcinog. Mutagen.* **19,** 109–119.
35. Wiklund, S. J. and Agurell, E. (2003) Aspects of design and statistical analysis in the Comet assay. *Mutagenesis* **18,** 167–175.
36. Singh, N. P., Stephens, R. E., and Schneider, E. L. (1994) Modifications of alkaline microgel electrophoresis for sensitive detection of DNA damage. *Int. J. Radiat. Biol.* **66,** 23–28.
37. Singh, N. P. and Stephens R. E. (1997) Microgel electrophoresis: sensitivity, mechanisms, and DNA electrostretching. *Mutat. Res.* **383,** 167–175.
38. Speit, G., Trenz, K., Schütz, P., Rothfuss, A., and Merk, O. (1999) The influence of temperature during alkaline treatment and electrophoresis on results obtained with the comet assay. *Toxicol. Lett.* **110,** 73–78.
39. Kasamatsu, T., Kohda, K., and Kawazoe, Y. (1996) Comparison of chemically induced DNA breakage in cellular and subcellular systems using the comet assay. *Mutat. Res.* **369,** 1–6.
40. Hartmann, A. and Speit, G. (1997) The contribution of cytotoxicity to effects seen in the alkaline comet assay. *Toxicol. Lett.* **90,** 183–188.
41. Hartmann, A., Kiskinis, E., Fjaellman, A., and Suter, W. (2001) Influence of cytotoxicity and compound precipitation on test results in the alkaline comet assay. *Mutation Res.* **497,** 199–212.
42. Henderson, L., Wolfreys, A., Fedyk, J., Bourner, C., and Windebank, S. (1998) The ability of the comet assay to discriminate between gentoxins and cytotoxins. *Mutagenesis* **13,** 89–94.
43. Olive, P. L. and Banath, J. P. (1995) Sizing highly fragmented DNA in individual apoptotic cells using the comet assay and a DNA crosslinking agent. *Exp. Cell. Res.* **221,** 19–26.
44. Kiffe, M., Christen, P., and Arni, P. (2003) Characterization of cytotoxic and genotoxic effects of different compounds in CHO K5 cells with the comet assay (single-cell gel electrophoresis assay). *Mutat Res.* **537,** 151–168.
45. Meintieres, S., Nesslany, F., Pallardy, M., and Marzin, D. (2003) Detection of ghost cells in the standard alkaline comet assay is not a good measure of apoptosis. *Environ. Mol. Mutagen.* **41,** 260–269.
46. Rundell, M. S., Wagner, E. D., and Plewa, M. J. (2003) The comet assay: genotoxic damage or nuclear fragmentation? *Environ. Mol. Mutagen.* **42,** 61–67.

21

Fast Micromethod DNA Single-Strand-Break Assay

Heinz C. Schröder, Renato Batel, Heiko Schwertner,
Oleksandra Boreiko, and Werner E. G. Müller

Summary

The Fast Micromethod is a convenient and quick fluorimetric microplate assay for the assessment of DNA single-strand breaks and their repair. This method measures the rate of unwinding of cellular DNA on exposure to alkaline conditions using a fluorescent dye which preferentially binds to double-stranded DNA, but not to single-stranded DNA or protein. The advantages of this method are that it requires only minute amounts of material (30 ng of DNA or about 3000 cells per single well), it allows simultaneous measurements of multiple samples, and it can be performed within 3 h or less (for one 96-well microplate). The Fast Micromethod can be used for the routine determination of DNA damage in cells and tissue samples after irradiation, exposure to mutagenic and carcinogenic agents, or chemotherapy.

Key Words: Alkali-labile sites; DNA damage; DNA repair; DNA single-strand breaks; DNA unwinding; environmental monitoring; Fast Micromethod; fluorescent dye; γ-rays; genotoxicity; HeLa cells; lymphocytes; microplate assay; UV-C light; X-rays.

1. Introduction

Cellular DNA is subject to damage by various physical and chemical agents including ionizing radiation, ultraviolet (UV) light and certain mutagenic chemicals. Besides damage caused by environmental factors, DNA lesions can occur spontaneously or be produced by oxygen radicals generated by endogenous metabolic reactions. Ionizing radiation (e.g., γ-rays and X-rays) causes predominantly DNA single-strand breaks, and much less frequently double-strand breaks, alkali-labile sites, and various oxidized purines and pyrimidines *(1)*. DNA single-strand breaks can also be generated by certain chemical agents or by repair endonucleases during excision repair of apurinic/apyrimidinic (AP) sites generated by ionizing radiation, and UV-induced cyclobutane pyrimidine

From: *Methods in Molecular Biology: DNA Repair Protocols: Mammalian Systems, Second Edition*
Edited by: D. S. Henderson © Humana Press Inc., Totowa, NJ

dimers and pyrimidine (6–4) pyrimidone photoproducts *(1)*. Because unrepaired DNA damage can lead to mutations, cancer, and cell death, DNA repair is essential for survival of all organisms.

Several methods have been introduced to measure DNA damage and repair activity, including the single cell gel electrophoresis (comet) assay *(2,3)*, fluorometric analysis of the DNA unwinding (FADU) *(4,5)*, and alkaline filter elution (AFE) *(6,7)*. The comet assay allows determination of DNA damage in single cells but not in tissue samples. The FADU method measures DNA denaturation in alkaline solution and requires termination of the unwinding reaction by neutralization after a given time period; this method measures the fluorescence caused by binding of the fluorescent dye ethidium bromide to the remaining double-stranded DNA but does not allow continuous measurements of the kinetics of DNA unwinding; about $1-2 \times 10^6$ cells are needed for each single determination. The AFE method is based on the different filtration rates of damaged DNA compared to undamaged DNA unwound by alkali *(6,7)*. DNA molecules with increased numbers of single-strand breaks consist of shorter strands after denaturation by alkali and are eluted from the filters faster than longer-strand DNA molecules. The AFE procedure is one of the most sensitive methods for measuring single-strand breaks. However, this procedure is time consuming, complicated, and only allows simultaneous analysis of limited numbers of samples; therefore it is less suitable for routine measurements.

To overcome these problems, a simple and quick method for the routine assessment of DNA single-strand breaks has been developed *(8)*. This method, called Fast Micromethod,[1] is a fluorescence-based microplate assay that can be used for measurement of DNA damage and repair in both human and animal cell and tissue samples. So far, the Fast Micromethod has been evaluated in studies on DNA integrity in cancer patients undergoing radiotherapy *(9)* and in bioindicator organisms or cells exposed to environmental genotoxins *(10–13)*.

The method is based on the ability of the commercially available fluorescence dye PicoGreen to preferentially bind to double-stranded DNA, and not to single-stranded DNA, RNA, and proteins *(14)*. This makes it possible to directly measure denaturation of double-stranded DNA without prior separation from single-stranded DNA or protein. Binding to double-stranded DNA occurs with high-fluorescence yields even in the presence of urea, high pH, or high ionic strength.

The principle of the assay is based on the fact that the hydrogen bonds in the DNA double helix are destabilized in alkaline solutions near pH 12. The kinetics of the unwinding of the two strands of the double-stranded DNA depends

---

[1] Patents/patent applications: DE 19724781, DE 19933078, and PCT/EP99/01545.

on the length of the molecule, but also on the number of single-strand breaks and alkali-labile sites present in the DNA. DNA with single-strand breaks and alkali-labile sites unwinds faster than DNA that is not damaged. The use of PicoGreen (which preferentially stains double-stranded but not single-stranded DNA) allows the remaining (not unwound) double-stranded DNA to be quantified. By this way, the Fast Micromethod can determine the frequency of single-strand breaks in the DNA.

The procedure consists of a simple lysis step of the cell or tissue sample directly on a microplate in the presence of urea, sodium dodecylsulfate (SDS), and EDTA at pH 10, followed by a DNA denaturation step at pH 12.4 after addition of NaOH solution to the same wells. The kinetics of DNA denaturation is determined by monitoring the decrease in fluorescence of the complex formed by PicoGreen (already present in the lysing solution) and the double-stranded DNA over a time period of at least 20 min. DNA denaturation starting immediately after the pH is raised from 10 to 12.4 is followed using a fluorescence microplate reader. Because of the faster unwinding of DNA with single-strand breaks, lower fluorescence values are measured, in the course of the unwinding reaction, for samples containing damaged DNA (single-strand breaks or DNA lesions that are converted to single-strand breaks at alkaline pH, such as alkali-labile sites, or transient breaks formed during incomplete excision repair).

The Fast Micromethod has been validated in experiments using both human and animal cell lines after irradiation or treatment with model genotoxins, as well as lymphocytes from persons exposed to irradiation. The results revealed that the method is also suitable for clinical applications, for example, for determination of DNA damage following radiotherapy of cancer patients *(9)*. Measurement of the individual radiosensitivity of patients based on peripheral blood mononuclear cells (PBMCs) may be used as a pretherapeutic test prior to radiotherapy, allowing assessment of the optimal dose for radiotherapy *(15)*. The Fast Micromethod has also been applied for measurement of DNA single-strand breaks in HeLa cells induced by γ-rays from a research reactor *(16)*. Using this method, the effects of irradiation in people exposed to fallout of the Chernobyl nuclear power plant accident has been determined *(17)*.

A study of the induction of DNA single-strand breaks in HeLa cells following exposure to ^{137}Cs γ-rays, UV-C light, and 4-nitroquinoline-*N*-oxide (NQO) revealed a dose-dependent increase in the strand scission factor (SSF) values from 0.010 (98% double-stranded DNA) for the negative control to 0.701 (20% double-stranded DNA) for 500 cGy (γ-rays; detection limit: 8 cGy), from 0.019 (96% double-stranded DNA) for the negative control to 1.196 (6% double-stranded DNA) for 1000 J·m^{-2} (UV-C light; detection limit: 10 J·m^{-2}) and from 0.003 (99% double-stranded DNA) for the negative control to 0.810 (16%

double-stranded DNA) for 0.5 μM NQO (detection limit: 0.03 μM), respectively *(18)*.

The Fast Micromethod can also be applied for measurements of DNA damage in solid tissues, for example, liver and muscle from mice *(8)* or fish liver *(13)*. In addition, this method can be used for determination of DNA lesions in lower taxa, for example, marine invertebrates like mussels *(19)* and sponges *(11,20)*, thus making this assay useful for assessment of genotoxic effects of pollution on aquatic organisms.

The sensitivity and precision of the Fast Micromethod is similar to (or even better than) that of the comet assay (performed according to **ref. 3**) concerning various DNA damaging agents including γ-rays (^{137}Cs), UV-C light and NQO *(18)*. The major advantages of the Fast Micromethod compared to comet assay are: (1) it is performed in microplates, thus allowing simultaneous analysis of large numbers of samples; and (2) it can be performed within 3 h or less. In addition, this method—like the comet assay—requires only minute amounts of material (approx 30 ng of DNA per well, corresponding to approx 3000 cells or approx 25 μg of tissue). Therefore, the Fast Micromethod is particularly useful if only small amounts of material are available or a large number of samples has to be analyzed within a short time period. Moreover, this method can be applied to frozen samples (e.g., biopsy material stored in liquid nitrogen). Because of its simplicity and speed even for analysis of multiple samples, Fast Micromethod is a serious alternative to conventional assays for DNA damage, especially for routine automation.

2. Materials [2]

2.1. Fast Micromethod

1. 96-Well microplate (black) for fluorescence measurement (Nunc).
2. Micropipets with handling volumes of 20, 25, 250, and 1000 μL.
3. Ice bath (optional).
4. Aluminum foil.
5. Fluorescence enzyme-linked immunosorbent assay (ELISA) plate reader with a 480-nm excitation filter and a 520-nm emission filter (e.g., Fluoroskan II, Labsystems, Helsinki, Finland).
6. Stopclock.
7. Fluorescent dye stock solution (solution A): PicoGreen dsDNA quantitation reagent (Molecular Probes). This solution must be stored at −20°C. (*See* **Note 1.**)
8. Calcium- and magnesium-free phosphate-buffered saline (Ca/Mg-free PBS; solution B): 137 mM NaCl, 2.7 mM KCl, 4.3 mM Na$_2$HPO$_4$, 1.5 mM KH$_2$PO$_4$.
9. Lysing solution (solution C): 9.0 M urea, 0.1% SDS, 0.2 M EDTA, pH 10 with NaOH. (*See* **Note 2.**)

---

[2] The Fast Micromethod DNA single-strand break assay is also available as a commercial kit (*BIOTEC*marin GmbH, Mainz, Germany).

10. Lysing solution supplemented with PicoGreen (solution D): 20 µL of the original stock dye/mL of lysing solution (solution C). Prepare shortly before use. (*See* **Note 3**.)
11. EDTA solution (solution E): 20 mM EDTA.
12. NaOH stock solution (solution F): 1.0 M NaOH, 20 mM EDTA. Dilute this solution with EDTA solution (solution E) before use. (*See* **below**.)
13. Working NaOH solution (solution G). This solution should be prepared freshly before use. Mix 2 mL of NaOH stock solution (solution F) with 18 mL of EDTA solution (solution E) and start pH checking and adjustment as follows. Add 5.0 mL of the resulting NaOH solution to 0.5 mL of Ca/Mg-free PBS (solution B) plus 0.5 mL of lysing solution (solution C) and check the pH. The pH should be 12.40 ± 0.02. If the pH is too high add some drops of NaOH stock solution (solution F) to the mixture of EDTA solution (solution E) and NaOH stock solution (solution F) and check the pH after addition to Ca/Mg-free PBS (solution B) and lysing solution (solution C) as **above**. If the pH is too low add some drops of EDTA solution (solution E) to the mixture of EDTA solution (solution E) and NaOH stock solution (solution F) and check the pH after addition to Ca/Mg-free PBS (solution B) and lysing solution (solution C) as above. Repeat these steps if necessary. Usually five adjustments are needed to achieve the right pH. The pH meter should be calibrated before each experiment using pH standards. (*See* **Note 4.**)
14. TE buffer (solution H): 10 mM Tris-HCl, pH 7.4, 1 mM EDTA.
15. Calf thymus DNA (Sigma).
16. pH standards for the calibration procedure: phosphate buffer, pH 7.00, and phosphate–NaOH buffer, pH 12.00.

2.2. Working With Cells

2.2.1. Cell Lines

1. HeLa S3 cells (American Type Culture Collection).
2. Mouse DBA/2 lymphoblasts L5178Y (American Type Culture Collection).
3. Cell culture flasks (Falcon).
4. Culture plates, for example, 96-well microplates, flat bottom (for dose–response experiments with chemical compounds; Nunc) or six-well plates, flat bottom (for irradiation experiments; Nunc).
5. RPMI 1640 medium (Gibco/ Invitrogen).
6. Fetal calf serum (FCS; Biochrom).
7. Trypsin–EDTA solution (for use with endothelial cell cultures; Sigma).
8. Neubauer chamber.

2.2.2. Peripheral Blood Mononuclear Cells

1. Ficoll Histopaque, for example, Histoprep (Sigma) containing Ficoll and metrizoic acid, or Ficoll-Paque Plus (Amersham), a Ficoll-sodium diatrizoate solution of the proper density and osmolarity for isolation of PBMCs.

2. Anticoagulant (e.g., 2.7% EDTA solution).
3. Syringes and needles.
4. 15-mL Centrifuge tubes.
5. Alternatively, special centrifuge tubes for isolation of PBMCs (e.g., Vacutainer CPT tubes, Becton Dickinson; or Uni-Sep tubes, Novamed).
6. Centrifuge with swinging bucket rotor and tube adapters (capable of generating 1500g, relative centrifuge force).
7. Ca/Mg-free PBS.
8. **Items 3–6**, and **8, Subheading 2.2.1.**

2.3. DNA Repair Measurements

Cell culture materials and equipment (*see* **Subheadings 2.2.1. and 2.2.2.**).

2.4. Standard Cell DNA

Bull sperm (approx 1×10^9 cells/mL) in stabilizing solution (containing 6% glycerol).

2.5. Working With Tissue Samples

1. Dimethyl sulfoxide (DMSO).
2. Homogenization buffer: Ca/Mg-free PBS (solution B) supplemented with 10% DMSO or TE/DMSO buffer (solution H supplemented with 10% DMSO).
3. DNA standard: 1 mg/mL of DNA (calf thymus) in TE–DMSO buffer (4°C).
4. Fluorescent dye YOYO (Molecular Probes). Add 5 µL of the original YOYO solution to 500 µL of 25% DMSO–water, and store 40-µL aliquots at –20°C.
5. YOYO working solution: Prepare this solution freshly by adding 960 µL of TE–DMSO buffer to the 40-µL aliquots of the fluorochrome in 25% DMSO–water.

3. Methods

3.1. Fast Micromethod

A scheme of the procedure is shown in **Fig. 1**. For monitoring of DNA unwinding, a fluorescence ELISA plate reader is required. The use of an ice bath is optional. Analysis of DNA damage can also be performed at room temperature. (*See* **Note 5.**)

1. Add 25 µL of cell suspension (150,000 cells/mL or 3000 cells/25 µL; this corresponds to approx 30 ng of DNA in human cells) in Ca/Mg-free PBS (solution B) or TE buffer (solution H) or different dilutions of a standard cell DNA (only necessary for assays that do not include nontreated samples, for example, nonirradiated control samples) into the wells of a black 96-well-microplate (**Fig. 1A,B**). Fluorescence blanks contain 25 µL of Ca/Mg-free PBS (solution B) or TE buffer (solution H) instead of sample and are processed in the same way as the samples. Each sample as well as the blank are measured in at least four replicates. (*See* **Notes 6** and **7.**)

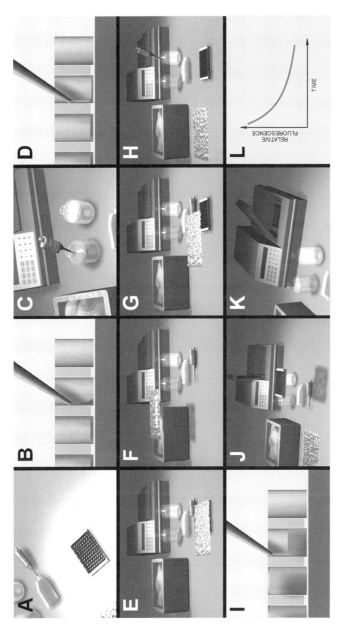

Fig. 1. Fast Micromethod DNA single-strand-break assay. Schematic presentation of the procedure. *Cell lysis step:* (**A**) 25 μL of cell suspension (or diluted standard cell DNA) are taken and (**B**) pipetted into the wells of the microplate. (**C**) Then 25 μL of lysing solution supplemented with PicoGreen (solution D) are taken and (**D**) added to the wells of the microplate containing the cells. (**E**) The plate is covered with aluminum foil (for light protection) and (**F**) incubated on ice for 40–60 min. *DNA denaturation (DNA unwinding) step:* (**G**) After the incubation period, the microplate is removed. (**H**) 250 μL of freshly prepared NaOH solution (adjusted to pH 12.4; solution G) is taken and (**I**) this solution is added to the lysate. (**J**) The plate is immediately transferred to the fluorescence microplate reader and (**K,L**) measurement at 480 nm excitation and 520 nm emission starts for up to 40 min. (Adapted from **ref. *16*.**)

2. Cell lysis: Add slowly 25 µL of lysing solution supplemented with the fluorescent dye (solution D) to the wells of the microplate without mixing or shaking (**Fig. 1C,D**). (*See* **Note 8**.)
3. Allow cell lysis to occur by incubating the microplate in the dark at room temperature or on ice for 40–60 min (**Fig. 1E–G**). Protect from light to prevent damage of DNA, for example, by covering the microplate with aluminum foil. Do not mix or shake.
4. DNA unwinding: Add 250 µL of freshly prepared and pH-adjusted working NaOH solution (solution G) to the lysates in the wells of the microplate (**Fig. 1H,I**). By addition of this solution, the pH value is raised to 12.4 and DNA denaturation (DNA unwinding) is initiated. (*See* **Notes 9** and **10**.)
5. Immediately thereafter, start fluorescence measurements at an excitation wavelength of 480 nm and an emission wavelength of 520 nm for up to 40 min (**Fig. 1J–L**). Fluorescence measurements are performed, depending on the type of fluorescence ELISA plate reader, every 30 s (also longer intervals are possible, e.g., 5 min) for at least 20 min. The first measurement of the samples is performed at time 0 as soon as working NaOH solution (solution G) has been added. (*See* **Note 11**.)

3.2. Working With Cells

3.2.1. Cell Lines and Treatment

1. Use exponentially growing cells. Cultures of human HeLa cells are maintained in RPMI 1640 medium with 10 mM N-2-hydroxyethylpiperazine-N'-2-ethanesulfonic acid (HEPES) and 10% FCS. The cells are kept in a fully humidified atmosphere and 5% CO_2–air at 37°C. Mouse DBA/2 lymphoblasts L5178Y are grown in RPMI 1640 medium with $NaHCO_3$ and 10% FCS at 37°C with 5% CO_2.
2. Count the cells. In each single well of the microplate, exactly the same amount of cells is needed (3000 cells/25 µL of Ca/Mg-free PBS or TE buffer).
3. Exposure to irradiation: Irradiate the cells (10^4–10^5 cells/mL) with different doses of γ-rays generated by a ^{137}Cs source (e.g., a Gammacell 2000 device, Mølsgaard Medical), or with different doses of UV-B light (e.g., Vilber Lourmat UV-B lamp VL215.M; peak at 312 nm), or UV-C light (e.g., Stratalinker™ 1800 UV Crosslinker; Stratagene), or using a full solar spectrum lamp allowing sun light simulation (e.g., SOL 500 lamp from Dr. Hönle AG, Planegg, Germany with filters H1 (from about 320 nm) or H2 (from about 295 nm), on ice or ice/water. After irradiation place the cells at 4°C, adjust to 150,000 cells/mL (by centrifugation and suspension in TE buffer), and analyze. (*See* **Note 12**.)
4. Exposure to chemical agents: Incubate the cells (10^4–10^5 cells/mL) in the absence or presence of different concentrations of compound, for example, the model mutagen NQO (**Fig. 2**), in culture medium for an appropriate time period at 37°C, dilute, and keep as described above. NQO is known to induce DNA single-strand breaks, besides formation of covalently bound adducts and intercalation of the quinoline between DNA base pairs *(21,22)*. If the compound is dissolved in organic solvent (e.g., DMSO), add also this solvent to the control (the maximum

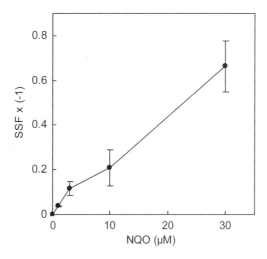

Fig. 2. DNA damage in HeLa cells after incubation with NQO. HeLa cells (1.5×10^5 cells/mL) were incubated in the absence or presence of different concentrations of NQO for 1.5 h. The frequency of single-strand breaks is expressed as SSF (unwinding period, 20 min). Nontreated cells were used as control. The mean values (± standard deviations) from six replicate determinations are shown. (Data from **ref. *16*.**)

DMSO concentration should not exceed 5 µL/mL). Using NQO and HeLa cells, a significant ($p < 0.01$) increase in SSF compared to the control and an almost linear dose–response curve are obtained in the concentration range between 1 and 30 µ*M* NQO. (*See* **Note 13**.)

The unwinding curves of DNA from L5178y mouse lymphoma cells and the calculated SSF values vs concentration for a compound, 2-aminoanthracene, requiring activation by liver S-9 fraction, are shown in **Fig. 3A,B**. The SSF values were calculated after an unwinding period of 20 min using the DNA of nontreated cells as a control.

3.2.2. Peripheral Blood Mononuclear Cells and Treatment

PBMCs are isolated from EDTA blood or blood samples (*see* **Notes 14** and **15**) containing an anticoagulating agent (heparin, EDTA, or citrate) by metrizoate-Ficoll centrifugation *(23)*, preferably using—for convenience and shortening of the procedure—special centrifuge tubes (e.g., Vacutainer CPT tube, Becton Dickinson; *see* **Note 16**). PBMCs are separated from erythrocytes and granulocytes by centrifugation through the preformed density gradient inside the tube.

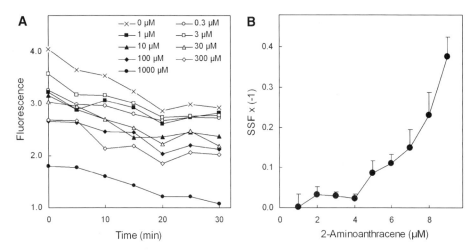

Fig. 3. DNA damage in L5178y mouse lymphoma cells induced by 2-aminoanthracene as determined by the Fast Micromethod. L5178y cells (3×10^5 cells/mL) were incubated with different amounts of compound for 2 h in the presence of S-9 fraction (Moltox post-mitochondrial supernatant; Molecular Toxicology, Inc., Boone, NC). (**A**) Unwinding curves after addition of alkaline solution. (**B**) SSF × (–1) values calculated after an unwinding period of 20 min using the DNA of nontreated cells as control. Error bars represent standard deviations of four replicate measurements. (Data from **ref. 9**.)

1. Follow the procedure described in **ref. 23** or given in the instruction manuals of commercial suppliers of Ficoll Histopaque; commercial reagents may contain Ficoll-sodium metrizoate (e.g., Histoprep; Sigma) or Ficoll-sodium diatrizoate solutions (e.g., Ficoll-Paque Plus; Amersham). Alternatively, use special centrifuge tubes (e.g., Vacutainer CPT tube) for isolation of PBMCs. Centrifuge the Vacutainer tube containing the cooled blood at 1500g for 15 min at room temperature. With the help of the gel, the leukocytes are separated from the erythrocytes. (*See* **Note 16**.)
2. Remove approximately half of the plasma above the mononuclear cell layer by aspiration.
3. Collect the cell layer with a Pasteur pipet and transfer the PBMCs into a fresh tube. Place the tube on ice, in order to prevent repair of broken DNA strands.
4. Count the cells using a Neubauer chamber and adjust the cell suspension to a cell density of 1.5×10^5 cells/mL by suspension in Ca/Mg-free PBS (solution B) or TE buffer (solution H). This ensures a DNA concentration of 30 ng/mL or 3500–4000 cells/25 µL, which is required for the fluorescence assay.
5. Distribute aliquots of the cell suspension (25 µl, approx 30 ng of DNA) in six replicates for each assay into the wells of a black 96-well microplate.

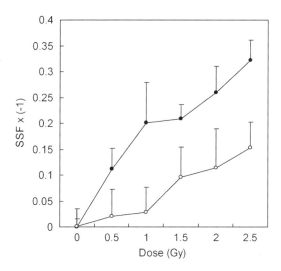

Fig. 4. Dose dependence of DNA damage and repair in PBMCs after irradiation with ^{60}Co γ-rays as determined by Fast Micromethod. PBMCs from one healthy individual were irradiated with the indicated doses and DNA was analyzed either immediately (●) or after 2 h of repair in culture medium (○). Results are expressed as SSF compared to control (nonirradiated PBMC; dose: 0 Gy). Each value represents the mean (± standard deviation) from six replicate determinations.

6. Irradiation with ^{60}Co or X-rays. Irradiate PBMC isolated from EDTA blood or heparin-containing blood samples, with different doses of ^{60}Co γ-rays or X-rays (high-energy photons) produced by a linear accelerator (e.g., as used in radiotherapy of cancer patients). Cool the samples before, during and after irradiation on ice, and analyze within 10 min (*See* **Note 12**.)

In the experiment shown in **Fig. 4**, human PBMC isolated from EDTA blood were exposed to different ^{60}Co γ-ray doses and subsequently DNA damage and repair were determined. A strong increase in the rate of DNA unwinding [increase in SSF × (–1) value], which is indicative for DNA damage, is already seen in cells irradiated with 0.5 Gy of ^{60}Co. The amount of single-strand breaks and alkali-labile sites, expressed as SSF, increases almost linearly in the dose range 0.5–2.5 Gy. A similar increase in the rate of DNA unwinding was also detected after irradiation of the cells using a linear accelerator; at 1 Gy and higher, this increase becomes highly significant *(15)*. To investigate DNA repair, the cells after irradiation were incubated further for 2 h at 37°C in 5% CO_2. The radiation-induced DNA strand breaks were partially repaired during the incubation period (**Fig. 4**). (*See* **Notes 17** and **18**.)

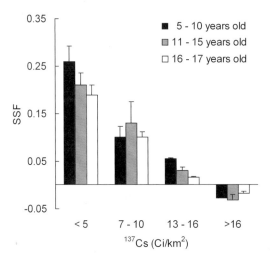

Fig. 5. DNA integrity in PBMCs from children of different age groups in relation to the level of soil contamination by radiocesium in the area of residence. Three age groups were selected: 5–10; 11–15; and 16–17 yr old. (Data from **ref. 17**.)

7. DNA damage in PBMC from radiotherapy patients. The Fast Micromethod can also be used for the assessment of DNA strand breaks produced in vivo in PBMC from patients undergoing radiotherapy. Collect EDTA blood or heparin-containing blood (2 mL) just before and immediately after irradiation of the patients, and immediately transfer to the tubes used for isolation of PBMC. Centrifuge and resuspend the PBMC as above. Determine cell counts using a Neubauer chamber. (*See* **Notes 19** and **20**.)

The Fast Micromethod has been used in a long-term study of the occurrence of DNA damage in PBMCs from a total number of 209 children living in areas of Belarus contaminated by radionuclides after the Chernobyl accident *(17)*. The children were divided into four groups, depending on the level of soil contamination by radiocesium in the area of residence: 5–7, 7–10, 13–16, and >16 Ci/km². The results revealed that the frequency of DNA strand breaks (SSF value) increased with the level of contamination by ^{137}Cs in all age groups studied (**Fig. 5**). The SSF values also depended on the age (period of exposure) of the children studied.

3.3. DNA Repair Measurements

1. Irradiate cell lines or PBMC as described in **Subheadings 3.2.1.** and **3.2.2.**
2. Determine DNA damage in cell lines after subsequent incubation of the cells in medium plus serum for different periods of time.

3. For determination of DNA repair in PBMCs, culture the cells in RPMI 1640 medium supplemented with 10–15% FCS at 37°C with a 5% CO_2 humidified atmosphere. Analyze the DNA immediately after irradiation and after different time periods in cell culture medium. Count the cells using a Neubauer chamber.

3.4. Standard Cell DNA (Bull Sperm)

1. Dilute one volume of original bull sperm (approx 1×10^9 cells/mL) with four volumes of stabilizing solution (containing 6% glycerol) to make a stock solution. Store at −20°C.
2. Add 960 µL of PBS (Ca/Mg-free) to 40 µL of standard cell DNA stock solution.
3. To obtain the standard curve, dilute 1:2, 1:4, 1:8 and so forth, to give a total of eight dilutions.
4. After measurement, select the standard cell DNA sample with the nearest fluorescence value (approx 20% higher than zero-time value of the control sample).
5. Add 960 µL of PBS (Ca/Mg-free) to 40 µL of standard cell DNA in the original tube.
6. Mix and subsequently dilute 1:1.
7. Repeat this step seven times (eight solutions with different concentrations of sample cell DNA are obtained).
8. Store the diluted stock solutions at 4°C.

3.5. Working With Tissue Samples

The method can also be used for determination of DNA damage in tissue samples if homogenized prior to thawing in a buffer containing an antifreeze such as DMSO and in the presence of liquid nitrogen (*see* **Note 21**). The following modifications are made.

1. Freeze the tissue in liquid nitrogen and store at −80°C. The chosen tissue should be immediately frozen in liquid nitrogen after treatment.
2. Homogenize the tissue (10 mg) in 1 mL of TE buffer (solution H) supplemented with 10% (v/v) DMSO in a mortar with a pestle precooled with liquid nitrogen. (Wear eye protection.)
3. Dilute the homogenate 1:10 with TE buffer (solution H).
4. The tissue samples to be analyzed should contain equal amounts of DNA. Determine the concentration of DNA using the fluorescent dye YOYO as follows:
    a. Dilute the DNA standard to 1 µg of DNA/mL in TE–DMSO buffer (three subsequent 1:10 dilutions).
    b. Prepare the working DNA solutions: 50–500 ng of DNA/mL in TE–DMSO buffer.
    c. Pipet 50 µL of diluted sample (or working DNA solutions) to each well, then add 50 µL of YOYO working solution.
    d. Add 250 µL of TE–DMSO to each well, and measure the fluorescence using a fluorescence microplate reader (e.g., Fluoroscan version 4.0: excitation, 485 nm; emission, 538 nm).

e. Calculate the amount of DNA in the samples with the help of the standard curve.
5. Adjust to 30 ng of DNA/25 µL with TE buffer (solution H).
6. Place 25 µL of the diluted homogenate in a well on the microplate. The amount of tissue which is required for one well of the microplate is 25 µg/well. (*See* **Note 22**.)
7. Add slowly 25 µL of lysing solution containing the fluorescent dye PicoGreen (solution G) and allow lysis of tissue for 1 h on ice in the dark.
8. Proceed with DNA unwinding at pH 12.40 and measurement of fluorescence as described in **items 4** and **5**, **Subheading 3.1**.
9. Measure each sample as well as the blank in multiple (usually six) replicates. Fluorescence blanks contain 25 µL of TE buffer instead of sample together with 25 µL of PicoGreen lysing solution (solution D) and 250 µL of NaOH solution (solution G).

3.6. Calculations

Results are expressed as strand scission factor (SSF) values which are calculated as the $\log_{10}$ of the ratio of the percentage of double-stranded DNA (dsDNA) from treated and control samples, respectively. (*See* **Note 18**.)

1. Correct for blank readings by subtracting the blank values (measured in wells containing 25 µL Ca/Mg-free PBS [Solution B] or TE buffer [Solution H] instead of samples) from all the values.
2. Set the mean fluorescence value of the (not yet) unwound DNA of the control (e.g., untreated cells) at time point 0 of denaturation at 100% dsDNA; calculate all the other values according to this value.
3. Calculate SSF according to the following equation:

   $$SSF = \log (\% \text{ dsDNA in the sample} / \% \text{ dsDNA in the control})$$

   SSF values are usually calculated after a denaturation period of 20 min. Negative values for SSF are indicative of an increased frequency of DNA strand breaks/alkali-labile sites. (*See* **Notes 23** and **24**.)
4. For practical reasons multiply SSF by –1 in graphical presentations (to obtain positive values).
5. Determine significant differences between means by one-way analysis of variance (ANOVA).

4. Notes

1. The fluorescent dye PicoGreen contains neither hazardous (above 1%) nor carcinogenic (above 0.1%) components as defined in 29 CFR 1910.1200 (OSHA Hazard Communication Standard).
2. Some components of lysing solution (solution C) may precipitate in the refrigerator, and the solution may become turbid. Warm the lysing solution before use at room temperature (20–25°C) until it becomes clear (usually 5–10 min). Store it again in the refrigerator after use.

3. Because the fluorescence dye PicoGreen is already present in lysing solution (and able to interact with the dsDNA despite the high urea concentration), a continuous monitoring of DNA unwinding under alkaline conditions is possible. This cannot be done using the FADU method because the fluorochrome ethidium bromide used in that method does not stain DNA in the presence of alkali and the unwinding reaction must be stopped (neutralization step) before the dye is added to the DNA.
4. The pH of the reaction mixtures (DNA denaturation step) in the microplate wells is not substantially different from the pH determined according to the described procedure. The precision of the pipetting routines (use of single and/or multipipets) should regularly be checked for routine determinations (*see* Costar EIA Performance Protocols). The coefficient of variance (CV) should be <2%.
5. A number of factors may affect DNA unwinding and the results obtained by the Fast Micromethod. These factors include cell density, pH, and ionic strength during DNA unwinding under alkaline conditions. The rate of DNA unwinding may depend on cell type or species. For example, marine invertebrates (e.g., mussels) have shorter DNA molecules resulting in higher rates of DNA unwinding. The optimal conditions for measuring DNA strand breaks in gills of *Mytilus galloprovincialis* (homogenate) were at 100 ng/mL of DNA and at pH 11.5 during DNA denaturation (maximum differences of slopes and SSF values of DNA unwinding kinetics) *(19)*.
6. The number of cells (or the amount of DNA) per well should be the same in single experiments. Using mouse L5178Y cells a linear dose response to γ-irradiation for up to 500 rad was found with 5000 cells/well, while a reduction of cell number (2000 cells/well) resulted in a higher sensitivity and a linear response for up to 15 rad *(8)*. One reason for this effect may be the influence of cellular proteins on denaturation kinetics even in the presence of the high concentrations of urea *(8)*. It is recommended to use 2500–3000 cells/well, at which normally reproducibility is high and standard deviations are low, but in certain cases the optimal number of cells must be determined.
7. The sensitivity of the assay allows detection of as little as 25 pg/well of DNA standard (calf thymus) in fluorescence microplate readers (e.g., Fluoroscan II reader; Labsystems). The increase in fluorescence is proportional to the concentration of the dsDNA (linear range) up to 35 ng/well (300 μL) at pH 12.4. The percentage contribution of single-stranded DNA to overall fluorescence is usually less than 10%. Therefore, corrections for fluorescence values of single-stranded DNA are not necessary if the amount of DNA does not exceed 30 ng/well (about 3000 cells/well).
8. Alternatively, first lysing solution and then cell suspension can be added. This has the advantage of avoiding DNA repair on the plates between the pipetting steps.
9. The influence of minute pH differences between the single wells is negligible if experiments are designed to fit on one microplate.

10. The kinetics of DNA denaturation depends on the size of the DNA and on small changes in the range of pH 11.0–12.0 *(24,25)*. Under standard assay conditions DNA denaturation starts at pH 12.1–12.2 and is very fast at pH 12.6 *(8)*. At the recommended pH 12.4, DNA unwinding rates are suitable for routine experiments and occur within <1 h.
11. Since the SSF values are usually calculated after a denaturation period of 20 min, the procedure can be simplified by measuring only at time 0 and at 20 min. However, this does not allow recognition of possible disturbances in DNA unwinding. Sometimes shorter unwinding periods are more appropriate, for example, in case of DNA from lower invertebrates.
12. Even keeping samples on ice may not result in a total suppression of DNA repair (*see* **ref. 26**).
13. Because of the faster unwinding of DNA with single-strand breaks, lower fluorescence values are measured for the treated (irradiated) samples in the course of the unwinding reaction.
14. Human blood samples should be handled as being potentially infected (e.g., hepatitis viruses or HIV).
15. The venous puncture should be carried out under sterile conditions, for example, with the assistance of a Vacutainer (Becton Dickinson), before the test. The sample must be immediately stored on ice until the test takes place.
16. Alternatively, freshly collected EDTA blood samples (1.6 mg of potassium-EDTA per milliliter of blood) are cooled on ice and PBMCs are isolated by centrifugation (1000*g* for 10 min at 8°C) through centrifuge tubes with a polysucrose–sodium metrizoate layer (density 1.077 g/mL) sequestered in the bottom of the tubes by plastic inserts (Uni-Sep centrifuge tubes from Novamed, Jerusalem, Israel). Under these conditions which allow isolation of PBMCs at reduced temperature within 15 min or less, postirradiation DNA repair can be neglected (<10%).
17. The curvilinear nature of the dose–response curves could be explained by the wide range of doses tested.
18. Like other assays that are based on DNA unwinding under alkaline conditions the Fast Micromethod cannot differentiate between single-strand breaks, alkali-labile sites, and transient breaks formed during repair of DNA damage. Alkali-labile sites in DNA are mostly abasic (apurinic/apyrimidinic) sites that are converted to single-strand breaks at alkaline pH *(27)*. Also DNA adducts may lead to alkali-labile sites.
19. A number of points need to be considered in studying human cells (e.g., PBMCs). Both the basal levels and the levels of irradiation-induced DNA single-strand breaks in human cells (e.g., PBMCs) are age dependent *(9,15)*. Also smoking can cause considerable DNA damaging effects as revealed in human lymphocytes using the comet assay *(28)*. In addition, irradiation-induced DNA damage depends on the cellular content of DNA repair enzymes *(29,30)* and DNA repair capacity may vary inter-individually *(31)*. Studies on cancer patients revealed interindividual differences in the level of irradiation-induced DNA single-strand breaks *(9)*.

20. Isolation of lymphocytes is a critical step when DNA from patients receiving radiotherapy is studied. To keep DNA repair as low as possible special centrifuge tubes can be used, which allow the rapid isolation of PBMCs within less than 15 min at low temperature (8°C). (*See* **Note 16**.)
21. Generation of artificial DNA strand breaks by analytical procedures is one major problem in analyzing DNA integrity. Homogenization of tissue samples in the presence of DMSO has been shown to cause only negligible amounts of DNA damage, and does not significantly lower the sensitivity of the method *(32)*. Thus human biopsy samples or samples from bioindicator organisms can be frozen in liquid nitrogen and kept at –80ºC until analysis without significant loss of DNA integrity.
22. It is recommended to determine the DNA content of the homogenates and to apply equalized amounts (based on DNA) of the homogenates to the microplate.
23. Positive values for SSF [i.e., negative values for SSF × (–1)] may be caused by formation of DNA crosslinks (reduction of the rate of DNA unwinding compared to untreated control).
24. Like other tests for DNA damage based on DNA unwinding under alkaline conditions Fast Micromethod cannot differentiate between single-strand breaks and alkali-labile sites. The latter lesions are converted into single-strand breaks by alkali.

Acknowledgments

This work was supported by grants from the Commission of the European Communities (EVK3-CT-1999-00005 and ICA2-CT-2000-10047), the Bundesministerium für Bildung und Forschung (Project 0312309 and WTZ-Ukraine/Belarus) and from Deutsche Krebshilfe (Dr. Mildred Scheel Stiftung; 10-1015-Schr 1).

References

1. Friedberg, E. C., Walker, G. C., and Siede, W. (1995) *DNA Repair and Mutagenesis*. American Society for Microbiology Press, Washington, DC.
2. Fairbairn, D. W., Olive, P. L., and O'Neill, K. L. (1995) The comet assay: a comprehensive review. *Mutat. Res.* **339,** 37–59.
3. Singh, N. P., McCoy, M. T., Tice, R. R., and Schneider E. L. (1988) A simple technique for quantitation of low levels of DNA damage in individual cells. *Exp. Cell Res.* **175,** 184–191.
4. Birnboim, H. C. (1990) Fluorimetric analysis of DNA unwinding to study strand breaks and repair in mammalian cells. *Methods Enzymol.* **186,** 550–555.
5. Birnboim, H. C. and Jevcak, J. J. (1981) Fluorimetric method for rapid detection of DNA strand breaks in human white blood cells produced by low doses of radiation. *Cancer Res.* **41,** 1889–1892.
6. Kohn, K. W. (1991) Principles and practice of DNA filter elution. *Pharmacol. Ther.* **49,** 55–77.

7. Kohn, K. W., Erickson, L. C., Ewig, R. A., and Friedman, C. A. (1976) Fractionation of DNA from mammalian cells by alkaline elution. *Biochemistry* **15**, 4629–4637.
8. Batel, R., Jakšić, Ž., Bihari, N., et al. (1999) A microplate assay for DNA damage determination (Fast Micromethod) in cell suspensions and solid tissues. *Anal. Biochem.* **270**, 195–200.
9. Elmendorff-Dreikorn, K., Chauvin, C., Slor, H., et al. (1999) Assessment of DNA damage and repair in human peripheral blood mononuclear cells using a novel DNA unwinding technique. *Cell. Mol. Biol.* **45**, 211–218.
10. Müller, W. E. G., Batel, R., Lacorn, M., et al. (1998) Accumulation of cadmium and zinc in the marine sponge *Suberites domuncula* and its potential consequences on single-strand breaks and on expression of heat-shock protein: A natural field study. *Mar. Ecol. Prog. Ser.* **167**, 127–135.
11. Schröder, H. C., Batel, R., Lauenroth, S., et al. (1999) Induction of DNA damage and expression of heat shock protein HSP70 by polychlorinated biphenyls in the marine sponge *Suberites domuncula* Olivi. *J. Exp. Mar. Biol. Ecol.* **233**, 285–300.
12. Schröder, H. C., Hassanein, H. M. A., Lauenroth, S., et al. (1999) Induction of DNA strand breaks and expression of HSP70 and GRP78 homolog by cadmium in the marine sponge *Suberites domuncula*. *Arch. Environm. Contam. Toxicol.* **36**, 47–55.
13. Schröder, H. C., Batel, R., Hassanein, H. M. A., et al. (2000) Correlation between the level of the potential biomarker, heat-shock protein, and the occurrence of DNA damage in the dab, *Limanda limanda*: a field study in the North Sea and the English Channel. *Mar. Environm. Res.* **49**, 201–215.
14. Haugland, R. P. (1996) *Handbook of Fluorescent Probes and Research Chemicals*, 6th Ed. Molecular Probes Inc., Eugene, OR.
15. Chauvin, C., Heidenreich, E., Elmendorff-Dreikorn, K., Slor, H., Kutzner, J., and Schröder, H. C. (1999) Lack of correlation between apoptosis and DNA single-strand breaks in X-irradiated human peripheral blood mononuclear cells in the course of ageing. *Mech. Ageing Dev.* **106**, 117–128.
16. Hassanein, H. M. A., Müller, C. I., Schlösser, D., Kratz, K.-L., Senyuk, O. F., and Schröder, H. C. (2002) Measurement of DNA damage induced by irradiation with γ-rays from a TRIGA Mark II research reactor in human cells using Fast Micromethod. *Cell. Mol. Biol.* **48**, 385–391.
17. Titov, L. P., Chernoshey, D. A., Kirylchik, E. J., Rytik, P. G., Melnov, S. B., and Wischnyakov, M. (2002) Stability and unstability of genome, and immune status of children living in territories contaminated by radiocesium of the Republic of Belarus, in: *Role of Anthropogenic and Natural Pathogens in Formation of Infectious and Non-infectious Human Diseases. Medical-ecological Aspects.* Proceedings of International Conference, Minsk, October 8–9, 2002, NESSI, Minsk, Belarus, pp. 493–500.
18. Bihari, N., Batel, R., Jakšić, Ž., Müller, W. E. G., Waldmann, P., and Zahn, R. K. (2002) Comparison between the Comet assay and Fast Micromethod for measuring DNA damage in HeLa cells. *Croatia Chem. Acta* **75**, 793–804.
19. Jakšić, Ž. and Batel, R. (2003) DNA integrity determination in marine invertebrates by Fast Micromethod. *Aquat. Toxicol.* **65**, 361–376.

20. Batel, R., Fafandjel, M., Blumbach, B., et al. (1998) Expression of the human XPB/ERCC-3 excission repair gene-homolog in the sponge *Geodia cydonium* after exposure to ultraviolet irradiation. *Mutat. Res.* **409,** 123–133.
21. Chan, A. C. and Walker, I. G. (1976) Reduced DNA repair during differentiation of a myogenic cell line. *J. Cell Biol.* **70,** 685–691.
22. Nagao, M. and Sugimura, T. (1976) Molecular biology of the carcinogen, 4-nitroquinoline 1-oxide. *Adv. Cancer Res.* **23,** 131–169.
23. Boyum, A. (1968) Isolation of mononuclear cells and granulocytes from human blood. *Scand. J. Clin. Lab. Invest.* **21 (Suppl. 97),** 77–89.
24. Erickson, L. C., Ross, W. E., and Kohn, K. W. (1979) Isolation and purification of large quantities of DNA replication intermediates by pH step alkaline elution. *Chromosoma* **74,** 125–139.
25. Yamanishi, D. T., Bowden, G. T., and Cress, A. E. (1987) An analysis of DNA replication in synchronized CHO cells treated with benzo[a]pyrene diol epoxide. *Biochim. Biophys. Acta* **910,** 34–42.
26. Bock, C., Dittmar, H., Gemeinhardt, H., Bauer, E., and Greulich, K. O. (1998) Comet assay detects cold repair of UV-A damages in a human B-lymphoblast cell line. *Mutat. Res.* **408,** 111–120.
27. Wani, A. A. and D'Ambrosio, S. M. (1986) Specific DNA alkylation damage and its repair in carcinogen-treated rat liver and brain. *Arch. Biochem. Biophys.* **246,** 690–698.
28. Piperakis, S. M., Petrakou, E., and Tsilimigaki, S. (2000) Effects of air pollution and smoking on DNA damage of human lymphocytes. *Environ. Mol. Mutagen.* **36,** 243–249.
29. Bohr, V. A. and Evans, M.K. (1990) Heterogeneity of DNA repair: implications for human disease and oncology. *Pediatr. Hematol. Oncol.* **7,** 47–69.
30. Bohr, V. A., Evans, M. K., and Fornace, A. J. (1989) Biology of disease: DNA repair and its pathogenetic implications. *Lab. Invest.* **61,** 143–161.
31. Celotti, L., Ferraro, P., and Biasin, M. R. (1992) Detection by fluorescence analysis of DNA unwinding and unscheduled DNA synthesis, of DNA damage and repair induced in vitro by direct-acting mutagens on human lymphocytes. *Mutat. Res.* **281,** 17–23.
32. Zahn, R. K., Jaud, S., Schröder, H. C., and Zahn-Daimler, G. (1996) DNA status in brain and heart as prominent co-determinant for life span? Assessing the different degrees of DNA damage, damage susceptibility, and repair capability in different organs of young and old mice. *Mech. Ageing Dev.* **89,** 79–94.

22

^{32}P-Postlabeling DNA Damage Assays
PAGE, TLC, and HPLC

Shinya Shibutani, Sung Yeon Kim, and Naomi Suzuki

Summary

^{32}P-Postlabeling analysis is a powerful technique for detecting, identifying, and quantifying DNA adducts induced by mutagens or carcinogens. The method involves enzymatic digestion of the DNA sample to nucleoside 3′-monophosphates, and partial purification of the adducted nucleotides followed by their 5′-labeling with ^{32}P. For analysis of DNA adducts, polyethyleneimine–cellulose thin-layer chromatography (TLC) plates have traditionally been used to resolve ^{32}P-labeled DNA adducts (^{32}P-postlabeling/ TLC analysis). However, the TLC procedure is time consuming and labor intensive. To expedite analyses, we recently devised a ^{32}P-postlabeling protocol that utilizes nondenaturing polyacrylamide gel electrophoresis (PAGE) and permits multiple DNA samples to be run on a single gel (^{32}P-postlabeling/PAGE analysis). Using this method, the detection limit for 5 µg of DNA is approx 7 adducts/10^9 nucleotides, similar to that for ^{32}P-postlabeling/TLC. For still higher sensitivity and resolution, high-performance liquid chromatography (HPLC) combined with a radioisotope detector system (^{32}P-postlabeling/HPLC analysis) can be used to increase the detection limit to approx 3 adducts/10^{10} nucleotides. Here we describe all three ^{32}P-postlabeling techniques.

Key Words: DNA adduct; DNA damage; gel electrophoresis; HPLC; ^{32}P-postlabeling; TLC.

1. Introduction

^{32}P-Postlabeling analysis has been widely used for detecting a variety of DNA adducts induced by endogenous and exogenous mutagens or carcinogens *(1–3)*. This technique has also been applied to explore the genotoxic effects of large numbers of anticancer drugs *(4)*, antibiotics *(5)*, estrogens *(6–9)* and antiestrogens *(10,11)*, and to identify the carcinogenic substance contained in Chinese herbs *(12)*. The basic methods, including enzymatic digestion of the DNA

From: *Methods in Molecular Biology: DNA Repair Protocols: Mammalian Systems, Second Edition*
Edited by: D. S. Henderson © Humana Press Inc., Totowa, NJ

sample and labeling the adducted nucleotides with ^{32}P, have not changed significantly in more than 20 yr since the technique was first described *(1)*. Traditionally, polyethyleneimine (PEI)-cellulose thin-layer chromatography (TLC) plates are generally used to resolve ^{32}P-labeled DNA adducts two-dimensionally using several different buffer conditions (^{32}P-postlabeling/TLC analysis) *(1,13,14)*; however, using this technique, only one ^{32}P-labeled sample can be analyzed per TLC plate and the migration of DNA adducts is variable from plate to plate. Separation by TLC is time consuming and labor intensive. To expedite analyses, we devised a ^{32}P-postabeling method that uses nondenaturing 30% polyacrylamide gel electrophoresis (PAGE) to resolve adducted nucleotides (^{32}P-postlabeling/PAGE analysis) *(15)*. The major advantages of this technique are: (1) multiple DNA samples can be loaded on a single gel along with standard markers; (2) DNA adducts can be resolved in only a few hours; and (3) exposure to ^{32}P during handling is minimized. The detection limit for both ^{32}P-postlabeling/TLC and ^{32}P-postlabeling/PAGE analyses is approx 7 adducts/10^9 nucleotides. However, greater sensitivity (down to approx 3 adducts/10^{10} nucleotides) can be achieved using high-performance liquid chromatography (HPLC) combined with a radioisotope detector (^{32}P-postlabeling/HPLC analysis) *(2,11,16)*. Although many modified procedures have been published, we describe general methods for three different ^{32}P-postlabeling techniques for analysis of DNA adducts. A flow diagram for these procedures is presented in **Fig. 1**.

For each method, the steps for enzymatic digestion and ^{32}P-labeling of DNA are the same (**Fig. 1A**). First, naked DNA or DNA isolated from cells is digested with micrococcal nuclease and spleen phosphodiesterase to produce unmodified nucleoside 3'-monophosphates ($dN_{3'p}$) and adducted nucleoside 3'-monophosphates ($dX_{3'p}$). Second, nuclease P1 is used to selectively dephosphorylate $dN_{3'p}$s since $dX_{3'p}$s are generally resistant to the enzyme. This step, therefore, enriches for adducted nucleotides, which in the next step are selectively 5'-labeled with T4 polynucleotide kinase and [γ-^{32}P]ATP. Labeled adducted nucleotides (now biphosphates, [32]$PdX_{3'p}$) are then resolved by either TLC (**Fig. 1B**), PAGE (**Fig. 1C**) or HPLC (**Fig. 1D**) and their radioactivity measured either by liquid scintillation counting and phosphorimaging (TLC and PAGE) or inline detection (HPLC).

2. Materials

2.1. Isolation of DNA From Cultured Cells and Animal and Human Tissues

1. 1.0% Sodium dodecyl sulfate (SDS), 10 mM ethylenediaminetetraacetic acid (EDTA), 20 mM Tris-HCl, pH 7.4; store at 4°C.
2. RNase A (Worthington Biochemical Corp., Freehold, NJ) (*see* **Note 1**).

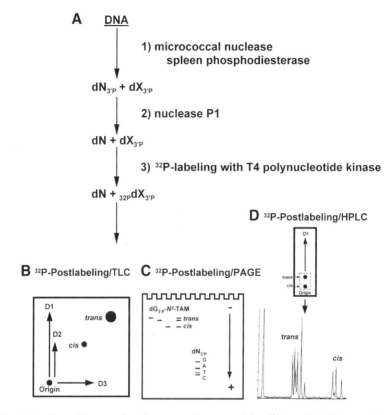

Fig. 1. A flow diagram for the procedures used for ^{32}P-postlabeling analyses.

3. RNase T1 (Worthington Biochemical Corp.).
4. Proteinase K (Sigma-Aldrich, St. Louis, MO).
5. Tris-saturated phenol (Roche Molecular Biochemicals); store at 4°C (*see* **Note 2**).
6. Chloroform (molecular biology grade).
7. Absolute ethanol (100% pure).
8. 1X Saline sodium citrate (SSC): 0.15 M NaCl, 1 mM EDTA, 0.015 M sodium citrate, pH 7.2; store at 4°C.
9. Polytron homogenizer.

2.2. Enzymatic Digestion of DNA Sample

1. Buffer: 8 mM CaCl$_2$, 17 mM sodium succinate buffer, pH 6.0; store at 4°C.
2. Micrococcal nuclease (Worthington Biochemical Corp.).
3. Spleen phosphodiesterase II (Worthington Biochemical Corp.) (*see* **Note 3**).
4. Nuclease P1 (Roche Molecular Biochemicals).
5. 1-Butanol (molecular biology grade).

2.3. Labeling Adducted Nucleotides With ^{32}P

1. [γ-^{32}P]ATP (specific activity, >6000 Ci/mmol); store at –20°C (*see* **Note 4**).
2. 3′-Phosphatase-free T4 polynucleotide kinase (Roche Molecular Biochemicals); store at –20°C (*see* **Note 5**).
3. Potato apyrase (Sigma-Aldrich).
4. 10 mM Spermidine aqueous solution; store at –20°C
5. 50 mM Dithiothreitol (DTT) aqueous solution; store at –20°C.
6. 10X Linker-kinase buffer: 100 mM MgCl$_2$, 700 mM Tris-HCl, pH 7.6; store at 4°C.
7. 10X Formamide dye: 10 mg of bromophenol blue and 10 mg of xylene cyanol FF in 10 mL of formamide.

2.4. ^{32}P-Postlabeling/TLC Analysis

1. Polyethylenimine (PEI)-cellulose thin-layer plates (Machery-Nagel; Duren, Germany).
2. Chromatography paper (Whatman Grade 3MM Chr.).
3. D1 buffer: 1.7 M Sodium phosphate buffer, pH 6.0.
4. D2 buffer: 1.1 M Lithium formate, 2.7 M urea, pH 3.5.
5. D3 buffer: 0.48 M LiCl, 0.3 M Tris-HCl, 5.1 M urea, pH 8.0.
6. β-PhosphorImager.

2.5. ^{32}P-Postlabeling/PAGE Analysis

1. 40% Acrylamide solution: 38 g of acrylamide and 2 g of N,N'-methylene *bis*-acrylamide in a final volume of 100 mL of distilled water; store at 4°C.
2. 10X TBE buffer: 2.24 M boric acid, 25.5 mM EDTA, 1 M Tris-base, pH 7.0.
3. 10% Ammonium persulfate aqueous solution. Make fresh and store for weeks at 4°C.
4. N,N,N',N'-Tetramethylethylenediamine (TEMED); store at 4°C.
5. Electrophoresis apparatus.

2.6. Measurement of ^{32}P-Labeled DNA Adducts

1. β-Phosphorimager.
2. X-ray film.
3. X-ray film developer.
4. Scintillation liquid.
5. Scintillation vials.
6. β-Liquid scintillation counter.

2.7. Determination of ^{32}P-Labeled-DNA Adducts by HPLC

1. 4 M Pyridinium formate, pH 4.3.
2. 0.2 M Ammonium formate, pH 4.0.
3. Acetonitrile–methanol (6:1 v/v).
4. Thermo Hypersil BDS C$_{18}$ analytical column (0.46 × 25 cm, 5 µm, Thermo Hypersil; Bellefonte, PA).

3. Methods

3.1. Isolation of DNA From Cultured Cells or Tissues

When the cultured cells or animals are exposed to a compound or its metabolites, the DNA can be extracted from the cells or tissues by using **steps 1–5**. When the DNA is directly incubated with a compound or its metabolites using cytosol, microsome, or an enzyme, the DNA should be extracted, following **steps 3–5**.

1. Suspend the cells (10^7–10^8) or tissue (10–100 mg) in 1.5 mL of ice-cold 1.0% SDS, 10 mM EDTA, 20 mM Tris-HCl, pH 7.4, in a polypropylene tube, and homogenize for 30 s at 0°C using a Polytron homogenizer.
2. Incubate the homogenate at 37°C for 30 min with RNase (300 µg) and RNase T1 (50 U), followed by incubation in 750 µg of Proteinase K for 30 min.
3. Add an equal volume of Tris-saturated phenol to the reaction mixture, mix well for 30 s at room temperature using a vortex, and centrifuge at 1600g for 10 min. If the organic (lower) and aqueous (upper) phases are not well separated, centrifuge again at a higher speed and/or for a longer time. Transfer the aqueous phase carefully to a fresh tube; discard the interface and organic phases. Add an equal volume of Tris-saturated phenol–chloroform (1:1 v/v), mix for 30 s and centrifuge. Transfer again the aqueous phase to a fresh tube. Add an equal volume of chloroform, mix for 30 s, and centrifuge. Transfer the aqueous phase to a fresh tube.
4. To extract the DNA, add three volumes of ice-cold absolute ethanol, mix well for 30 s, centrifuge at 14,000g for 10 min, and remove the supernatant. Dissolve the precipitate in 100 µL of distilled water, add 500 µL of ice-cold absolute ethanol, mix well, and centrifuge; repeat this process once.
5. Repeat **steps 2–4** to avoid contamination by the RNA.
6. Dissolve the purified DNA in 1 mL of 0.01X SSC. Estimate the concentration of the DNA using a UV spectrophotometer (50 µg = 1.0 OD$_{260nm}$). Approximately 100 µg of DNA can be extracted from 10^8 cells or 100 mg of the tissue; the recovery of DNA varies depending on the cell type and organs used. Store the DNA sample at –70°C.

3.2. Enzymatic Digestion of DNA Sample

This process enriches for the adducted nucleotides that result from enzymatic digestion of the DNA (*see* **Note 6**).

1. Incubate 5 µg of DNA at 37°C overnight (approx 16 h) with 1 µL of micrococcal nuclease (1.5 U) and 1 µL of spleen phosphodiesterase (0.1 U) in 50 µL of 17 mM sodium succinate buffer, pH 6.0, containing 8 mM CaCl$_2$ (*see* **Note 7**).
2. Add 1 µL of nuclease P1 (1.0 U) into the reaction mixture and incubate at 37°C for 1 h.

3. Evaporate the sample to dryness using a SpeedVac.

If adducted nucleotides such as tamoxifen-derived DNA adducts *(10,11)* are efficiently extracted by 1-butanol, the following butanol fractionation can be used to minimize the contamination of $dN_{3'p}$ and to enrich the $dX_{3'p}$ in the DNA digest. This additional procedure minimizes the background of $dN_{3'p}$ during ^{32}P-postlabeling analysis, which in turn increases the detection limit.

4. Dissolve the digested DNA in 100 μL of distilled water, and extract twice with 200 μL of butanol. Centrifuge at 14,000g for 5 min at room temperature and transfer the top layer (butanol phase) to a fresh tube. Back-extract the butanol fraction with 50 μL of distilled water, centrifuge, and remove the bottom layer (aqueous phase). Evaporate the remaining butanol fractions to dryness and use for adduct analysis.

3.3. Labeling Adducted Nucleotides With ^{32}P

Wild-type T4 polynucleotide kinase is generally used to label the $dX_{3'p}$ with ^{32}P. Since the wild-type enzyme has 3'-phosphatase activity, the 3'-monophosphate is, in some cases, removed from the $dX_{3'p}$, resulting in an inefficient labeling with ^{32}P. To avoid this, it is essential at this step that the T4 polynucleotide kinase be free of 3'-phosphatase activity (*see* **Note 5**).

1. Dissolve the digested DNA or authentic standard in 16 μL of distilled water, 3 μL of 10X linker-kinase buffer, pH 7.6, 3 μL of 50 mM DTT, and 3 μL of 10 mM spermidine in an Eppendorf tube, and incubate at 37°C for 40 min with 3 μL of [γ-^{32}P]ATP (10 μCi/μL) and 2 μL of 3'-phosphatase free T4 polynucleotide kinase (10 U/μL).
2. To decompose nonreacted [γ-^{32}P]ATP, add 1 μL of potato apyrase (50 mU/μL) and incubate at 37°C for an additional 30 min.
3. Evaporate the reaction mixture to dryness under vacuum.

While handling [γ-^{32}P]ATP, minimize exposure to ^{32}P by wearing protective clothing, gloves, and body dosimeters. Use Plexiglas shielding and monitor the work area with a Geiger counter. ^{32}P-labeled waste materials should be discarded following appropriate safety procedures.

3.4. ^{32}P-Postlabeling/TLC Analysis

To separate the ^{32}P-labeled adduct in two dimensions on a PEI-cellulose TLC plate, several buffers are required that depend on the individual adducts (*see* **Note 8**). We describe here the typical example used for separation of tamoxifen–DNA adducts *(12)*.

1. Cut the PEI-cellulose TLC plate (20 × 20 cm^2) into four plates (10 × 10 cm^2). Wash the plates using distilled water (approx 500 mL) for 30 min at room temperature with shaking, and then allow to dry.
2. Staple a paper wick (10 × 15 cm^2) at the top of the TLC plate.

3. Dissolve the ³²P-labeled nucleotides in 5 µL of distilled water, apply the sample at the corner 2 cm from the bottom and 2 cm from left side of the TLC plate (10 × 10 cm²) using a capillary tube and a hair dryer (*see* **Note 9**). Rinse the sample tube with 3 µL of distilled water and apply onto the TLC plate. Discard the sample tube after testing for radioactivity using a Geiger counter.
4. TLC plates are developed using D1 buffer in a glass chamber overnight (approx 16 h) at room temperature (*see* **Note 10**). Remove the paper wick and staples, wash the TLC plate twice using 500 mL of distilled water with shaking, and dry the plate.
5. Staple a new paper wick (10 × 10 cm²) at the top of the TLC plate, and develop in the same direction with D2 buffer in a glass chamber for approx 3 h at room temperature (*see* **Fig. 1B**). After that, remove the paper wick, wash the TLC plate twice using 500 mL of distilled water with shaking, and dry the plate.
6. Staple a new paper wick (10 × 10 cm²) to the right side of the TLC plate, and develop further in D3 buffer at a right angle to the previous direction of development in a glass chamber for approx 3 h at room temperature (*see* **Fig. 1B**). After that, remove the paper wick, wash the TLC plate twice using 500 mL of distilled water with shaking, and dry the plate.
7. The position of adducts on the TLC plate can be established by exposing to a β-PhosphorImager screen or by autoradiography. The time of exposure for β-PhosphorImager or X-ray film varies depending on the radioactivity of the targeted adducts; lower radioactivity requires overnight exposure.

3.5. ³²P-Postlabeling/PAGE Analysis

³²P-Postlabeling/PAGE analysis can be applied to any DNA adducts using nondenaturing 30% polyacrylamide gel under the following experimental conditions. The detection limit of DNA adducts is approx 7 adducts/10^9 bases. A typical ³²P-postlabeling/PAGE used for analysis of tamoxifen-induced DNA adducts is presented in **Fig. 2**.

3.5.1. Preparation of Polyacrylamide Gel

1. Mix 60 mL of 40% polyacrylamide solution, 10 mL of distilled water and 10 mL of 10X TBE buffer, pH 7.0.
2. Before pouring the gel, add 1.0 mL of 10% ammonium persulfate and 35 µL of TEMED.
3. Pour the solution between the glass plates (35 × 42 × 0.04 cm), which have been taped together. Then insert the comb in the top.
4. When the gel has polymerized, carefully remove both the comb and tape from the glass plates.
5. Set the glass plates–gel sandwich into the gel apparatus and fill the upper and bottom tanks with 1X TBE buffer, and flush out the wells.
6. Before loading the ³²P-labeled samples, run the electrophoresis for at least 30 min at 1200–1400 V, 20–50 mA.

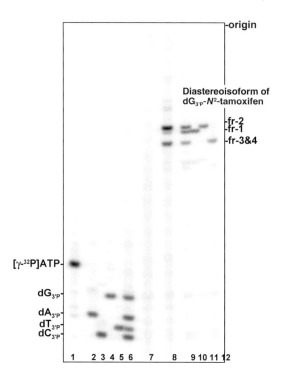

Fig. 2. ^{32}P-Postlabeling/PAGE analysis of antiestrogen-derived DNA adducts. Standard markers; **1**, [γ-^{32}P]ATP (approx. 0.1 μCi/33 fmol); **2**, dA$_{3'P}$; **3**, dC$_{3'P}$; **4**, dG$_{3'P}$; **5**, dT$_{3'P}$; **6**, a mixture of four dN$_{3'P}$; **9**, a mixture of *trans*- and *cis*-isoforms of dG-N^2-tamoxifen; **10**, a *trans*-isoform of dG-N^2-tamoxifen (fr-1); **11**, a *trans*-isoform of dG-N^2-tamoxifen (fr-2); **12**, a mixture of *cis*-isoforms of dG-N^2-tamoxifen (fr-3 and -4). **7**, Untreated DNA; **8**, DNA treated with tamoxifen α-sulfate.

Unpolymerized acrylamide is a neurotoxin. Wear gloves when preparing and handling polymerized acrylamide gels.

3.5.2. Performance of Gel Electrophoresis

1. Dissolve the ^{32}P-labeled sample in 3 μL of distilled water and 2 μL of 10X formamide dye. Dilute 1 μL of [γ-^{32}P]ATP (10 μCi/μL) with 198 μL of distilled water. Take 2 μL of the diluted [γ-^{32}P]ATP and mix well with 3 μL of 10X formamide dye.
2. Apply the ^{32}P-labeled samples and the diluted [γ-^{32}P]ATP standard using a 10 to 20 μL pipet. Run the electrophoresis for approx 5 h at 1200–1800 V, 20–50 mA. Stop the electrophoresis when the xylene cyanol (upper dye) is approx 14 cm from the top of the gel; the position of ^{32}P-labeled dC$_{3'P}$ is approx 5 cm from the bottom. (*See* **Fig. 2**, lanes 3 and 6).

32P-Postlabeling Analysis

3. Wrap the gel in plastic wrap. Determine the position of ^{32}P-labeled adducts by a β-phosphorimager analysis or by autoradiography.

3.6. Measurement of the Level of ^{32}P-Labeled DNA Adducts

The level of DNA adducts resolved on the TLC plate or nondenaturing polyacrylamide gel can be estimated by the following procedures.

3.6.1. Level of DNA Adducts on a PEI-Cellulose TLC Plate

1. Mark the adduct positions by placing the TLC plate on top of the developed X-ray film. When the origin spot on the TLC plate is adjusted to that of the X-ray film, the adduct positions on the TLC plate can be easily determined.
2. Scrape the radioactive adducts from the TLC plate, put them into the scintillation vials and mix well with 4 mL of scintillation liquid.
3. Measure the radioactivity using a β-liquid scintillation counter and compare with a known amount of [γ-^{32}P]ATP.

Relative adduct levels (RAL) are calculated using the following equation:

$$RAL = \frac{\text{adducted nucleotides (dpm or cpm)}}{dN_{3'P} \text{ constituted DNA (pmol)} \times \text{specific activity of } [\gamma\text{-}^{32}P] \text{ATP (dpm or cpm/pmol)}}$$

For example, (total dpm in adducts)/ 2.02×10^{11} dpm, assuming that 5 µg of DNA represented 1.52×10^4 pmol of $dN_{3'P}$ and the specific activity of the [γ-^{32}P]ATP is 1.33×10^7 dpm/ pmol. The specific activity of [γ-^{32}P]ATP is corrected according to the extent of decay (the half-life of ^{32}P is 14.29 d).

3.6.2. Level of DNA Adducts on a Nondenaturing Polyacrylamide Gel

1. Subject the ^{32}P-labeled samples to a nondenaturing gel along with the diluted concentration of [γ-^{32}P]ATP (10 µCi/µL) as described in the **step 1** in **Subheading 3.5.2**.
2. After completion of PAGE, measure the integrated values of adducts using β-phosphorimager and compare with that of the 100-fold dilution of [γ-^{32}P]ATP used. When the integrated values are beyond the linear response range, shorter exposure of ^{32}P-labeled products should be used to determine the radioactivity. RAL can be determined by the equation shown in **Subheading 3.6.1**.

3.7. Determination of ^{32}P-Labeled-DNA Adducts by HPLC

To increase the resolution and the detection limit of DNA adducts, HPLC connected inline to a radioisotope detector system can be used for ^{32}P-postlabeling analysis. Prior to subjecting ^{32}P-labeled samples to HPLC, ^{32}P-labeled adducts should be partially purified by one of the following procedures using either PEI-cellulose TLC or nondenaturing polyacrylamide gel electrophoresis (*see* **Note 11**).

3.7.1. Partial Purification of Adducts by PEI–Cellulose TLC

1. Develop the ^{32}P-labeled samples on TLC plates, following **steps 1–4**, **Subheading 3.4**.
2. Follow **steps 1** and **2** in **Subheading 3.6.1.**, and scrape the ^{32}P-labeled adducts remaining on the TLC plate into 1.5-mL microcentrifuge tubes.
3. Extract the ^{32}P-labeled adducts using 0.5 mL of 4 M pyridinium formate, pH 4.3, overnight at room temperature. After centrifugation (14,000g for 5 min), evaporate the supernatant to dryness.

3.7.2. Partial Purification of Adducts by Nondenaturing PAGE

1. Electrophorese the ^{32}P-labeled samples as described in **Subheading 3.5.**
2. Put the wrapped gel on a solid support (e.g., a used X-ray film) and tape together.
3. Staple the wrapped gel and unexposed X-ray film together in a dark room, and place in a cassette. The exposure time depends on the radioactivity of ^{32}P-labeled adducts.
4. Develop the film using an X-ray film developer.
5. Line up the staple holes on the wrapped gel and the developed X-ray film with pins. Mark the adduct positions.
6. Cut the ^{32}P-labeled material from the gel, put in a fresh tube, and extract (no need to crush the gel) using 1 mL of distilled water with shaking overnight at room temperature.
7. Centrifuge the sample at 14,000g for 5 min, transfer the supernatant in a fresh tube, and evaporate to dryness.

The extraction efficiency of DNA adducts from TLC or PAGE can be estimated as follows: for example, the radioactivity of 1/20 volume of the supernatant fraction (S) or the whole precipitate (P) is mixed with scintillation liquid (4 mL) in a scintillation vial and measured using a β-liquid scintillation counter. The extraction efficiency of DNA adducts is estimated using the following equation:

$$\text{Recovery } (\%) = \frac{S \times 20 \text{ (dpm)}}{S \times 20 + P \text{ (dpm)}} \times 100$$

3.7.3. ^{32}P-Postlabeling/HPLC

To resolve DNA adducts, such as tamoxifen-derived DNA adducts *(16)*, partially purified ^{32}P-labeled products are injected into a Hypersil BDS C$_{18}$ analytical column (0.46 × 25 cm, 5 μm), eluted at a flow rate of 1.0 mL/min using a linear gradient of 0.2 M ammonium formate and 20 mM H$_3$PO$_4$, pH 4.0, containing 20 to 30% acetonitrile–methanol (6:1, v/v) for 40 min, 30–50% acetonitrile–methanol (6:1, v/v) for 5 min, followed by an isocratic condition of 50% acetonitrile–methanol (6:1, v/v) for 15 min. The radioactivity is monitored using a radioisotope detector connected to a HPLC instrument (*see* **Note 12**). A typical ^{32}P-postlabeling/HPLC chromatogram for the analysis of tamoxifen-derived DNA adducts is presented in **Fig. 3**.

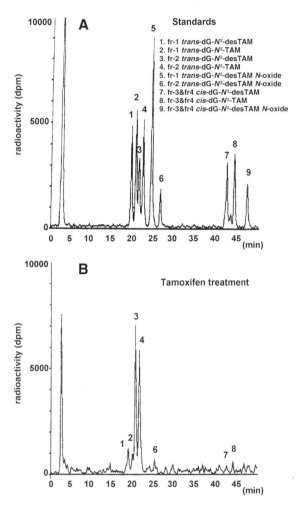

Fig. 3. ^{32}P-Postlabeling/HPLC analysis of anti-estrogen-derived DNA adducts in an animal. (**A**) Standards of *trans*- and *cis*-isoforms of dG-N^2-tamoxifen, dG-N^2-N-desmethyltamoxifen and dG-N^2-tamoxifen N-oxide adducts. (**B**) Hepatic DNA from rats treated with tamoxifen.

3.8. Quantification of DNA Adducts

The relative adduct levels are calculated by using the specific radioactivity of [γ-^{32}P]ATP labeled adducted nucleotides as described in **Subheading 3.6.** In most of cases, the adduct level is underestimated because of the incomplete DNA digestion, inefficiency of adduct labeling by T4 polynucleotide kinase,

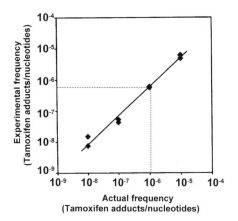

Fig. 4. Comparison of experimental and actual frequencies of antiestrogen-derived DNA adducts.

and loss of adducted nucleotides during the enrichment procedure *(17)*. When site-specifically modified oligodeoxynucleotides containing a target DNA adduct are available, such an oligomer can be used as an internal standard in order to determine the accurate level of DNA adducts. For example, oligodeoxynucleotides containing a single tamoxifen-derived DNA adduct (5'TCC TCCTCXCCTCTC, where X is the adducted site) can be prepared by post-synthetic methods *(18)* or by phosphoramidite chemical synthesis *(19)*. The concentration of oligomer can be determined, based on the extinction coefficient at 260 nm *(20)*. When 0.152–152 fmol (0.743-743 pg) of this oligomer are mixed with 5 µg of purified calf thymus DNA (15,200 pmol of $dN_{3'p}$), the actual level of tamoxifen adducts in the mixture is 1 adduct/10^8 nucleotides to 1 adduct/10^5 nucleotides. Such standard DNA can be used to determine the recovery of adducts and to quantify adducts using ^{32}P-postlabeling analyses. A typical standard curve is presented in **Fig. 4**. The amount of adducts detected increases linearly depending on the amount of the oligomer used. The recovery of adducts is 56% of the actual amount present; therefore, the actual level of adducts can be estimated by dividing the experimental values by 56% (*see* **Note 13**).

4. Notes

1. Commercially available RNase often contains DNase. Decontamination can be achieved by heating RNase solution at 95°C for 10 min.
2. The pH of Tris-saturated phenol should be 7.8–8.0 because the DNA partitions into the phenol phase under acidic conditions.

3. Some of commercially available spleen phosphodiesterases contain 3'-phosphatase activity. Therefore, we use a spleen phosphodiesterase II from Worthington Biochemical Corp.
4. Highly radioactive [γ-^{32}P]ATP (approx 6000 Ci/mmol) can be purchased from several companies; however, the stated concentration of ATP is not always accurate. Therefore, [γ-^{32}P]ATP should be obtained from companies that routinely determine the ATP concentration before each shipment.
5. The 3'-phosphatase activity of wild-type T4 polynucleotide kinase depends on the company and their lots. We have found Roche to be a good source of 3'-phosphatase-free T4 polynucleotide kinase.
6. When the adduct ($dX_{3'p}$), such as 8-hydroxy-2'-deoxyguanosine *(21)*, is not resistant to nuclease P1, $dX_{3'p}$ should be isolated from the DNA digest using HPLC or TLC.
7. Generally, 5 µg of DNA is used for the analysis. When the level of DNA adducts is expected to be low, the amounts of DNA analyzed can be increased to 50 µg. In such cases, 10-fold higher amounts of micrococcal nuclease (15 U) and spleen phosphodiesterase (1.0 U) should be used for DNA digestion using the same volume (50 µL) of sodium succinate buffer. The same amount of nuclease P1 (1.0 U) can be applied even though amounts of DNA have been increased.
8. Extra precaution should be taken to select the appropriate buffer depending on the DNA adduct to be analyzed. Search publications describing ^{32}P-postlabeling analyses used for related compounds and drugs to find the best buffer conditions.
9. To have the highest separation of ^{32}P-labeled adducts, the applied spot should be as small as possible.
10. The level of the buffer in the glass chamber should be <0.3 cm to avoid diffusion of ^{32}P-labeled adducts into the buffer. To check if ^{32}P-labeled adducts were lost in the buffer, measure the radioactivity of the buffer using a Geiger counter.
11. ^{32}P-labeled sample may be applied directly to HPLC without partial purification. However, the high background from free ^{32}P may reduce the resolution and detection limit of DNA adducts.
12. To adapt the HPLC system to resolve other DNA adducts, the gradient of acetonitrile:methanol (6:1 v/v) and/or pH of 0.2 *M* ammonium formate, 20 m*M* H_3PO_4 may be modified.
13. Simply, known amounts of standard DNA (1 adduct/10^7 nucleotides or 1 adduct/10^6 nucleotides, respectively) are prepared by mixing 1.52 or 15.2 fmol of oligomer with 5 µg of purified calf thymus DNA (15,200 pmol of dNs) and are analyzed by ^{32}P-postlabeling analyses together with samples containing an unknown amount of adduct. When compared with the standards, the accurate level of adducts in the samples can be determined.

Acknowledgments

This work was supported by Grants ES09418 and ES04068 from the National Institute of Environmental Health Sciences.

References

1. Randerath, K. and Randerath, E. (1993) Postlabeling methods—an historical review, in: *IARC Scientific Publications*, No 124, Lyon, France, pp. 3–9.
2. Anonymous (1993) In: Postlabeling methods for detection of DNA adducts, in: *IARC Scientific Publications*, No.124, (Phillips, D. H., Castegnaro, M., and Bartsch, H., eds.) Lyon, France, pp. 1–379.
3. Anonymous (1994) DNA adducts: identification and biological significance, in: *IARC Scientific Publications*, No. 125, Lyon, France, pp. 1–478.
4. Pluim, D., Maliepaard, M., van Waardenburg, R. C., Beijnen, J. H., and Schellens, J. H. (1999) ^{32}P-postlabeling assay for the quantification of the major platinum-DNA adducts. *Anal. Biochem.* **275**, 30–38.
5. Reddy, M. V. and Randerath, K. (1987) ^{32}P-analysis of DNA adducts in somatic and reproductive tissues of rats treated with the anticancer antibiotic, mitomycin C. *Mutat Res.* **179**, 75–88.
6. Gladek, A. and Liehr, J. G. (1989) Mechanism of genotoxicity of diethylstilbestrol in vivo. *J. Biol. Chem.* **264**, 16,847–16,852.
7. Han, X., Liehr, J. G., and Bosland, M. C. (1995) Induction of a DNA adduct detectable by ^{32}P-postlabeling in the dorsolateral prostate of NBL/Cr rats treated with estradiol-17 beta and testosterone. *Carcinogenesis* **16**, 951–954.
8. Seraj, M. J., Umemoto, A., Tanaka, M., Kajikawa, A., Hamada, K., and Monden, Y. (1996) DNA adduct formation by hormonal steroids in vitro. *Mutat. Res.* **370**, 49–59.
9. Yasui, M., Matsui, S., Santosh Laxmi, Y. R., et al. (2003) Mutagenic events induced by 4-hydroxyequilin in supF shuttle vector plasmid propagated in human cells. *Carcinogenesis* **24**, 911–917.
10. Shibutani, S., Suzuki, N., Terashima, I., Sugarman, S. M., Grollman, A. P., and Pearl, M. L. (1999) Tamoxifen-DNA adducts detected in the endometrium of women treated with tamoxifen. *Chem. Res. Toxicol.* **12**, 646–653.
11. Shibutani, S., Ravindernath, A., Suzuki, N., et al. (2000) Identification of tamoxifen-DNA adducts in the endometrium of women treated with tamoxifen. *Carcinogenesis* **21**, 1461–1467.
12. Stiborová, M., Fernando, R. C., Schmeiser, H. H., Frei, E., Pfau, W., and Wiessler, M. (1994) Characterization of DNA adducts formed by aristolochic acids in the target organ (forestomach) of rats by ^{32}P-postlabeling analysis using different chromatographic procedure. *Carcinogenesis* **15**, 1187–1192.
13. Randerath, K., Reddy, M. V., and Gupta, R. C. (1981) ^{32}P-labeling test for DNA damage. *Proc. Natl. Acad. Sci. USA* **78**, 6126–6129.
14. Reddy, M. V. and Randerath, K. (1986) Nuclease P1-mediated enhancement of sensitivity of ^{32}P-postlabeling test for structurally diverse DNA adducts. *Carcinogenesis* **7**, 1543–1551.
15. Terashima, I., Suzuki, N., and Shibutani, S. (2002) ^{32}P-Postlabeling/polyacrylamide gel electrophoresis analysis: application to the detection of DNA adducts. *Chem. Res. Toxicol.* **15**, 305–311.

16. Shibutani, S., Suzuki, N., Santosh Laxmi, Y. R., et al. (2003) Identification of tamoxifen-DNA adducts in monkeys treated with tamoxifen. *Cancer Res.* **63,** 4402–4406.
17. Phillips, D. H., Farmer, P. B., Beland, F. A., et al. (2000) Methods of DNA adduct determination and their application to testing compounds for genotoxicity. *Environ. Mol. Mutagen.* **35,** 222–233.
18. Terashima, I., Suzuki, N., and Shibutani, S. (1999) Mutagenic potential of α-(N^2-deoxyguanosinyl)tamoxifen lesions, the major DNA adducts detected in endometrial tissues of patients treated with tamoxifen. *Cancer Res.* **59,** 2091–2095.
19. Santosh Laxmi, Y. R., Suzuki, N., Dasaradhi, L., Johnson, F., and Shibutani, S. (2002) Preparation of oligodeoxynucleotides containing a diastereoisomers of α-(N^2-2'-deoxyguanosinyl)tamoxifen by phosphoramidite chemical synthesis. *Chem. Res. Toxicol.* **15,** 218–225.
20. Cantor, C. R., Warshaw, M. M., and Shapiro, H. (1970) Oligonucleotide interactions. 3. Circular dichroism studies of the conformation of deoxyoligonucleotides. *Biopolymers* **9,** 1059–1077.
21. Zeisig, M., Hofer, T., Cadet, J., and Möller, L. (1999) ^{32}P-postlabeling high-performance liquid chromatography (^{32}P-HPLC) adapted for analysis of 8-hydroxy-2'-deoxyguanosine. *Carcinogenesis* **20,** 1241–1245.

23

Electrophoretic Mobility Shift Assays to Study Protein Binding to Damaged DNA

Vaughn Smider, Byung Joon Hwang, and Gilbert Chu

Summary

The electrophoretic mobility shift assay (EMSA) can be used to identify proteins that bind specifically to damaged DNA. EMSAs detect the presence of key DNA repair proteins, such as ultraviolet (UV)-damaged DNA binding protein, which is involved in nucleotide excision repair, and Ku and DNA-PKcs, which are involved in double-strand break repair. This chapter describes EMSA protocols for detecting proteins that bind to UV-damaged DNA, cisplatin-damaged DNA, and DNA ends. The chapter also describes variations of the EMSA that can be used to obtain additional information about these important proteins. The variations include the reverse EMSA, which can detect binding of ^{35}S-labeled protein to damaged DNA, and the antibody supershift assay, which can define the composition of protein–DNA complexes.

Key Words: DNA-PK; double-strand break repair; global genomic repair; Ku; nonhomologous end-joining; nucleotide excision repair; UV-DDB; V(D)J recombination.

1. Introduction

DNA repair pathways must include proteins that recognize and bind to damaged DNA. The search for such proteins has been facilitated by the use of electrophoretic mobility shift assays (EMSAs), which were first used to detect transcription factors that bind to specific DNA sequences *(1,2)*. To study DNA repair, EMSAs have been adapted to detect proteins that bind to specific DNA lesions or structural features.

In an EMSA, proteins bound to DNA can be resolved as distinctly migrating complexes by nondenaturing polyacrylamide gel electrophoresis. To analyze crude cellular extracts for proteins involved in DNA repair, a ^{32}P-labeled DNA probe is prepared so that it contains the DNA lesion or DNA structure of inter-

From: *Methods in Molecular Biology: DNA Repair Protocols: Mammalian Systems, Second Edition*
Edited by: D. S. Henderson © Humana Press Inc., Totowa, NJ

est. To mask the effects of nonspecific DNA binding proteins, the DNA probe is incubated with cell extract in the presence of an excess of competitor DNA that does not contain the DNA structure.

For any EMSA, several general principles must be followed to ensure proper detection of the appropriate protein–DNA complexes. The procedure for extracting proteins from the cell nuclei must be optimized for the target protein. A cocktail of protease inhibitors must be present, particularly for high molecular weight proteins that may be vulnerable to degradation. The salt concentration of the extraction buffer must be high enough to dissociate the target protein from DNA, but not so high that interfering components are also extracted. The electrophoresis conditions must be optimized to obtain a well defined mobility shift: in particular, the salt concentration of the electrophoresis buffer can be adjusted to permit stable formation of the specific protein–DNA complex while minimizing the formation of nonspecific complexes. Use of a minigel apparatus permits resolution of the desired complexes in about 30 min; end-labeling of the probe DNA with ^{32}P permits autoradiographic detection of the complexes in a few hours. In our laboratory, gels are poured in lots of 13 and stored for up to 3 wk at 4°C so that experiments can be done with minimal setup time.

Once a complex is detected, its specificity for damaged DNA must be confirmed. If the mobility-shifted complex contains a protein rather than some other biochemical molecule, complex formation should be sensitive to the addition of proteases. If the protein binds specifically to damaged DNA, a complex of the same mobility should not form with undamaged probe DNA. Alternatively, unlabeled competitor DNA with and without damage can be compared for their ability to compete away the labeled complex. The spectrum of DNA structures recognized by the protein can be determined by testing different competitor DNAs.

Of course, the discovery of a protein that binds to damaged DNA does not prove that it is involved in DNA repair. To establish biological significance, we have found it fruitful to screen mutant cell lines for abnormalities in binding activity. Successful screening depends on procedures described below for rapidly making cell extracts from small numbers of cells.

This approach has been used to study a number of DNA repair proteins. In this chapter, we will describe our protocols for studying two such proteins (**Fig. 1**): UV-DDB (UV-damaged DNA binding protein), which is involved in the nucleotide excision repair of UV-induced damage *(3)*, and DNA dependent protein kinase (DNA-PK), which is involved in DNA double-strand break repair and V(D)J recombination *(4)*.

In the case of UV-DDB, the probe DNA was a linear DNA fragment damaged by exposure to UV-radiation, and the competitor DNA consisted of unla-

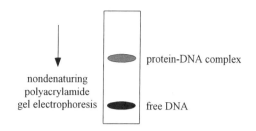

	UV-DDB	**Ku**
Structures bound by protein	bulky DNA adducts	DNA ends
Probe DNA	UV-damaged DNA	linear DNA
Competitor DNA	undamaged DNA	supercoiled DNA

Fig. 1. EMSA for studying UV-DDB and Ku. UV-damaged DNA probe for UV-DDB or linear DNA probe for Ku may be incubated with cell extracts in the presence of the competitor DNA to block nonspecific binding. After incubation, the protein–DNA complex is resolved from the free DNA probe by nondenaturing gel electrophoresis.

beled linear DNA that was left undamaged *(5)*. A complex with a reasonably well-defined mobility was detected even though UV-radiation introduced DNA lesions throughout the length of the DNA probe. This phenomenon was previously reported for the binding of a sequence-specific DNA binding protein to a series of DNA probes containing the binding sequence at different sites. In that case, the mobility of the protein–DNA complex was relatively insensitive to the position of the binding sequence *(6)*. The specificity of UV-DDB was established by showing that the binding activity was competed away by the addition of unlabeled UV-damaged competitor DNA to the binding reaction (**Fig. 2A**). UV-DDB was then purified *(7)* and partially sequenced in order to isolate a cDNA encoding a polypeptide of the expected molecular weight. To demonstrate that this polypeptide bound specifically to UV-damaged DNA, the cDNA was transcribed and translated in vitro so that it was labeled with ^{35}S-labeled methionine *(8)*. A reverse electrophoretic mobility shift assay was then used to show that the mobility of the labeled protein was specifically shifted by the addition of damaged DNA, but not be undamaged DNA (**Fig. 2B**).

To determine the biological relevance of UV-DDB, we screened a series of mutant cell lines known to be hypersensitive to UV radiation. A role in nucleotide excision repair was supported by the discovery that UV-DDB was absent in a subset of xeroderma pigmentosum group E cells, which are defective in

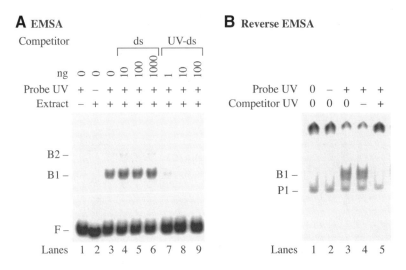

Fig. 2. EMSA and reverse EMSA for UV-DDB. (**A**) EMSA. Crude protein extracts from HeLa cells were incubated with f148 probe DNA in the presence of different amounts of unlabeled competitor DNA. The probe DNA was UV irradiated (lanes 1 and 3–9) or left intact (lane 2). Protein extract was omitted (lane 1) or added (lanes 2–9). Control binding reactions omitted competitor DNA (lanes 1–3). Competition was carried out with intact double-stranded DNA (ds) (lanes 4–6), UV-irradiated double-stranded DNA (UV-ds) (lanes 7–9). (**B**) Reverse EMSA. The fraction of ^{35}S-labeled DDB1 protein that bound to a UV-DNA cellulose column was assayed in a binding reaction with unlabeled f148 DNA. Probe DNA was omitted (lane 1), added as undamaged DNA (lane 2), or added as UV-damaged DNA (lanes 3–5). Unlabeled plasmid competitor DNA was either omitted (lanes 1–3), added as undamaged DNA (lane 4), or added as UV-damaged DNA (lane 5). Some of the ^{35}S-labeled DDB1 protein is retained in the well of the gel because it was purified from a UV-DNA cellulose column, so that some high molecular weight UV-damaged DNA co-eluted as a complex with ^{35}S-labeled DDB1 protein.

this repair pathway (*5*). Furthermore, UV-DDB also recognized competitor DNA carrying intrastrand DNA crosslinks induced by the anticancer drug cisplatin, and levels of UV-DDB were increased in extracts from cells selected for resistance to cisplatin (*9*). The biological relevance was established by showing that microinjection of XPE cells with purified UV-DDB restored DNA repair to the cells (*10*). More specifically, the ectopic expression of UV-DDB in hamster cells enhanced the global genomic repair of cyclobutane dimers (*11*). Global genomic repair refers to the removal of DNA lesions from nontranscribed regions of the genome and from the nontranscribed strand in transcribed regions of the genome.

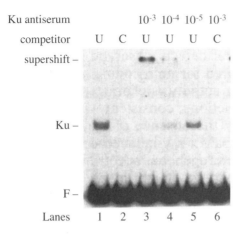

Fig. 3. EMSA and antibody supershift for Ku. Nuclear extract from the hamster cell line AA8 was incubated in the presence of radiolabeled f148 probe and uncut plasmid (U) DNA competitor in lane 1, or cut plasmid (C) in lane 2. The Ku antiserum was included in lanes 3–5 in dilutions ranging from 10^{-3} to 10^{-5} in the presence of uncut plasmid (U). The supershifted complex is specific for DNA ends as shown in lane 6, where cut plasmid (C) is the competitor.

In the case of Ku, the probe was an undamaged linear DNA fragment, and the competitor DNA was supercoiled plasmid DNA (**Fig. 3**). This assay system detected a protein–DNA complex, which we originally denoted as DNA end-binding (DEB) factor *(12)*. Binding specific for DNA ends was established by showing that the binding activity was competed away by the addition of unlabeled competitor plasmid DNA cleaved with any one of several different restriction enzymes, producing DNA ends with 5′ overhanging, 3′ overhanging, or blunt ends. The identification of DEB factor as Ku protein was first established by a useful adjunct to the EMSA, in which incubation of the binding reactions with anti-Ku antibodies produced a supershift of the original protein–DNA complex *(13)*.

To determine the biological relevance of DEB factor, we screened a series of mutant cell lines hypersensitive to ionizing radiation, reasoning that a DNA end-binding protein might be involved in the repair of the double-strand breaks produced by ionizing radiation. This hypothesis was confirmed by the discovery that DEB factor was absent in three cell lines from X-ray complementation group 5, which is defective in double-strand-break repair *(13)*. Strikingly, these cells were also defective in V(D)J recombination, the pathway that generates immunological diversity by cleaving and rearranging the immunoglobulin

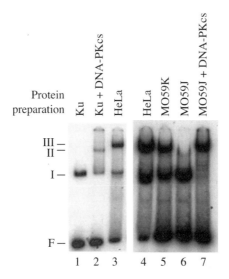

Fig. 4. EMSA for complex of DNA-PKcs and Ku bound to DNA. Radiolabeled f32 probe was incubated with purified protein consisting of 11 ng of Ku (lane 1), or 11 ng of Ku plus 35 ng of DNA-PKcs (lane 2). In addition, the probe was incubated with 500 ng of nuclear extract prepared from HeLa cells (lanes 3 and 4), M059K cells derived from a human glioma (lane 5), or M059J cells lacking DNA-PKcs derived from the same glioma (lanes 6 and 7). In lane 7, the M059J extract was supplemented with 50 ng of purified DNA-PKcs. The gel shows migration of the free probe (F) and protein–DNA complexes containing Ku (I), Ku and DNA-PKcs (II), and Ku, DNA-PKcs, and an unidentified factor (III).

genes in B-cells and the T-cell receptor genes in T-cells. Biological relevance was established by showing that transfection of the mutant cells with a cDNA expression vector for the 86-kDa subunit of Ku restored both ionizing radiation resistance and V(D)J recombination *(14,15)*.

Upon binding to DNA ends, Ku is capable of translocating inward to allow binding of additional Ku molecules. Ku also recruits the DNA-dependent protein kinase catalytic subunit (DNA-PKcs) to the DNA ends, thus forming a higher order complex with DNA (**Fig. 4**). (DNA-PK consists of its catalytic subunit DNA-PKcs and the heterodimeric Ku protein.) To detect binding by DNA-PKcs, we modified the EMSA by shortening the length of the radiolabeled DNA probe to 32 base pairs (bp) and eliminating competitor DNA. Indeed, higher order complexes were detected with either purified proteins or cell extracts. Interestingly, cell extracts generate a higher order complex with a slightly lower mobility than seen with purified proteins (compare lanes 2 and 3 in **Fig. 4**). Supershift assays with antibodies demonstrate that this complex contains both Ku and DNA-PKcs (data not shown). Extracts from M059J cells, which lack DNA-PKcs, do not support the formation of the higher order complex, but supplementation of the mutant extracts with purified DNA-PKcs restores the higher order complex (compare lanes 6 and 7 in **Fig. 4**). Finally, DNA-PKcs is capable of binding to DNA ends in the absence of Ku. Indeed, when the proteins are incubated with a short 32 bp oligonucleotide in the absence of competitor DNA, the EMSA is capable of detecting complexes containing DNA-PKcs. These complexes include DNA bound to DNA-PKcs, or Ku in a complex with DNA-PKcs *(16)*.

EMSAs have been used to identify other structure-specific proteins. The randomly damaged UV-irradiated DNA probe, which was used to detect UV-DDB in human extracts, failed to detect a corresponding factor in yeast extracts, but instead detected photolyase, which repairs UV-induced cyclobutane pyrimidine dimers by photoreactivation *(17)*. Like UV-DDB, yeast photolyase recognized both UV-damaged and cisplatin crosslinked DNA. Although UV-DDB expression was associated with resistance to cisplatin, yeast photolyase conferred sensitivity to cisplatin *(18)*. DNA probes with other structures have detected additional proteins. An oligonucleotide containing a single GT mismatch was used to identify the GT binding protein that has now been shown to be involved in mismatch repair *(19)*. An oligonucleotide containing a single intrastrand cisplatin crosslink was used to search extracts for damage-specific DNA binding proteins that proved to be the HMG-1 and HMG-2 proteins *(20,21)*. In this case, the proteins were not involved in DNA repair, but rather interfered with the nucleotide excision repair machinery and conferred sensitivity to cisplatin *(22)*.

In conclusion, the EMSA has proven to be extremely useful for studying DNA repair. Our EMSA protocols for UV-DDB, Ku, and DNA-PKcs are described in detail, because they remain useful for studying these proteins further, and because they are paradigms for how the EMSA might be used to discover new repair proteins or gain new insights into DNA repair in the future.

2. Materials

2.1. Probe DNA

2.1.1. Preparation of f148 and f32 Probe

1. f148: pRSVcat plasmid.
2. f32: oligonucleotides f32-1 (GTAGTTGTACCCGATGGTGGACGCGTGC) and f32-2 (GGCCGCACGCGTCCATCGGGTACAACTAC).
3. *Hin*dIII and *Pvu*II: 10 U/µL (New England Biolabs, Beverly, MA).
4. Agarose (Kodak-IBI, Rochester, NY).
5. 1X TBE: 89 mM Tris-base, 89 mM borate, 2 mM EDTA.

6. Ethidium bromide solution: 2 µg/mL, dissolved in distilled water and filtered through 0.4-µm filter.
7. Long-wavelength UV light source (366 nm): Model UVL-56 (Ultra-Violet Products, Inc., San Gabriel, CA).
8. IBI-electroeluter: Model 46000 (IBI-Kodak, Rochester, NY).
9. Polyethylene tubing: 1.6 mm inner diameter.
10. Salt solution: 10 µL of 0.5% bromophenol blue and 1 mL of 10 M ammonium acetate.
11. 18-gage needle.
12. Razor blade.
13. Isobutanol.
14. 100% Ethanol.
15. 80% Ethanol.
16. TE: 10 mM Tris-HCl, pH 8.0, and 1 mM EDTA.

2.1.2. Labeling of f148 or f32 Probe (Klenow Method)

1. 10X Klenow buffer: 100 mM Tris-HCl, pH 7.5, 50 mM MgCl$_2$, and 75 mM dithiothreitol (DTT).
2. 10 mM dATP, dGTP, and dTTP.
3. [α-^{32}P]dCTP: 3000 Ci/mmol, 10 mCi/mL (Amersham, Arlington Heights, IL).
4. DNA polymerase I, Klenow fragment: 5 U/µL (New England Biolabs, Beverly, MA).

2.1.3. Labeling of f148 Probe (Exonuclease III/Klenow Method)

1. **Items 1–4** of **Subheading 2.1.2.**
2. Exonuclease III: 100 U/µL, diluted in TE (New England Biolabs, Beverly, MA).

2.2. Cell Extracts

2.2.1. Whole Cell Extract

1. Phosphate-buffered saline (PBS): 1 g/L of D-glucose, 36 mg/L of sodium pyruvate, 36 mg/L of calcium phosphate, and 36 mg/L of magnesium phosphate.
2. Lysis buffer: 700 mM NaCl, 1 mM EGTA, 1 mM EDTA, 10 mM β-glycerophosphate, 2 mM MgCl$_2$, 10 mM KCl, 1 mM sodium vanadate, 1 mM phenylmethylsulfonyl fluoride (PMSF), 1 mM DTT, 0.1% Nonidet P-40 (NP-40), 10 µg/mL of each pepstatin, leupeptin, and aprotinin.
3. Bradford solution (Bio-Rad, Hercules, CA).

2.2.2. Cytoplasmic/Nuclear Extracts

1. PBS: *see* **item 1, Subheading 2.2.1.**
2. Buffer A without NP-40: 10 mM N-2-hydroxyethylpiperazine-N'-2-ethanesulfonic acid (HEPES)–KOH, pH 7.4, 1.5 mM MgCl$_2$, 10 mM KCl, 1 mM DTT, and 1 mM EDTA.
3. Buffer A with NP-40: 10 mM HEPES–KOH, pH 7.4, 1.5 mM MgCl$_2$, 10 mM KCl, 1 mM DTT, 1 mM EDTA, and 1.0% (w/v) NP-40.

EMSAs to Study DNA Repair Proteins 331

4. Buffer C: 20 mM HEPES–KOH, pH 7.4, 20% glycerol (v/v), 500 mM NaCl, 1.5 mM MgCl$_2$, 0.2 mM EDTA, and 0.5 mM DTT.
5. 100X Protease inhibitor cocktail: 100 mM PMSF, 10 mg/mL of pepstatin A, 10 mg/mL of leupeptin, and 10 mg/mL of aprotinin.

2.3. Preparation of Nondenaturing Gels

1. Minigel plates: 0.75 mm thickness, 80 × 100 mm (Hoefer, San Francisco, CA).
2. Gel caster: SE215 Mighty-Small multiple gel caster (Hoefer).
3. Acrylamide–*bis*-acrylamide: 29:1, w/w (Molecular biology grade, Sigma, St. Louis, MO).
4. 5X TGE buffer: 50 mM Tris-HCl, pH 8.5, 380 mM glycine, and 2 mM EDTA. (*See* **Note 1**.)
5. Ammonium persulfate: 10% of fresh solution.
6. *N*,*N*,*N'*,*N'*-Tetramethylethylenediamine (TEMED).
7. Gel combs: 10- or 15-lane combs with 0.75 mm thickness (Hoefer).

2.4. EMSA for Studying UV-DDB

2.4.1. UV-Damaged f148 DNA

1. Germicidal lamp: G15T8 (General Electric, Cleveland, OH).
2. Radiometer/photometer: Model IL1350 (International Light Inc., Newburyport, MA).

2.4.2. Competitor DNA to Mask Nonspecific Binding

1. Poly(dI-dC) (Amersham Biosciences).
2. Salmon sperm DNA: Sheared by passing through an 18-gage needle with pressure.
3. Linear plasmid DNA: pRSVcat plasmid digested with *Pvu*II and extracted with phenol–chloroform, and precipitated in ethanol.

2.4.3. Competitor DNAs to Test Specificity of Binding

1. Supercoiled pRSVcat plasmid.
2. AP treatment buffer: 0.01 M sodium citrate, 0.1 M NaCl, pH 5.0.
3. Pt treatment buffer: 3 mM NaCl, 1 mM NaH$_2$PO$_4$, pH 7.4.
4. *cis*-Diamminedichloroplatinum (*cis*-DDP, Sigma, St. Louis, MO): For a stock solution, dissolve in Pt treatment buffer and store at –20°C in the dark.
5. 100% Ethanol.
6. *N*-Methyl-*N'*-nitro-*N*-nitrosoguanidine (MNNG, Sigma): dissolve to 15 mM in ethanol and store at –20°C.
7. MNNG treatment buffer: 10 mM Tris-HCl, pH 7.5, and 50 mM NaCl.

2.4.4. Binding Assay

1. 5X Binding buffer: 12 mM HEPES–KOH, pH 7.9, 60 mM KCl, 5 mM MgCl$_2$, 4 mM Tris, 0.6 mM EDTA, 1 mM DTT, and 12% glycerol (v/v).

2. Bovine serum albumin (BSA): 10 mg/mL.
3. Plasmid DNA: pRSVcat, 1 mg/mL.
4. Poly(dI-dC)/salmon sperm DNA: 1 and 0.5 mg/mL.
5. Whole cell or nuclear extract: 0.1–0.5 mg/mL.
6. Loading dye: 0.025% bromophenol blue in 5X binding buffer.
7. Whatman 3MM paper (Whatman, Clifton, NJ).
8. X-ray film: Kodak XAR-5 (IBI-Kodak, Rochester, NY).

2.5. Variation on EMSA: Reverse Electrophoretic Mobility Shift Assay

2.5.1. In Vitro Synthesis of ^{35}S-Labeled DDB1

1. DDB1 plasmid: cloned into pCDNA3 vector (Invitrogen, Carlsbad, CA).
2. Rabbit reticulocyte lysate (Promega, Madison, WI).
3. ^{35}S-labeled methionine: 1000 Ci/mmol, 50 mCi/mL (Amersham, Arlington Heights, IL).
4. T7 RNA polymerase buffer: 40 mM Tris-HCl, pH 7.9, 6 mM MgCl$_2$, 2 mM spermidine, 10 mM DTT, 0.5 mM ATP, 0.5 mM CTP, 0.5 mM GTP, and 0.5 mM UTP.
5. T7 RNA polymerase: 10 U/µL (Promega, Madison, WI).
6. Amino acid mixture minus methionine: 1 mM (Promega, Madison, WI).
7. RNasin ribonuclease inhibitor: 40 U/µL (Promega, Madison, WI).
8. RNase-free water: treated with 0.1% diethyl pyrocarbonate (DEPC) at 37°C overnight and autoclaved at 120°C for 20 min.

2.5.2. Purification of Active ^{35}S-Labeled DDB

1. UV-DNA cellulose resin: prepared as described previously (7).
2. Buffer D: 10 mM HEPES–KOH, pH 7.9, 2 mM EDTA, 2 mM DTT, 0.01% NP-40 (v/v).

2.5.3. Reverse Electrophoretic Mobility Shift Assay

1. 5X binding buffer: 12 mM HEPES–KOH, pH 7.9, 60 mM KCl, 5 mM MgCl$_2$, 4 mM Tris, 0.6 mM EDTA, 1 mM DTT, and 12% glycerol (v/v).
2. BSA: 10 mg/mL.
3. Plasmid DNA: pRSVcat, 1 mg/mL.
4. Poly(dI-dC)/salmon sperm DNA: 1 and 0.5 mg/mL.
5. ^{35}S-labeled DDB1.
6. Loading dye: 0.025% bromophenol blue in 5X binding buffer.
7. 50% Methanol and 10% acetic acid.
8. Amplify (Amersham, Arlington Heights, IL).
9. Whatman 3MM paper (Whatman, Clifton, NJ).
10. Baby powder.
11. Whatman 3MM paper (Whatman, Clifton, NJ).
12. X-ray film: Kodak XAR-5 (IBI-Kodak, Rochester, NY).

2.6. EMSA for Studying Ku or DNA-PKcs
2.6.1. Probe
Materials for purifying and radiolabeling the f148 and f32 probe are described in **Subheading 2.1.**

2.6.2. Competitor DNA Against Nonspecific Binding
Supercoiled pRSVcat plasmid: purified by CsCl centrifugation *(23)*.

2.6.3. Competitor DNAs to Test Specificity of Binding
1. Poly(dA) or poly(dT) (Amersham Biosciences).
2. Double-stranded linear DNA: unlabeled f148 DNA.
3. Single-stranded circular DNA: phage M13 DNA (New England BioLabs, Beverly, MA).
4. Supercoiled pRSVcat plasmid: purified by CsCl centrifugation *(23)*.

2.6.4. Ku Binding Assay
1. 5X binding buffer: 12 mM HEPES–KOH, pH 7.9, 60 mM KCl, 5 mM MgCl$_2$, 4 mM Tris, 0.6 mM EDTA, 1 mM DTT, and 12% glycerol (v/v).
2. Supercoiled plasmid DNA: pRSVcat, 1 mg/mL.
3. Protein extract: 0.1–0.5 mg/mL.
4. Loading dye: 0.025% bromophenol blue in 5X binding buffer.
5. 4% Polyacrylamide gel.
6. Whatman 3MM paper (Whatman, Clifton, NJ).
7. X-ray film: Kodak XAR-5 (IBI-Kodak, Rochester, NY).

2.6.5. DNA-PKcs Binding Assay
1. 5X binding buffer: 50 mM Tris-HCl, pH 7.8, 5 mM EDTA, 500 mM NaCl, 25% glycerol, 1 mM DTT.
2. Protein extract: 0.1–5.0 mg/mL.
3. Loading dye: 0.025% bromophenol blue in 1X binding buffer.
4. 4% Polyacrylamide gel.
5. Whatman 3MM paper (Whatman, Clifton, NJ).
6. X-Ray film: Kodak XAR-5 (IBI-Kodak, Rochester, NY).

2.6.6. Variation on EMSA: Antibody Supershift
1. Anti-Ku antibody: HT serum from a human autoimmune patient with polymyositis–scleroderma overlap syndrome *(13)*.
2. 1% BSA.

3. Methods
3.1. Probe DNA
3.1.1. Preparation of f148 Probe
The 148-bp DNA fragment (f148) was isolated from bacterial chloramphenicol acetyltransferase gene.

1. Digest pRSVcat plasmid with *Hin*dIII and *Pvu*II restriction enzymes.
2. Separate the digested DNA fragments on a 1.5% agarose gel.
3. Stain the gel with ethidium bromide solution and visualize the 148-bp DNA fragment (f148) with long-wavelength UV light. Long-wavelength UV is used instead of short-wavelength UV to prevent damage to the DNA.
4. Excise the band containing the f148 fragment with a razor blade.
5. Elute the f148 DNA fragment from the gel slice by using an IBI-electroeluter, as described below.
6. Soak the gel slice containing f148 in 0.2X TBE buffer for 5–10 min.
7. Preclear the electroeluter chamber by electrophoresis in 0.2X TBE for 20 min at 150 V. Carefully remove air bubbles in the chamber.
8. Place gel slice in circular receptacle. Surround the gel slice with 0.2X TBE buffer but do not cover the gel slice. Keep the valve up.
9. Flush the V-channel with 0.2X TBE buffer by using an 18-gage needle on a 1-mL syringe with polyethylene tubing to protect electroeluter device.
10. Underlay 125 µL of salt solution into the V-channel.
11. Close the cover to move the valve to the intermediate position.
12. Run the electroeluter at 150 V; watch for migration of the DNA band out of the gel slice with the hand-held long-wavelength UV light.
13. Stop the electroelution when all of the DNA fragments have left the gel slice (approx 10–20 min).
14. Push the valve to the lowest position.
15. Withdraw the contents in the V channel and rinse the channel with 100 µL of the salt solution.
16. Extract the eluate twice with 400 µL of isobutanol, which will remove the contaminated dyes and reduce the volume.
17. Measure the final volume of the eluate and add two volumes of 100% ethanol and precipitate at –20°C for 2 h.
18. Centrifuge the ethanol precipitate at 13,000g for 20 min at 4°C.
19. Wash the pellet with cold 80% ethanol solution.
20. Suspend the pellet in TE solution and measure the concentration *(23)*.

3.1.2. Preparation of f32 Probe

1. Mix 300 µL of 5 ng/µL of f32-1 with 300 µL of 5 ng/µL f32-2 in TE buffer.
2. Heat the mixture of oligonucleotides to 100°C for 2 min.
3. Anneal the oligonucleotides by turning off the heat source and allowing the mixture to cool to room temperature.

3.1.3. Labeling of f148 or f32 Probe (Klenow Method) (see **Note 3**)

1. Set up reaction mixture at room temperature as follows:
   a. 10X Klenow buffer   1 µL
   b. f148 or f32 DNA fragment (5 ng/µL)   4 µL
   c. 10 mM dATP   0.5 µL
   d. 10 mM dGTP   0.5 µL

EMSAs to Study DNA Repair Proteins

    e. 10 mM dTTP   0.5 μL
    f. [α-^{32}P]dCTP (10 μCi/μL)   1 μL
    g. Klenow (5 U/μL)   1 μL
    h. Distilled water   to 10 μL

2. Incubate at room temperature for 20 min.
3. Inactivate the Klenow by incubating the reaction mixture at 65°C for 10 min.
4. Purify the labeled f148 from unincorporated nucleotides by using spin-column chromatography *(23)*.

3.1.4. Labeling of f148 Probe (Exonuclease III/Klenow Method) (see **Note 3**)

1. Set up exonuclease III digestion reaction as below:
    a. 10X Klenow buffer   2 μL
    b. f148 DNA fragment (5 ng/μL)   4 μL
    c. Exonuclease III (0.2 U/μL, diluted in TE)   1 μL
    d. Distilled water   to 10 μL
2. Incubate at room temperature for 10 min.
3. Inactivate the exonuclease III by incubating the reaction mixture for 10 min at 65°C.
4. Add the following to the cooled reaction mixture at room temperature:
    a. 10 mM dATP   1 μL
    b. 10 mM dGTP   1 μL
    c. 10 mM dTTP   1 μL
    d. [α-^{32}P]dCTP (10 μCi/μL)   1 μL
    e. Klenow (5 U/μL)   1 μL
    f. Distilled water   to 20 μL
5. Incubate at 37°C for 30 min.
6. Inactivate the Klenow by incubating the reaction mixture at 65°C for 10 min.
7. Purify the labeled f148 from unincorporated nucleotides by using spin-column chromatography *(23)*.

3.2. Preparation of Cell Extracts

3.2.1. Whole Cell Extract

1. Harvest 2×10^6 cells from culture dishes in 1 mL of ice-cold PBS.
2. Pellet the cells by centrifugation for 1 min at 13,000g.
3. Resuspend the pellet in 30 μL of lysis buffer.
4. Incubate the lysates at 4°C for 30 min with gentle shaking.
5. Centrifuge the lysates at 13,000g for 30 min at 4°C.
6. Save the supernatant at –80°C. (*See* **Note 4**.)
7. Measure the protein concentration by a modification of the Bradford method *(24)*.

3.2.2. Cytoplasmic/Nuclear Extracts (see **Notes 5** and **6**)

1. Harvest 2×10^6 adherent cells from culture dishes in 1 mL of ice-cold PBS.
2. Wash the cells once with 500 μL of 1X PBS.

3. Add 10 μL of protease inhibitor cocktail to 1 mL of buffer A without NP-40. Wash the cells once with 500 μL of buffer A without NP-40.
4. Add 1 μL of protease inhibitor cocktail to 100 μL of buffer A with NP-40. Resuspend the pellets in 20 μL of buffer A with NP-40.
5. Incubate the suspensions at 4°C for 10 min with gentle shaking.
6. Centrifuge the lysates at 13,000g for 5 s at 4°C.
7. Save the supernatant as cytoplasmic extract at –80°C. (*See* **Note 4**.)
8. Wash the pellet in 300 μL of buffer A without NP-40 by gentle pipetting. Microfuge at 13,000g for 5 s.
9. Add 10 μL of protease inhibitor cocktail to 1 mL of buffer C. Resuspend the pellet from **step 8** in 75 μL of buffer C.
10. Incubate the suspensions at 4°C for 20 min with gentle shaking.
11. Centrifuge the nuclear lysates at 13,000g for 15 min at 4°C.
12. Save the supernatant as nuclear extract at –80°C.
13. Measure the protein concentration by the method of Bradford *(24)*.

3.3. Preparation of Nondenaturing Gel (see **Note 7**)

1. Assemble 13 sets of 0.75-mm thick minigels in a gel caster.
2. Prepare 4% polyacrylamide gel solution as follows:
   a. Acrylamide–*bis*-acrylamide (29:1)   14 mL
   b. 5X TGE buffer   20 mL
   c. Distilled water   65 mL
   d. Ammonium persulfate (10%)   1 mL
   e. TEMED   120 μL
3. Gently mix the gel solution and immediately pour to the top of the assembled gel caster with caution to avoid bubbles.
4. Tap the gel caster gently several times to remove the bubbles in the gel caster.
5. Insert the comb into the top of each gel.
6. Leave at room temperature for at least 4 h.
7. Disassemble the gel caster and wrap each gel with Saran Wrap and store at 4°C until use. Be careful to remove residual polyacrylamide that adheres to the outside of the individual gels—this can interfere with efficient electrophoresis. (*See* **Note 8**.)

3.4. EMSA for Studying UV-DDB

3.4.1. UV Irradiation of the f148 Probe (see **Notes 9** and **10**)

The labeled f148 is damaged by UV radiation at a DNA concentration of 0.2 μg/mL with a germicidal lamp at a flux of 10.4 J/m^2/s for total doses of 100–5000 J/m^2.

3.4.2. Competitor DNA Against Nonspecific Binding (see **Note 11**)

Crude cellular extracts have nonspecific DNA binding proteins, including DNA end binding proteins. Thus, excess unlabeled linear double-stranded

DNA is included in the reaction mixture. The amount of the competitor DNA is determined empirically because different extracts require different amounts of competitor DNA. A mixture of poly(dI-dC) and salmon-sperm DNA (2:1, w/w) is generally used.

3.4.3. Competitor DNAs to Test Specificity of Binding (see **Note 12**)

The binding specificity for UV-damaged DNA and single-stranded DNA is measured by competition assay. The single-stranded competitor DNA is prepared by heating linear double-stranded DNA at 100°C for 5 min and then cooling rapidly in ice water. UV-damaged single-stranded DNA is prepared by UV irradiating single-stranded DNA.

The chemically damaged competitor DNA is prepared as described previously *(25)*. Apurinic DNA is prepared by incubating DNA (300 µM DNA phosphate) in AP treatment buffer at 70°C for different times. Approximately one purine base is released every 4 min under these reaction conditions. Cisplatin damaged DNA is prepared by incubating supercoiled plasmid (300 µM DNA phosphate) with *cis*-DDP at 37°C for 12–18 h in the dark. The DNA is purified by ethanol precipitation. MNNG is incubated with DNA (300 µM DNA phosphate) for 24 h at 37°C in MNNG treatment buffer. The DNA is purified by ethanol precipitation.

3.4.4. Binding Assay

1. Assemble the reaction mixture as follows:
   a. 5X Binding buffer — 2 µL
   b. UV-^{32}P-f148 probe — 0.2 or 0.02 ng
   c. BSA (10 mg/mL) — 0.3 µL
   d. Plasmid DNA (1 mg/mL) — 2 µL
   e. Poly(dI-dC)/salmon sperm DNA (1/0.5 mg/mL) — 0.5–2 µL
   f. Distilled water — to 8 µL
   g. Whole cell or nuclear extract (0.1–0.5 mg/mL) — 2 µL
2. Add whole cell or nuclear extract last.
3. Incubate the reaction mixture at room temperature for 30 min.
4. Add 2 µL of loading dye to the reaction mixture and gently mix together.
5. Resolve the protein–DNA complexes by nondenaturing gel electrophoresis at 10 V/cm at room temperature.
6. Dry the gel on Whatman 3MM paper and expose to X-ray film at –80°C (**Fig. 2A**).

3.5. Variation on EMSA: Reverse Electrophoretic Mobility Shift Assay

Specific binding of UV-DDB to UV-damaged DNA probe can also be visualized by a reverse mobility shift assay with labeled UV-DDB and unlabeled UV-f148 probe (**Fig. 2B**).

3.5.1. In Vitro Synthesis of ^{35}S-labeled DDB1

UV-DDB consists of a 125-kDa subunit, DDB1, and a 48-kDa subunit, DDB2. Labeled DDB1 protein is synthesized by transcribing the DDB1 cDNA from a T7 promoter with T7 RNA polymerase and translating the mRNA in a rabbit reticulocyte lysate in the presence of [^{35}S]methionine.

1. Assemble the in vitro transcription/translation reaction mixture as follows:
   a. pCDNA3 (DDB1) plasmid (1 mg/mL)   1 µL
   b. Rabbit reticulocyte lysate   25 µL
   c. [^{35}S]methionine   4 µL
   d. T7 RNA polymerase buffer   2 µL
   e. T7 RNA polymerase   1 µL
   f. Amino acid mixture minus methionine   1 µL
   g. RNasin ribonuclease inhibitor   1 µL
   h. RNase-free water   to 50 µL
2. Incubate at 30°C for 120 min.
3. Store the in vitro translation products at –80°C.

3.5.2. Purification of Active ^{35}S-Labeled DDB1 Protein

All of the DDB1 protein synthesized in vitro is not active to bind UV-damaged DNA. Thus, the active fraction is purified through UV-DNA cellulose affinity chromatography.

1. Pack 10 µL of UV-DNA cellulose resin in a 200 µL Pipetman tip and equilibrate the column with buffer D.
2. Load the in vitro translated ^{35}S-labeled DDB1 proteins onto the column (50 µL of the reticulocyte lysate).
3. Wash the column extensively with 600 µL of buffer D containing 200 mM NaCl.
4. Elute the bound ^{35}S-labeled DDB1 protein with buffer D containing 1 M NaCl.
5. Store the fractions at –80°C.

3.5.3. Reverse Electrophoretic Mobility Shift Assay

1. Assemble the reaction mixture as follows:
   a. 5X binding buffer   2 µL
   b. UV-f148 probe   1 ng
   c. BSA (10 mg/mL)   0.3 µL
   d. Plasmid DNA (1 mg/mL)   2 µL
   e. Poly (dI-dC)/salmon sperm DNA (1/0.5 mg/mL)   0.5 µL
   f. Distilled water   to 8 µL
   g. ^{35}S-labeled DDB1   2 µL
2. Incubate the reaction at room temperature for 30 min.
3. Add 2 µL of loading dye to the reaction mixture and gently mix together.
4. Resolve the protein–DNA complex by nondenaturing gel electrophoresis at 10 V/cm at room temperature.

5. Fix the gel in 50% methanol and 10% acetic acid for 2 h.
6. Soak the gel in Amplify for 30 min.
7. Dry the gel on Whatman 3MM paper.
8. Treat the sticky surface of the dried gel with baby powder, which allows direct contact of the gel with the X-ray film.
9. Expose the gel to X-ray film at –80°C.

3.6. EMSA for Studying Ku

3.6.1. Probe

Prepare the f148 probe as described in **Subheading 3.1.1.** and radiolabel the probe as described in **Subheading 3.1.3.** (*See* **Notes 13** and **14**.)

3.6.2. Competitor DNA Against Nonspecific Binding (see **Note 15**)

Crude nuclear extracts have proteins that are not specific for DNA ends that may bind to the labeled f148 probe. To minimize the effects of these proteins on the EMSA, supercoiled plasmid is included in the binding reaction. Because the supercoiled pRSVcat plasmid contains the f148 fragment but has no free ends, it is a good choice for being a nonspecific competitor. In general, human cell extracts require 2 µg of nonspecific competitor in a 10 µL reaction. Rodent cell extracts, however, require only 50–100 ng of nonspecific competitor.

3.6.3. Competitor DNAs to Test Specificity of Binding

Competition assays are performed using single-stranded linear DNAs (poly[dA] or poly[dT]), double-stranded linear DNA (unlabeled f148), single-stranded circular DNA (phage M13), or supercoiled double-stranded DNA (pRSVcat). These competitor DNAs are included in the binding reaction in concentrations ranging from 0.2 to 200 ng. Competitors containing free DNA ends (unlabeled f148) or DNA containing single-stranded to double-stranded DNA transitions (phage M13) compete for Ku binding activity.

3.6.4. Binding Assay

1. Prepare the reaction mixture as follows:
   a. 5X binding buffer   2 µL
   b. f148 probe   0.2 ng
   c. Supercoiled plasmid   (human extracts)   2 µg
                             (rodent extracts)   or 50–100 ng
   d. Distilled water   to 8 µL
   e. Nuclear extract   0.5 µg
2. Incubate the reaction for 5 min at room temperature.
3. Add 2 µL of loading dye to the reaction and mix by gentle pipetting.

4. Resolve the protein–DNA complexes by nondenaturing gel electrophoresis at 10 V/cm at room temperature.
5. Dry the gel on 3MM Whatman paper and expose to X-ray film at –80°C.

3.6.5. Variation on EMSA: Antibody Supershift

Anti-Ku antibodies may be added to the binding reaction to supershift the Ku/DNA complex. Serial dilutions of the antibodies are made in 1% BSA and added to the reaction mix above, prior to adding the protein extract (**Fig. 3**).

3.7. EMSA for Studying DNA-PKcs

3.7.1. Probe

Prepare the f32 probe as described in **Subheading 3.1.2.** and radiolabel the probe as described in **Subheading 3.1.3.**

3.7.2. Competitor DNA

We have found that shortening the length of the DNA probe allows a complex containing DNA-PKcs to form on DNA ends in the absence of competitor DNA. (*See* **Note 16**.) In addition, certain competitor DNAs, including poly(dA) and supercoiled plasmid, result in the supershift of the DNA-PKcs-containing complex. This may result from the ability of DNA-PKcs to synapse different DNA molecules.

3.7.3. Binding Assay (see **Notes 17** and **18**)

1. Prepare the reaction mixture as follows:
   a. 5X binding buffer 2 μL
   b. f132 probe 0.2 ng
   c. Distilled water to 8 μL
   d. Nuclear extract (human extracts) 0.5 μg
      (rodent extracts) or 2.0 μg
   e. NaCl to a final concentration of (human extracts) 200 mM
      (rodent extracts) or 400 mM
2. Incubate the reaction for 5 min at room temperature.
3. Add 2 μL of loading dye to the reaction and mix by gentle pipeting.
4. Resolve the protein–DNA complexes by nondenaturing gel electrophoresis at 10 V/cm at room temperature.
5. Dry the gel on 3MM Whatman paper and expose to X-ray film at –80°C (**Fig. 4**).

3.7.4. Variation on EMSA: Antibody Supershift

Anti-DNA-PKcs antibodies may be added to the binding reaction to supershift or disrupt the DNA-PKcs containing complexes. Serial dilutions of the antibodies are made in 1% BSA and added to the reaction mix above, prior to adding the protein extract.

4. Notes

1. The high-salt buffers TGE and TBE were superior to low-salt buffers in these EMSAs because the high-salt buffers reduced nonspecific binding more efficiently.
2. When labeling the DNA probe with Klenow fragment, the radiolabeled nucleotide used to fill in the 5' overhang of the DNA should be complementary to an internal base of the 5' overhang. Inefficient labeling may occur if the radiolabeled nucleotide is complementary to the last base of the 5' overhang because of exonuclease activity in the Klenow fragment.
3. The specific activity of the probe can be increased by using the exonuclease III/ Klenow method (**Subheading 3.1.4.**). This can be helpful in detecting UV-DDB in rodent cells, where levels are substantially lower than in human cells.
4. Extracts may be stored at 4°C for 1–3 d, but should be stored at –80°C for longer periods of time. Extracts should be stored in aliquots to minimize repeated freeze-thawing. Thawing of the extracts should be done on ice to minimize protease activity which may occur at higher temperatures.
5. The extraction procedure can be modified by using different salt concentrations in buffer C or changing the NP-40 concentration.
6. The presence of NP-40 in the extraction buffer (buffer C) can inhibit extraction of some proteins. Washing the nuclei following the treatment with buffer A containing NP-40 removes the NP-40 and overcomes this effect (**Subheading 3.2.2., step 8**). In addition, larger proteins may be affected by proteases to a greater extent than smaller proteins, making it critical that protease inhibitors be present at all steps in the extraction procedure following the lysis in buffer A.
7. Lowering the acrylamide concentration and decreasing the length of the DNA probe can facilitate the detection of larger DNA–protein complexes (like DNA-PKcs) by permitting the migration of these complexes into the gel.
8. Conduction leaks in the gel can lead to aberrant migration of the free probe and protein/DNA complexes. This can be minimized by cleaning areas around the gel and gel box where salt buildup can occur.
9. In detecting UV-DDB, increasing the UV dose on the probe DNA can be utilized to increase the number of lesions per DNA molecule. Higher order complexes corresponding to multiple UV-DDB binding events can be visualized at these higher doses *(7)*.
10. When cyclobutane pyrimidine dimers were removed from a UV-damaged DNA probe by treatment with purified photolyase, the binding was decreased compared to the untreated UV-damaged DNA probe. This suggests that UV-DDB recognizes at least some cyclobutane dimers *(7)*.
11. Because these EMSAs rely on the structure-specific binding properties of the DNA repair proteins, it is important that the competitor DNAs be free of contaminating structures which might compete for binding activity.
12. UV-DDB recognizes DNA damaged by several agents including UV irradiation, but direct binding was observed only to the UV-damaged double-stranded DNA probe. The binding of UV-DDB to apurinic sites, cisplatin adducts, nitrogen

mustard adducts, and single-stranded DNA was detected by showing that these forms of damaged DNA could compete with UV-damaged DNA for binding to UV-DDB *(7,25)*. UV-DDB may have low affinities for these forms of DNA, so that only the competition assay but not the direct binding assay detected binding.

13. Increasing the length of the radiolabeled DNA probe can be utilized to detect multiple Ku binding events. This is possible because Ku is able to bind DNA and translocate along the molecule. Thus, longer DNA lengths permit more Ku molecules to bind and produce a "ladder" pattern in the EMSA *(12)*.
14. Decreasing the length of the DNA probe can allow a decrease in the amount of competitor DNA used in the EMSA for Ku.
15. The competitor used in EMSA for Ku should be supercoiled DNA which is free of contaminating nicked or linear DNA. Ku binds strongly to free DNA ends and has been reported to bind nicks, so competitors with these structures will obscure Ku binding activity.
16. There should not be competitor DNA in EMSAs to detect DNA-PKcs. In addition, the probe length should be short. We have found 32 bp to be optimal in EMSAs to detect DNA-PKcs.
17. DNA-PKcs levels are lower in rodent extracts than human extracts, thus the concentration of extract used in an EMSA should be increased from 0.5 μg (human extract) to 2.0 μg (rodent extract).
18. Since DNA-PKcs EMSAs do not utilize competitor DNA, we have tested the effect of different salt concentrations on nonspecific probe binding and DNA-PKcs complex formation. DNA-PKcs complexes can be detected with NaCl concentrations as high as 500 mM, with higher salt decreasing nonspecific binding. For rodent extracts, which use more protein extracts, we recommend increasing the NaCl concentration to 400 mM.

References

1. Garner, M. M. and Revzin, A. (1981) A gel electrophoresis method for quantifying the binding of proteins to specific DNA regions: application to the components of the *E. coli* lactose operon regulatory system. *Nucleic Acids Res.* **9,** 3047–3059.
2. Fried, M. and Crothers, D. M. (1981) Equilibrium and kinetics of lac repressor-operator interactions by polyacrylamide gel electrophoresis. *Nucleic Acids Res.* **9,** 6505–6525.
3. Tang, J. and Chu, G. (2002) Xeroderma pigmentosum complementation group E and UV-damaged DNA-binding protein. *DNA Repair* **1,** 601–616.
4. Smider, V. and Chu, G. (1997) The end-joining reaction in V(D)J recombination. *Semin. Immunol.* **9,** 189–197.
5. Chu, G. and Chang, E. (1988) Xeroderma pigmentosum group E cells lack a nuclear factor that binds to damaged DNA. *Science* **242,** 564–567.
6. Singh, H., Sen, R., Baltimore, D., and Sharp, P. (1986) A nuclear factor that binds to a conserved sequence motif in transcriptional control elements of immunoglobulin genes. *Nature* **319,** 154–158.

7. Hwang, B. J. and Chu, G. (1993) Purification and characterization of a protein that binds to damaged DNA. *Biochemistry* **32,** 1657–1666.
8. Hwang, B. J., Liao, J., and Chu, G. (1996) Isolation of a cDNA encoding a UV-damaged DNA binding factor defective in xeroderma pigmentosum group E cells. *Mutat. Res.* **362,** 105–117.
9. Chu, G. and Chang, E. (1990) Cisplatin-resistant cells express increased levels of a factor that recognizes damaged DNA. *Proc. Natl. Acad. Sci. USA* **87,** 3324–3327.
10. Keeney, S., Eker, A. P. M., Brody, T., Vermeulen, W., Bootsma, D. and Hoeijmakers, J. H. J. (1994) Correction of the DNA repair defect in xeroderma pigmentosum group E by injection of a DNA damage-binding protein. *Proc. Natl. Acad. Sci. USA* **91,** 4053–4056.
11. Tang, J., Hwang, B. J., Ford, J. M., Hanawalt, P. C., and Chu, G. (2000) Xeroderma pigmentosum p48 gene enhances global genomic repair and suppresses UV-induced mutagenesis. *Mol. Cell* **5,** 737–744.
12. Rathmell, W. K. and Chu, G. (1994) A DNA end-binding factor involved in double-strand break repair and V(D)J recombination. *Mol. Cell. Biol.* **14,** 4741–4748.
13. Rathmell, W. K. and Chu, G. (1994) Involvement of the Ku autoantigen in the cellular response to DNA double-strand breaks. *Proc. Natl. Acad. Sci. USA* **91,** 7623–7627.
14. Smider, V., Rathmell, W. K., Lieber, M., and Chu, G. (1994) Restoration of X-ray resistance and V(D)J recombination in mutant cells by Ku cDNA. *Science* **266,** 288–291.
15. Taccioli, G. E., Gottlieb, T. M., Blunt, T., et al. (1994) Ku80: product of the XRCC5 gene and its role in DNA repair and V(D)J recombination. *Science* **265,** 1442–1445.
16. Hammarsten, O. and Chu, G. (1998) DNA-dependent protein kinase: DNA binding and activation in the absence of Ku. *Proc. Natl. Acad. Sci. USA* **95,** 525–530.
17. Patterson, M. and Chu, G. (1989) Evidence that xeroderma pigmentosum cells from complementation group E are deficient in a homolog of yeast photolyase. *Mol. Cell. Biol.* **9,** 5105–5112.
18. Fox, M., Feldman, B., and Chu, G. (1994) A novel role for DNA photolyase: binding to drug-induced DNA damage is associated with enhanced cytotoxicity in yeast. *Mol. Cell. Biol.* **14,** 8071–8077.
19. Jiricny, J. (1994) Colon cancer and DNA repair: have mismatches met their match? *Trends Genetics* **10,** 164–168.
20. Donahue, B.A., Augot, M., Bellon, S. F., et al. (1990) Characterization of a DNA damage-recognition protein from mammalian cells that binds specifically to intrastrand d(GpG) and d(ApG) DNA adducts of the anticancer drug cisplatin. *Biochemistry* **29,** 5872–5880.
21. Toney, J., Donahue, B. A., Kellett, P. J., Bruhn, S. L., Essigmann, J. M., Lippard, S. J. (1989) Isolation of cDNAs encoding a human protein that binds selectively to DNA modified by the anticancer drug *cis*-diamminedichloroplatinum(II). *Proc. Natl. Acad. Sci. USA* **86,** 8328–8332.

22. Brown, S., Kellet, P., and Lippard, S. (1993) Ixr1, a yeast protein that binds to platinated DNA and confers sensitivity to cisplatin. *Science* **261,** 603–605.
23. Sambrook, J., Fritsch, E., and Maniatis, T. (1989) *Molecular Cloning: A Laboratory Manual.* Cold Spring Harbor Laboratory Press, Plainview, NY.
24. Bradford, M. M. (1976) A rapid and sensitive method for the quantitation of microgram quantities of protein utilizing the principle of protein-dye binding. *Anal. Biochem.* **72,** 248–254.
25. Payne, A. and Chu, G. (1994) Xeroderma pigmentosum group E binding factor recognizes a broad spectrum of DNA damage. *Mutat. Res.* **310,** 89–102.

24

Construction of MMR Plasmid Substrates and Analysis of MMR Error Correction and Excision

Huixian Wang and John B. Hays

Summary

We describe simple and efficient construction of mismatch repair (MMR) substrates, by generation of gapped plasmids using one sequence-specific nicking endonuclease (N.*Bst*NBI), ligation of synthetic oligomers into the gaps, and introduction of defined single nicks for initiation of MMR excision using a second such endonuclease (N.*Alw*I). We further describe measurement of completed mismatch correction and a sensitive quantitative assay for MMR excision intermediates. These methods can be easily adapted for construction of substrates containing defined DNA lesions, for analysis of MMR responses to DNA damage and for studies of other DNA repair pathways.

Key Words: DNA lesion; error correction; excision; mismatch repair; N.*Alw*I; N.*Bst*NBI.

1. Introduction

Indispensable for biochemical studies of mismatch repair (MMR) in vitro are highly purified DNA substrates containing defined DNA mismatches and, typically for eukaryotic MMR assays, specific nicks for initiation of excision. Previously, such substrates have been prepared from single-stranded and double-stranded phage DNA whose sequences were identical, except for a base substitution or insertion/deletion at a unique site. The technically demanding and time-consuming procedures included preparation of single-stranded circular phage DNA, preparation and linerization of double-stranded (RF) phage DNA, denaturation and reannealing of the two DNAs, and final purification of the mismatched substrate. The typically low initial yields of phage DNA and the multiple preparation steps worked to reduce yields of purified substrates. Since covalently closed (supercoiled) DNA was never isolated, adventitious

From: *Methods in Molecular Biology: DNA Repair Protocols: Mammalian Systems, Second Edition*
Edited by: D. S. Henderson © Humana Press Inc., Totowa, NJ

nicks introduced into the relatively large DNA molecules (6 kilobase pairs) during the multiple preparation steps remained present. Previously, MMR error correction endpoints have been assayed as restored restriction endonuclease recognition sites or reverted mutations in reporter genes. Such assays require that base mismatches be in special DNA-sequence contexts, and these assays cannot be used to analyze MMR responses to DNA lesions, except in special cases.

DNA is frequently damaged by physical and chemical agents in the environment, by certain drugs, by endogenous oxyradicals, and by spontaneous depurination, depyrimidination and deamination. The resultant loss, modification, or covalent adduction of DNA bases can have cytotoxic, mutagenic, or carcinogenic consequences that severely threaten human health. Currently, research in many laboratories aims to understand how cells process particular DNA lesions, in order to more precisely define the hazards posed by particular genotoxins, to better delineate individual variations in susceptibility and, where possible, to point toward remedies. Furthermore, better understanding of how some DNA lesions impact tumor or pretumorous cells more severely than normal cells may increase the efficacy of cancer therapy.

We describe here a simple procedure for preparation of MMR substrates, and a general but highly sensitive and quantitative MMR excision assay. These methods are applicable to DNA mismatches in any sequence context. Our techniques can be easily adapted to incorporate oligomers containing specific DNA lesions into substrates for analysis of MMR responses to DNA damage, and to studies of other DNA repair processes.

The protocol describes construction of (G/T)-mismatch-containing substrate s19CPDC(g/t) from plasmid pUC19CPDC, and homoduplex control substrate s19CPD(a/t) from plasmid pUC19CPD, by a new method: direct production of defined gaps in high-copy-number plasmids using one sequence-specific nicking enzyme, ligation of mismatch-creating synthetic oligomers into these gaps, and introduction of a defined nick using a second such enzyme. Besides assays for in vitro MMR error correction in the presence of exogenous deoxyribonucleoside triphosphates (dNTPs), this protocol further describes analysis of MMR excision in the absence of dNTPs by annealing of radiolabeled probes to gaps in DNA intermediates, generated here from substrates s19CPDC(g/t) and s19CPD(a/t).

2. Materials

1. *Ase*I digestion buffer: 50 mM Tris-HCl, pH 7.9, 100 mM NaCl, 10 mM MgCl$_2$, 1 mM dithiothreitol (DTT).
2. *Ahd*I digestion buffer: 20 mM Tris-acetate, pH 7.9, 50 mM potassium acetate, 10 mM magnesium acetate; 1 mM DTT.

3. CFS buffer: 2% caffeine, 50% formamide, 1.0 M NaCl, 10 mM Tris-HCl, pH 8.0, 1 mM EDTA.
4. Ligation buffer: 40 mM Tris-HCl, pH 7.8, 10 mM MgCl$_2$, 1 mM DTT, 0.5 mM ATP.
5. MMR buffer: 20 mM Tris-HCl, pH 7.6, 1.5 mM ATP, 1 mM glutathione, 0.1 mM in each of four dNTPs, 5 mM MgCl$_2$, 110 mM KCl, 500 µg/mL of bovine serum albumin (BSA).
6. Stop solution: 25 mM EDTA, 0.67% sodium dodecyl sulfate (SDS), 90 µg/mL of Proteinase K.
7. 100% and 70% ethanol.
8. 10 mg/mL of ethidium bromide solution.
9. Restriction endonucleases: *Ahd*I; *Ase*I; N.*Alw*I; N.*Bst*NBI.

3. Methods
3.1. Plasmids Used to Generate Substrates

Plasmid pUC19Y was derived from pUC19 by removing all GAGTC and GGATC sequences (recognition sites for endonucleases N.*Bst*NBI and N.*Alw*I, respectively). The first C:G base pair in the *Ear*I site at position 2488 of the original pUC19 sequence is base pair 1 in pUC19Y and its derivatives. Nucleotides (nt) in the sense-strands and the complementary antisense strands are correspondingly numbered 1, 2 ... and 1′, 2′ ..., respectively. (Sense and antisense strands are defined by the pUC19 *bla* gene). Plasmids pUC19CPD and pUC19CPDC (**Fig. 1**) were derived from pUC19Y, which contains two *GAGTC* sites, separated by 32 nt on the antisense strand, and unique endonuclease N.*Alw*I sites, on the sense strands *(1)*.

3.2. Generation and Purification of Gapped Plasmids
3.2.1. Large-Scale Preparation of Plasmid DNA

1. Transform *E. coli* strain SCS110 (*dam*⁻, *endA*⁻), obtained from Clontech, with plasmid pUC19CPDC and pUC19CPD, by the standard CaCl$_2$-mediated technique *(2)*. Select appropriate colonies for further propagation of plasmids (*see* **Note 1**).
2. Purify plasmid DNA from 3 L of overnight LB broth by alkaline lysis and at least two rounds of isopycnic sedimentation in cesium chloride plus ethidium bromide according to standard methods *(3)* (*see* **Note 2**).

3.2.2. Generation of Gapped Plasmids

1. Incubate 400 µg of DNA with 300 U of endonuclease N.*Bst*NBI at 55°C for about 2 h to completely nick the DNA, that is, to cleave at one or both of the two available N.*Bst*NBI sites. Add another 300 U of N.*Bst*NBI endonuclease to these completely nicked DNA mixtures and continue incubation at 55°C for an additional 4 h, to maximize double nicking (*see* **Note 3**). Monitor the progress of

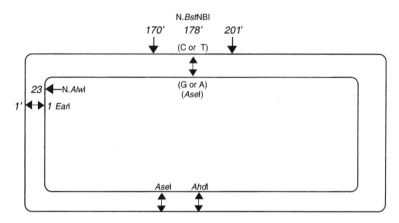

Fig. 1. Schematic representation of plasmids pUC19CPD and pUC19CPDC. Base pair number (1/1′) shown here corresponds to the first base pair in the *Ear*I endonuclease site at position 2488 in pUC19, the inside strand collinear with *bla* gene. Additional numbering is clockwise as shown. Approximate sites of restriction endonucleases cleaving both DNA strands (*Ase*I, *Ahd*I) and the exact cleavage sites for single strand nicking endonucleases N.*Alw*I and N.*Bst*NBI are indicated. Plasmids pUC19CPD and pUC19CPDC encode T/A and C/G base pair, respectively, at position 178′/178.

nicking by gel electrophoresis of an aliquot of digestion mixture at various times during the first 2 h of digestion.
2. Add to the digestion mixture a 50-fold molar excess of 32-nt oligomer (complementary to the sequence flanked by nicks), and incubate at 85°C for 5 min to denature plasmid DNA at the doubly nicked region, thus removing the 32-mer antisense-strand oligomer and creating a 32-nt gap. Anneal the displaced 32-mer to the added (excess) complementary 32-mer by slowly cooling samples to room temperature (*see* **Note 4**).
3. Remove the 32-base pairs (bp) oligoduplex and the excess single-stranded oligomers, by four passages through a Centricon-100 filter (Millpore). Reduce the sample to 10% of its original volume in each passage, reconstituting the retentate to the original volume using TE buffer between each passage.

3.2.3. Purification of Gapped Plasmid by BND-Cellulose Chromatography (see **Note 5**)

1. Prepare BND cellulose resin as a 50% suspension in TE buffer containing 0.3 *M* NaCl, according to the published method *(4)*.
2. Incubate the purified (concentrated) DNA from the Centricon-100 filtration (*see* **Subheading 3.2.2., step 3**) with 20 mL of BND-cellulose resin slurry for 10 min at room temperature, while mixing on a Barnstead/Thermolyne Labquake (model 400110) rotator.

3. Load the entire DNA–resin mixture into a 30-mL (2.5 × 7.5 cm) column, yielding a settled bed volume of 8–9 mL.
4. Wash the column with 100 mL of TE buffer containing 1.0 M NaCl at a flow rate of 1 mL/min, elute gapped plasmid DNA with 15 mL of CFS buffer, collect (1-mL) fractions and pool the plasmid-containing fractions. Assay for the presence of putative gapped (faster-migrating) plasmid DNA by agarose gel electrophoresis and ethidium bromide staining.
5. Dialyze against three changes of TE buffer at 4°C, and concentrate the DNA to 100 µg/mL, by Centricon-100 filtration, if necessary.

3.3. Construction of Substrates for In Vitro MMR Studies

1. Mix a 10-fold molar excess of oligomer 5'GCG GAT *ATT AAT* GTG ACG GTA GCG AGT CGC TC3' with 50 µg of gapped pUC19CPDC or gapped pUC19CPD plasmid DNA, incubate at 80°C for 5 min in ligation buffer, and slowly cool the mixtures down to room temperature to anneal the oligomer into the gapped regions of the plasmids (*see* **Note 6**). The boldface thymine (nt 178') creates a G/T mismatch (pUC19CPDC) or an A/T pair (pUC19CPD) depending on the gapped plasmid. Changing the oligomer sequences at different positions and inserting them into these or other gapped plasmids generates different (unique) mismatches.
2. Add to the annealing mixture fresh DTT and ATP to final concentrations of 1.0 mM, plus 100 U of T4 DNA ligase, and incubate overnight at 16°C to ligate the annealed oligomer into the gaps.
3. Analyze a small aliquot of the ligation mixture by 1% agarose gel electrophoresis in the presence of 1 µg/mL of ethidium bromide. The ligated and unligated products migrate as supercoiled and nicked DNA, respectively. Determine ligation efficiency by comparing the intensity of the supercoiled DNA band to the summed intensities of the supercoiled and nicked DNA bands. Ligation efficiency should at least be 50%. Note that nicked DNA binds more ethidium bromide than supercoiled DNA, so equal-amount DNA bands stain differently.
4. Separate ligated from unligated nicked plasmid DNA, by equilibrium cesium chloride ultracentrifugation in the presence of ethidium bromide at 336,000g (65,000 rpm) for 6 h in a Beckman VTi-80 rotor, recover the DNA, remove the ethidium bromide by several butanol extractions, and eliminate the CsCl by dialysis against three changes of TE buffer.
5. Use nicking endonuclease (N.*Alw*I) to cleave one strand of dsDNA four nt 3' to the recognition sequence (GGATCNNNN↓; unmethylated adenines only) in the mismatched or homoduplex plasmid products of **step 4**. The single nicks in substrates sCPD(a/t)n, sCPDC(g/t)n are at nt position 23 (*see* **Note 7**).

3.4. MMR Error Correction in Cell-Free Extracts

1. Prepare nuclear extracts from Hela cells as previously described *(5)*.
2. Assemble on ice standard 15-µL MMR reaction mixtures containing 75 fmol (100 ng) sCPDC(g/t)n substrate and 100 µg of nuclear extract in MMR buffer.

Add nuclear extracts as the last component and immediately transfer to 37°C to initiate the reactions.
3. Incubate the reaction mixtures at 37°C for 12 min and terminate the reactions by adding 30 μL of stop solution. Incubate the reactions at 37°C for an additional 15 min.
4. Purify the DNA by two extractions with equal volumes of phenol, and precipitate the DNA with ethanol.
5. Resuspend the precipitated DNA in 15 μL of *Ase*I digestion buffer and incubate with 4 U of *Ase*I endonuclease and 1 μg of RNase A for 2 h at 37°C. Correction of (G/T) mismatches to (A/T) restores a second site for *Ase*I endonuclease, which thus cuts the products into fragments of 0.8 and 1.2 kb.
6. Separate the digested products by 1% agarose gel electrophoresis in TAE buffer and visualize DNA bands by ethidium bromide staining. Capture the gel image using an UV-7500 or similar imaging system and quantitatively analyze the bands using Imagequant software. Repair yield equals the ratio of the summed intensities of the 0.8- and 1.2-kb fragments to the total of this sum plus the intensity of the 2.0-kb band corresponding to singly cut (uncorrected) DNA.

3.5. Analysis of MMR Excision Intermediates

1. Assemble standard MMR reaction mixtures containing nicked DNA substrates, mismatched or homoduplex (control), as described in **Subheading 3.4., step 2.**, but without exogenous dNTPs. Almost no DNA resynthesis occurs in the absence of exogenous dNTPs, so excision gaps generated by MMR excision remain unfilled (*see* **Note 8**).
2. Incubate the reaction mixtures at 37°C for 7 min. Extract and precipitate the DNA as described **above**. For kinetic analyses, remove individual samples at desired times.
3. Linearize the substrate DNA by digesting with 4 U of *Ahd*I endonuclease plus 1 μg of RNase A in 15 μL of *Ahd*I digestion buffer, for 2 h at 37°C.
4. Add 0.5 pmol of a particular ^{32}P-labeled oligomer probe to the digestion mixture and incubate the mixture at 85°C for 5 min; then slowly cool down to room temperature to anneal the radiolabeled probe to complementary single strand DNA in the MMR excision gaps. Label the oligomer probe (typically 30 nt) with [γ^{32}P]ATP in the presence of T4 polynucleotide kinase according to the standard method. To compare excision at different locations in the substrate DNA, select oligomers corresponding to DNA sequences that will have similar melting temperature and base compositions and add unlabeled oligomer as necessary to adjust each probe solution to the same specific radioactivity.
5. Separate annealed products from free oligomers by 1.0% agarose gel electrophoresis in TAE buffer (40 m*M* Tris-acetate, 1 m*M* EDTA, pH 8.0). Measure intensities of the DNA bands in agarose gels by ethidium staining. Dry the gels and measure radioactivity in each lane by PhosphorImaging. Use DNA band intensities (*see* **Subheading 3.4., step 5**) to normalize radioactivity measurements for any variations in DNA recovery and loading.

6. Subtract from normalized excision signals the background excision signals generated using the same probes with control homoduplex DNA substrate.

4. Notes

1. Some randomly picked colonies from transformation will have a significant fraction of plasmid DNA in dimers or higher multimers. Screen DNA "minipreps" from half a dozen colonies by agarose gel electrophoresis and choose colonies whose "minipreps" show little multimerized (slow-moving) plasmid DNA.
2. The Qiagen kit for plasmid purification does not work well for DNA preparation from *dam*⁻ bacterial strain, in our hands. One round of cesium chloride isopycnic sedimentation sometimes is not sufficient to produce high-purity plasmid preparation.
3. The critical first step in preparation of gapped DNA is dual nicking of DNA by endonuclease N.*Bst*NBI (or another site-specific nicking enzyme). Since there is no simple way to assay the progress of dual nicking, some practical way to estimate the endonuclease concentration and incubation time needed to approach complete dual incision is required. We use enough N.*Bst*NBI endonuclease to completely transform form I into form II DNA (incised at least once) in 2 h at 55°C, then add an equal amount of fresh N.*Bst*NBI endonuclease and continue the incubation for 4 h. This yields at least 80% doubly nicked DNA: after partial denaturation and reannealing in the presence of 50-fold excess oligomer (complementary to the oligomer produced by dual incision), less than 20% of the endonuclease-N.*Bst*NBI-treated starting material can still be ligated to covalently closed circles (under conditions where a different singly-nicked plasmid DNA is ligated with at least 95% efficiency).
4. One major limitation of this method for incorporation of oligomers into specific gaps is the requirement for a GAGTCNNNN sequence at the 3′ end of the oligomers, to specify the 3′ N.*Bst*NBI-endonuclease nick. A second limitation is the need for a gap large enough to make purification of gapped plasmid with BND-cellulose efficient. If a lesion-containing oligomer must be shorter, to facilitate HPLC purification, for example, or is not compatible with a GAGTCNNNN sequence, the lesion-containing oligomer can be ligated to an appropriate second oligomer "arm" on its 3′ side, using a "bottom-strand" scaffold complementary to both oligomers, to produce oligomer fitting a gap large enough so that the recipient can be efficiently prepared. The oligomer can be radiolabeled, or can incorporate fluorescent–dye-tagged or biotin-tagged nucleotides to facilitate various analyses.

   The upper limit on gap lengths will be a function of the melting temperature (T_m) of the corresponding gap oligomer. As long as the T_m is below 90°C, the nick-flanked gap region, but not the remainder of the plasmid will rapidly denature during the heating and reannealing process. (Base-stacking at the nicks can be taken into account by calculating the T_m for an oligomer containing two more nucleotides. Even when the thermodynamic T_m approaches the infinite-chain value, denaturation kinetics may differ enough for selective removal of the oligomer digestion product during a brief heating period.) We have recently prepared plasmids containing 66-nt gaps, using a plasmid with four N.*Bst*NBI sites in a 66-bp region (H. Wang and J. B. Hays, unpublished).

5. BND-cellulose column purification is crucial for separation of gapped molecules from nicked-DNA contaminants. The yield of gapped molecules from this step is about 60% of the starting material, that is, about 50% of the total doubly nicked DNA. The longer the gap, the higher the recovery and purity of gapped molecules eluted from BND-cellulose, presumably because of stronger initial binding to the resin. Our best yields have been 25% with a 22-nt and about 50% with a 30- or 32-nt gap. It is convenient to simply increase the total input DNA if more gapped plasmid DNA is needed.

   DNA treated exhaustively with N.*Bst*NBI endonuclease, but not subjected to denaturation and reannealing in the presence of competing antisense 30-mer, is not retained by BND-cellulose: less than 0.5% of the input DNA remains to be eluted with the CFS-buffer after prior washing with 1 M NaCl. If the 20% of the original total DNA that appeared not to have been doubly nicked (i.e., that could still be ligated after the heating-annealing step [*see* **Note 3**]) was washed through with similar (>99.5%) efficiencies, then the singly nicked material subsequently eluted would be <0.005 × (20%), that is, <1/100, of the yield of gapped DNA. For most biochemical studies of repair of DNA-lesion-containing substrates, contamination with lesion-free plasmid at levels of <1% should be acceptable. If trace amounts of lesion-free contaminants are problematic, for example, when the end points are lesion-bypass efficiencies or lesion-targeted mutation spectra, backgrounds could be reduced further by more exhaustive initial N.*Bst*NBI digestion to increase the percentage of doubly nicked plasmid. If a recognition site for a restriction endonuclease unable to cleave single-stranded DNA were incorporated into the sequence where the gap is generated, the dsDNA contaminants in gapped plasmid preparation could be removed by linearization with that restriction enzyme and digestion with exonuclease V (RecBCD).

6. Purification of synthesized oligomers used for incorporation is highly recommended to increase ligation efficiency and thus final yield.

7. The commercial availabilities of several other nicking endonucleases make it possible to use one of these instead of N.*Alw*I, once a particular recognition sequence is uniquely present at the appropriate position. Thus, preparation of DNA from *dam*⁻ bacteria (lacking the d[GATC]-adenine methylation that blocks N.*Alw*I digestion) would not be necessary, and two rounds of cesium chloride isopycnic sedimentation in the initial purification of plasmid DNA may not be needed.

8. Where desired, one can demonstrate that DNA intermediates generated in the absence of exogenous dNTPs are genuine intermediate in error correction, thus indirectly identifying them as gapped DNA. Incubate or mock-incubate the purified intermediate DNA with 2 U of DNA pol I (Klenow) plus all four dNTPs (17 μM each) in 15 µL of *Ase*I digestion buffer, for 20 min at room temperature. Terminate the gap-filling reaction by heating at 85°C for 20 min and assay error cor-

rection: add 4 U of *Ase*I endonuclease and 1 µg of RNase A to the mixture after cooling and continue the incubation at 37°C for 2 h. Separate the digestion products by 1% agarose gel electrophoresis and measure band intensities as described. We find yields of corrected product obtained in this way to be nearly identical to product yields from identical reactions in complete MMR buffer; mock-incubated intermediate DNA molecules are resistant to *Ase*I digestion at presumed gap positions.

References

1. Wang, H. and Hays, J. B. (2002) Mismatch repair in human nuclear extracts: Quantitative analysis of excision of nicked mismatched DNA substrates, constructed by a new technique employing synthetic oligonucleotides. *J. Biol. Chem.* **277,** 26,136–26,142.
2. Seidman, C. E., Sheen, J., and Jessen, T. (1998) Introduction of plasmid DNA into cells, in: *Current Protocols in Molecular Biology.* Vol. 1, pp. 1.8.1–1.8.10.
3. Heilig, J. S., Elbing, K. L., and Brent, R. (1998) Large-scale preparation of plasmid DNA, in: *Current Protocols in Molecular Biology.* Vol. 1, pp. 1.7.1–1.7.16.
4. Gillam, I., Millward, S., Blew, D., von Tigerstrom, M., Wimmer, E., and Tenner, G. M. (1967) The separation of soluble ribonucleic acids on benzoylated diethylaminoethyl cellulose. *Biochemistry* **6,** 3043–3056.
5. Challberg, M. D. and Kelly, T. J. (1979) Adenovirus DNA replication in vitro: origin and direction of Daughter strand synthesis. *J Mol Biol.* **135,** 999–1012.

25

Characterization of Enzymes That Initiate Base Excision Repair at Abasic Sites

Walter A. Deutsch and Vijay Hegde

Summary

Abasic sites in DNA arise under a variety of circumstances, including destabilization of bases through oxidative stress, as an intermediate in base excision repair, and through spontaneous loss. Their persistence can yield a blockade to RNA transcription and DNA synthesis and can be a source of mutations. Organisms have developed an enzymatic means of repairing abasic sites in DNA that generally involves a DNA repair pathway that is initiated by a repair protein creating a phosphodiester break ("nick") adjacent to the site of base loss. Here we describe a method for analyzing the manner in which repair endonucleases differ in the way they create nicks in DNA and how to distinguish between them using cellular crude extracts.

Key Words: AP endonucleases; AP lyases; base excision repair (BER); DNA abasic sites; DNA damage; DNA oligonucleotides; oxidative stress; tetrahydrofuran.

1. Introduction

Apurinic/apyrimidinic (AP) abasic sites arise in DNA under a variety of circumstances that, taken together, make them one of the more common lesions found in DNA. For example, free radicals can interact with DNA bases leading to their destabilization and ultimate loss. Free radical attack on DNA can also lead to DNA base modifications, some of which are known to be removed by *N*-glycosylases, resulting in the formation of an AP site as part of the base excision repair (BER) pathway *(1)*. Even in the absence of environmental factors, DNA bases are known to be spontaneously lost *(2)*, leaving behind abasic lesions that, if left unrepaired, can be mutagenic and can also form a blockade to RNA transcription *(3)*.

To illustrate the importance of abasic sites, all organisms that have thus far been tested have the ability to repair these sites by creating an incision adjacent

From: *Methods in Molecular Biology: DNA Repair Protocols: Mammalian Systems, Second Edition*
Edited by: D. S. Henderson © Humana Press Inc., Totowa, NJ

to the AP site to initiate the repair process. The major class of AP endonuclease, at least quantitatively, is one that hydrolytically cleaves 5′ and adjacent to an abasic site, producing nucleotide 3′-hydroxyl and 5′-deoxyribose-5-phosphate termini *(4)*. Examples of this activity are the exonuclease lll of *E. coli* and the human multifunctional APE/ref-1 *(5,6)*.

Another kind of activity that acts on abasic sites is part of the BER pathway that is initiated by *N*-glycosylases directed toward a modified or nonconventional base in DNA. Often, these *N*-glycosylases also possess AP lyase activity that cleaves DNA 3′ to an abasic site via a β-elimination reaction to leave a 3′ 4-hydroxy-2-pentenal-5-phosphate. An example of this type of activity is that possessed by *E. coli* endonuclease lll (endo III), a broad-specificity *N*-glycosylase/AP lyase used for the repair of oxidative damage to DNA *(7,8)*. In some cases, *N*-glycosylases/AP lyases not only cleave DNA via a β-elimination reaction but also are capable of carrying out a second ∂-elimination incision. This results in the removal of the AP site and the formation of a one-nucleotide gap bordered by 3′- and 5′-phosphate termini. The β, ∂-elimination reaction can be concerted, as appears to be the case for the *E. coli* formamidopyrimidine glycosylase (Fpg), which repairs oxidative DNA damage, primarily in the form of 8-oxoguanine (8oxoG) *(9,10)*. On the other hand, repair of 8oxoG by the *Drosophila* S3 *N*-glycosylase/AP lyase activity has been concluded to occur in two distinct steps, catalyzing a ∂-elimination reaction on a second encounter with the lesion after first dissociating from the AP substrate when the β-elimination reaction was completed *(11)*.

There are several ways to monitor enzyme activity on abasic sites present in naturally occurring or synthesized DNA substrates. Originally, we utilized a [^{3}H]-labeled supercoiled phage DNA in a filter-binding assay that accurately measured DNA nuclease activity *(12)*, but this technique was hampered by the tedious and time-consuming preparation of the substrate DNA. Moreover, this assay could only quantify the cleavage of a preprepared abasic DNA substrate, not identify the type of cleavage event unless it was followed up by DNA synthesis to determine whether the incision created a productive 3′-OH terminus *(12,13)*. Recently, we have turned to an assay that utilizes a 5′-end-labeled DNA duplex oligonucleotide containing a single abasic site. After reaction with an AP endonuclease or AP lyase, the products of the reaction are separated on a polyacrylamide gel. Based on the migration of the cleaved product, one can easily visualize the type of strand cleavage possessed by a DNA repair endonuclease by autoradiography *(11,14,15)*. Quantitation of the product(s) formed can be performed either by video densitometric analysis of autoradiograms, or by Phosphorimager analysis and scanning of dried gels. Importantly in this assay, different types of cleavage events generate unique visual images of the products formed by autoradiography; the assay is also adaptable in most

AP Endonuclease Assay

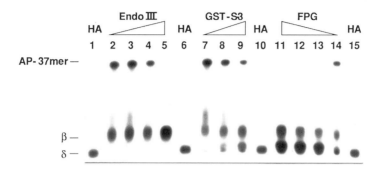

Fig. 1. Mechanisms of nuclease action on abasic site-containing DNA. Reactions contained 1 pmol of AP 37-mer and were incubated for 30 min at 37°C with *E. coli* endonuclease III (Endo III; lanes 2–5) at protein amounts of 100, 150, 200, and 400 pg, respectively, glutathione-*S*-transferase (GST)-conjugated *Drosophila* S3 (lanes 7–9) at 20, 40, and 80 pg, respectively, or *E. coli* formamidopyrimidine glycosylase (Fpg; lanes 11–14) at 160, 120, 80, and 80 pg, respectively; lanes 1, 6, 10, and 15 contain the products of hot alkali (HA; piperidine) treatment of the AP 37-mer. The reaction products were separated on a 16% polyacrylamide DNA sequencing gel. The electrophoretic mobilities of the uncleaved apurinic/apyrimidine (AP) 37-mer and DNA cleavage products corresponding to β- and ∂-elimination reactions are indicated. (Adapted with permission from **ref. *11*.**)

cases to the use of both highly purified enzymes as well as cruder preparations.

For the assay described here, a synthetic oligonucleotide is utilized that is 37 bp in length (37-mer). Within the 37-mer is a single uracil (U) residue placed at position 21 during the synthesis of the oligonucleotide. After 5′-end-labeling and gel purification of the single-stranded U-containing oligonucleotide, the complementary strand is annealed to create a duplex 37-mer (*see* **Note 1**). This forms a substrate for uracil-DNA glycosylase *(16)*, which liberates the nonconventional base and forms an abasic site in its place.

Alternatively, tetrahydrofuran can be placed at position 21 within the 37-mer. Tetrahydrofuran is an analog of an abasic site and represents a productive substrate for a hydrolytic AP endonuclease *(17)*. It is, however, refractive to cleavage by AP lyases, therefore making it a convenient substrate for measuring hydrolytic AP endonuclease activity in crude preparations that would ordinarily be compromised by AP lyase activity.

To demonstrate the utility of the oligonucleotide assay, the products of three different AP lyases acting on an abasic oligonucleotide DNA substrate are presented in **Fig. 1**. In each case, the migration pattern of the reaction products provides direct information on the type of DNA termini produced by each enzyme. The hot alkali control (HA) provides a landmark for the production of a β, ∂-elimination reaction that reflects the production of a 5'- and 3'-phosphoryl terminus. DNA fragments containing a terminal phosphoryl group migrate faster than those of the same length that lack a terminal phosphate. As can be seen in **Fig. 1**, the reaction products are completely distinct for each of the enzymes tested. Endonuclease lll produces a β-elimination product, regardless of the amount of protein added to the assay, as shown in lane 5, where the substrate is totally consumed yet yields only a single product. For *Drosophila* S3, a β-elimination product is also evident at low protein concentrations, yet higher amounts of protein yield what is clearly a ∂-elimination product as well. This suggests that S3 is dissociating from the abasic substrate once the original β-elimination reaction is completed and on a second encounter is then cleaving the remaining AP site via a ∂-elimination reaction. This is in contrast to what is observed for *E. coli* Fpg, which produces equal amounts of β- and ∂-elimination products, regardless of protein concentration (**Fig. 1**) or time of incubation (not shown). This indicates that Fpg is remaining bound to the AP substrate as it carries out both incision activities.

Another utility of the oligonucleotide assay is that it is amendable to analyzing 5'-acting AP endonucleases in crude extracts without concern over the contribution of contaminating AP lyases that could make interpretation of the actual products formed difficult. This is accomplished by switching from an authentic AP site created in the oligonucleotide substrate by U incorporation to a tetrahydrofuran analog of an AP site that is refractive to cleavage to AP lyases. As seen in **Fig. 2**, the 5'-acting hydrolytic human APE/ref-1 is clearly capable of acting on a 37-mer synthetic DNA substrate with a tetrahydrofuran spacer (lanes 9 and 10), yet the same substrate is totally refractive to cleavage by the AP lyase activity possessed by *Drosophila* S3 (lanes 6 and 7). The same preparation of *Drosophila* S3 is, however, active on a DNA substrate containing a single 8oxoG residue (lanes 2 and 3).

2. Materials

All solutions should be made using molecular biology-grade reagents and sterile distilled water.

2.1. 5'-End-Labeling and Purification of Oligonucleotides Containing a Single Tetrahydrofuran or U Residue

The oligonucleotides used in our studies are commercially prepared to our specifications and contain the nonconventional bases U or 8oxoG (Operon Technologies, Alameda, CA) or the abasic spacer tetrahydrofuran (Genosys, Pittsburgh, PA). The single-stranded oligos are deprotected and purified by spin-column chromatography (Gibco-BRL, Grand Island, NY). The individual

AP Endonuclease Assay

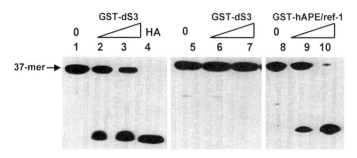

Fig. 2. Activity of GST-dS3 and GST-hAPE/ref-1 on an 8oxoG site and a tetrahydrofuran spacer-containing DNA substrate. Lane 1, 8oxoG 37-mer alone. Lanes 2 and 3, 1 pmol of 8oxoG 37-mer was incubated with 0.2 pmol (lane 2) and 0.4 pmol (lane 3) of purified GST-dS3 at 37°C for 30 min, and the products were separated on a 16% polyacrylamide DNA sequencing gel and analyzed by autoradiography. Lane 4, hot alkali (HA) treatment to generate a β-, ∂-elimination product. Lane 5, 1 pmol tetrahydrofuran-containing 37-mer alone and incubated with 0.2 pmol and 0.4 pmol GST-dS3 (lanes 6 and 7, respectively). Lanes 8–10 are 1 pmol tetrahydrofuran-containing 37-mer alone and incubated with 0.4 pmol and 0.8 pmol purified GST-hAPE/ref-1, respectively.

single-stranded and purified oligonucleotides are then resuspended in distilled water to 10 pmol/μL.

1. 10 U/μL of T4 polynucleotide kinase (Stratagene, La Jolla, CA).
2. 10X T4 polynucleotide kinase buffer: 700 mM Tris-HCl, pH 7.6, 100 mM MgCl$_2$, 50 mM dithiothereitol (DTT), 1 mM spermidine-HCl.
3. [γ-^{32}P]ATP, 10 mCi/mL, 6000 Ci/mmol (Amersham, Arlington Heights, IL).
4. 10X Annealing buffer: 100 mM Tris-HCl, pH 7.6, 100 mM MgCl$_2$, 10 mM EDTA.
5. Loading buffer: 50% glycerol, 0.5% bromophenol blue, 0.5% xylene cynanol.
6. Phenol, molecular biology grade, neutralized, and equilibrated with 10 mM Tris-HCl, pH 8.0, 1 mM EDTA.
7. Phenol–chloroform–isoamyl alcohol mixture (25:24:1 by volume).
8. 40% Acrylamide stock: 38:2 acrylamide/*bis*-acrylamide in 100 mL of distilled water.
9. 10X TBE: 890 mM Tris-borate, 20 mM EDTA, pH 8.0.
10. Nondenaturing 20% polyacrylamide gel (per 100 mL): 50 mL of 40% acrylamide stock, 10 mL of 10X TBE, 500 μL of 10% ammonium persulfate, 60 μL of N,N,N',N'-tetramethylethylenediamine (TEMED), distilled H$_2$O to 100 mL final volume.
11. Centrex MF-0.4 microcentrifuge tubes (Schleicher & Schuell, Keene, NH).
12. 35 mM N-2-hydroxyethylpiperazine-N'-2-ethanesulfonic acid (HEPES)–KOH, pH 7.4.

2.2. Cleavage of Uracil-Containing Oligonucleotides

1. Uracil-DNA glycosylase (Epicentre, Madison, WI).
2. 10X Uracil-DNA glycosylase buffer: 200 mM Tris-HCl, pH 8.0, 10 mM EDTA, 10 mM DTT, 100 µg/mL bovine serum albumin (BSA).
3. 10 mM HEPES–KOH, pH 7.4.

2.3. Enzymatic Reactions and Electrophoresis

1. Abasic oligonucleotides (see **Subheadings 3.1.–3.3.**).
2. Purified AP endonuclease or AP lyase (Trevigen, Gaithersburg, MD) *(11)*.
3. 10X *Drosophila* S3 (dS3) buffer: 300 mM HEPES, pH 7.4, 500 mM KCl, 10 mg/mL BSA, 0.5% Triton X-100, 10 mM DTT, 5 mM EDTA.
4. 10X *E. coli* Endo lll buffer: 150 mM KH$_2$PO$_4$, pH 6.8, 100 mM EDTA, 100 mM β-mercaptoethanol, 400 mM KCl.
5. 10X *E. coli* Fpg buffer: 150 mM HEPES, pH 7.5, 500 mM KCl, 10 mM β-mercaptoethanol, 5 mM EDTA.
6. 10X Human APE/ref-1 buffer: 500 mM HEPES, pH 7.5, 500 mM KCl, 10 µg/mL of BSA, 100 mM MgCl$_2$, 0.5% Triton X-100.
7. 1 M Piperidine.
8. Denaturing polyacrylamide gel: 16% polyacrylamide solution, 7 mM urea, 1X TBE.
9. Formamide loading buffer: 96% formamide, 0.05% xylene cyanol, 0.05% bromophenol blue, 10 mM EDTA.
10. 15% methanol: 10% acetic acid solution.
11. Whatman 3MM paper.
12. X-ray film (Kodak XAR-5) or Phosphorimager.

3. Methods

Characterization of endonucleases that act at abasic sites existing in DNA can be divided into several parts: first, a 5'-radiolabeled synthetic oligonucleotide containing either a single tetrahydrofuran residue, or one containing a single deoxyU residue, is prepared. Single-stranded oligos are then annealed to their nonradioactive complementary oligonucleotide (*see* **Note 1**). The duplexes are then purified and subsequently further processed by a uracil-DNA glycosylase so as to form an abasic site if necessary. This, or the tetrahydrofuran-containing oligonucleotide, is then employed as a substrate for enzyme reactions using proteins known, or suspected, to act on abasic sites. Upon completion of the enzymatic assays, the reaction products are then separated on a DNA sequencing gel.

3.1. 5' End-Labeling and Purification of Oligonucleotides

Bacteriophage T4 polynucleotide kinase is used to catalyze the transfer of the γ-phosphate of ATP to the 5'-hydroxyl terminus of the oligonucleotide.

The following procedure produces sufficient quantities of 5′-end-labeled duplex oligonucleotides for several enzymatic reactions (*see* **Note 2**).

1. Prepare 5′-end labeling reaction mixtures in a 0.5-mL microcentrifuge tube containing the following: 3 µL of [^{32}P]ATP, 4 µL 10X kinase buffer, 2 µL of oligonucleotide, 2 µL (2 U) of polynucleotide kinase, and 2 µL distilled water.
2. Incubate the reaction mixture for 30 min at 37°C.
3. Extract the reaction mixture once with phenol/chloroform/isoamyl alcohol.
4. Mix for 1 min, and then centrifuge at 12,000g for 3 min at room temperature in a microcentrifuge. Transfer the aqueous supernatant to a new tube. Add 2.5 volumes of ethanol, mix, and store the tube at –20°C for 1 h.
5. Recover the oligos by centrifugation at 12,000g for 15 min at 4°C in a microcentrifuge. Remove the supernatant, and leave the tube open at room temperature until all the ethanol has evaporated.
6. Dissolve the pellet in 20 µL of distilled water.

3.2. Annealing Reaction (see **Note 1**)

1. Mix together in a microcentrifuge tube the following: 20 µL of labeled oligos (1 pmol/mL), 4 µL complementary strand, 4 µL of 10X annealing buffer, and 4 µL of distilled water to 40 µL final volume.
2. Incubate the annealing mixture at 75°C for 10 min.
3. Slowly cool to room temperature.
4. Add 10 µL of loading buffer and mix well.
5. Separate the labeled duplex oligonucleotides on a 20% nondenaturing polyacrylamide gel and then subject to autoradiography.
6. Excise the band corresponding to the labeled duplex oligos from the gel, and transfer to a Centrex MF-0.4 microcentrifuge tube.
7. Crush the acrylamide gel into small pieces against the wall of the tube, add 200 µL of 35 mM HEPES–KOH, pH 7.4, to the tube, and incubate for 5 h to overnight at 4°C to elute the labeled oligos from the gel. (Typically, ≥95% of labeled oligo is eluted.)
8. Collect the duplex oligos by centrifugation at 4000g for 3 min in a microcentrifuge.

3.3. Uracil Excision

1. Uracil-DNA glycosylases are used to hydrolyze the *N*-glycosidic bond between the deoxyribose sugar and uracil base. The reaction mixture is as follows: 10 pmol of labeled uracil-containing duplex oligo, 4 µL 10X uracil-DNA glycosylase buffer, 1 µL of uracil-DNA glycosylase (1 U/µL), and distilled water to 40 µL final volume.
2. Incubate the reaction for 20 min at 37°C. Extract the reaction mixture once with phenol/chloroform/isoamyl alcohol and precipitate the DNA as described in **Subheadings 3.1., steps 3–5**.
3. Dissolve the purified labeled duplex oligonucleotide in 10 mM HEPES–KOH, pH 7.4. It can be stored at 4°C for up to 1 wk.

3.4. Enzymatic Reaction

1. Mix together in a microcentrifuge tube: approx 1 pmol of γ-^{32}P abasic oligonucleotide (typically 10,000 cpm), 1 µL of 10 X reaction buffer, X µL of enzyme, and distilled water to 10 µL final volume. Incubate at 37°C for the desired time.
2. Stop the reactions by heating at 75°C for 10 min. Add 2 µL of formamide loading buffer, heat for 4 min at 90°C, cool on ice, and then load immediately on a denaturing polyacrylamide gel.

3.5. Hot Alkali Treatment

1. Add 90 µL of 1 M piperidine to 10 µL of 5′-end-labeled abasic oligonucleotide, and incubate for 30 min at 90°C.
2. Lyophilize to dryness using a Speed Vac, and redissolve the pellet in 20 µL of distilled water.
3. Repeat lyophilization step twice more in order to remove all the piperidine.
4. Dissolve the remaining pellet in 50 µL of formamide loading buffer, heat for 4 min at 90°C, cool on ice, and then load immediately on a denaturing polyacrylamide gel.

3.6. Analysis of Endonuclease Activity by Denaturing Gels

1. Load an equal amount of radioactivity (about 5000 cpm) per lane on a pre-electrophoresed 16% denaturing polyacrylamide gel (*see* **Note 3**).
2. Electrophorese in 1X TBE buffer at 45-W constant power until the bromophenol dye front is near the bottom of the gel.
3. Remove the gel plates, pry apart, and transfer the gel to a bath containing 15% methanol and 10% acetic acid for 20 min.
4. With the gel still attached to the glass plate, place a similar sized piece of Whatman 3MM paper on top of the gel, and then carefully peel off the 3MM paper with the gel attached to it.
5. Cover the gel with plastic wrap (Saran Wrap), and dry under vacuum at 80°C for 45 min.
6. Expose the dried gel to X-ray film at –70°C for 12–16 h with an intensifying screen. Alternatively, PhosphorImager cassettes can be used for the same length of time, but at room temperature.

4. Notes

1. One advantage of the oligonucleotide assay is that the sequence can be manipulated so as to determine whether enzyme activity is affected by surrounding DNA bases either adjacent to, or opposite, the target site.
2. Caution should be taken in the preparation of ^{32}P-labeled oligonucleotides when planning assays over a sustained period. We have found that regardless of the method of storage of prepared oligos, degradation products begin to appear in our controls that are presumably owing to radioactive decay that splinters the oligos into smaller fragments. Generally, after 2 wk unused oligos are of little use because of such fragmentation.

3. We have used gels containing 20% polyacrylamide, but to maximize the separation of a β- and ∂-elimination product, 16% gels are preferred.

References

1. Friedberg, E. C., Walker, G. C., and Siede, W. (1995) *DNA Repair and Mutagenesis.* American Society of Microbiology Press, Washington, D.C.
2. Lindahl, T. and Nyberg, B. (1972) Rate of depurination of native deoxyribonucleic acid. *Biochemistry* **11,** 3610–3618.
3. Loeb, L. A. and Preston, B. D. (1986) Mutagenesis by apurinic/apyrimidinic sites. *Annu. Rev. Genet.* **20,** 201–230.
4. Doetsch, P. W. and Cunningham, R. P. (1990) The enzymology of apurinic/apyrimidinic endonucleases. *Mutat. Res.* **236,** 173–201.
5. Demple, B. and Harrison, L. (1994) Repair of oxidative damage to DNA: enzymology and biology. *Annu. Rev. Biochem.* **63,** 915–948.
6. Demple, B., Herman, T., and Chen, D. S. (1991) Cloning and expression of APE, the cDNA encoding the major human apurinic endonuclease: definition of a family of DNA repair enzymes. *Proc. Natl. Acad. Sci. USA* **88,** 11,450–11,454.
7. Dizdaroglu, M., Laval, J., and Boiteux, S. (1993) Substrate specificity of the *Escherichia coli* endonuclease III: excision of thymine- and cytosine-derived lesions in DNA produced by radiation-generated free radicals. *Biochemistry* **32,** 12,105–12,111.
8. Kow, Y. W. and Wallace, S. S. (1987) Mechanism of action of *Escherichia coli* endonuclease III. *Biochemistry* **26,** 8200–8206.
9. Dodson, M. L., Michaels, M., and Lloyd, R. S. (1994) Unified catalytic mechanism for DNA glycosylases. *J. Biol. Chem.* **269,** 32,709–32,712.
10. Bailly, V., Verly, W. G., O'Conner, T., and Laval, J. (1989) Mechanism of DNA strand nicking at apurinic/apyrimidinic sites by *Escherichia coli* [formamidopyrimidine] DNA glycosylase. *Biochem. J.* **262,** 581–589.
11. Yacoub, A., Augeri, L., Kelley, M. R., Doetsch, P. W., and Deutsch, W. A. (1996) A *Drosophila* ribosomal protein contains 8-oxoguanine and abasic site DNA repair activities. *EMBO J.* **15,** 2306–2312.
12. Spiering, A. L. and Deutsch, W. A. (1986) *Drosophila* apurinic/apyrimidinic DNA endonucleases. Characterization of mechanism of action and demonstration of a novel type of enzyme activity. *J. Biol. Chem.* **261,** 3222–3228.
13. Warner, H. R., Demple, B. F., Deutsch, W. A., Kane, C. M., and Linn, S. (1980) Apurinic/apyrimidinic endonucleases in repair of pyrimidine dimers and other lesions in DNA. *Proc. Natl. Acad. Sci. USA* **77,** 4602–4206.
14. Yacoub, A., Kelley, M. R., and Deutsch, W. A. (1996) *Drosophila* ribosomal protein PO contains apurinic/apyrimidinic endonuclease activity. *Nucleic Acids Res.* **24,** 4298–4303.
15. Deutsch, W. A. and Yacoub, A. (1999) Characterization of DNA strand cleavage by enzymes that act at abasic sites in DNA. *Methods Mol. Biol.* **113,** 281–288.
16. Lindahl, T. (1980) Uracil-DNA glycosylase from *Escherichia coli. Methods Enzymol.* **65,** 284–295.

17. Wilson, D. M. 3rd, Takeshita, M., Grollman, A. P., and Demple, B. (1995) Incision activity of human apurinic endonuclease (Ape) at abasic site analogs in DNA. *J. Biol. Chem.* **270,** 16,002–16,007.

26

Base Excision Repair in Mammalian Cells

Yoshihiro Matsumoto

Summary

A rapid, convenient and safe in vitro assay system for base excision repair is described. Whole cell extracts are prepared by detergent-based cell lysis and provide a vigorous activity of AP site repair. A circular DNA substrate is used for detection of both DNA polymerase β-dependent and proliferating cell nuclear antigen (PCNA)-dependent pathways. Repaired and unrepaired DNA substrates are separated by agarose gel electrophoresis as a linear DNA molecule and a nicked circular molecule, respectively, and detected by staining with SYBR Green I. This assay system does not require radioactive substrates or nucleotides, and provides a sensitivity in which 10 ng of a DNA substrate per reaction is sufficient for quantitative repair analysis.

Key Words: AP site; base excision repair; DNA polymerase β; PCNA; SYBR Green.

1. Introduction

Various types of base modifications in chromosomal DNA are induced not only by exposure to chemical and physical DNA damaging agents, but also occur during physiologically normal cellular metabolism. All living organisms from bacteria to humans are facilitated to correct the majority of such DNA lesions by a mechanism called base excision repair (BER: *see* **refs.** *1,2* for review). This repair mechanism processes the damaged DNA by sequential reactions: removal of the modified base resulting in formation of an apyrimidinic/apurinic (AP) site; incisions of the phosphate backbone at its 5′ and 3′ sides leading to excision of the AP site (and its adjacent nucleotides); DNA synthesis; and ligation. Biochemical studies of in vitro repair systems derived from *Xenopus laevis* oocytes and mammalian cells have revealed that BER proceeds not in a single pathway but in multiple alternative pathways *(3,4)*. In one classification of BER, two major pathways are distinguished as either

DNA-polymerase β (pol β)-dependent or proliferating cell nuclear antigen (PCNA)-dependent. Another classification divides BER into a short-patch repair that replaces only a single nucleotide and a long-patch repair that replaces two or more nucleotides. Furthermore, recent studies have demonstrated that the pol β-dependent pathway and the PCNA-dependent pathway are not mutually exclusive but crisscross each other *(5,6)*. Which BER pathway employed may depend on the type of modified base and/or on cellular conditions such as cell cycle stage, but the mechanisms are not yet fully elucidated.

In this chapter, methods of construction of circular DNA substrates, preparation of cell extracts and a repair assay for in vitro BER analysis are described. The method for extract preparation is based on detergent-induced cell lysis. The AP-site repair activity of the extracts prepared by this method is significantly higher than the activity of those prepared by the method of Manley et al. *(7,8)*. Furthermore, the repair assay using extracts thus prepared demonstrated contribution of both pol β-dependent and PCNA-dependent pathways to AP site repair *(8)*. Another advantage of this method is its simplicity that allows one to prepare extracts simultaneously from many different cultures. In addition, it can be applied to any size culture, from six-well plates to large culture bottles. The protocol described here includes some modifications of the original protocol *(8)*.

The method of repair assay described employs a circular DNA substrate containing a defined lesion at a specific position *(9)*. This type of DNA substrate allows investigation of the PCNA-dependent repair pathway, which is not functional on linear DNA fragments *(8)*, as well as the pol β-dependent pathway. After the repair reaction, the circular DNA substrate is treated with AP endonuclease (following digestion with damage-specific DNA glycosylase, if necessary) and subsequently with a restriction enzyme whose recognition sequence is interrupted by the defined lesion. Restriction enzymes generally cannot introduce a double-strand break at the recognition sequence that is disrupted by an incised AP site. Consequently, the unrepaired substrate will remain in a nicked circular form whereas the repaired substrate will be converted into linear DNA (**Fig. 1**). These DNA substrates can be easily separated by regular agarose gel electrophoresis.

Another advantage of this repair assay is that the protocol does not use radioisotope for detection of repair products. Instead, both unrepaired and repaired DNA substrates are detected by staining with SYBR Green in an agarose gel. SYBR Green provides a significantly higher sensitivity than ethidium bromide and allows detection of less than 0.1 ng of DNA per band. As a typical repair assay includes 10 ng of the DNA substrate as a starting material, 1% of the DNA substrate can be detected as either repaired or unrepaired product. Furthermore, although quantitation of the covalently closed circular DNA by stain-

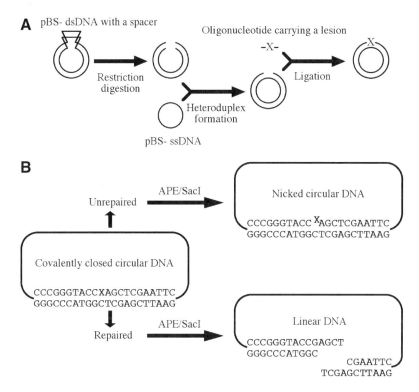

Fig. 1. (**A**) Strategy for preparation of cccDNA carrying a single lesion. (**B**) Schematic diagram of an in vitro assay of AP site repair. An AP site is designated as X. Although the unpaired DNA substrate is represented by the input substrate in this figure, other repair intermediates whose repaired strand has not been ligated are also in a nicked circular form. Therefore, only the completely repaired products are converted into a linear form.

ing with a fluorescent dye often results in underestimation due to limited binding of the dye to the torsion-restricted DNA, the linear DNA (repaired substrate) and the nicked circular DNA (unrepaired substrate) are equally stained with the dye. Thus the protocol described here provides a rapid, convenient and safer way for studying BER of mammalian cells in vitro.

2. Materials
2.1. Construction of Abasic Site-Containing cccDNA

1. Double-stranded DNA (dsDNA) of a modified pBS- vector which carries a spacer fragment of several hundred base pairs between *Ava*I and *Eco*RI sites (*see* **Note 1**). pBS- is a phagemid vector from Stratagene.

2. Single-stranded DNA (ssDNA) of pBS- (without a spacer fragment). A large-scale preparation of ssDNA of pBS- can be performed by scaling up the method described by Trower *(10)* to 200 mL of culture. We use F'-Tet^r-transduced HI1006 *(11)* as a host strain to promote a high yield of ssDNA. The preparation thus obtained usually contains only a small quantity of ssDNA from the helper phage. An alternative protocol for ssDNA preparation is described by Svoboda and Vos *(12)*.
3. Oligonucleotide, 5'-CCGGGTACCXAGCTCG-3' (X is either 3-hydroxy-2-hydroxymethyltetrahydrofuran or deoxyuridine; *see* **Note 2**). A phosphoramidite derivative of 3-hydroxy-2-hydroxymethyltetrahydrofuran is available for automated oligonucleotide synthesis under the name of dSpacer from Glen Research Corporation, Sterling, VA.
4. Restriction enzymes: *Eco*RI and *Ava*I (New England Biolabs, 10 U/µL). (*See* **Note 3**.)
5. 10X *Eco*RI digestion buffer: 500 mM Tris-HCl, pH 8.0, 100 mM MgCl$_2$, 1 M NaCl. Store at –20°C.
6. 10 mg/mL bovine serum albumin (BSA). Store at –20°C.
7. Phenol–chloroform–isoamyl alcohol (25:24:1). Store at 4°C.
8. TE buffer: 10 mM Tris-HCl, pH 7.5, 1 mM EDTA, pH 8.0.
9. 3 M Sodium acetate, pH 5.2.
10. 100% Ethanol.
11. 10X Annealing buffer: 100 mM Tris-HCl, pH 7.5, 1.5 M NaCl.
12. T4 polynucleotide kinase (Invitrogen, 10 U/µL).
13. 5X Phosphorylation buffer: 350 mM Tris-HCl, pH 7.6, 50 mM MgCl$_2$, 500 mM KCl, 5 mM β-mercaptoethanol (Invitrogen). Store at –20°C.
14. 10 mM ATP. Store at –20°C.
15. T4 DNA ligase (New England Biolabs, 400 U/µL).
16. 10X Ligation buffer: 500 mM Tris-HCl, pH 7.5, 100 mM MgCl$_2$, 10 mM ATP, 100 mM DTT, 0.25 mg/mL of BSA (New England Biolabs). Store at –20°C.
17. Cesium chloride (powder).
18. 10 mg/mL of ethidium bromide.
19. Cesium chloride solution: 1 g of cesium chloride per 1 mL of TE buffer.
20. Cesium chloride-saturated isopropanol. Isopropanol is mixed 1:1 (v/v) with the cesium chloride-saturated water. The upper phase is the organic layer.
21. Centricon-30 (Amicon).
22. TBE buffer: 44.5 mM Tris base, 44.5 mM boric acid, 1 mM EDTA.
23. Agarose gel electrophoresis equipment with 1% agarose gel in TBE buffer with 0.5 µg/mL of ethidium bromide.
24. Uracil-DNA glycosylase (Epicenter or Perkin Elmer; *see* **Note 4**).

2. 2. Preparation of Whole Cell Extracts

Whole cell lysis buffer: 40 mM N-2-hydroxyethylpiperazine-N'-2-ethanesulfonic acid (HEPES)-NaOH, pH7.5, 500 mM NaCl, 10% glycerol, 0.1% NP-40, proteinase inhibiter mix (e.g., Complete Mini, EDTA-free from Roche:

add to the solution immediately before use). Prepare this solution from sterilized stock solutions and precool on ice before use.

2.3. Repair Assay

1. Covalently closed circular DNA containing a synthetic AP site in a unique *Sac*I site that is prepared as described above (**Fig. 1**). An alternative method for construction of similar DNA substrates is described by Frosina et al. (**ref. 13**; *see also* Chapter 27).
2. 1 M HEPES–NaOH, pH 7.5.
3. 1 M $MgCl_2$.
4. 0.1 M Dithiothreitol (DTT): Store in small aliquots at −20°C.
5. 0.1 M ATP: Store in small aliquots at −20°C.
6. 1 mM Deoxyribonucleoside triphosphate (dNTP) mixture: 1 mM dATP, 1 mM dCTP, 1 mM dGTP, 1 mM TTP. Store in small aliquots at −20°C.
7. 0.1 M NAD: Store in small aliquots at −20°C.
8. 0.5 M phosphocreatine: Store in small aliquots at −20°C.
9. 2 U/µL of creatine phosphokinase: Store in small aliquots at −20°C.
10. Stop solution: 10 mM Tris-HCl, pH 7.5, 300 mM sodium acetate, 10 mM EDTA, pH 8.0, 0.5% SDS.
11. Phenol–chloroform–isoamyl alcohol (25:24:1).
12. 100% Ethanol.
13. APE/RNase solution: 40 mM HEPES–NaOH, pH7.5, 100 mM NaCl, 10 mM $MgCl_2$, 5 µg/mL of AP endonuclease (*see* **Note 5**), 0.1 mg/mL of RNase A. Prepare from stock solution immediately before use. 20 mg/mL of RNase A is heated for 15 min in boiling water to inactivate contaminating DNase activity and stored in small aliquots at −20°C.
14. *Sac*I (New England Biolabs, 20 U/µL).
15. 6X Agarose gel loading solution: 60 mM EDTA, pH 8.0, 30% glycerol, 0.6% sodium dodecylsulfate (SDS), 0.25% bromophenol blue.
16. TBE buffer: 44.5 mM Tris base, 44.5 mM boric acid, 1 mM EDTA.
17. 1X SYBR Green I/TBE buffer: Dilute SYBR Green I (Molecular Probes) 10,000-fold in TBE buffer before use.
18. Agarose gel electrophoresis equipment.
19. UV transilluminator: a standard type for ethidium bromide (300 nm) can be used for SYBR Green I.
20. Digital camera and appropriate image processing software.

3. Methods
3.1. Construction of cccDNA Carrying a Single Lesion

1. Digest 100 µg of dsDNA of the modified pBS- with 30 U of *Ava*I and 30 U of *Eco*RI in 200 µL *Eco*RI digestion buffer (20 µL of 10X *Eco*RI digestion buffer, 2 µL of 10 mg/mL BSA; adjust the final volume with H_2O) at 37°C for 4 h.
2. Deproteinize by phenol–chloroform extraction. Back-extract with 100 µL of TE buffer, and combine the two aqueous phases.

3. Add 100 µg of ssDNA of pBS- to the linearized dsDNA.
4. Precipitate DNA with 1/10 volume of 3 M sodium acetate and 2.5 volume of ethanol.
5. Rinse the precipitated DNA with ethanol to completely remove the salt-containing supernatant.
6. Dissolve the DNA in 400 µL of H_2O.
7. Incubate the DNA solution at 72°C for 10 min.
8. Add 44 µL of 10X annealing buffer to the DNA solution while keeping it at 72°C.
9. Cool the sample gradually to room temperature. We use a water bath with refrigerating circulation. It takes approx 1 h to reach room temperature.
10. Run a small aliquot of the sample (0.2 µg of DNA) in 1% agarose gel electrophoresis. Heteroduplex molecules will appear as an additional band with a slow mobility similar to that of nicked circular plasmid DNA (**Fig. 2**, lane 3; *see* **Note 6**). If heteroduplex molecules have not been formed (*see* **Note 7**), go to **step 11** and then repeat **steps 5–10**.
11. Precipitate the DNA with sodium acetate and ethanol, and rinse the pellet with ethanol.
12. Dissolve the DNA in 100 µL of H_2O.
13. While waiting for heteroduplex formation, phosphorylate 300 pmol of the oligonucleotide with 10 U of T4 polynucleotide kinase and ATP (final 0.1 mM) in 30 µL of phosphorylation buffer (6 µL of 5X phosphorylation buffer; adjust the final volume with H_2O) at 37°C for 1 h.
14. Ligate the DNA and the phosphorylated oligonucleotide with 1200 U of T4 DNA ligase (*see* **Note 8**) in 200 µL of ligation buffer (20 µL of 10X ligation buffer; adjust the final volume with H_2O) at 15°C for several hours or overnight.
15. Run a small aliquot of the sample (0.2 µg of DNA) in a 1% agarose gel electrophoresis. If ligation is complete, the small spacer fragment disappears (**Fig. 2**, lane 4).
16. Recover DNA by phenol–chloroform–isoamyl alcohol extraction and ethanol precipitation.
17. Dissolve the precipitated DNA in 200 µL of TE buffer.
18. Dissolve 3.8 g of cesium chloride in 3.5 mL of TE buffer.
19. Mix the ligated DNA, the cesium chloride solution, and 0.1 mL of 10 mg/mL ethidium bromide.
20. Transfer the DNA mixture to a tube for the VTi65 rotor (Beckman).
21. Fill up the tube with the cesium chloride solution, and seal the tube (*see* **Note 9**).
22. Centrifuge the tube at 246,000g for more than 12 h. After centrifugation, cccDNA molecules will form a minor lower band visualized with a long-wave (366 nm) UV illuminator, while linear and nicked DNA molecules form a major upper band.
23. Withdraw the cccDNA with a 3-mL syringe from the side of the tube.
24. Extract ethidium bromide from the recovered DNA sample with the cesium chloride-saturated isopropanol. Repeat the extraction at least four times or until

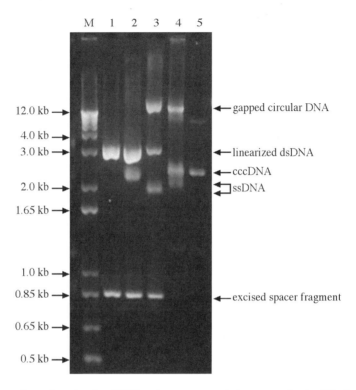

Fig. 2. Construction of a circular DNA substrate containing a synthetic AP site. Small aliquots of DNA from several steps in substrate preparation were analyzed by agarose gel electrophoresis. Lane 1, dsDNA after digestion with *Ava*I and *Eco*RI; lane 2, mixture of ssDNA and linear dsDNA; lane 3, the sample after heteroduplex formation; lane 4, the sample after ligation; lane 5, the final product after purification by ultracentrifugation in CsCl/ethidium bromide. 0.2% of the total sample at each step was loaded. Lane M, molecular weight markers.

ethidium bromide is completely removed as judged with the long-wave UV illuminator.
25. Concentrate the DNA solution in a Centricon-30 by centrifugation with the JA-20 rotor (Beckman) at 3700g for 15–30 min until the volume decreases to <0.1 mL.
26. Add 0.5 mL of TE buffer to the DNA solution in the Centricon-30, and concentrate it in the same manner. Repeat three times to remove the cesium chloride, which otherwise tends to precipitate with ethanol.
27. Transfer the DNA solution to a 1.5-mL microtube.
28. Precipitate the DNA with sodium acetate and ethanol.
29. Dissolve the precipitated DNA in 100 µL of TE buffer.

30. Measure the concentration of the DNA at an absorbance of 260 nm. Alternatively, digest a small aliquot of the purified DNA with *Eco*RI and compare its intensity in a 1% agarose gel with those of 10–100 ng of a standard DNA. The yield of the cccDNA is usually 20–30 μg.
31. Store the cccDNA at 4°C. For formation of a natural AP site, the uracil-containing DNA should be treated with uracil-DNA glycosylase immediately before starting the repair assay (*see* **Subheading 3.3.**).

3.2. Preparation of Whole-Cell Extracts

1. Grow cells in a six-well plate (growth area, 9.5 cm^2/well) until they are 50–70% confluent (*see* **Note 10**). In the case of mouse 3T3 cells, if 2×10^5 cells per well are plated, the cells will be ready for the next step approx 24 h after plating.
2. Remove the media. Wash the cells with ice-cold phosphate-buffered saline (PBS). All the following procedures should be carried out with ice-cold solutions on ice or at 4°C.
3. Add 0.3 mL of ice-cold whole cell lysis buffer per well. Leave the plate on ice for 5 min. Occasional shaking of the plate facilitates detaching cells.
4. Transfer the cell lysate into a 1.5-mL tube. Leave the tube on ice for 5 min.
5. Spin at 16,000*g* for 5 min at 4°C. Recover the supernatant without touching the cloudy viscous part at the bottom.
6. Distribute the recovered extract into small aliquots (e.g., 50 μL), quickly freeze in dry ice, and store at –80°C.
7. Measure the protein concentration of the extract. The protein concentration should be >1 mg/mL (*see* **Note 11**).

3.3. Repair Assay

1. Mix the following reagents (the volumes are provided as per reaction and should be scaled up for experiments involving multiple samples): 0.8 μL of 1 *M* HEPES–NaOH, pH 7.5, 0.4 μL of 1 *M* MgCl$_2$, 0.4 μL of 0.1 *M* DTT, 0.8 μL of 0.1 *M* ATP, 0.8 μL of 1 m*M* dNTP mixture, 0.8 μL of 0.1 *M* NAD, 3.2 μL of 0.5 *M* phosphocreatine, 0.5 μL of 2 U/μL of creatine phosphokinase, 10 ng of the AP-site containing circular DNA, and H$_2$O to adjust the volume per reaction to 30 μL (*see* **Note 12**). Preincubate at 25°C.
2. Dilute the whole cell extract with whole cell lysis buffer in a 1.5-mL tube to provide 10 μg of protein in 10 μL of buffer. Preincubate at 25°C.
3. Start the repair reaction by adding the reagent mix made at **step 1** (30 μL) to the diluted extract made at **step 2** (10 μL). Incubate at 25°C for an appropriate time (usually 10 min to several hours).
4. Stop the repair reaction by adding 160 μL of stop solution to the reaction mixture.
5. Recover the DNA by phenol–chloroform–isoamyl alcohol extraction and ethanol precipitation. Remove the residual supernatant completely.
6. Dissolve the precipitated DNA in 10 μL of APE/RNase solution. Incubate the sample at 37°C for 30 min.

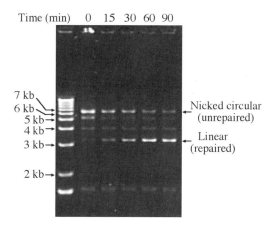

Fig. 3. Time course of synthetic AP site repair in a 3T3 whole cell extract. The repair reaction was conducted for 0, 15, 30, 60, or 90 min at 25°C, and the reaction products were analyzed by agarose gel electrophoresis. The linear DNA was quantitated as 6% at 0 min, 29% at 15 min, 51% at 30 min, 80% at 60 min, and 83% of the total DNA substrate at 90 min.

7. Add 0.2 µL of SacI (20 U/µL) to the sample (see **Note 13**). Incubate the sample at 37°C for 1 h.
8. Add 2 µL of 6X agarose gel loading solution to the sample. Incubate the sample at 50°C for 10 min (see **Note 14**).
9. Load 6 µL of the sample onto a 1% agarose gel in TBE buffer (no ethidium bromide in the gel or running buffer) and electrophorese the sample at 3.8 V/cm for 2 h (see **Note 6**).
10. Stain the gel in 1X SYBR Green I /TBE buffer with gentle rocking for at least 1 h in the dark.
11. Take a picture of the stained gel over a UV transilluminator with a digital camera.
12. Measure the intensity of DNA bands using an appropriate image processing software. A typical result is shown in **Fig. 3**.

4. Notes

1. Insertion of a spacer fragment into a phagemid vector is optional. However, this modification is helpful in checking for complete digestion by restriction enzymes and for ligation efficiency (the small spacer fragment should disappear after ligation).
2. The position of the AP site in the DNA substrate used in this protocol is different from the original construct described previously *(9)*. This new construct mimics an AP site formed by depurination of a guanine, the most abundant spontaneous

lesion. Another modified base can be introduced instead of the AP site into the circular DNA substrate, if its phosphoramidite derivative is available. In this case, the sample should be treated with a specific DNA-glycosylase prior to the incubation in APE/RNase solution at **step 6** of **Subheading 3.3**.

3. Restriction enzymes from commercial sources are sometimes contaminated with AP endonuclease and/or exonuclease. Try to use enzymes of the best quality, avoid overdigestion, and perform control experiments to rule out contamination.
4. Make sure that uracil-DNA glycosylase is not contaminated with AP lyase.
5. AP endonuclease used here is a recombinant human enzyme purified from the bacterial overexpression system as described by Xanthoudakis et al. *(14)*. It is also commercially available (e.g., from Trevigen).
6. The relative mobility of the gapped circular DNA to the linear DNA depends on electrophoretic field strength. The electrophoresis shown in **Fig. 2** was performed at 3.8 V/cm.
7. Most cases when heteroduplex molecules are not formed result from residual salt after ethanol precipitation which may interfere with denaturation of dsDNA.
8. The unit definition of T4 DNA ligase from New England Biolabs is different from those from most other companies. One cohesive end ligation unit (New England Biolabs) is equivalent to 0.015 Weiss (ATP-PP exchange) units.
9. It is critical to prepare the DNA/cesium chloride mixture at the accurate density. When the sample is filled up in a Beckman Optiseal tube (cat. no. 362185), the total weight including a tube and a plug should be 9.0 ± 0.05 g.
10. Although this protocol is made for the cells grown in a six-well plate, the size of cell culture can be changed. In that case, multiply the volume of the solutions by the relative size of the culture.
11. The protein concentration is measured by the Bradford method using a commercial kit. In the author's laboratory, bovine γ-globulin is used as a standard, while many other laboratories use BSA instead. However, since BSA provides unusually high absorbance in the Bradford assay *(15)*, use of BSA as a standard may result in underestimation of the protein concentration.
12. Final concentrations of the reagents in a reaction mixture after addition of the extracts are: 30 mM HEPES–NaOH, pH 7.5, 10 mM MgCl$_2$, 125 mM NaCl, 2.5% glycerol, 0.025% NP-40, 1 mM DTT, 2 mM ATP, 20 mM each dNTP, 2 mM NAD, 40 mM phosphocreatine, 1 U of creatine phosphokinase, 10 ng of the DNA substrate, and 10 µg of protein from the extract in a 40-µL reaction.
13. If the original construct *(9)* is used for repair assay, *Acc*65I or *Kpn*I instead of *Sac*I should be used for digestion at **step 7** of **Subheading 3. 3**. If different constructs in which another unique restriction enzyme site is interrupted are used, it is recommended to confirm in advance that the restriction enzyme cannot introduce a double-strand break at the recognition site disrupted by an incised AP site.
14. This step is critical for dissociation of nucleases from DNA before gel electrophoresis. If this step is skipped, DNA bands in agarose gel will become blurred and inappropriate for quantitation.

Acknowledgments

The author thanks C. C. Stobbe for critical reading of the manuscript. This work is supported by National Institutes of Health Grants CA06927, CA63154, and an appropriation from the Commonwealth of Pennsylvania.

References

1. Hickson, I. D. (ed.) (1997) *Base Excision Repair of DNA Damage*. Landes Bioscience, Austin, TX.
2. Dianov, G. L., Sleeth, K. M., Dianova, I. I., and Allinson, S. L. (2003) Repair of abasic sites in DNA. *Mutation Res.* **29**, 157–163.
3. Matsumoto, Y., Kim, K., and Bogenhagen, D. F. (1994) Proliferating cell nuclear antigen-dependent abasic site repair in *Xenopus laevis* oocytes: an alternative pathway of repair of base excision DNA repair. *Mol. Cell. Biol.* **14,** 6187–6197.
4. Frosina, G., Fortini, P., Rossi, O., et al. (1996) Two pathways for base excision repair in mammalian cells. *J. Biol. Chem.* **271,** 9573–9578.
5. Prasad, R., Dianov, G. L., Bohr, V. A., and Wilson, S. H. (2000) FEN1 stimulation of DNA polymerase beta mediates an excision step in mammalian long patch base excision repair. *J. Biol. Chem.* **275,** 4460–4466.
6. Kedar, P. S., Kim, S. J., Robertson, A., et al. (2002) Direct interaction between mammalian DNA polymerase beta and proliferating cell nuclear antigen. *J. Biol. Chem.* **277,** 31,115–31,123.
7. Manley, J. L., Fire, A., Samuels, M., and Sharp, P. A. (1983) *In vitro* transcription: Whole cell extract. *Methods Enzymol.* **101,** 568–582.
8. Biade, S., Sobol, R. W., Wilson, S. H., and Matsumoto, Y. (1998) Impairment of proliferating cell nuclear antigen-dependent apurinic/apyrimidinic site repair on linear DNA. *J. Biol. Chem.* **273,** 898–902.
9. Matsumoto, Y. (1999) Base excision repair assay using *Xenopus laevis* oocyte extracts. *Methods Mol. Biol.* **113,** 289–300.
10. Trower, M. K. (1996) Preparation of ssDNA from phagemid vectors. *Methods Mol. Biol.* **58,** 363–366.
11. Matsumoto Y., Shigesada, K., Hirano, M., and Imai, M. (1986) Autogenous regulation of the gene for transcription termination factor rho in *Escherichia coli*: localization and function of its attenuators. *J. Bacteriol.* **166,** 945–958.
12. Svoboda, D. L. and Vos, J. M. (1999) Assays of bypass replication of genotoxic lesions in mammalian disease and mutant cell-free extracts. *Methods Mol. Biol.* **113,** 555–576.
13. Frosina, G., Cappelli, E., Fortini, P., and Dogliotti, E. (1999) In vitro base excision repair assay using mammalian cell extracts. *Methods Mol. Biol.* **113,** 301–315.
14. Xanthoudakis, S., Miao, G., Wang, F., Pan, Y. C., and Curran, T. (1992) Redox activation of Fos-Jun DNA binding activity is mediated by a DNA repair enzyme. *EMBO J.* **11,** 3323–3335.
15. Instruction manual for Bio-Rad Protein Assay.

27

In Vitro Base Excision Repair Assay Using Mammalian Cell Extracts

Guido Frosina, Enrico Cappelli, Monica Ropolo,
Paola Fortini, Barbara Pascucci, and Eugenia Dogliotti

Summary

Base excision repair (BER) is the main pathway for removal of endogenous DNA damage. This repair mechanism is initiated by a specific DNA glycosylase that recognizes and removes the damaged base through N-glycosylic bond hydrolysis. The generated apurinic/apyrimidinic (AP) site can be repaired in mammalian cells by two alternative pathways which involve either the replacement of one (short patch BER) or more nucleotides (long patch BER) at the lesion site. This chapter describes a repair replication assay for measuring BER efficiency and mode in mammalian cell extracts. The DNA substrate used in the assay is either a randomly depurinated plasmid DNA or a plasmid containing a single lesion that is processed via BER (for example a single AP site or uracil residue). The construction of a single lesion at a defined site of the plasmid genome makes the substrate amenable to fine mapping of the repair patches, thus allowing discrimination between the two BER pathways.

Key Words: AP site; base excision repair; in vitro repair assay; mammalian cell extracts; 8-oxoguanine; uracil.

1. Introduction

Base excision repair (BER) is a major cell repair mechanism which corrects a broad range of DNA lesions (for a review, see **ref. 1**). BER deals with DNA damage generated not only by environmental genotoxins like ionizing radiation, alkylating agents and oxidative reagents, but also by endogenously produced oxygen radicals and other reactive species. Therefore, its correct functioning is very important for genome stability and cell viability *(2,3)*. The primary pathway for BER involves the recognition by a DNA glycosylase of the damaged base followed by cleavage of the N-glycosyl bond to generate an

From: *Methods in Molecular Biology: DNA Repair Protocols: Mammalian Systems, Second Edition*
Edited by: D. S. Henderson © Humana Press Inc., Totowa, NJ

apurinic/apyrimidinic (AP) site. The AP site is then recognized by an endonuclease. The major AP endonucleases cleave hydrolytically the phosphodiester bond on the 5′ side of the AP site. A phosphodiesterase then excises the generated 5′ deoxyribose phosphate terminus to leave a single nucleotide gap. This gap can then be filled by a DNA polymerase (polymerase β in mammalian cells) and the nick is sealed by a DNA ligase. In eukaryotes, an alternative BER pathway that involves the replacement of more than a single residue is also present *(4–6)*. Repair synthesis, which is dependent upon proliferating cell nuclear antigen (PCNA) occurs on the 3′ side of the damaged residue and involves the replacement of two to six nucleotides. It is likely that these two repair mechanisms have evolved to repair structurally distinct lesions and/or to operate in different cell cycle stages (for a review, *see* **ref. 7**).

The knowledge of human nucleotide excision repair has been greatly improved by the development and use of a cell-free in vitro repair assay *(8; see* Chapter 29). This same methodology has been successfully applied in the past few years to the analysis of the BER process *(4–6,9,10)*. This chapter details a BER synthesis assay using mammalian cell extracts and, as DNA substrate, a plasmid containing a single AP site or lesions that are repaired *via* AP site formation (e.g., uracil). AP sites are the typical BER lesions since they are generated as intermediates during the repair process itself. For many research questions randomly depurinated plasmid DNA, which is easier to produce, can be used as an alternative repair substrate. However, only DNA containing a single lesion at a defined site is amenable to fine mapping of the repair patches, thus allowing discrimination between the two BER pathways.

2. Materials
2.1. Construction of DNA Substrates (see **Notes 1** and **2**)

1. Single-stranded circular DNA. Any plasmid which contains the origin of replication of single-stranded phage DNA is suitable. We currently use the phagemid pGEM-3Zf (+) (Promega) (**Fig. 1**) to produce single-stranded (+) pGEM-3Zf DNA. Materials listed below (**items 2–4**) are required to prepare single-stranded circular DNA from this plasmid. A method for preparing single-stranded DNA is described in **ref. 11**.
2. M13K07 helper phage DNA (Promega).
3. Kanamycin stock solution (10 mg/mL) dissolved in distilled water. Sterilize by filtration and store in aliquots at –20°C.
4. 2X YT bacterial growth medium: 10 g of yeast extract, 16 g of Bacto-tryptone and 5 g of NaCl/L, dissolved in water and autoclaved.
5. Primers for in vitro DNA replication:
   a. An oligonucleotide that contains a single uracil residue (e.g., 5′GATCCTCTAGAGUCGACCTGCA3′).
   b. A control oligonucleotide (e.g., 5′GATCCTCTAGAGTCGACCTGCA3′) (Fig. 1).

Base Excision Repair Assay

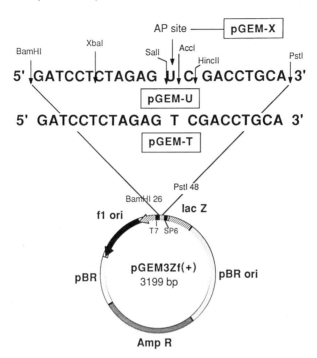

Fig. 1. Scheme of the circular duplex DNA molecules used as substrates. pGEM-T and pGEM-U plasmids were obtained by priming single-stranded (+) pGEM-3Zf DNA with the indicated oligonucleotides and performing in vitro DNA synthesis as described. The single AP site-containing plasmid, pGEM-X, was obtained by incubation of pGEM-U with UDG. The positions of the AP site and the restriction sites in its proximity, which may be utilized for fine mapping of the repair patch, are also indicated. Reprinted with permission from **ref. 5**.

    c. An oligonucleotide which contains a single uracil residue within the recognition sequence of *Sal*I (e.g., 5′GATCCTCTAGA<u>U</u>TCGACCTGCAGGC ATGCA3′) (**Fig. 4**). These oligonucleotides are prepared by automated DNA synthesis.
6. T4 Polynucleotide-DNA-kinase (PNK) (New England Biolabs).
7. Stock solutions for kinasing of primers:
    a. 1 M Tris-HCl, pH 7.6.
    b. 0.2 M MgCl$_2$.
    c. 1 M Dithiothreitol (DTT). Autoclave solutions a and b, and store at –20°C. Filter-sterilize solution c, and store at –20°C.
8. TE (pH 8.0): 10 mM Tris-HCl, pH 8.0, 1 mM EDTA. Sterilize by autoclaving.
9. Sephadex G-50 (medium) dissolved in TE, pH 8.0 (*see* **Note 3**). Store at 4°C in a screw-cap bottle.

10. 10X Annealing buffer: 200 mM Tris-HCl, pH 7.4, 20 mM MgCl$_2$, 0.5 M NaCl. Sterilize by filtration, and store in aliquots at –20°C.
11. 10X in vitro synthesis buffer: 175 mM Tris-HCl, pH 7.4, 37.5 mM MgCl$_2$, 215 mM DTT, 7.5 mM ATP and 0.4 mM each dNTP. Store in aliquots at –20°C (*see* **Note 4**).
12. [γ-^{32}P]ATP aqueous solution 250 μCi, 370 MBq/mL.
13. 0.1 M ATP stock solution. ATP is dissolved in distilled water. Sterilize by filtration and store in aliquots at –80°C.
14. T4 gene 32 protein (single-stranded DNA binding protein), T4 DNA polymerase holoenzyme and T4 DNA ligase (Roche).
15. *Escherichia coli* uracil-DNA-glycosylase (UDG) and endonuclease III (Nth) (*see* **Note 5**).
16. 10X *E. coli* uracil-DNA-glycosylase and endonuclease III buffer: 70 mM HEPES–KOH, pH 7.8, 50 mM 2-mercaptoethanol, 20 mM Na$_2$EDTA, 350 mM NaCl. Sterilize by filtration and store in aliquots at –20°C.
17. Solid cesium chloride.
18. Ethidium bromide solution (10 mg/mL): Ethidium bromide is dissolved in distilled water. Store in aliquots at 4°C in dark bottles. Ethidium bromide is a potent mutagen. Wear gloves and a mask when weighing it out.
19. 1-Butanol saturated with water.
20. 7 M Ammonium acetate. Dispense into aliquots and sterilize by autoclaving.
21. 3 M Sodium acetate, pH 5.2. Dispense into aliquots and sterilize by autoclaving.
22. Phenol–chloroform–isoamyl alcohol (25:24:1) saturated with TE, pH 8.0.
23. Chloroform–isoamyl alcohol (24:1).
24. Ethanol 100%.
25. 1% Agarose gel.

2.2. Preparation of Extracts

2.2.1. Preparation of Whole-Cell Extracts ("Manley" Type)

1. Hypotonic lysis buffer, pH 7.9: 10 mM Tris-HCl, 1 mM EDTA, 5 mM DTT, 0.5 mM spermidine, 0.1 mM spermine. Store at 4°C.
2. Sucrose–glycerol buffer, pH 7.9: 50 mM Tris-HCl, 10 mM MgCl$_2$, 2 mM DTT, 25% sucrose (molecular biology grade), 50% glycerol (Fluka). Store at 4°C.
3. Dialysis buffer, pH 7.9: 25 mM N-2-hydroxyethylpiperazine-N'-2-ethanesulfonic acid (HEPES)–KOH, 100 mM KCl, 12 mM MgCl$_2$, 1 mM EDTA, 17% glycerol (Fluka), 2 mM DTT (add just before use), pH to 7.9 with 5 M KOH. Store at 4°C.
4. Protease inhibitors (all from Sigma): (a) 500 mM (87 mg/mL) phenylmethylsulfonyl fluoride (PMSF) dissolved in acetone, store at –20°C in a dark glass bottle; (b) 5 mg/mL of pepstatin A, chymostatin, antipain and aprotinin, each dissolved in sterile 20% dimethylsulfoxide, store at –20°C; (c) 5 mg/mL of leupeptin dissolved in sterile water; store at –20°C. Protease inhibitors are highly toxic: wear gloves and manipulate with care.
5. Solid ammonium sulfate.

2.2.2. Preparation of Nuclear Extracts

1. Buffer H: 10 mM HEPES–KOH, pH 7.9, 1.5 mM MgCl$_2$, 10 mM KCl, 0.5 mM DTT (add just before use), 0.5 mM PMSF (add just before use).
2. Buffer N20: 20 mM HEPES–KOH, pH 7.9, 25% glycerol, 1.5 mM MgCl$_2$, 0.2 mM EDTA, 20 mM NaCl, 0.5 mM DTT (add just before use), 0.5 mM PMSF (add just before use).
3. Buffer N200: 20 mM HEPES–KOH, pH 7.9, 25% Glycerol, 1.5 mM MgCl$_2$, 0.2 mM EDTA, 200 mM NaCl, 0.5 mM DTT (add just before use), 0.5 mM PMSF (add just before use).
4. Buffer C (dialysis buffer): 20 mM HEPES–KOH, pH 7.9, 20% Glycerol, 50 mM KCl, 0.2 mM EDTA, 0.5 mM DTT (add just before use), 0.5 mM PMSF (add just before use).

   Additional protease inhibitors (Sigma):
   a. 5 mg/mL of pepstatin A, chymostatin and aprotinin, each dissolved in sterile 20% dimethy sulfoxide. Store at –20°C;
   b. 5 mg/mL of leupeptin dissolved in sterile water. Store at –20°C.

2.2.3. Preparation of Crude Extracts ("Tanaka" Type)

1. Phosphate-buffered saline (PBS).
2. Buffer I: 10 mM Tris-HCl, pH 7.8, 200 mM KCl.
3. Buffer II: 10 mM Tris-HCl, pH 7.8, 200 mM KCl, 2 mM EDTA, 40% of glycerol, 0.2% Nonidet P-40, 2 mM DTT, 0.5 mM PMSF, 10 µg/mL of aprotinin, 5 µg/mL of leupeptin, and 1 µg/mL of pepstatin.

2.3. In Vitro Repair Assay

1. 5X reaction buffer: 25 mM MgCl$_2$, 200 mM HEPES–KOH, pH 7.8, 2.5 mM DTT, 10 mM ATP, 100 µM dGTP, 100 µM dCTP, 100 µM TTP, 100 µM dATP, 200 mM phosphocreatine, 1.8 mg/mL of bovine serum albumin. Store in aliquots at –80°C for up to 2 mo.
2. Creatine phosphokinase (CPK) (type I, Sigma), 2.5 mg/mL dissolved in 5 mM glycine, pH 9.0, 50% glycerol. Store in aliquots at –20°C.
3. TE (pH 7.8): 10 mM Tris-HCl, pH 7.8, 1 mM EDTA. Sterilize by autoclaving.
4. pGEM-X and pGEM-T plasmid substrates (**Fig. 1**) dissolved in TE, pH 7.8, at 100 ng /µL. Store in aliquots at –80°C.
5. α-^{32}P-labeled deoxynucleotides ([α-^{32}P]TTP or [α-^{32}P]dCTP when pGEM-X is used as substrate), 3000 Ci/mmol, 10 mCi/mL, aqueous solution. Use a recently labeled deoxynucleotide to maximize the sensitivity of the repair assay. Perform all the steps behind a shield for ^{32}P and wear protective clothing (gloves and glasses) for β-emitting radioisotopes.
6. 3 M KCl, sterile solution, store at 4°C.
7. Appropriate α-^{32}P-5'-end-labeled DNA size markers (stored at –20°C).
8. Whole cell extracts (stored at –80°C or in liquid nitrogen).

9. RNase A: Dissolve at a concentration of 2 mg/mL in 10 mM Tris-HCl, pH 7.5, and 15 mM NaCl. Heat for 15 min at 100°C and allow to cool slowly to room temperature. Store in aliquots at –20°C.
10. Proteinase K. Dissolve at a concentration of 2 mg/mL in H$_2$O. Store in aliquots at –20°C.
11. Restriction enzymes with suitable 10X reaction buffers.
12. Denaturing loading buffer: 80% formamide, 0.1% xylene cyanol, and 0.1% bromophenol blue.
13. Polyacrylamide gel electrophoresis equipment.
14. PhosphorImager or liquid scintillation counter.

3. Methods
3.1. Construction of DNA Substrates Containing a Single Lesion

1. Phosphorylate both the control and the uracil-containing oligonucleotides by using PNK according to standard procedure (*see* **Note 6**). Remove the free ATP by chromatography on, or centrifugation through, small columns of Sephadex G-50.
2. Anneal the phosphorylated oligonucleotide to single-stranded (ss) pGEM-3Zf(+) DNA by performing the following program in a thermal cycler: 75°C for 5 min; from 75°C to 30°C over 30 min. Assemble each annealing reaction in a tube appropriate for the thermal cycler (usually a 0.5-mL microcentrifuge tube), as follows (*see* **Note 7**):
   a. Phosphorylated primer (40 ng/µL)   10.5 µL
   b. ss pGEM-3Zf(+) (200 ng/µL)   10 µL
   c. 10X Annealing buffer   3 µL
   d. Distilled water   6.5 µL
3. Spin down the samples and assemble the in vitro replication reactions by adding in each tube:
   a. 10X synthesis buffer   3.6 µL
   b. T4 Gene 32 protein (5 mg/mL)   0.5 µL
   c. T4 DNA ligase (1 U/µL)   1 µL
   d. T4 DNA polymerase (1 U/µL)   1 µL
4. Incubate 5 min on ice, then for 5 min at room temperature and finally for 90 min at 37°C.
5. Transfer and pool the replication reactions in a 1.5-mL microcentrifuge tube. Purify the DNA by standard procedure. Briefly stated, extract the DNA with an equal volume of Phenol–chloroform–isoamyl alcohol (25:24:1) saturated with TE, followed by extraction with an equal volume of chloroform–isoamyl alcohol (24:1). Precipitate the DNA by adding one-fourth volume of 7 M ammonium acetate and two volumes of cold pure ethanol. Keep the samples at –80°C overnight.
6. Centrifuge for 30 min at top speed (approx 15,000g) in an Eppendorf centrifuge at 4°C and wash the pellet with 70% ethanol. Dry the pellet using a SpeedVac for 2–3 min and resuspend it in 30 µL of TE, pH 8.0.

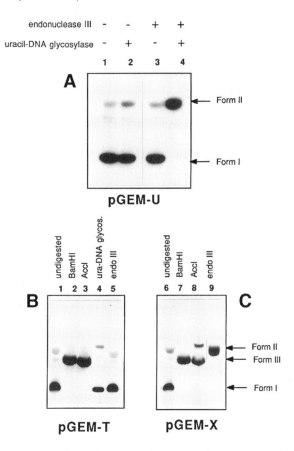

Fig. 2. Characterization of the single AP site-containing substrate. (**A**) Lane 1: construct containing a single uracil residue, pGEM-U; lane 2: after incubation with UDG; lane 3: after incubation with Nth; lane 4: after incubation with UDG followed by incubation with Nth. (**B**) Lane 1: construct containing the control oligonucleotide, pGEM-T; lane 2: after digestion with *Bam*HI (20 U); lane 3: after digestion with *Acc*I (20 U); lane 4: after incubation with UDG; lane 5: after incubation with Nth. (**C**) Lane 6: construct containing a single AP site, pGEM-X; lane 7: after digestion with *Bam*HI (20 U); lane 8: after digestion with *Acc*I (20 U); lane 9: after incubation with Nth. (Reprinted with permission from **ref. 5**.)

7. Analyze the replication products by 1% agarose gel electrophoresis (*see* **Note 8**).
8. Purify the closed circular duplex DNA molecules by equilibrium density gradient centrifugation with cesium chloride (*see* **Note 9**).
9. Incubate the plasmid molecules containing a single uracil residue (pGEM-U, **Fig. 2A**) with *E. coli* uracil-DNA glycosylase to create circular duplex DNA containing a single AP site (pGEM-X; **Fig. 2C**; *see* **Note 10**).

3.2. Construction of Radioactive DNA Substrates Containing a Single Lesion

Radioactive single lesion-containing plasmid DNA molecules are constructed essentially as described in the previous section. Only the modified steps are reported in the following:

1. Phosphorylate both the control and the modified oligonucleotides by using PNK and [γ-^{32}P]ATP nucleotide according to standard procedure (*see* **Note 11**).
2. Assemble each annealing reaction in a tube as follows:
   a. ^{32}P-labeled oligonucleotide   33 ng
   b. ss pGEM-3Zf(+)   110 ng
   c. 10X Annealing buffer   3 µL
   d. H$_2$O   up to 30 µL
3. Perform **steps 3–6** as in **Subheading 3.1.**
4. Analyze the replication products on a 1% agarose gel. Dry the gel and expose it to X-ray film or analyze by Instant Imager. At least 90% of covalently closed circular molecules is expected (*see* **Note 12**).

3.3. Preparation of Extracts (*see* Notes 13–19)

3.3.1. Preparation of Whole-Cell Extracts ("Manley" Type)

1. Grow cells (*see* **Note 18**) in monolayers up to 0.6 – 1.2 × 10^9 cells in 850 cm^2 roller bottles. Harvest the cells when they are just confluent. In the case of CHO cells (*see* **Note 19**) one roller bottle yields approx 1 × 10^8 cells. Alternatively, cells growing in suspension, like lymphoblastoid cell lines, may be used. Harvest cells from 1–2 L of an exponentially growing culture when the cell density is 6–8 × 10^5/mL.
2. Wash the monolayers twice with sterile PBS. Detach cells by incubation with 0.25% trysin/PBS (Gibco). Add complete growth medium and pellet the cells by 15 min centrifugation at approx 320*g* 1500 rpm. Resuspend the cells in 150 mL of cold (4°C) PBS and pellet again.
3. Resuspend the cells in PBS and count them. Spin at 1500 rpm for 15 min. Carefully remove the supernatant. Measure the packed cell volume (pcv). This is usually about 1 mL for approx 0.6 × 10^9 cells. At this step, the pellet can be frozen on dry ice and stored at –80°C.
4. Resuspend the cells in 4 pcv of hypotonic lysis buffer containing the following protease inhibitors: per 1 mL pcv, 5 µL of 87 mg/mL of PMSF, 0.5 µL each of 5 mg/mL of leupeptin, pepstatin, chymostatin, antipain, and aprotinin. Leave the cells swelling for 20 min on wet ice. If the pellet had been frozen, first add the hypotonic lysis buffer containing the protease inhibitors and, keeping the tube on ice, resuspend the cell pellet.
5. Pour the cell suspension into a 20-mL glass homogenizer with a Teflon pestle and homogenize on ice with 20 strokes.

All the following steps are performed in a cold room (4°C).

6. Pour the homogenate into a glass beaker (50 mL for 1 mL of pcv) on ice over a magnetic stirrer. Stir very slowly (about 60 rpm).
7. Add dropwise 4 pcv of sucrose/glycerol buffer. Mix it well.
8. Add dropwise 1 pcv of saturated ammonium sulfate. The viscosity of the solution gradually increases. Stirring must be as slow as possible to avoid fragmentation of high molecular weight cellular DNA that should sediment in the next ultracentrifugation step. Stir for 30 min.
9. Pour (do not pipet) the viscous solution into polyallomer tubes (Beckman) for an SW41 or SW55 rotor. The tubes must be filled to the top, otherwise they will collapse in the next ultracentrifugation step.
10. Centrifuge for 3 h at $235,000g$ (37,000 rpm) with SW41 rotor or at $214,000g$ (42,000 rpm) with SW55 rotor. For larger extract volumes use Ti60 rotor at $226,000g$ (41,000 rpm) and thick-wall polycarbonate tubes. Set the temperature at 2°C.
11. Carefully remove the tubes from the centrifuge. With a Pasteur pipet discard aggregates that lie on the meniscus. With another Pasteur pipet remove the supernatant leaving the last 1 mL (The latter fraction contains high molecular weight DNA that has not fully pelletted). Measure the volume of supernatant (usually 6–7 mL for 1 mL of pcv).
12. Transfer the supernatant to a 30-mL Corex tube or polyallomer tube for SW28 with a magnetic stirring bar in the tube. Stir on ice at normal speed (5–10 revolutions/s). Slowly add 0.33 g of pure solid ammonium sulfate per mL of supernatant. When it is dissolved add 10 µL of 1 M NaOH per gram of ammonium sulfate added. Continue stirring for 30 min.
13. Remove the magnetic stirring bar. Centrifuge for 20 min at $10,500g$ (8,000 rpm) in an HB-4 rotor (when using Corex tubes) or in a SW28 rotor ($11,500g$) (when using polyallomer tubes) at 2°C.
14. Remove the supernatant with a Pasteur pipet, leaving the pellet as dry as possible.
15. Resuspend the pellet in a small amount of dialysis buffer (0.05 volume of the high-speed supernatant). Do not try to dissolve the pellet, just collect the suspension (which should be thick and milky) by scraping with the end of a 1-mL Gilson tip, not pipetting up and down.
16. Dialyze for 1.5–2 h in 500 mL of extract dialysis buffer. Change the buffer and dialyze for another 10 h. Do not dialyze longer, as significant amounts of precipitate will form.
17. Remove the dialysate and centrifuge in Eppendorf tubes. Centrifuge for 10 min at approx $16,000g$ (14,000 rpm) to remove any precipitate. Transfer the supernatant to a fresh tube on ice.
18. Dispense 50-µL aliquots into conical cryotubes. Freeze immediately on dry ice and store in liquid nitrogen or at –80°C. In the latter case, a progressive decrease in repair activity will be observed starting 6 mo after preparation.
19. For an active extract, recover about 1 mL of extract per 1 mL of pcv. The protein concentration is 10–15 mg/mL.

3.3.2. Preparation of Nuclear Extracts

1. Harvest at least 60×10^6 exponentially growing cells.
2. Resuspend the cells in 7.5 mL of buffer H.
3. Spin the cells 15 min at approx 320g (1500 rpm) at 2°C. Remove the supernatant.
4. Resuspend the pellet in 4.0 mL of buffer H. Allow to swell on ice 20 min. During this time, add the following protease inhibitors: 1 µL each of 5 mg/mL of leupeptin, pepstatin A, chymostatin, and aprotinin.
5. Lyse in a 10-mL homogenizer with 30 strokes on ice.
6. Collect nuclei by centrifuging 15 min at approx 320g (1500 rpm).
7. Keep the supernatant (supernatant A).
8. Resuspend the nuclei in 80 µL of buffer N20. Measure the total volume with a micropipet.
9. Transfer the suspension to small flat-bottomed tube ("bijoux" type) containing a magnetic stirring bar and kept on a petri dish with ice. Stir slowly (about 60 rpm). Add dropwise an equal volume of buffer N200. Allow 30 min.
10. Transfer the suspension to a 1.5 mL Eppendorf tube. Centrifuge at 20,800g (14,000 rpm) in a refrigerated Eppendorf centrifuge (2°C) for 45 min.
11. Dialyse the supernatant against 400 mL of buffer C. Change the dialysis buffer after 1 h. Allow 1 more hour.
12. Recover the supernatant from the dialysis bag and centrifuge for 30 min at 20,800g (14,000 rpm) in a refrigerated Eppendorf centrifuge.
13. Aliquot the supernatant into conical cryotubes. Freeze immediately on dry ice. Store in liquid nitrogen.

3.3.3. Preparation of Crude Extracts ("Tanaka" Type)

1. Harvest at least 30×10^6 exponentially growing cells.
2. Wash the cells three times with PBS.
3. Resuspend the cells in buffer I at a concentration of 50×10^6 cells/mL.
4. Add an equal volume of buffer II.
5. Stir the cell suspension by rocking for 1 h at 4°C. Then centrifuge at 20,800g (14,000 rpm) for 10 min at 2°C.
6. Dispense the supernatant into aliquots and store in liquid nitrogen.

3.4. In Vitro Repair Assays (see Notes 20–30)

Two alternative experimental approaches are described, both based on the use of single lesion-containing plasmids. The first one (**Subheading 3.4.1.**), involves the use of radiolabeled dNTPs to quantify and map repair synthesis at the lesion site. The second one (**Subheading 3.4.2.**) involves the use of radiolabeled plasmid molecules containing the lesion of interest. In this second case the repair reaction is monitored by quantifying the relative percentages of the different plasmid forms after gel electrophoresis.

**Table 1
Basic Composition of a Single Repair Reaction**

Component	(µL)
5X reaction buffer	10
CPK (2.5 mg/mL)	1
[α-^{32}P]dNTP (10 µCi/µL)	0.2
H$_2$O	4.8
pGEM-X or pGEM-T (100 ng/µL)	3
cell extract	Variable [a]
dialysis buffer	Variable [b]
TE, pH 7.8	Up to 50

[a] Usually 20–200 µg protein are used.
[b] The total amount of dialysis buffer should be the same in all reactions.

3.4.1. Repair Synthesis by a Plasmid Assay

1. Premix appropriate amounts of 5X reaction buffer and CPK in order to run two more reactions than needed. Each 50-µL reaction requires 10 µL of 5X buffer and 1 µL of CPK. Keep these on ice.
2. Add to the mix 3 M KCl to have a total final salt concentration of 70 mM KCl. Remember to take into account the contribution to KCl concentration given by the extracts and eventually by other buffers used.
3. Assemble on dry ice the DNA substrates (pGEM-X and pGEM-T, kept at –80°C) and cell extracts (kept at –80°C or in liquid nitrogen) and keep frozen on dry ice.
4. Mix the appropriate amounts of labeled deoxynucleotide ([α-^{32}P]TTP or [α-^{32}P]dCTP in the case of pGEMX) (*see* **Notes 22–24**) and H$_2$O. Each reaction includes 0.2 µL (2 µCi) of labeled deoxynucleotide and 4.8 µL of H$_2$O. Make a mix for two more reactions than needed.
5. Add the components to microfuge tubes on wet ice in the following order (**Table 1**; *see* **Note 25**):
   a. Extract buffer (the amount depends on the amount of extract added, up to a total volume of 20 µL) to the bottom of the tubes.
   b. TE, pH 7.8 (up to a total volume of 50 µL).
   c. Plasmid DNA to the bottom of the tubes.
   d. 11 µL of 5X reaction buffer/CPK/KCl to the side of the tubes.
   e. 5 µL of [α-^{32}P]dNTP/H$_2$O to the side of the tubes.
6. Quick-thaw the cell extract, add the appropriate amounts, and vortex-mix.
7. Briefly centrifuge the tubes in a microcentrifuge.

8. Incubate at 30°C for 1 h (*see* **Note 26**). Stop the reactions by adding EDTA (20 mM final concentration) on wet ice (*see* **Note 27**).
9. Briefly centrifuge the tubes in a microcentrifuge.
10. Add 2 µL of 2 mg/mL of RNase A (80 µg/mL final concentration). Incubate at 37°C for 10 min.
11. Add 3 µL of 10% SDS (0.5% final concentration) and 6 µL of 2 mg/mL Proteinase K (190 µg/mL final concentration). Incubate at 37°C for 30 min.
12. Extract and precipitate the DNA as in **Subheading 3.1., step 5**.
13. Pellet the DNA by centrifugation in a microcentrifuge at 20,800g (14,000 rpm), 4°C.
14. Gently remove the ethanol with a Gilson micropipet keeping the tip as far as possible from the DNA pellet. Leave 100–200 µL of ethanol above the pellet.
15. Wash the pellet by adding 1 mL of cold (–20°C) 70% ethanol.
16. Centrifuge at 20,800g (14,000 rpm), 30 min, 4°C.
17. Carefully remove the ethanol rinse without disturbing the DNA pellet.
18. Dry the DNA pellet at room temperature. Do not overdry the pellet because it will become difficult to redissolve.
19. Incubate the DNA with the appropriate restriction endonuclease according to manufacturer's instructions (*see* **Note 25**).
20. Extract and precipitate DNA as in **Subheading 3.1., step 5**.
21. Resuspend the precipitate in 4 µL of Tris–EDTA. Vortex-mix and then add 10 µL of denaturing loading buffer to the tubes. Vortex-mix the DNA.
22. Heat the DNA samples at 95°C for 2 min.
23. Load onto a 15% polyacrylamide gel containing 7 M urea in 90 mM Tris-borate/ 2 mM EDTA, pH 8.8 .
24. Electrophorese the samples at 30 mA for 60–90 min at room temperature.
25. Fix the gel in 10% methanol, 10% acetic acid for 20 min.
26. Dry the gel and expose it to X-ray film in a cassette with intensifying screens. Store the gel at –80° C. The results of such an analysis are shown in **Fig. 3**.
27. Quantitate the repair incorporation by phosphorimager analysis or electronic autoradiography. Alternatively, excise the band from the gel and count it in a liquid scintillation counter.

3.4.2. Repair Kinetics by a Plasmid Assay

The DNA lesion of interest is constructed within the recognition sequence of a restriction endonuclease which is unable to cleave when the lesion is present on DNA. Lesion removal can be monitored either by incubation with the specific repair enzyme (e.g., *E. coli* uracil DNA glycosylase in the case of uracil) or by restoration of restriction endonuclease sensitivity.

1. Premix appropriate amounts of 5X reaction buffer and CPK in order to run two more reactions than needed. Each 50-µL reaction requires 10 µL of 5X buffer and 1 µL of CPK. Keep on ice.

Base Excision Repair Assay

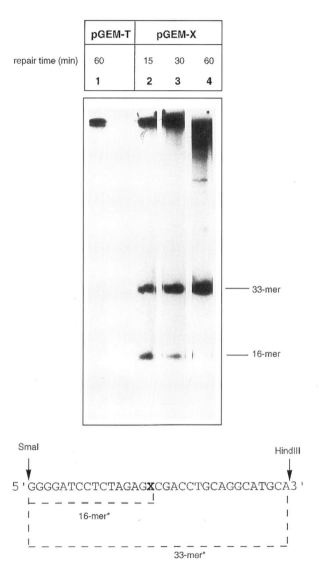

Fig. 3. Kinetics of repair replication and ligation of a single abasic site by CHO-9 whole-cell extracts. (Top) Autoradiograph of a denaturing polyacrylamide gel. Repair replication was performed in the presence of [α-^{32}P]TTP. pGEM-T (lane 1) or pGEM-X (lanes 2–4) were digested with SmaI and HindIII to release the 33-mer, which originally contained the AP site. pGEM-X was incubated with cell extracts from 15 (lane 2), 30 (lane 3), and 60 (lane 4) min. Unligated repair products arising from the single nucleotide insertion pathway (16-mer) are visible at the bottom of the gel after 15 (lane 2) and 30 (lane 3) min of repair time. (Bottom) Scheme of the expected repair products.

2. Assemble the repair reaction adding the components to microcentrifuge tubes on wet ice in the following order:
   a. 5X reaction buffer   10 µL
   b. CPK (2.5 mg/mL)   1 µL
   c. Labeled DNA (2 ng/µL)   2 µL
   d. Cell extracts   variable [a]
   e. TE, pH 7.8   up to 50 µL

   [a] Usually 20–40 µg of nuclear cell extracts and 100–150 µg of whole cell extracts are used.

3. Briefly vortex-mix and centrifuge the tubes.
4. Incubate at 30°C for the appropriate time.
5. Stop the reaction by adding EDTA to a final concentration of 20 mM on wet ice.
6. Perform **steps 9–18** as in **Subheading 3.4.1**.
7. Incubate the samples with the lesion-specific repair enzyme or with the appropriate restriction endonucleases according to the manufacturer's instructions.
8. Analyse the samples by gel electrophoresis on a 1.2% agarose gel.
9. Measure the relative yield of the different plasmid forms by electronic autoradiography.

The results obtained by using HeLa cell nuclear extracts on uracil-containing plasmid molecules are shown in **Fig. 4** (*see* **Notes 29** and **30**).

4. Notes

1. DNA substrates containing multiple AP sites can also be used *(9)*. This type of DNA substrate is much easier to produce, although it does not allow discrimination between the two BER pathways. In brief, depurinated plasmid DNA can be obtained as follows:
   a. Incubate plasmid DNA in 1X depurination buffer (10 mM sodium citrate, 100 mM sodium chloride, pH 5.0) at 70°C for various times (e.g., 15, 30, 45, and 60 min).
   b. Precipitate the DNA by standard procedure.
   c. Isolate pure supercoiled plasmid forms by equilibrium density gradient centrifugation with cesium chloride (*see* **Note 9**).
   d. Measure the number of AP sites by digestion of the plasmid DNA with an AP-endonuclease, such as *E. coli* endonuclease III (Nth protein), followed by

Fig. 4. *(Opposite page)* Repair assay with duplex plasmid molecules containing a single uracil residue. **(A)** Scheme of the oligonucleotide used to construct the plasmid pGEM U:C is indicated. **(B)** Autoradiograph of agarose gels: lane 1: construct containing the control oligonucleotide, pGEM-T; lane 2: pGEM-T after 120 min of incubation at 30°C, no extract; lane 3: pGEM-T after digestion with *Sal*I; lane 4: pGEM-T after digestion with UDG and Nth; lane 5: pGEM-U:C construct; lane 6: pGEM-U:C after digestion with *Sal*I; lane 7: pGEM-U:C after digestion with UDG and Nth. *(Continued)*

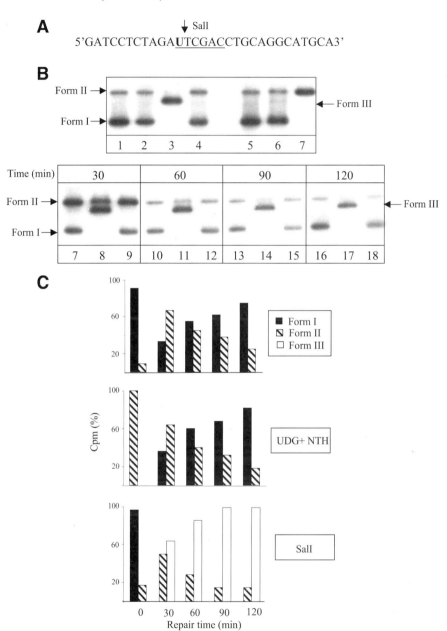

Fig. 4. *(Continued)* **(B)** Lanes 7, 10, 13, and 16: pGEM-U:C after incubation with HeLa nuclear extracts for increasing periods of time; lanes 8, 11, 14, and 17 followed by *Sal*I digestion; lanes 9, 12, 15, and 18 followed by digestion with UDG and Nth. **(C)** The relative percentages of form I, II, and III are obtained by electronic autoradioagraphy of the gel.

1% agarose gel electrophoresis and densitometric scanning of the relative amounts of supercoiled and nicked forms. The number of abasic sites is calculated by Poisson analysis.

   e. Store the depurinated plasmid DNA in aliquots at –80°C.
2. The standard alkaline lysis method for DNA plasmid preparation should be avoided since the alkaline lysis step might induce alkali-labile sites which would increase the background incorporation. The method of choice for plasmid preparation involves a neutral lysis step by SDS *(10)*.
3. Slowly add 30 g of Sephadex G-50 (medium) to 250 mL of TE, pH 8.0, and make sure the powder is well dispersed. Autoclave and allow to cool to room temperature. Decant the supernatant and replace with an equal volume of TE, pH 8.0.
4. Sterile solutions should be used to prepare the in vitro replication buffer. 100 mM sterile stock solutions of dNTPs can be purchased from Amersham Biosciences.
5. *E. coli* uracil-DNA-glycosylase and endonuclease III are commercially available.
6. Phosphorylation of the primer.
   a. Mix:

1 M Tris-HCl	3 µL
0.2 M MgCl$_2$	1.5 µL
1 M DTT	1.5 µL
1 mM ATP	13 µL
T4 PNK (20 U/ µL)	1 µL
22-mer (1 µg/µL)	2 µL
H$_2$O	8 µL

   b. Incubate at 37°C for 45 min and at 65°C for 10 min to inactivate PNK.
7. In our experimental system, the optimal molar ratio of phosphorylated primer (22-mer) to single-stranded plasmid DNA (3.2 kb) is 30:1. This ratio should be established experimentally on a case by case basis.
8. A successful in vitro replication reaction produces almost 90% closed circular duplex molecules. Ten percent of the replication products are usually molecules containing nicks or short gaps (which run as form II). Occasionally, aberrant replication products, which migrate faster than closed circular duplex DNA (form I), are detected on the gel. To avoid artifacts, it is extremely important to isolate closed circular molecules from the replication products.
9. Closed circular DNA can be purified by centrifugation in cesium chloride–ethidium bromide gradients by standard procedure *(12)*. We suggest:
   a. To use polyallomer Quick-Seal centrifugation tubes (13 × 51mm) for a Beckman Type-65 rotor.
   b. To avoid loading more than 30 µg of DNA per tube.
   c. To check carefully the final density of the cesium chloride solution (0.5 g added to 10 mL, η= 1.3905).
   d. To centrifuge at 290,000*g* (55,000 rpm) for 15 h at 18°C (using a VTi 65 rotor). Following collection of the closed circular plasmid DNA from the tube, remove the ethidium bromide by water-saturated butanol extraction and

dialyse the aqueous phase against several changes of TE, pH 8.0. Concentrate the DNA by butanol extraction to a final volume of 200–300 μL and then precipitate the DNA by adding 1/10 volume of 3 M sodium acetate and two volumes of pure cold ethanol. Leave the sample at –80°C for at least 30 min, and recover the DNA by centrifugation at 20,800g (14,000 rpm) for 30 min. Resuspend the pellet in an appropriate volume of TE, pH 8.0, and run the sample in a 1% agarose gel. The relative amount of nicked circular forms (form II) should not exceed 5% of the total DNA molecules.

10. To confirm that these plasmid molecules contain a single abasic site, digest a small aliquot with an AP-endonuclease (*see* **Note 1**). A complete conversion from closed circular (form I) to nicked (form II) forms should be observed (*see* **Fig. 2C**). Conversely the plasmid molecules containing the control oligonucleotide (pGEM-T) should not be cleaved (**Fig. 2B**).
11. End-labeling reaction of the oligonucleotide.
    a. Mix:
       10 pmol of oligonucleotide
       1X T4 PNK buffer
       50 μCi of [γ^{32}P]ATP (usually 5 μL from the commercial source)
       10 U T4 PNK (New England Biolabs)
       H$_2$O up to 50 μL of final volume
       At least 15–20 × 10^6 cpm/μL should be expected (1 × 10^{10} cpm/μg oligonucleotide).
12. No purification by cesium chloride gradient is required.
13. The procedure for preparation of whole cell extracts is basically that described by Manley et al. *(13,14)* for transcription-competent mammalian cell extracts with minor modifications *(8,15)*.
14. The procedure for preparation of crude extracts is basically that described by Tanaka et al. *(16)* with minor modifications *(17)*.
15. All amounts and volumes indicated in **Subheadings 2** and **3** refer to a pcv of 1 mL usually obtained with approx 0.6 × 10^9 cells. We often find it more convenient to prepare more extracts and we usually start with a pcv of 2 mL obtained by pelleting 1–1.2 × 10^9 cells. In this case all amounts and volumes should be doubled.
16. Rinse all glassware used during the cell extract preparation with PBS, pH 7.6, just before use.
17. Add DTT to buffers (hypotonic lysis, sucrose–glycerol and dialysis buffer) immediately before use. Some degradation products of DTT that form during prolonged storage may partially inactivate cell extracts.
18. In order to make active extracts, cells should be mycoplasma-free, healthy and exponentially growing.
19. CHO-9 cells and its derivative mutants grow very well in F10/DMEM 1:1 with 10% fetal calf serum. Inoculate in each roller bottle a minimum of 30 × 10^6 cells.
20. The procedure for the in vitro repair assay is basically that described in Wood et al. *(8,15)* with minor modifications *(5,9)*. The protocol described here refers to the use of plasmid DNA containing a single lesion as DNA substrate.

21. A simplified version of the in vitro BER assay involves the use of DNA substrates containing multiple AP sites *(9)*. In brief, for each reaction, 300 ng of heat-depurinated plasmid (e.g., pUC12, 2.7 kb) (*see* **Note 1**) are mixed with 300 ng of control undamaged plasmid of a different size (e.g., pBR322, 4.3 kb) and incubated with cell extracts as described in **Subheading 3.4.**, the only exception being the use of [α-^{32}P]dATP as the labeled deoxynucleotide. This is because a preferential loss of purines is produced after heating. The plasmid DNAs are purified by standard procedure and linearized with a restriction endonuclease that cuts at a single site. Damaged and undamaged plasmids are then separated overnight by 1% agarose gel electrophoresis in the presence of ethidium bromide (0.5 mg/mL). Repair synthesis is quantified by densitometric scanning of the gel autoradiograph. The amount of [α-^{32}P]dAMP incorporated is corrected for the amount of recovered DNA as measured by densitometric scanning of the photographic negative of the gel.
22. Unless otherwise indicated, perform all operations in microcentrifuge tubes in racks on wet ice.
23. Two BER pathways have been identified in mammalian cells: a single-nucleotide insertion pathway, which is catalyzed by DNA polymerase β *(10,18)*, and a proliferating cell nuclear antigen (PCNA)-dependent pathway, which involves the resynthesis of 2–6 nucleotides 3′ to the lesion *(4–6)*. With pGEM-X, the use of [α-^{32}P]TTP allows one to monitor mainly short-patch BER (the single AP site is located opposite an adenine and one-gap filling reactions are the major repair route for this lesion), whereas [α-^{32}P]dCTP is suggested for measuring the long-patch BER (cytosine is the most represented base 3′ to the abasic site) (**Figs. 1 and 3**).
24. Better reproducibility can be obtained if fresh batches of α-^{32}P-labeled deoxynucleotides are used. Stabilized aqueous solutions (such as the redivue Amersham preparations) are also recommended.
25. Take care that all buffers (e.g., dialysis buffer in which extracts are dissolved) are present in equal amounts in all the reactions.
26. An incubation time of 1 h is usually appropriate for studying both the short- and long-patch BER pathways. The latter pathway is slower than the former but both reactions are completed within 60 min. However, because of a certain degree of variability in the repair activity of different extract preparations, we suggest running a preliminary kinetic experiment (15-, 30-, 60-, 180-min incubation times) when setting up the system. This kind of experiment is very informative and may help in choosing the most suitable incubation time.
27. At this step the samples can be stored in Plexiglas box at –80°C for 24–48 h if necessary.
28. In the case of pGEM-X incubation of the repaired plasmid DNA with *Sma*I and *Hin*dIII restriction endonucleases releases a fragment of 33 bp which originally contained the abasic site (**Fig. 3**). This restriction fragment can be easily analyzed by 15% denaturing polyacrylamide gel electrophoresis. The occurrence of complete repair is shown by the appearance on the gel autoradiograph of a band

corresponding to a 33-mer. Unligated products containing one replaced nucleotide migrate as 16 bp oligonucleotides (**Fig. 3**). Appropriate DNA size markers can be obtained by 5′-end-labeling oligonucleotides having the same sequence as the restriction fragment.
29. The repair kinetics, as monitored by *Sal*I digestion, is faster than that detected by uracil DNA glycosylase/Nth cleavage (**Fig. 4C**). These results are likely explained by *Sal*I cleavage not only of fully repaired molecules but also of repair intermediates (i.e., duplex molecules containing abasic sites and/or or nicks).
30. A similar experimental approach is described in Allinson et al. *(19)*. In this case the repaired plasmid molecules were digested with restriction endonucleases to generate a fragment originally containing the lesion of interest. The reaction products were then analyzed by electrophoresis in a 10% denaturing polyacrylamide gel: This allowed the identification of the repair intermediates generated during the repair process (i.e., repair blocked after incision or after addition of the first nucleotide during repair synthesis). In the plasmid assay described in **Subheading 3.4.2.** both repair intermediates would generate form II plasmid molecules.

Acknowledgments

This work was supported by grants of Compagnia di S. Paolo (Programma Oncologia), Italian Association for Cancer Research (AIRC) and Ministry of University and Research, Fondo Investimenti Ricerca Base (FIRB).

References

1. Wood, R. D. (1996) DNA repair in eukaryotes. *Annu. Rev. Biochem.* **65,** 135–167.
2. Sobol, R. W., Horton, J. K., Kuhn, R., et al. (1996) Requirement of mammalian DNA polymerase β in base excision repair. *Nature* **379,** 183–186.
3. Xanthoudakis, S., Smeyne, R. J., Wallace, J. D., and Curran, T. (1996) The redox/DNA repair protein, Ref-1, is essential for early embryonic development in mice. *Proc. Natl. Acad. Sci. USA* **93,** 8919–8923.
4. Matsumoto, Y., Kim, K., and Bogenhagen, D. F. (1994) Proliferating cell nuclear antigen-dependent abasic site repair in *Xenopus laevis* oocytes: an alternative pathway of base excision DNA repair. *Mol. Cell. Biol.* **14,** 6187–6197.
5. Frosina G., Fortini, P., Rossi, O., et al. (1996) Two pathways for base excision repair in mammalian cells, *J. Biol.Chem.* **271,** 9573–9578.
6. Klungland, A. and Lindahl, T. (1997) Second pathway for completion of human DNA base excision repair: reconstitution with purified proteins and requirement for DNase IV (FEN1). *EMBO J.* **16,** 3341–3348.
7. Dogliotti, E., Fortini, P., Pascucci, B., and Parlanti, E. (2001) The mechanism of switching among multiple BER pathways. *Prog. Nucleic Acids Res. Mol. Biol.* **68,** 3–27.
8. Wood, R. D., Robins, P., and Lindahl, T. (1988) Complementation of the xeroderma pigmentosum DNA repair defect in cell-free extracts. *Cell* **53,** 97–106.

9. Frosina, G., Fortini, P., Rossi, O., Carrozzino, F., Abbondandolo, A., and Dogliotti, E. (1994) Repair of abasic sites by mammalian cell extracts. *Biochem. J.* **304,** 699–705.
10. Kubota, Y., Nash, R., Klungland, A., Schar, P., Barnes, D., and Lindahl, T. (1996) Reconstitution repair of DNA base excision-repair with purified human proteins: interaction between DNA polymerase β and the XRCC1 protein. *EMBO J.* **15,** 6662–6670.
11. Svoboda, D. L. and Vos, J.-M. H. (1999) Assays of bypass replication of genotoxic lesions in mammalian disease and mutant cell-free extracts. *Methods Mol. Biol.* **113,** 555–576.
12. Sambrook, J., Fritsch, E. F., and Maniatis, T. (1989) *Molecular Cloning: A Laboratory Manual.* Cold Spring Harbor Laboratory Press, Plainville, NJ.
13. Manley, J. L., Fire, A., Cano, A., Sharp, P. A., and Gefter, M. L. (1980) DNA-dependent transcription of adenovirus genes in a soluble whole-cell extract. *Proc. Natl. Acad. Sci. USA* **77,** 3855–3859.
14. Manley, J. L., Fire, A., Samuels, M., and Sharp, P. A. (1983) In vitro transcription: whole-cell extract. *Methods Enzymol.* **101,** 568–582.
15. Wood, R. D., Biggerstaff, M., and Shivji, M. K. K. (1995) Detection and measurement of nucleotide excision repair synthesis by mammalian cell extracts in vitro. *Methods* **7,** 163–175.
16. Tanaka, M., Lai, J. S., and Herr, W. (1992) Promoter-selective activation domains in Oct-1 and Oct-2 direct differential activation of an snRNA and mRNA promoter. *Cell* **68,** 755–767.
17. Biade, S., Sobol, R. W., Wilson, S. H., and Matsumoto, Y. (1998) Impairment of proliferating cell nuclear antigen-dependent apurinic/apyrimidinic site repair on linear DNA. *J. Biol. Chem.* **273,** 898–902.
18. Dianov, G., Price, A., and Lindahl, T. (1992) Generation of single-nucleotide repair patches following excision of uracil residues from DNA. *Mol. Cell. Biol.* **12,** 1605–1612.
19. Allinson, S. L., Dianova, I. I., and Dianov, G. L. (2001) DNA polymerase β is the major dRP lyase involved in repair of oxidative base lesions in DNA by mammalian cell extracts. *EMBO J.* **23,** 6919–6926.

28

Biochemical Assays for the Characterization of DNA Helicases

Robert M. Brosh, Jr. and Sudha Sharma

Summary

Helicases are ubiquitous enzymes that disrupt complementary strands of duplex nucleic acid in a reaction dependent on nucleoside-5'-triphosphate hydrolysis. Helicases are implicated in the metabolism of DNA structures that are generated during replication, recombination, and DNA repair. Furthermore, an increasing number of helicases have been linked to genomic instability and human disease. With the growing interest in helicase mechanism and function, we have set out to describe some basic protocols for biochemical characterization of DNA helicases. Protocols for measuring ATP hydrolysis, DNA binding, and catalytic unwinding activity of DNA helicases are provided. Application of these procedures should enable the researcher to address fundamental questions regarding the biochemical properties of a given helicase, which would serve as a platform for further investigation of its molecular and cellular functions.

Key Words: ATPase; DNA metabolism; DNA unwinding; helicase; molecular motor.

1. Introduction

DNA motor proteins represent a ubiquitous family of enzymes that translocate along nucleic acid polymers (for review, *see* **ref. *1***). DNA helicases are a specialized class of molecular motor proteins that disrupt complementary strands of duplex DNA in a reaction dependent on nucleoside-5'-triphosphate hydrolysis *(2–5)*. Evidence suggests that helicases play important roles in nucleic acid metabolism, and a growing number of helicases have been linked to human disease *(6–8)*. Understanding the mechanism of helicase-catalyzed DNA unwinding and the DNA substrate requirements for efficient helicase action may provide insight to the types of specific nucleic acid structures in replication, recombination or DNA repair that the helicase acts upon.

From: *Methods in Molecular Biology: DNA Repair Protocols: Mammalian Systems, Second Edition*
Edited by: D. S. Henderson © Humana Press Inc., Totowa, NJ

A number of methods have been developed to analyze the mechanism and function of prokaryotic and eukaryotic DNA helicases. The directionality of movement of a given helicase along ssDNA has been inferred from strand displacement assays using a linear partial duplex DNA substrate containing two radiolabeled oligonucleotides that are annealed to the very proximal opposite ends of a long ssDNA molecule *(9,10)*. Preferential release of one of the two oligonucleotides on such a substrate is used to define the polarity of movement of the helicase along the bound ssDNA residing between the duplexes. However, formal proof for unidirectional translocation of a helicase along a DNA molecule is warranted, and has been provided recently for the PcrA helicase *(11,12)*. Although directional unwinding by a helicase can be inferred from the result of a strand displacement assay using a directionality DNA substrate, the helicase may not require a preexisting ssDNA tail of defined polarity for either loading or initiation of unwinding. This principle becomes important for a variety of DNA replication or repair intermediates (e.g., replication fork, 5′ flap, D-loop, Holliday Junction) that the helicase may act upon in vivo. A number of helicases prefer a forked duplex compared to a duplex with a ssDNA tail for unwinding (e.g., **refs.** *13–15*); however, this is not a universal property of all DNA helicases. The DNA substrate specificity of a helicase may be conferred by its DNA binding preference *(16)* or by DNA structural elements that serve to occlude one of the single strands of the duplex from the central channel of the helicase *(15)*. A useful tool for probing the importance of DNA structural elements for helicase unwinding function is a DNA substrate with a specifically positioned steric block such as a biotin-streptavidin complex whose strong affinity and size can block a helicase when it is positioned on the single strand on which the enzyme translocates *(14)*. However, certain DNA helicases have the ability to displace streptavidin from biotin-labeled oligonucleotides *(17,18)*, suggesting a role for removing proteins during cellular processes such as DNA repair.

The length dependence for duplex DNA unwinding is an important property of DNA helicases *(10)*. By testing duplex DNA substrates of increasing lengths, information pertaining to the processivity of a DNA helicase can be determined. Whereas some helicases are very processive and can unwind thousands of base pairs of duplex DNA, other helicases only efficiently unwind fairly short duplexes (<50 bp). The presence of an auxiliary factor can greatly alter the ability of a helicase to unwind duplex DNA, and in some cases enable a relatively non-processive helicase to act in a processive fashion and unwind very long duplexes. An example of a protein auxiliary factor for a DNA helicase is the human single-stranded DNA binding protein Replication Protein A (RPA). RPA interacts with WRN *(19,20)*, BLM *(21)*, and RECQ1 *(22)* helicases and stimulates their DNA unwinding activities. Since RPA has been shown to have roles in DNA replication, recombination and repair, these helicases are likely to function with the single stranded DNA binding protein during one or more of these processes.

In addition to classical B-form DNA duplex substrates, certain helicases can unwind DNA:RNA hybrids or RNA duplexes (e.g., **ref. 23**). Specialized DNA helicases can efficiently unwind alternate DNA structures including triplexes (e.g.,**ref. 24**) and tetraplexes (e.g., **ref. 16**) which can be formed by sequences that are widely distributed throughout the human genome. Such alternate DNA structures that deviate from the canonical Watson-Crick base-pairing potentially interfere with cellular processes such as replication or transcription, and may give rise to genomic instability. At least some of the genomic instability detected in human RecQ disorders may be a consequence of the absence of the respective DNA helicase to resolve these structures.

DNA damage may also hinder pathways of nucleic acid metabolism in vivo. Helicases are likely to encounter DNA lesions during the cellular processes of replication or DNA repair. Generally speaking, helicase inhibition by helix-distorting lesions is strand-specific, that is, a DNA adduct inhibits DNA unwinding catalyzed by a helicase when the adduct is positioned in the strand that the helicase translocates upon, but to a significantly lesser degree when the adduct is on the opposite strand to the helicase translocating strand (for review, *see* **ref. 25**). A classic example of strand-specific inhibition was reported for Rad3, a 5' to 3' DNA helicase implicated in nucleotide excision repair and transcription *(26)*. Stereochemistry and orientation of the adduct can impact the ability of the adduct to deter DNA unwinding, as demonstrated for the WRN helicase *(27)*. Information on the mechanism of DNA unwinding by a helicase can be obtained by in vitro helicase studies using DNA substrates with site-specific adducts.

Up to this point, we have focused on DNA structural elements that might be important for efficient DNA unwinding catalyzed by a helicase. However, solution reaction conditions are also an important factor in the optimization of helicase activity. Recently, it was reported that *E. coli* RecQ helicase activity is sensitive to the ratio of free magnesium ion to ATP concentration with an optimal ratio of 0.8 and a free magnesium ion concentration of 50 μM *(28)*. In addition, *E. coli* RecQ helicase activity displayed a sigmoidal dependence on ATP concentration, suggesting multiple interacting ATP sites *(28)*. Like *E. coli* RecQ, WRN helicase activity increases with Mg^{2+}:ATP ratios up to 1; however, greater Mg^{2+}:ATP ratios of up to 4 do not significantly further increase or decrease WRN unwinding (our unpublished data). These results and others indicate that the reaction conditions for helicase catalyzed unwinding should be optimized for each DNA helicase since the solution factors may uniquely affect the structural and functional properties of the helicase in question.

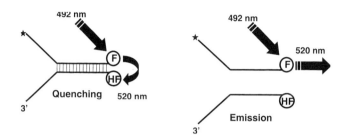

Fig. 1. Schematic representation of a typical fluorescent DNA substrate design. A forked duplex flanked by noncomplementary 5′ and 3′ ssDNA tails is shown. The length of the ssDNA tails and duplex region can be optimized for the helicase under study according to its DNA substrate preference and processivity, respectively. Fluorescein (F) is covalently attached to the 3′ blunt end and hexachlorofluorescein (HF) is attached to the 5′ blunt end of the DNA substrate. The schematic depicts how, when in close proximity, HF quenches the signal emitted from F upon excitation. Once the DNA substrate is unwound, the emission from F upon excitation is free to be detected because the HF is no longer in close proximity with F. The asterisk (*) indicates the position of the 5′ ^{32}P label for the radiolabeled DNA substrate which can be used in radiometric assays to substantiate the unwinding results from the fluorometric assay.

To better understand helicase mechanism and function, kinetic analyses of helicase-catalyzed unwinding of duplex DNA using fluorescence stopped-flow instrumentation is valuable since unlike classic radiometric assays, the data is collected continuously throughout the reaction in real time. The fluorometric assay uses the principle of fluorescence resonance energy transfer (FRET) to observe the unwinding of duplex DNA *(29)*. FRET occurs between a donor (e.g., fluorescein) and acceptor (e.g., hexachlorofluorescein) covalently attached to the complementary strands of the duplex substrate, and upon separation of the complementary strands, F and HF are no longer in close proximity and the fluorescence emission from F excitation can be detected by a photosensor (**Fig. 1**). Unwinding data from fluorometric assays can be substantiated using chemical quench flow kinetic analyses of data from radiometric helicase assays. Fast kinetics of helicase activity can be studied under two sets of conditions: (1) single-turnover, in which the DNA substrate is saturated with an excess of enzyme; (2) pre-steady state, in which the enzyme is saturated with an excess of DNA substrate. Results from single-turnover and pre-steady state kinetic analyses can be used to detect lag phase or initial burst kinetics of helicase-catalyzed DNA unwinding, respectively, under a given set of conditions *(30–32)*. The kinetic mechanism for the formation of the active monomeric or multimeric helicase-DNA complex or the kinetic step size in sequential unwinding mechanisms can be determined from such analyses.

The focus of this chapter for characterizing helicase-catalyzed DNA unwinding will be the use of radiometric helicase assays, particularly for those applications to determine DNA structural elements of the helicase substrate important for efficient helicase activity. The protocols for radiometric helicase assays are relatively easy to implement and interpret, enabling the researcher to gain some useful insight to mechanistic aspects of the helicase reaction as they relate to properties of the DNA substrate and characterization of the enzyme's unwinding activity on DNA substrates that might be acted upon in DNA metabolism. In addition, biochemical assays to detect and quantitate DNA binding and ATP hydrolysis by a helicase are described.

2. Materials

2.1. Radiolabeled Duplex DNA Substrate

1. T4 Polynucleotide kinase (10,000 U/mL) and 10X kinase buffer (New England Biolabs; NEB).
2. [γ-^{32}P]ATP (>5000 Ci/mmol; Amersham Biosciences).
3. Polyacrylamide gel electrophoresis (PAGE)-purified oligonucleotides (Midland Certified Reagent Company, Loftstrand Technologies, or vendor of choice).
4. Microspin G-25 Columns (Amersham Biosciences, cat. no. 27-5325-01).
5. 1 M NaCl.

2.2. Radiolabeled Oligonucleotide-Based Holliday Junction Substrate

1. **Items 1–4, Subheading 2.1.** For item 3, *see* **Note 1**.
2. 10% polyacrylamide gel: To 26 mL of ddH$_2$O is added 10 mL of 40% acrylamide (19:1 acrylamide:*bis*-acrylamide; Bio-Rad), 4 mL of 10X TBE (Molecular Biology Grade, Quality Biological Inc.), 0.3 mL of 10% ammonium persulfate, and 30 µL of *N,N,N',N'*-tetramethylethylenediamine (TEMED). Mix and pour the solution to make a 1.5-mm thick 18 cm × 16 cm gel.
3. Vertical Slab Gel Electrophoresis Unit (The Sturdier Model SE400, Hoefer Scientific Instruments) and Power Supply.
4. 2X Helicase load buffer: 36 mM EDTA, 50% glycerol, 0.08% bromophenol blue, 0.08% xylene cyanol.
5. Sterile razor blades.
6. Kodak X-ray film (XOMAT AR or BioMax MS) and film cassette.
7. Dialysis tubing (Spectra/Por, MWCO: 6000–8000) that has been preequilibrated in 0.5X TBE.
8. Weighted dialysis clips (Spectrum).
9. Complete Counting Scintillation Cocktail (3a70B, Research Product International Corp.) and 20-mL scintillation vials.

2.3. Radiometric Helicase Assay

1. 12% Polyacrylamide gel: same as **Subheading 2.2., item 2** except that 12 mL of 40% acrylamide and 24 mL of ddH$_2$O are used. *See* **Note 2**.

2. Vertical slab gel electrophoresis unit and power supply (**Subheading 2.2., item 3**).
3. 5X Reaction buffer: 150 mM HEPES, pH 7.4, 25% glycerol, 200 mM KCl, 500 μg/mL of bovine serum albumen (BSA), 5 mM MgCl$_2$. *See* **Note 3**.
4. 10 mM ATP (dilute 100 mM ATP (ultrapure >98%, Amersham Biosciences) 1:10 in ddH$_2$O).
5. 2X Helicase load buffer, *see* **Subheading 2.2., item 4**. *See* **Note 4**.
6. Radiolabeled DNA substrate (*see* **Subheadings 3.1., 3.2.**)
7. PhosphorImager plates (Molecular Dynamics).

2.4. Nitrocellulose Filter DNA Binding Assay

1. Nitrocellulose filters (0.45 μm, 25 mm diameter, Whatman).
2. Vacuum manifold (Millipore, cat. no. XX2702550).
3. Blunt forceps.
4. 5X Reaction buffer (*see* **Subheading 2.3., item 3**). *See* **Note 3**.
5. Radiolabeled DNA substrate (*see* **Subheadings 3.1., 3.2.**).
6. Counting scintillation cocktail (3a70B, Research Product International Corp.) and 20-mL scintillation vials.

2.5. Gel-shift DNA Binding Assay

1. 5% polyacrylamide gel: same as **Subheading 2.2., item 2** except that 5 mL of 40% acrylamide and 31 mL of double-distilled water (ddH$_2$O) are used. *See* **Note 5**.
2. Vertical slab gel electrophoresis unit and power supply (*see* **Subheading 3.2., step 3**).
3. 5X DNA binding reaction buffer: 125 mM Tris-HCl, pH 7.5, 100 mM NaCl, 10 mM MgCl$_2$, 500 μg/mL of BSA.
4. Radiolabeled DNA substrate (*see* **Subheadings 3.1., 3.2.**).
5. Native loading buffer: 74% glycerol, 0.01% xylene cyanol, 0.01% bromophenol blue.
6. PhosphorImager plates (Molecular Dynamics).

2.6. ATPase Assay

1. Baker-flex Cellulose PEI sheets (J.T. Baker). To prepare a PEI cellulose sheet for thin-layer chromatography (TLC), cut the 20 cm × 20 cm sheet in half with scissors. With a pencil and ruler, draw a line 1 cm from the bottom of a half sheet. Make small tick-marks along the line at 1-cm intervals with a pencil. These tick-marks will serve as positions for spotting aliquots of quenched ATPase reaction mixtures (*see* **Subheading 3.6., step 5**).
2. 5X Reaction buffer (*see* **Subheading 2.3., item 3**). *See* **Note 3**.
3. DNA effector: M13 ssDNA (New England Biolabs), poly(dT) or poly(dA) (Amersham Biosciences), single stranded or double stranded oligonucleotide (*see* **Subheading 2.1., item 3**).
4. [^{3}H]ATP (Amersham Biosciences) or [γ-^{32}P]ATP (Perkin Elmer Life Sciences).
5. ATPase stop solution: 6.7 mM rADP, 6.7 mM rATP, 33 mM EDTA. To prepare, add equal volumes of 20 mM ATP, 20 mM ADP, and 100 mM EDTA. 100

DNA Helicase Biochemical Assays

mM ATP and ADP stocks, prepared by dissolving powder stocks (Calbiochem), can be diluted to 20 mM. 500 mM EDTA (Research Genetics) can be diluted to 100 mM.

6. TLC solvent: 1 M formic acid (Sigma), 0.8 M LiCl for the [^{3}H]ATP hydrolysis assay, or 0.75 M KH$_2$PO$_4$ for the [γ-^{32}P]ATP hydrolysis assay.
7. TLC tank.
8. Blunt forceps.
9. Short wavelength UV handheld lamp (for [^{3}H]ATP hydrolysis assay).
10. Complete Counting Scintillation Cocktail (3a70B, Research Product International Corp.) and 20-mL scintillation vials for measurement of [^{3}H]ATP hydrolysis or PhosphorImager plate (Molecular Dynamics) for [γ-^{32}P]ATP hydrolysis.

3. Methods

3.1. Preparation of Radiolabeled Two- or Three-Stranded Duplex DNA Substrate

1. Radiolabel the gel-purified oligonucleotide at its 5′ end by adding 10 pmol of oligonucleotide (1 µL, 10 pmol/µL) to a microfuge tube containing 13 µL H$_2$O, 2 µL of 10X T4 Kinase buffer, and 3 µL of [γ-^{32}P]ATP (> 5000 Ci/mmol). Add 1 µL of T4 Polynucleotide Kinase (New England Biolabs). Mix the contents and briefly pulse down in a microfuge. Incubate the reaction mixture at 37°C for 60 min. Inactivate the kinase reaction by heating at 65°C for 10 min.
2. To remove the unincorporated [γ-^{32}P]ATP, load the 20-µL reaction mixture onto a prespun MicroSpin G-25 column. Centrifuge at 735g for 1 min, and collect the eluate.
3. For the annealing reaction, add 25 µL of H$_2$O, 2.5 µL of 1 M NaCl, and 25 pmol (2.5 µL, 10 pmol/mL) of complementary oligonucleotide to the tube containing 20 µL of ^{32}P-labeled oligonucleotide. (*See* **Note 6**.) Incubate the mixture in a boiling water bath for 5 min and then briefly pulse down the tube's contents. Immediately, transfer the tube to a 65°C water bath. On transfer, turn off the thermostat of the water bath and allow the oligonucleotides to anneal upon cooling to room temperature over a period of 4–5 h.
4. For three-stranded duplex DNA substrates where an upstream primer is present, add 50 pmol (2 µL, 25 pmol/µL) of upstream primer to the tube containing annealed duplex DNA mixture (50 µL). Incubate the mixture in a 37°C water bath for 60 min. After incubation, turn off the thermostat of the water bath and allow the upstream oligonucleotide to anneal to the duplex DNA substrate upon cooling to room temperature over a period of 3 h. (*See* **Note 7**.)
5. Store the mixture containing annealed duplex DNA substrate shielded at 4°C until required. The final concentration of the DNA substrate is 200 fmol/µL and can be diluted to 10 fmol/µL in substrate dilution buffer (10 mM Tris-HCl, pH 8.0, 50 mM KCl, 1 mM EDTA) for addition (1 µL, 10 fmol/µL) to helicase reaction mixtures (*see* **Subheading 3.3., step 2**) prior to the experiment.

3.2. Preparation of Four-Stranded Oligonucleotide-Based Holliday Junction

1. Radiolabel the gel-purified oligonucleotide X12-1 (*see* **Note 1**) at its 5′ end by adding 200 ng of oligonucleotide X12-1 (1 µL, 200 ng/µL) to a microfuge tube containing 4.5 µL of H_2O, 1 µL of 10X T4 kinase buffer, and 2.5 µL of [γ-^{32}P]ATP (>5000 Ci/mmol). Add 1 µL of T4 polynucleotide kinase (10 U/µL). Mix the contents and briefly pulse down in a microfuge. Incubate the reaction mixture at 37°C for 60 min. Inactivate the kinase reaction by adding 1 µL of 0.5 M EDTA and heating at 65°C for 20 min.
2. To remove the unincorporated [γ-^{32}P]ATP, load the 10-µL reaction mixture onto a prespun MicroSpin G-25 column. Centrifuge at 735g for 1 min, and collect the eluate.
3. Remove 0.5 µL of the eluate, add to 49.5 µL of TE, and save for determination of specific activity.
4. For the annealing reaction, add 1000 ng of oligonucleotides X12-2, X12-3, and X12-4 (1 µL, 1000 ng/µL) (*see* **Note 1**) to the tube containing 9.5 µL of ^{32}P-labeled X12-1. Incubate the mixture at 90°C for 5 min, 65°C for 10 min, and then 37°C for 10 min. After 10 min of incubation, turn off the thermostat of the water bath and allow oligonucleotides to anneal on cooling to room temperature over a period of 2–3 h.
5. Pour a nondenaturing 10% polyacrylamide gel (1.5 mm thick, 40 cm long) with wells of 8 mm width. (*See* **Subheading 2.2.**, item 2)
6. To the 12.5 µL of annealed HJ(X12-1) mixture, add an equal volume of 2X helicase load buffer.
7. Carefully load the annealing reaction mixture in helicase load buffer (two lanes, 12.5 µL per lane) onto a nondenaturing 10% (19: 1 acrylamide: bisacrylamide) polyacrylamide gel. Electrophorese the samples at 125 V in 1X TBE for 4 h.
8. Dismantle the gel apparatus at the end of the run and place the glass plate horizontally on the bench behind a shield. Discard the running buffer into ^{32}P liquid waste. Separate the glass plates with a plastic wedge and allow the gel to stick to one glass plate. Carefully blot off excess liquid surrounding the gel with a paper towel and cover the gel bound to one glass plate with Saran Wrap.
9. Expose the gel to X-ray film for approx 2–5 min. Process by autoradiography.
10. Carefully align the film with the gel and identify the desired band representing the four-stranded HJ structure. The four-stranded synthetic HJ will have migrated less than 2 cm into the gel and represents the slowest migrating species on the gel. Mark the desired band, and excise it from the rest of the gel using a razor blade.
11. Reexpose the gel lacking the excised band to verify that the radiolabeled band was removed.
12. Carefully place the gel slice in a dialysis bag (Spectra/Por, MWCO: 6000–8000) that has been preequilibrated in 0.5X TBE. Add 0.5 mL 0.5X TBE to the bag containing gel slice. Clamp with weighted dialysis clips (Spectrum). Place the clamped dialysis bag in a horizontal gel electrophoresis unit containing 0.5X

DNA Helicase Biochemical Assays

TBE. Electrophorese the sample at 60 V for 4 h at room temperature. At the end of the 4 h, reverse the current for 1 min.
13. Remove the radioactive liquid sample containing the HJ(X12-1) and store shielded at 4°C until needed.
14. Remove 1 μL of HJ(X12-1) sample and place in scintillation vial with 3 mL of scintillation cocktail (Vial A). Place 1 μL of the diluted eluate of the ^{32}P-labeled X12-1 oligonucleotide (**step 3**) in a scintillation vial that contains 3 mL of scintillation cocktail (vial B). Count Vials A and B in a liquid scintillation counter using a ^{32}P program. Use the cpm value from Vial B to determine the specific activity and calculate the concentration (ng/μL) of HJ(X12-1) using the cpm value from Vial A.
15. Convert the concentration of HJ(X12-1) to pmol/μL using the molecular weight of HJ(X12-1).

3.3 Radiometric Helicase Assay

1. Pour a nondenaturing 12% polyacrylamide gel (1.5 mm thick) with wells of 5 mm width. *See* **Subheading 2.3., item 1**.
2. Reaction mixtures are set up in 20-μL aliquots. Add 1 μL of appropriately diluted helicase enzyme to a prechilled microfuge tube sitting on ice containing 4 μL of 5X helicase reaction buffer, 10 fmol of DNA duplex substrate (1 μL, 10 fmol/μL), 2 μL of 10 m*M* ATP (or the specified nucleoside triphosphate, final concentration 1 m*M* nucleoside triphosphate), and 14 μL of H$_2$O. (*See* **Note 3**.)
3. Mix the reaction mixture briefly, pulse down in a microfuge briefly, and incubate in a 37°C (or desired temperature) water bath for 15 min (or the appropriate time).
4. Quench the reaction by adding 20 μL of 2X helicase load buffer containing a 10-fold excess of unlabeled oligonucleotide with the same sequence as the labeled strand. (*See* **Note 8**.)
5. Carefully load the products of the helicase reactions onto nondenaturing 12% (19: 1 acrylamide–*bis*-acrylamide) polyacrylamide gels. Electrophorese the samples at 180 Volts in 1X TBE for 2 h or the appropriate time. (*See* **Note 2**.)
6. Dismantle the gel apparatus at the end of the run and place the glass plate horizontally on the bench. Separate the glass plates with a plastic wedge and allow the gel to stick to one glass plate. Carefully blot off excess liquid surrounding the gel with a paper towel and cover the gel bound to one glass plate with Saran Wrap.
7. Expose the gel to a PhosphorImager screen for 5–15 h. Alternatively, expose the gel to X-ray film overnight at –20°C and process by autoradiography the following day.
8. Visualize radiolabeled DNA species in the polyacrylamide gel using a PhosphorImager. Examples of phosphorimages of native gels showing the WRN helicase products from a reaction containing 5' flap DNA substrate or a synthetic oligonucleotide-based Holliday Junction are shown in **Fig. 2** and **3**, respectively. The identity of the reaction products can be confirmed by resolving control

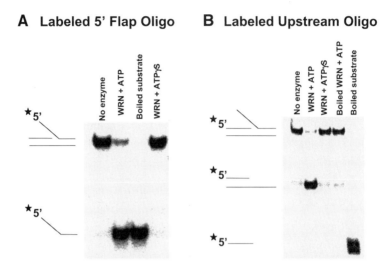

Fig. 2. Werner syndrome helicase (WRN) unwinds a 5′ ssDNA flap substrate. A 26 nt 5′ flap substrate (*see* text for details) with a 5′ ^{32}P label on the 5′ flap oligonucleotide (**A**) or the 25-mer upstream oligonucleotide (**B**) was incubated with WRN in helicase reaction buffer (30 m*M* HEPES, pH 7.6, 5% glycerol, 40 m*M* KCl, 0.1 mg/mL of BSA, 8 m*M* MgCl$_2$) containing 10 fmol of duplex DNA substrate (0.5 n*M* DNA substrate concentration) and 2 m*M* ATP where indicated. Helicase reactions were initiated by the addition of WRN and then incubated at 37°C for 15 min. Reaction mixtures (20 µL) were quenched with 20 µL of 2X helicase load buffer (36 m*M* EDTA, 50% glycerol, 0.08% bromophenol blue, 0.08% xylene cyanol) containing a 10-fold excess of unlabeled oligonucleotide with the same sequence as the labeled strand. The products of the helicase reactions were resolved on nondenaturing 12% (19:1 acrylamide–*bis*-acrylamide) polyacrylamide gels. Radiolabeled DNA species were visualized using a PhosphorImager. (**A**) Lane 1, no enzyme control; lane 2, 3.8 n*M* WRN + ATP; lane 3, heat-denatured DNA substrate control; lane 4, 3.8 n*M* WRN + ATPγS. (**B**) Lane 1, no enzyme control; lane 2, 3.8 n*M* WRN + ATP; lane 3, 3.8 n*M* WRN + ATPγS; lane 4, heat-denatured WRN protein control + ATP; lane 5, heat-denatured DNA substrate control.

samples of the one-, two-, or three-stranded radiolabeled DNA structures on the polyacrylamide gel. (*See* **Note 9**.)
9. Quantitate helicase reaction products using the ImageQuant software (Molecular Dynamics).
10. Calculate percent helicase substrate unwound using the following formula: percent unwinding = $100 \times (P/(S + P)$, where P is the product and S is the residual substrate. Determine values of P and S by subtracting background values in controls having no enzyme and heat-denatured substrate, respectively (*See* **Note 9**).

DNA Helicase Biochemical Assays

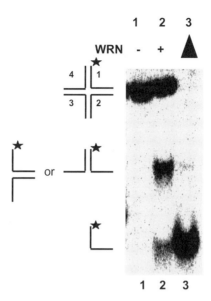

Fig. 3. WRN helicase unwinds a synthetic Holliday Junction structure. An oligonucleotide-based Holliday Junction substrate (HJ(X12-1); see text for details) was incubated with WRN (12 nM) in HJ unwinding buffer (20 mM Tris-HCl, pH 7.5, 2 mM MgCl$_2$, 0.1 mg/mL of BSA, 1 mM DTT) containing 2.5 fmol HJ(X12-1) and 2 mM ATP where indicated. HJ unwinding reaction (20 μL) was initiated by the addition of WRN and then incubated at 37°C for 15 min, stopped by the addition of 20 μL of 50 mM EDTA, 40% glycerol, 0.1% bromophenol blue, 0.1% xylene cyanole, and electrophoresed on nondenaturing 10% (19:1 acrylamide–bis-acrylamide) polyacrylamide gels. Radiolabeled DNA species were visualized using a PhosphorImager. Filled triangle, heat-denatured HJ(X12-1) substrate control.

An example of the quantitative data representation for unwinding of a series of related DNA substrates by WRN helicase is shown in **Fig. 4**.

3.4. DNA Nitrocellulose Filter Binding Assay

1. Pretreat nitrocellulose filters (0.45 μm; Whatman) by boiling in deionized distilled water for 20 min. Store for up to 1 d in helicase reaction buffer.
2. Calculate dilutions of protein to be tested for DNA binding. Set up DNA binding reaction mixtures (20 μL) as described in **Subheading 3.3., steps 2** and **3** in the absence of ATP. (*See* **Note 10**.)
3. At the end of incubation (15 min), dilute the reaction mixture with 1 mL of prewarmed (37°C) 1X helicase reaction buffer. (*See* **Note 11**.) Pass the diluted mixture over a nitrocellulose filter immobilized in a Millipore vacuum filtration apparatus at a flow rate of 4 mL/min.

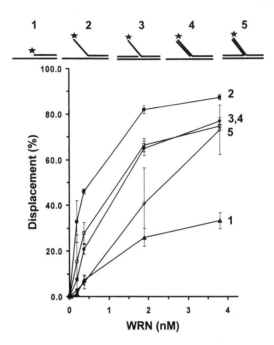

Fig. 4. WRN helicase activity on forked DNA substrates with dsDNA 5'- and/or 3'-tails. DNA substrates were incubated with increasing amounts of WRN in helicase reaction buffer (**Fig. 2A**) containing 10 fmol of duplex DNA substrate (0.5 nM DNA substrate concentration) and 2 mM ATP. Helicase reactions were initiated by the addition of WRN and then incubated at 37°C for 15 min. Quantitative analyses of WRN helicase data are shown.

4. Wash the filters with two 1-mL aliquots of warmed 1X helicase reaction buffer.
5. Turn off the vacuum and remove the cover of the manifold. Carefully remove the filters from the manifold using blunt edge forceps and place each filter onto the top of a scintillation vial to dry for approx 4 min.
6. Place the dried filters in scintillation vials with scintillation cocktail. Count in a liquid scintillation counter on a ^{32}P program.
7. Subtract background radioactivity (counts obtained from filters of control binding reactions lacking the helicase) from total radioactivity bound to the filter to determine counts of specifically bound DNA-protein complexes.
8. Calculate the fraction of the radiolabeled DNA substrate bound by the DNA helicase. (*See* **Note 12.**)

3.5. Gel Mobility Shift Assay for DNA Binding

1. Pour a nondenaturing 5% polyacrylamide gel (1.5 mm thick) with wells of 8-mm width. *See* **Subheading 2.5., item 1**. After the gel is polymerized (2–3 h), let it cool at 4°C.

DNA Helicase Biochemical Assays

2. Binding reaction mixtures are set up in 20-µL aliquots. Add 1 µL of appropriately diluted helicase enzyme to a prechilled microfuge tube sitting on ice containing 4 µL of 5X DNA binding reaction buffer, 1 µL of 10 mM dithiothreitol, 10 fmol of DNA duplex substrate (1 µL, 10 fmol/µL), 2 µL of 10 mM ATPγS (or the specified nucleotide, final concentration 1 mM), and 13 µL H$_2$O. (*See* **Note 3**.)
3. Mix the reaction mixture briefly, pulse down in a microfuge briefly, and incubate in a 37°C (or desired temperature) water bath for 15 min (or the appropriate time).
4. At the end of the incubation, add 3 µL of native loading buffer.
5. Carefully load DNA binding reaction mixtures onto nondenaturing 5% (19: 1 acrylamide–*bis*-acrylamide) polyacrylamide gels. (*See* **Note 5**.) Electrophorese the samples at 220 V in 1X TBE (or 0.5X TBE) for 2 h at 4°C.
6. Dismantle the gel apparatus and wrap the plate in Saran Wrap as described under **Subheading 3.3., step 6**.
7. Expose the gel to a phosphorimager screen for 5–15 h. Alternatively, expose the gel to X-ray film overnight at –20°C and process by autoradiography the following day.
8. Visualize the radiolabeled DNA species in a polyacrylamide gel using a Phosphor-Imager. *See* **Fig. 5** and **Notes 13–15**. Quantitate the fraction of DNA substrate bound by protein using the ImageQuant software (Molecular Dynamics).

3.6. ATPase Assay

1. Reaction mixtures are set up in 20-µL aliquots for K_m determination and 30-µL aliquots for k_{cat} determination. For simplicity, the 30-µL reaction volume will be described here. Add 2 µL of appropriately diluted helicase enzyme to a prewarmed microfuge tube containing 6 µL of 5X helicase reaction buffer, 3 µL of the desired DNA effector that has been appropriately diluted (e.g., M13 ssDNA circle, 0.1 mg/mL or 200 nM single-stranded or duplex oligonucleotide DNA molecules), 9.6 µL 2.5 mM [^{3}H]ATP (final concentration 0.8 mM) or [γ-^{32}P]ATP, and 9.4 µL of H$_2$O. Alternatively, DNA effector can be omitted and replaced with 3 µL of H$_2$O. (*See* **Note 3**.)
2. Mix the reaction mixture briefly, pulse down in a microfuge briefly, and incubate in a 37°C (or desired temperature) water bath for a fixed time (e.g., 10 min) to determine the K_m or a set of kinetic time points (e.g., 2, 4, 6, 8, 10 min) to determine k_{cat}.
3. At the end of the incubation, carefully remove 5-µL aliquots from the reaction mixture and add to 5-µL aliquots of ATPase stop solution (6.7 mM rADP, 6.7 mM rATP, 33 mM EDTA) that have been previously dispensed in labeled microfuge tubes. Mix and pulse down the tube contents in a microfuge. Repeat for all time points.
4. Spot 2–3 µL of quenched reaction mixture sample onto an appropriate position on a TLC sheet (*see* **Subheading 2.6., item 1**). Proceed with the next sample, spotting 2–3 µL each time. After an entire set of samples has been spotted consecutively, allow the TLC sheet to dry by air or with a blow dryer (low heat setting).

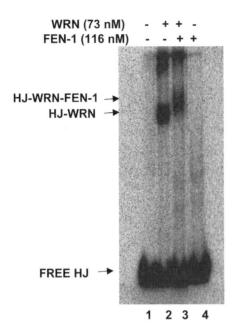

Fig. 5. WRN recruits FEN-1 to the Holliday Junction. WRN-HJ complex is supershifted by FEN-1. WRN (73 nM) and FEN-1 (116 nM) were incubated in HJ binding buffer (20 mM triethanolamine [pH 7.5], 2 mM MgCl$_2$, 0.1 mg/mL of BSA, 1 mM dithiothreitol) with 25 fmol of HJ(X12-1) and 1 mM ATPγS in a total volume of 20 μL at 24°C for 20 min. Protein–DNA complexes were fixed in the presence of 0.25% glutaraldehyde crosslinking agent for 10 min at 37°C. Protein–DNA complexes were analyzed on nondenaturing 5% polyacrylamide 0.5X TBE gels and visualized by PhosphorImage analysis.

5. Repeat the spotting (as described in **Subheading 3.6., step 4**), for each sample until the entire 10 μL has been spotted. This will require four or five sample applications.
6. Once the TLC plate has dried from the final application of samples, carefully dip the TLC plate into a glass baking dish containing a thin layer of deionized distilled water. Allow buffer to wick up toward the 1 cm horizontally marked line on the TLC plate. Prior to the water-front reaching the 1 cm line, carefully place the TLC plate in the tank containing a thin layer of 1 M formic acid–0.8 M LiCl for the [^{3}H]ATP hydrolysis assay or 0.75 M KH$_2$PO$_4$ for the [γ-^{32}P]ATP hydrolysis assay.
7. Allow the solvent front to reach 1 cm from the top of the TLC sheet for the [^{3}H]ATP assay. Carefully remove the TLC sheet from the tank and place the TLC sheet horizontally on the bench under a low intensity heat lamp. Do not overdry

the TLC plate as this will result in problems during the scraping step (samples will flake).

8. If [γ-^{32}P]ATP was used, the radioactivity on the TLC plate can be directly visualized on a PhosphorImager and quantitated using the ImageQuant software. The released ^{32}P$_i$ will have migrated significantly farther than [γ-^{32}P]ATP. If [^{3}H]ATP was used, then proceed with **step 9**.

9. Use a handheld short-wavelength UV lamp to visualize ADP and ATP spots which migrate to lower and higher positions, respectively, relative to the origin of spotting. Circle ADP and ATP spots with a pencil. Using a blunt-ended metal spatula, carefully scrape spots off and transfer with blunt-ended forceps to scintillation vials. Add 3 mL of scintillant and count using a ^{3}H program.

10. Subtract background radioactivity (counts obtained from ADP spots of control reactions lacking the helicase) from total radioactivity for each ADP spot to determine counts of ADP produced for each time reaction. Determine the specific activity of the ATP using the values of the ATP spots from the no enzyme control reactions and the calculated amount of ATP (4000 pmol in this case) in each spot. Using the specific activity term, convert the corrected ADP counts to pmol ADP. Initial rates of ATP hydrolysis (k_{cat}) can be determined by linear regression analyses of kinetic plots (pmol ADP produced versus time (min)). Experiments should be designed such that at a given helicase concentration, <20% of the substrate ATP is consumed in the reaction over the entire time course of the experiment.

4. Notes

1. Synthetic HJ(X12) is made by annealing four 50-mer oligonucleotides (X12-1, X12-2, X12-3, and X12-4) whose sequences are referenced by Mohaghegh et al., 2001*(33)*. HJ substrates with a 5′ ^{32}P label on oligonucleotides X12-1, X12-2, X12-3, or X12-4 are designated HJ(X12-1), HJ(X12-2), HJ(X12-3), or HJ(X12-4), respectively.

2. The final percentage polyacrylamide of the native gel depends on the size of the intact duplex DNA substrate and unwound radiolabeled product of the helicase reaction. For example, resolution of an unwound 44-mer from a forked 19 bp duplex flanked by 25 nt 3′ and 5′ ssDNA arms can be achieved by electrophoresis at 180 V for 2 h on a 12% polyacrylamide gel. The percentage polyacrylamide of the gel and running time for electrophoresis can be altered to optimize resolution of the intact substrate from the released strand. For M13 partial duplex substrates, the percentage acrylamide is generally lowered to 6–8% to insure that the M13 partial duplex migrates out of the well into the gel.

3. The components of the reaction buffer (pH, buffering agent, salt, stabilizing agent (BSA or glycerol), nucleoside 5′ triphosphate) should be optimized for the DNA helicase under investigation.

4. The 2X helicase buffer can be supplemented with 0.6% sodium dodecyl sulfate (SDS) to reduce helicase protein binding to the unwound DNA products of the helicase reaction.

5. To optimize resolution, the ratio of acrylamide–*bis*-acrylamide used for the preparation of the native gel can be modified. A 30:2 ratio could be tried.
6. The ratio of ^{32}P-labeled oligonucleotide to complementary oligonucleotide (1:2.5) ensures that the great majority of labeled oligonucleotide becomes annealed to complementary oligonucleotide.
7. To construct a synthetic replication fork structure with duplex leading and lagging strand arms, add 50 pmol (1 µL, 50 pmol/µL) of complementary leading and lagging strand primers to the tube containing annealed duplex DNA mixture (50 µL), incubate at 37°C for 1 h, and allow to slow cool to room temperature.
8. The presence of excess unlabeled oligonucleotide prevents reannealing of the unwound strand to its complementary strand.
9. For a control reaction, set up a helicase reaction mixture in which the enzyme is omitted from the incubation. This serves as the "no enzyme" control. A second control is the heat-denatured DNA substrate control in which a "no enzyme" control tube is incubated at 95°C for 5 min prior to loading.
10. Note that nucleotide-dependent conformational changes in protein structure may influence DNA binding properties of the helicase. Therefore, binding mixtures containing ATPγS, ADP, or no nucleotide should be tested to address the effect of nucleotide on DNA binding.
11. To reduce nonspecific binding of DNA to the filter, it may be useful to include BSA (50 µg/mL) in the 1-mL aliquot of 1X helicase reaction buffer and 1-mL aliquots of 1X Helicase Reaction Buffer used for wash steps (*see* **Subheading 3.4., step 4**).
12. The following formula, based on Scatchard analysis, can be used to analyze the data by Hill plot:

$$K_d = (1 - f)[\text{Pt}]/f \qquad (1)$$

$$\log [\text{Pt}] = \log (f/(1 - f)) + \log K_d. \qquad (2)$$

K_d is the dissociation constant of the DNA–protein complex, [Pt] is the total concentration of helicase protein present in the reaction, and f is the ratio of the amount of the bound DNA over the total amount of DNA present in the reaction. Plot the logarithm of [Pt] against the logarithm of $f/(1-f)$; the y-intercept represents the logarithm of K_d.

13. To trap protein–DNA complexes, it may be helpful to add a crosslinking agent. For example, incubate the protein–DNA complex mixture with 0.25% glutaraldehyde (final concentration) for 10 min at 37°C. Use triethanolamine in place of Tris as the buffering agent when glutaraldehyde is used for crosslinking.
14. Specific binding of the protein to the DNA substrate may be confirmed by the detection of a supershifted complex when an antibody against the DNA binding helicase protein is incubated with the protein–DNA complex mixture; however, if the DNA binding region and the antibody recognition (epitope) region overlap, a supershifted species may not be detected.
15. A supershifted complex of a helicase bound to its protein partner and the DNA substrate may be detected in some circumstances. A good example of this is the

WRN-FEN-1 interaction in which it was shown that WRN recruits FEN-1 to a Holliday Junction *(34)*. See **Fig. 5** for a representation of this result.

Acknowledgments

Because of page length limitations, we were not able to reference a number of papers in the literature from laboratories that developed various biochemical techniques for the study of DNA helicases. We wish to thank our colleagues in The Laboratory of Molecular Gerontology, National Institute on Aging, NIH for helpful discussion.

References

1. Singleton, M. R. and Wigley, D. B. (2003) Multiple roles for ATP hydrolysis in nucleic acid modifying enzymes. *EMBO J.* **22,** 4579–4583.
2. Caruthers, J. M. and McKay, D. B. (2002) Helicase structure and mechanism. *Curr. Opin. Struct. Biol.* **12,** 123–133.
3. Delagoutte, E. and von Hippel, P. H. (2002) Helicase mechanisms and the coupling of helicases within macromolecular machines. Part I: Structures and properties of isolated helicases. *Q. Rev. Biophys.* **35,** 431–478.
4. Hall, M. C. and Matson, S. W. (1999) Helicase motifs: the engine that powers DNA unwinding. *Mol. Microbiol.* **34,** 867–877.
5. Patel, S. S. and Picha, K. M. (2000) Structure and function of hexameric helicases. *Annu. Rev. Biochem.* **69,** 651–697.
6. Bachrati, C. Z. and Hickson, I. D. (2003) RecQ helicases: suppressors of tumorigenesis and premature ageing. *Biochem. J.* **374,** 577–606.
7. Brosh, R. M. Jr. and Bohr, V. A. (2002) Roles of the Werner syndrome protein in pathways required for maintenance of genome stability. *Exp. Gerontol.* **37,** 491–506.
8. Opresko, P. L., Cheng, W. H., and Bohr, V. A. (2004) At the junction of RecQ helicase biochemistry and human disease. *J. Biol. Chem.* **279,** 18,099–18,102. E-Pub 15023996.
9. Matson, S. W. (1986) *Escherichia coli* helicase II (urvD gene product) translocates unidirectionally in a 3′ to 5′ direction. *J. Biol. Chem.* **261,** 10,169–10,175.
10. Matson, S. W., Bean, D. W., and George, J. W. (1994) DNA helicases: enzymes with essential roles in all aspects of DNA metabolism. *Bioessays* **16,** 13–22.
11. Dillingham, M. S., Wigley, D. B., and Webb, M. R. (2002) Direct measurement of single-stranded DNA translocation by PcrA helicase using the fluorescent base analogue 2-aminopurine. *Biochemistry* **41,** 643–651.
12. Dillingham, M. S., Wigley, D. B., and Webb, M. R. (2000) Demonstration of unidirectional single-stranded DNA translocation by PcrA helicase: measurement of step size and translocation speed. *Biochemistry* **39,** 205–212.
13. Ahnert, P. and Patel, S. S. (1997) Asymmetric interactions of hexameric bacteriophage T7 DNA helicase with the 5′- and 3′-tails of the forked DNA substrate. *J. Biol. Chem.* **272,** 32,267–32,273.

14. Brosh, R. M. Jr., Waheed, J., and Sommers, J. A. (2002) Biochemical characterization of the DNA substrate specificity of Werner syndrome helicase. *J. Biol. Chem.* **277,** 23,236–23,245.
15. Kaplan, D. L. (2000) The 3′-tail of a forked-duplex sterically determines whether one or two DNA strands pass through the central channel of a replication-fork helicase. *J. Mol. Biol.* **301,** 285–299.
16. Sun, H., Karow, J. K., Hickson, I. D., and Maizels, N. (1998) The Bloom's syndrome helicase unwinds G4 DNA. *J. Biol. Chem.* **273,** 27,587–27,592.
17. Morris, P. D. and Raney, K. D. (1999) DNA helicases displace streptavidin from biotin-labeled oligonucleotides. *Biochemistry* **38,** 5164–5171.
18. Morris, P. D., Byrd, A. K., Tackett, A. J., et al. (2002) Hepatitis C virus NS3 and Simian virus 40 T antigen helicases displace streptavidin from 5′-biotinylated oligonucleotides but not from 3′-biotinylated oligonucleotides: evidence for directional bias in translocation on single-stranded DNA. *Biochemistry* **41,** 2372–2378.
19. Brosh, R. M. Jr., Orren, D. K., Nehlin, J. O., et al. (1999) Functional and physical interaction between WRN helicase and human Replication Protein A. *J. Biol. Chem.* **274,** 18,341–18,350.
20. Shen, J. C., Gray, M. D., Oshima, J., and Loeb, L. A. (1998) Characterization of Werner syndrome protein DNA helicase activity: directionality, substrate dependence and stimulation by Replication Protein A. *Nucleic. Acids. Res.* **26,** 2879–2885.
21. Brosh, R. M. Jr., Li, J. L., Kenny, M. K., et al. (2000) Replication Protein A physically interacts with the Bloom's syndrome protein and stimulates its helicase activity. *J. Biol. Chem.* **275,** 23,500–23,508.
22. Cui, S., Klima, R., Ochem, A., Arosio, D., Falaschi, A., and Vindigni, A. (2003) Characterization of the DNA-unwinding activity of human RECQ1, a helicase specifically stimulated by human Replication Protein A. *J. Biol. Chem.* **278,** 1424–1432.
23. Kikuma, T., Ohtsu, M., Utsugi, T., et al. (2004) Dbp9p, a member of DEAD box protein family, has DNA helicase activity. *J. Biol. Chem.* E-Pub. 15028736.
24. Brosh, R. M. Jr., Majumdar, A., Desai, S., Hickson, I. D., Bohr, V. A., and Seidman, M. M. (2001) Unwinding of a DNA triple helix by the Werner and Bloom syndrome helicases. *J. Biol. Chem.* **276,** 3024–3030.
25. Villani, G. and Tanguy, L. G. (2000) Interactions of DNA helicases with damaged DNA: possible biological consequences. *J. Biol. Chem.* **275,** 33,185–33,188.
26. Naegeli, H., Bardwell, L., and Friedberg, E. C. (1993) Inhibition of Rad3 DNA helicase activity by DNA adducts and abasic sites: implications for the role of a DNA helicase in damage-specific incision of DNA. *Biochemistry* **32,** 613–621.
27. Driscoll, H. C., Matson, S. W., Sayer, J. M., Kroth, H., Jerina, D. M., and Brosh, R. M. Jr. (2003) Inhibition of Werner syndrome helicase activity by benzo[c]phenanthrene diol epoxide dA adducts in DNA is both strand-and stereoisomer-dependent. *J. Biol. Chem.* **278,** 41,126–41,135.
28. Harmon, F. G. and Kowalczykowski, S. C. (2001) Biochemical characterization of the DNA helicase activity of the *Escherichia coli* RecQ helicase. *J. Biol. Chem.* **276,** 232–243.

29. Bjornson, K. P., Amaratunga, M., Moore, K. J., and Lohman, T. M. (1994) Single-turnover kinetics of helicase-catalyzed DNA unwinding monitored continuously by fluorescence energy transfer. *Biochemistry* **33,** 14,306–14,316.
30. Ali, J. A., Maluf, N. K., and Lohman, T. M. (1999) An oligomeric form of *E. coli* UvrD is required for optimal helicase activity. *J. Mol. Biol.* **293,** 815–834.
31. Lucius, A. L., Vindigni, A., Gregorian, R., et al. (2002) DNA unwinding step-size of *E. coli* RecBCD helicase determined from single turnover chemical quenched-flow kinetic studies. *J. Mol. Biol.* **324,** 409–428.
32. Nanduri, B., Byrd, A. K., Eoff, R. L., Tackett, A. J., and Raney, K. D. (2002) Pre-steady-state DNA unwinding by bacteriophage T4 Dda helicase reveals a monomeric molecular motor. *Proc. Natl. Acad. Sci. USA* **99,** 14,722–14,727.
33. Mohaghegh, P., Karow, J. K., Brosh, R. M. Jr., Bohr, V. A., and Hickson, I. D. (2001) The Bloom's and Werner's syndrome proteins are DNA structure-specific helicases. *Nucleic. Acids. Res.* **29,** 2843–2849.
34. Sharma, S., Otterlei, M., Sommers, J. A., et al. (2004) WRN Helicase and FEN-1 form a complex upon replication arrest and together process branch-migrating DNA structures associated with the replication fork. *Mol. Biol. Cell* **15,** 734–750.

29

Repair Synthesis Assay for Nucleotide Excision Repair Activity Using Fractionated Cell Extracts and UV-Damaged Plasmid DNA

Maureen Biggerstaff and Richard D. Wood

Summary
Methods are described for measuring nucleotide excision repair (NER) of damaged plasmid DNA using fractionated mammalian cell extracts. NER creates a single-stranded gap of approx 25–30 nt. Filling of this gap by repair synthesis can be monitored by the incorporation of radioactive nucleotides. We first describe the preparation of ultraviolet light (UV)-damaged and control plasmid DNA substrates and purification of their closed-circular forms. To increase the specificity for NER, plasmid molecules containing pyrimidine hydrates and other lesions sensitive to *Escherichia coli* Nth protein are eliminated. The preparation of whole cell extracts active in NER is described, both for cells grown as attached cultures and those grown in suspension. Cell extracts are partially purified on phosphocellulose to produce a fraction that can carry out the full NER reaction when combined with purified RPA and PCNA proteins. This enables NER to be quantified in an assay with exceptionally low background in nondamaged DNA.

Key Words: DNA repair synthesis; fractionation of cell extracts; nucleotide excision repair; plasmid DNA purification; proliferating cell nuclear antigen; protein purification; replication factor A.

1. Introduction
Mammalian cells remove carcinogenic damage caused to DNA by ultraviolet (UV) light and certain other mutagens mainly by using the pathway known as nucleotide excision repair (NER). This involves damage recognition, unwinding of the DNA around the site of damage, incision on either side of the lesion, removal of a fragment containing the lesion, and finally DNA synthesis and ligation to form a repair patch of approx 30 nucleotides. The use

of purified proteins and reconstituted systems has revealed the protein components that are essential for the core dual-incision reaction *(1,2)*.

As for NER in prokaryotes, the defining event is the excision of damage in a short oligonucleotide, by cleavage of the damaged strand on either side of a lesion in DNA. Adducts in DNA initially cause some intrinsic local distortion of the helix. This distortion appears to be the first structural feature recognized. Initial distortion recognition generally requires XPC-RAD23B (Rad4–Rad23 in yeast) which binds to distortions and facilitates the next key event *(3)*. This is the formation of an unwound DNA structure around a lesion in a preincision assembly of protein factors generating an "open complex." This opening confers single-stranded character to approx 25–30 DNA residues in the vicinity of the lesion *(4)*. Proteins necessary and sufficient to form a productive, open preincision complex are TFIIH, XPA, RPA, and XPG (TFIIH, Rad14, RPA, and Rad2 in yeast) *(5)*. The catalytic activity for opening the DNA strands is provided by TFIIH, which among its 10 subunits includes the two ATP-dependent helicases XPB and XPD (Ssl2 and Rad3 in yeast). This bubble-like intermediate is the substrate for cleavage by the structure-specific NER nucleases. As a result, XPG (Rad2 in yeast) cuts on the 3' side of the damaged DNA, and ERCC1-XPF enzyme (Rad10–Rad1 in yeast) cleaves on the 5' side *(6–10)*.

The dual incision event releases a fragment of approx 24–32 nt containing the damage and creates a single-stranded gap of this size. This gap is filled in a carefully controlled reaction *(11)* by a DNA polymerase holoenzyme complex that includes DNA polymerase δ or ε, the sliding clamp PCNA, and involves the PCNA loading factor RFC and the RPA protein. After this patching event, a DNA ligase then seals the remaining nick.

Repair synthesis of plasmid DNA can be measured as a consequence of these damage-specific incision and excision events. Plasmids damaged with physical or chemical agents that are known substrates for NER can be used. Most versatile are "whole-cell" extracts prepared using a procedure originally established to study transcription of RNA in vitro *(12,13)*. During incubation of the plasmid DNA with such extracts in the presence of deoxyribonucleoside triphosphates (one of which is radiolabeled), Mg^{2+}, ATP, and an ATP-regenerating system, repair synthesis patches are generated in the plasmid DNA at sites of NER. To distinguish this DNA synthesis mode from nonspecific background synthesis, a second undamaged plasmid of different size is included in the reaction mixtures as an internal control. Following incubation, plasmid DNA is recovered and linearized by cutting with a restriction enzyme. The two plasmids are resolved by gel electrophoresis. Bands are detected by autoradiography of the gels, and their intensity is quantified to calculate the incorporation of deoxynucleotides into each plasmid during DNA repair synthesis *(14,15)*.

These repair synthesis assays are useful for measuring the activity of individual purified DNA repair proteins *(16–19)* and to observe in vitro complementation between different repair-defective cell extracts *(20,21)*. The reaction can be resolved into incision and gap-filling stages by using fractionated cell extracts *(22)*. Other reagents such as antibodies, enzymes and inhibitors can be added to cell-free systems. DNA can be used with lesions arising from treatment with various types of chemicals or radiation, or containing a single lesion at a defined site. Repair synthesis has also been used as an assay with DNA assembled into nucleosomes *(23–25)*. Nucleosomes appear to suppress overall repair synthesis somewhat with UV-irradiated DNA, but this does not present a serious technical problem *(23,24,26)*.

The approach of measuring NER synthesis in cell extracts has been adapted to other eukaryotic cells, including *Saccharomyces cerevisiae (27,28)*, *Xenopus (25,29)*, and *Drosophila (30)*. Extracts from fresh human malignant lymphoid cells have also been used *(31)*. Other sources of cells such as tissue samples present difficulties caused by the diversity of cell types and cell-cycle states within the tissue sample, as well as regions of dying cells that can be sources of degradative enzymes.

Although the methods are straightforward, it is a considerable amount of work to set up the cell-free repair synthesis assay in a laboratory. Major investments of time and effort are required to grow sufficient quantities of healthy cells, maintain stocks of active extracts, and prepare a large amount of highly purified DNA substrate with low background.

For the purpose of assaying the activity of an individual NER protein it is necessary to have an extract from mutant cells lacking the corresponding activity. The extract is fractionated on a phosphocellulose column *(22,32)* using conditions which separate proliferating cell nuclear antigen (PCNA) and replication factor A (RPA) from the rest of the repair proteins (CFII). Purified RPA is added back at the incision stage, and purified PCNA for a short pulse at the synthesis stage during the assay. The use of CFIIs to assay for purified proteins results in much lower background repair synthesis in undamaged DNA than the procedure originally described with unfractionated extracts *(33,34)*. The DNA substrate is damaged plasmid and does not need to contain a single specific lesion, although it should be carefully prepared. Other, more detailed questions relating to the incision stage of nucleotide excision repair can be answered in vitro using dual incision assays such as those described in Chapter 30.

2. Materials
2.1. Plasmid Preparation
1. Luria broth (LB).
2. LB agar plates.
3. Ampicillin.

4. Qiagen DNA purification kit (Qiagen).
5. TE buffer: 10 mM Tris-HCl, pH 8.0, 1 mM EDTA.
6. Petri dishes (Sterilin).
7. Germicidal lamp (254 nm peak wavelength).
8. 10X Nth buffer: 0.4 M N-2-hydroxyethylpiperazine-N'-2-ethanesulfonic acid (HEPES)–KOH, pH 8.0, 1.0 M KCl, 5.0 mM EDTA. 5.0 mM dithiothreitol (DTT), and 2 mg/mL of bovine serum albumin (BSA) (nuclease-free; Gibco).
9. Nth protein (endonuclease III). (*See* **Note 1**.)
10. Cesium chloride (Gibco).
11. Ethidium bromide: 10 mg/mL (Sigma).
12. Syringes and needles.
13. Aluminum foil.
14. Quick-Seal centrifuge tubes (Beckman).
15. Ti60 rotor.
16. CsCl–isopropanol–TE: Dissolve 10 g of CsCl in 10 mL of TE, and then add 50 mL of isopropanol (an extra 1–2 mL of TE may be needed to allow the CsCl to go into solution).
17. Absolute ethanol.
18. Thick-walled 25 × 89 mm centrifuge tubes (Beckman).
19. SW28 ultracentrifuge rotor (Beckman).
20. Sucrose. UltraPure (Gibco).
21. Sucrose gradient buffer: 25 mM Tris-HCl, pH 7.5, 1 M NaCl, 5 mM EDTA.
22. Peristaltic pump.
23. Magnetic stirrer.
24. Ultraclear centrifuge tubes 25 × 89 mm (Beckman).
25. Agarose (molecular biology grade).
26. 70% Ethanol.

2.2. Cell Extract Preparation

1. 175-cm^2 Tissue-culture flasks (Falcon).
2. 850-cm^2 Roller bottles (Falcon cat. no. 3027).
3. Phosphate-buffered saline A (PBSA): 10 g/L NaCl, 0.25 g/L KCl, 0.25 g/L KH$_2$PO$_4$, 1.43 g/L of Na$_2$HPO$_4$.
4. 0.25% Trypsin (Gibco) and phosphate-buffered saline.
5. Fetal calf serum (FCS) (Gibco).
6. Hypotonic lysis buffer: 10 mM Tris-HCl, pH 8.0, 1 mM EDTA, 5 mM DTT.
7. Phenylmethylsulfonyl fluoride (PMSF): 500 mM in dry methanol (Sigma) or 4-(2-aminoethyl)benzenesulfonyl fluoride (AEBSF): 100 mM in H$_2$O (Calbiochem).
8. Leupeptin: 5 mg/mL in H$_2$O (Sigma).
9. Pepstatin: 5 mg/mL in dimethyl sulfoxide (DMSO) (Sigma).
10. Chymostatin: 5 mg/mL in DMSO (Sigma).
11. Aprotinin (Sigma).
12. Glass homogenizer with Teflon pestle.
13. Sucrose–glycerol buffer: 50 mM Tris-HCl, pH 8.0, 10 mM MgCl$_2$, 2 mM DTT, 25% sucrose (Gibco), 50% glycerol (Fluka).

Nucleotide Excision Repair

14. Ammonium sulfate (BDH Analar), finely powdered using a mortar and pestle.
15. Saturated ammonium sulfate solution in water, neutralized by adding 1 M NaOH after saturation.
16. Cell extract dialysis buffer: 25 mM HEPES–KOH, 0.1 M KCl, 12 mM MgCl$_2$, 1 mM EDTA, 17% glycerol (Fluka), 2 mM DTT (add just before use); adjust pH to 7.9 with 5 M KOH.
17. Phosphocellulose P11 (Whatman).
18. Buffer A: 25 mM HEPES–KOH, pH 8.0, 10% glycerol, 150 mM KCl, 1 mM EDTA, 2 mM DTT.
19. Chromatography column (1.6 cm diameter) to hold 5 mL of resin.
20. Amicon stirred pressure cell (Amicon).

2.3. Repair Reactions

1. 5X buffer for repair reactions: 200 mM HEPES, 25 mM MgCl$_2$, 2.5 mM DTT (Sigma), 10 mM ATP (Sigma) pH 5.2, 100 µM dGTP, 100 µM dCTP, 100 µM dTTP, 40 µM dATP (nucleotide triphosphates from Amersham Biosciences, 100 mM), 110 mM phosphocreatine (Sigma di-Tris salt), 1.7 mg/mL of BSA (nuclease-free, Invitrogen Life Technologies), adjust the pH with 1 M KOH to give pH 7.8 in the final reaction (set up a mock reaction to test this).
2. Creatine phosphokinase (CPK, rabbit muscle; Sigma), 2.5 mg/mL in 10 mM glycine, pH 9.0, and 50% glycerol. Store at –20°C.
3. [α-^{32}P] dATP (3000 Ci/ mmol; Amersham Biosciences).
4. 10% Sodium dodecyl sulfate (high quality, e.g., BDH Analar).
5. Proteinase K: 2 mg/mL (Sigma).
6. RNase A: 2 mg/mL (Sigma).
7. Phenol–chloroform–isoamyl alcohol (25:24:1) (Fluka).
8. 0.5 M EDTA, pH 8.0.
9. 7.5 M Ammonium acetate.
10. Absolute ethanol (–20°C).
11. RPA (*see* **Note 2**).
12. Whole-cell extract or CFII.
13. PCNA (*see* **Note 3**).
14. DNA mix: UV-damaged plasmid and a different-sized undamaged plasmid (control) (*see* **Note 4**).
15. *Bam*HI restriction endonuclease (Roche).
16. Kodak film and cassette with screens or phosphorimager cassette and Phosphor-Imager.

3. Methods

3.1. Preparation of DNA Substrate

3.1.1. Plasmid Purification

1. Make fresh transformants of each plasmid to be used in the assay in a strain of *E. coli* containing the *recA* and *endA* mutations (*see* **Note 4**). Grow overnight at 37°C in LB containing 25 µg/mL ampicillin.

2. Spread 100 µL of transformed cells on a freshly prepared LB agar plate containing 25 µg/mL of ampicillin, and incubate overnight at 37°C.
3. Inoculate one bacterial colony into 10 mL of LB + ampicillin. Incubate overnight at 37°C with shaking.
4. Transfer 2 mL of this culture into each of five 2-L flasks containing 500 mL of LB + ampicillin. Incubate overnight at 37°C with shaking.
5. Prepare plasmid by the Qiagen method using the appropriate-sized pack following the manufacturer's instructions (*see* **Note 5**).

3.1.2. UV Irradiation

1. Dilute the plasmid to 50 µg/mL in TE.
2. Small volumes of plasmid can be irradiated in 20 to 50 µL drops in a Petri dish. Pour larger amounts 15 mL at a time into 15-cm Petri dishes, and place on a rocker.
3. Remove the lid of the Petri dish, and irradiate at a dose rate of 0.5 $J/m^2/s$ with UV light (peak wavelength of 254 nm), while the dish is rocked to give an even fluence over the surface for 15 min. This gives a total of 450 J/m^2 (*see* **Note 6**).
4. Collect the irradiated plasmid and measure the volume.

3.1.3. Treatment With Nth Protein

1. Add 1/9 volume of 10X Nth protein buffer. Keep a 250-ng sample for gel electrophoresis.
2. Add the appropriate amount of Nth protein as calculated by titration (*see* **Note 1**).
3. Incubate at 37°C for 30 min followed by inactivation at 65°C for 2 min. Keep a 250-ng sample for gel electrophoresis.
4. Mix the 250-ng samples reserved from the unirradiated and irradiated plasmids, before and after Nth protein treatment and measure the amount of nicking by gel electrophoresis (*see* **Note 1**).

3.1.4. Cesium Chloride Equilibrium Centrifugation

1. Add 1.1 g of cesium chloride powder to the plasmid solution, and mix until fully dissolved. Carefully transfer 30-mL aliquots into 33 mL Beckman Quick-Seal tubes using a syringe with some tubing on the end. Wrap the tubes in aluminum foil, and add 3 mL of ethidium bromide at 10 mg/mL into the top using a syringe and needle. Seal the tubes according to the manufacturer's instructions (*see* **Note 7**).
2. Mix thoroughly by inverting the tubes, and centrifuge at 38,000 rpm in a Ti 60 rotor (145,000g) for 60 h at 18°C.
3. Remove the band of closed-circular plasmid using a syringe and needle. Work in the dark room under a safe light, and visualize the bands by brief illumination using a 365-nm lamp.
4. Recover the DNA by shaking in an equal volume of CsCl–isopropanol–TE solution in the dark under a safe light. Remove the top, pink layer.
5. Repeat the extraction two to three times until the solution is colorless.
6. Measure the volume and add 3 volumes of TE to each, and then two total volumes of ethanol.

7. Divide between several thick-walled 25 × 89 mm centrifuge tubes. Mix and leave at –20°C for at least 30 min.
8. Centrifuge at 72,000g (20,000 rpm) for at least 20 min in an SW28 rotor. Carefully wash the pellet in 70% ethanol and resuspend in TE.

3.1.5. Sucrose Density Centrifugation

1. Make up fresh solutions of 5 and 20% sucrose in sucrose gradient buffer.
2. Pour one gradient/500 µg of plasmid DNA. Place 40 mL of the 5% solution in a 100-mL beaker containing a stir bar, on a magnetic stirrer. Pour 40 mL of the 20% solution into another beaker, and pump the 20% solution into the beaker containing the 5% solution (stirring the 5% solution constantly) and at the same time pump two lines from the 5% solution into the bottom of two 25 × 89 mm ultraclear centrifuge tubes. Use a peristaltic pump set to deliver 3–4 mL/min (*see* **Note 8**). Leave at 4°C for at least 1 h.
3. Precool a rotor and buckets to 4°C. Carefully layer 1 mL of DNA (500 µg) in TE buffer onto the top of each gradient (*see* **Note 9**). Centrifuge at 113,000g (25,000 rpm) at 2°C in an SW28 rotor for 19 h.
4. Carefully remove the tubes from the buckets, and collect 1.5-mL fractions by pumping out from the bottom. Keep the gradients cool.
5. Load 10 µL of each fraction onto a 0.8% agarose gel containing 0.25% ethidium bromide (*see* **Note 10**). Place loading buffer in the end wells as a marker. Separate by electrophoresis at 100 V for 3–4 h.
6. The closed-circular DNA should be present in about four to five fractions in the lower third of the gradient (**Fig. 1**). Pool the fractions containing the closed circular DNA and no dimerized or nicked forms, and add two volumes of ethanol, chill for 10 min, and centrifuge at 72,000g (20,000 rpm) in an SW28 rotor. Rinse with 70% ethanol, centrifuge again, dry the pellet, and resuspend in 1 mL of TE buffer/gradient.

3.1.6. Preparation of DNA Mixture

Check that each plasmid has an acceptably low level of nicked DNA. Examine 250 ng of each one separately by electrophoresis on a 0.8% agarose gel containing 0.25 µg/mL ethidium bromide. Prepare a "DNA mix" by combining equal amounts of purified damaged and undamaged plasmid. For repair reactions the DNA concentration should be 50 µg/mL of each plasmid in TE (before combining). This is equivalent to 250 ng of each plasmid in a 10-µL aliquot. Check the activity of the mixture in a repair assay, using a repair-proficient cell extract. Store the DNA mix in aliquots at –80°C.

3.2. Preparation of Fractionated Cell Extracts
3.2.1. Growth and Harvesting of Fibroblasts

1. Grow fibroblasts in 175-cm^2 tissue-culture flasks or 850-cm^2 roller bottles turning at 10 rpm. Fractionated extracts will require about 12 roller bottles or 10^9 cells (*see* **Note 11**).

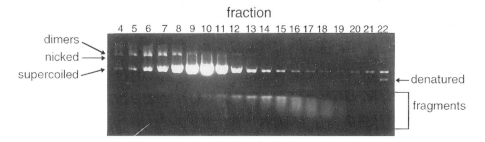

Fig. 1. Sucrose density gradient centrifugation of UV-damaged plasmid DNA. pBluescript KS+ plasmid DNA was isolated on a Qiagen giga column, UV-irradiated, Nth-treated and the supercoiled form isolated after cesium chloride equilibrium centrifugation. Five hundred micrograms of this were separated on a sucrose gradient; 1.6-mL fractions were collected, and 10 µL of each subjected to electrophoresis on a 0.8% agarose gel containing 0.25 µg/mL of ethidium bromide. Arrows indicate the form of DNA: Supercoiled plasmid DNA is the major species. Dimeric plasmid is present in fractions 4–8. Nicked forms migrate slightly faster than dimers and are mainly removed by the cesium chloride equilibrium centrifugation, but a small amount remains in fractions 11 and 12. Fragments of DNA can be seen in fractions 10–18 and it is important to remove as much of this material as possible from the preparation, since it inhibits repair synthesis *(40)*. Denatured forms occur occasionally as a result of the alkaline lysis step. Fractions 9, 10, and 11 were pooled and reprecipitated for use in the final DNA mix.

2. When the cells are just confluent, wash twice with 100 mL/bottle of sterile PBSA.
3. Decant the PBSA, and add 100 mL of 0.25% trypsin–PBSA until the cells round up and are almost ready to detach. Decant the trypsin solution.
4. Add 50 mL of medium + 15% FCS, and gently agitate until the cells detach. Centrifuge for 5 min at 165g (1000 rpm) in a Sorvall HB-4 rotor to pellet the cells gently.
5. Gently resuspend in 50 mL of medium + 15% FCS and pellet again.
6. Resuspend in 25 mL of cold PBSA (4°C), and pellet again.
7. Resuspend in 25 mL of cold PBSA, and transfer to a 30-mL Corex tube. Centrifuge at 370g (1500 rpm) for 10 min at 4°C. Carefully remove the supernatant, and measure the packed cell volume (pcv).

3.2.2. Growth and Harvesting of Lymphoblastoid Cells

1. Start with 1–2 L of actively growing cells in suspension at 6-8 × 10^5 cells/mL (*see* **Note 11**). For a fractionated extract, use 10 L or 10^9 cells.
2. Centrifuge at 1200 rpm (HB-4 rotor; 235g) for 10 min to pellet the cells and resuspend in cold PBSA. Combine the pellets and wash twice more.
3. After the final wash, carefully remove the supernatant, and measure the pcv.

3.2.3. Whole-Cell Extracts (see **Note 12**)

1. Resuspend the cells in 4 pcv of hypotonic lysis buffer and leave on ice for 20 min. Add per 1 mL of pcv: 5 µL of 87 mg/mL PMSF or 0.1 mM AEBSF (final concentration); 2 µL each of 5 mg/mL leupeptin, pepstatin, and chymostatin; and 50 µL of aprotinin.
2. Break the cells in a glass homogenizer with a Teflon pestle (about 20 strokes). Check by dye exclusion that the cells are broken (see **Note 13**).
3. Transfer the homogenate to a glass beaker in a tray of ice over a magnetic stirrer in the cold, and stir very slowly.
4. Gently add 4 pcv of sucrose–glycerol buffer, and mix carefully.
5. Slowly add 1 pcv of saturated ammonium sulfate solution (see **Note 14**).
6. Pour carefully into polyallomer tubes for an SW55 rotor (or SW50). Do not pipet (see **Note 15**).
7. Centrifuge for 3 h in a Beckman SW50 or SW55 rotor at 42,000 rpm or a Sorvall AH-650 rotor at 212,000g (42,500 rpm at r_{max}), at 2°C.
8. Carefully remove the supernatant with a Pasteur pipet, leaving the last 1 mL or 1 cm above the pellet in the tube (see **Note 16**). Measure the volume of supernatant (usually 6–7 mL/mL of pcv). Transfer to a 35-mL polyallomer tube on ice, with a magnetic stirring bar in the bottom of the tube.
9. Weigh out finely powdered solid ammonium sulfate, 0.33 g for each milliliter of supernatant. Slowly add this to the supernatant while stirring continously. When it is dissolved, add 10 µL of 1 M NaOH/g of ammonium sulfate added. Stir for 30 min more.
10. Centrifuge for 30 min (for a CFII) or 1 h (for a nonfractionated extract) at approx 20,000g (11,000 rpm at r_{max}) in a Sorvall HB-4 or HB-6 rotor at 4°C.
11. Remove the supernatant, leaving the pellet as dry as possible. Resuspend the pellet in just enough dialysis buffer to allow it to be drawn up into a 1-mL syringe (without the needle) or a cut-off 1-mL pipet tip. For a whole-cell extract, do not dilute in dialysis buffer, since this will yield an extract with very low protein concentration. The suspension should be very thick and milky.
12. Transfer the suspension into a prepared dialysis bag (e.g., Spectra/Por 2, molecular mass cutoff 12–14 kDa), and clamp it fairly tightly, avoiding trapping of bubbles inside the bag. Dialyze 1–2 h in 500 mL of extract dialysis buffer. For a CFII, dialyze into extract dialysis buffer containing 150 mM KCl. Dialyze in the cold, and then change the buffer and continue dialysis for 8–12 h in 2 L of fresh buffer. Alternatively, change the buffer three times, and dialyze for only 6 h.
13. Remove the dialysate, and centrifuge for 10 min in a cold microcentrifuge to remove any precipitate. Transfer the supernatant into a fresh tube, and mix by inverting the tube several times (do not vortex).
14. Snap-freeze in small aliquots at –80°C. Use each aliquot only once, and do not refreeze.
15. The yield should be about 5 mg of protein/roller bottle or liter of lymphoblast suspension, at 15–20 mg protein/mL of extract. At least 60 mg of whole-cell extract are needed to make a successful phosphocellulose fraction.

3.2.4. Phosphocellulose Fractionation (CFII)

1. Prepare phosphocellulose P11 (Whatman) according to the manufacturer's instructions. Pour 5 mL into a 1.6-cm diameter column in the cold, and equilibrate with buffer A at a speed of 20 mL/h.
2. Load 60 mg of extract protein, and wash in buffer A until the UV absorbance decreases and is steady (about two column-volumes).
3. Elute with buffer A containing 1 M KCl. Collect 1-mL fractions.
4. Pool the peak fractions.
5. Concentrate in an Amicon pressure cell concentrator for 2–3 h to about 1 mL (*see* **Note 17**).
6. Dialyze overnight into extract dialysis buffer.
7. Microcentrifuge for 10 min at full speed and aliquot into small (20-µL) amounts and snap-freeze. Store at –80°C. A good CFII should contain 8–15 mg of protein/mL.

3.3. Repair Reactions to Assay for Activity of Purified Proteins

3.3.1. Setting Up Repair Reactions

Carry out all operations on ice.

1. Make a premix A by combining 10 µL/reaction of 5X buffer + 1 µL per reaction of CPK.
2. Make a premix B: 0.2 µL of fresh (<1 wk after activity date) [α-^{32}P]dATP, 1.5 µL of 1 M KCl, and 7.3 µL of milli-Q water (*see* **Note 18**). For reactions with CFIIs add approx 150 ng of RPA (*see* **Note 2**). Adjust the KCl and water in premix B to a volume of 9 µL/reaction and the KCl to give 70 mM in a 50-µL reaction, taking into account the concentration of KCl in the extract and other added reagents. Make enough for two more reactions than needed. Carry out all procedures using appropriate radiological protection and monitoring.
3. Place 10 µL of DNA mix (250 ng of each plasmid) in the bottom of a 1.5-mL microcentrifuge tube.
4. Add 11 µL of premix A.
5. Add 9 µL of premix B.
6. Quick-thaw the extracts, and add 200 µg of extract protein, or 100 µg of CFII protein in 15 µL of extract dialysis buffer (*see* **Note 19**).
7. Add an appropriate amount of purified protein made up to 5 µL in buffer (if different from extract dialysis buffer), or 5 µL of buffer alone (*see* **Note 20**).
8. Centrifuge for a few seconds to collect the components at the bottom of the tube. Mix by vortexing, centrifuge again, and incubate for 3 h at 30°C in a heat block or water bath.
9. Reactions using CFIIs require the addition of PCNA (*see* **Note 3**) at this stage. Dilute the stock PCNA in its own buffer, add 25 ng to each reaction, and incubate for a further 10 min at 30°C. A "cold chase" of unlabeled dATP can be added at this stage if background signal in the damaged plasmid is a problem (*see* **Note 21**).

10. After incubation, stop the reactions by adding EDTA to 20 mM and store at −20°C or below.

3.3.2. Working Up Repair Reactions

1. To each reaction add 2 µL of RNase A (80 µg/mL final), and incubate at 37°C for 10 min.
2. Add 3 µL of SDS (0.5% final concentration) and 6 µL of Proteinase K (190 µg/mL final concentration) to each tube, mix, and incubate at 37°C for 30 min.
3. Extract the mixture with 50 µL of phenol:chloroform:isoamyl alcohol (25:24:1) and microcentrifuge at maximum speed for 5–10 min.
4. Remove the aqueous phase, and transfer to a fresh tube containing 25 µL of 7.5 M ammonium acetate. Mix and add 160 µL of absolute ethanol. Mix, and chill the tubes on dry ice for 15–30 min.
5. Microcentrifuge for 10 min at 4°C. Remove the supernatant with an ultrafine gel-loading tip, avoiding the DNA pellet. Add 200 µL of 70% ethanol, spin, and remove as much ethanol as possible.
6. Dry the DNA pellet in a SpeedVac for about 5 min until the liquid is removed, but do not overdry.
7. Linearize the DNA with an appropriate restriction enzyme. The plasmids have a unique *Bam*HI site and 10 U are used/50 µL reaction in the manufacturer's buffer. Incubate for 60 min at 37°C.
8. Add loading dye to the tubes, mix and load onto a 0.8% agarose gel in 1X TBE with both the gel and buffer containing 0.25 µg/mL ethidium bromide. Run the samples into the gel at 100 V for 15 min and then at 40 V overnight.
9. Photograph the gel with UV-B transillumination, digitally or using Polaroid type 57 film (f 11, 1 s), and type 55 film (f8, 30 s) to obtain a negative for quantification.
10. Dry the gel onto Whatman 3MM paper under vacuum at 80°C for about 1.5 h.
11. Put the dried gel into a cassette with intensifying screens and expose to preflashed X-ray film (*see* **Note 22**) and place at −80°C. Expose the film for 3–6 h for extracts and 6 h to overnight for CFIIs. (*See* **Fig. 2**.) Alternatively, put the gel into a PhosphorImager cassette, and read using the manufacturer's instructions.

3.3.3. Analysis of Results

1. Excise the band from the exposed gel (together with the attached filter paper) under UV light, and obtain the cpm in a scintillation counter. Cpm can also be obtained by using an appropriately calibrated PhosphorImager.
2. Measure the relative densities of the bands of both plasmids on the negative of the photograph of the ethidium bromide-stained gel.
3. Calculate the fmol of dAMP incorporated/reaction by dividing the cpm by 10 and by the correction factor for the activity date of the isotope, as read from a ^{32}P decay table. Normalize this for DNA recovery by dividing by the relative amount of DNA obtained from the density measurement of the ethidium bromide-stained gel (*see* **Note 23**).

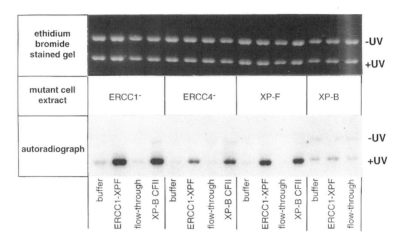

Fig. 2. Repair synthesis assay for the detection of ERCC1-XPF complex in pooled fractions from a mono S column. Pooled fractions from a mono S FPLC column were assayed for complementation of different excision repair mutant CFIIs from the following cells: Chinese hamster ovary 43-3B (ERCC1−); CHO UV41 (ERCC4−, equivalent to XPF); human XP-F; and CHO 27-1 (XPB). Each CFII was supplemented with Mono S buffer alone (buffer), pooled Mono-S fractions (ERCC1-XPF), flowthrough from the Mono S column (flowthrough), and a CFII from the mutant CHO 27-1 (XP-B CFII). A total of 100 µg of CFII protein or CFII + complementing protein were added in each lane to 250 ng of each of UV-damaged and undamaged plasmid DNA, supplemented with RPA and PCNA as described in the text.

4. Notes

1. The endonuclease III enzyme is expressed from *E. coli* and purified exactly according to the method of Asahara et al. *(35)*. It will be quite concentrated after purification (10–20 mg/mL) and must be titrated to establish the correct amount to use. Excessive enzyme treatment will cleave unirradiated DNA. It is important to inactivate the enzyme at 65°C for 5 min with gentle shaking when large volumes are being treated to prevent excess plasmid nicking. Titration of the enzyme can be carried out using 50 ng of irradiated plasmid (450 J/m^2) mixed with 250 ng of unirradiated plasmid. Add 1 µL of 10X Nth buffer and 1 µL of diluted enzyme and distilled water to a final volume of 10 µL. Incubate for 30 min at 37°C, and stop the reaction by heating to 65°C. Add loading buffer containing 1% SDS and examine by electrophoresis through a 0.8% agarose gel containing 0.25% ethidium bromide until separation of nicked (form II) and closed-circular plasmid DNA (form I) is evident (approx 4 h at 75 V for a 15 × 10 cm^2 gel). Photograph the gel with Polaroid type 55 film and scan the negative by densitometry to quantify each band. The nicked DNA has a higher molar fluorescence than closed-circular DNA by a factor of about 1.6 (this can vary for the staining conditions

and plasmid used). Calculate the fraction of nicked molecules for each plasmid by the formula:

Nicked fraction = (area form II/1.6) /[(area form I) + (area form II/1.6)] (1)

The average nicks per molecule can then be calculated from the nonnicked fraction (1 − nicked fraction) by the Poisson distribution, where the average nicks per circle = −ln (nonnicked fraction). Make a plot of the amount of enzyme vs average nicks per circle in UV-irradiated and nonirradiated DNA. Identify the amount of enzyme for which UV-specific nicking is complete, but for which minimal nicking of unirradiated DNA occurs. Use this dilution for the large plasmid preparation.

2. Human RPA can be produced as a recombinant protein in *E. coli* and purified according to the method of Henricksen et al. *(36)*. The normal yield is about 500 µg/mL in extract dialysis buffer containing 250 mM KCl. It is stored in aliquots at −80°C.

3. Recombinant human PCNA can be purified from an *E. coli* expression strain on Q-Sepharose, S-Sepharose, hydroxyapatite, and phenyl-Sepharose as described by Fien and Stillman *(37)*, except for the procedure for the final column. Protein is equilibrated and loaded onto the phenyl Sepharose column in buffer containing 25 mM Tris-HCl, pH 7.5, 1 mM EDTA, 0.01% NP-40, 10% glycerol, 2 mM benzamidine, 2 µM pepstatin A, 10 mM NaHSO$_3$, 1 mM PMSF, and 5 mM DTT containing 1.0 M ammonium sulfate. The protein is eluted from the column using a gradient in this buffer from 1.0 M ammonium sulfate and 0% ethylene glycol to 0 M ammonium sulfate and 50% ethylene glycol. PCNA elutes from the column in 0.35 M ammonium sulfate–32.5% ethylene glycol. At the end of this procedure, PCNA should be a single homogenous band of 36 kDa as detected by Coomassie blue staining of an SDS-polyacrylamide gel *(11)*. The final protein concentration after dialysis into extract dialysis buffer is about 600 µg/mL.

4. Any plasmids are theoretically suitable, but to obtain a high yield of DNA it is most convenient to use plasmids with relaxed copy control, such as those derived from pUC vectors. We use the 2.9-kb pBluescript KS+ (pBS, Stratagene) as the irradiated plasmid, and the same plasmid with a 0.8-kb insert in the *Eco*R1 site, pHM14 (3.7 kb) as the unirradiated control. Use of an *E. coli* strain that contains the *recA* mutation may help limit the formation of plasmid multimers. JM109 is a widely available *recA* strain that also contains an *endA* mutation which helps to reduce degradation of the plasmid DNA during cell lysis. pBS and pHM14 plasmids prepared from stocks of transformed JM109 cells stored at −20°C or −80°C can form dimers, which cannot be used in the repair assay. This should be avoided by using fresh transformants from a freshly plated-out culture.

5. Typical yields are 2 mg of plasmid/L of culture for the pBS plasmids. At least 10 mg of UV-irradiated plasmid are needed at this stage, that is, 5 L of culture. This is because subsequent treatment by Nth protein nicks UV-irradiated DNA at pyrimidine hydrates and the nicked DNA is separated from closed-circular by cesium chloride density gradient centrifugation. Since only closed-circular DNA

is used in the repair assay, up to 60% of the UV-irradiated plasmid is removed. Six milligrams of the unirradiated plasmid are sufficient.

6. UV-C irradiation at 254 nm produces roughly 1 pyrimidine dimer photoproduct (about 0.75 cyclobutane pyrimidine dimer and 0.25 [6–4] photoproduct)/1000 base pairs per 100 J/m^2. A relatively low fluence rate ensures a more even irradiation of the whole DNA sample.

7. Cesium chloride gradients are set up essentially as described in the *Current Protocols in Molecular Biology* manual *(38)*. We centrifuge in a Ti60 rotor at 38,000 rpm to achieve 145,000g. All operations are carried out in the dark when the DNA is in contact with ethidium bromide to reduce damage to the DNA. Wear gloves, face shield, and protective clothing when handling ethidium bromide, and dispose of it according to local safety regulations.

8. Pumping the solutions to the bottom of the tubes is best achieved using glass capillaries attached to the end of the pump tubing. These are placed in the gradient tubes and can be removed at the end by lifting them out smoothly and gently without disturbing the gradient.

9. When the DNA is loaded onto the gradient, it should be visible as a distinct layer. If it starts to sink into the gradient, this indicates that there is still some cesium chloride present in the DNA. The time of centrifugation is dependent on the plasmid and the density of the solutions. For our plasmids, 19 h give good separation of nicked and supercoiled DNA.

10. When loading aliquots of fractions onto an agarose gel, there is no need to add loading solution containing dye to every fraction, since the sucrose in the fractions makes them dense enough to stay at the bottom of the wells.

11. We have prepared extracts from as few as 10^8 cells, but starting with 10^9 cells is much easier and more likely to be successful. Cells that are overgrown yield extracts with very poor repair activity. The viability of the cells should be measured by exclusion of a dye, such as trypan blue or nigrosin, and should be 98% or greater. Fibroblast cells should be just confluent when harvested, since they should be as healthy as possible for good extracts.

12. These methods for whole cell extract preparation are adapted from those of Manley et al. *(12,39)*.

13. The cells should stay intact during incubation with hypotonic lysis buffer and only break during Dounce homogenization.

14. After about half of this is added, the viscosity increases dramatically. Stir very slowly to avoid shearing the DNA (about 1 turn/s). Continue stirring for about 30 min.

15. The resulting solution is very viscous and must be poured into the tubes gently. This may take a little practice!

16. The last 1 mL is usually viscous, because it contains DNA that has not fully pelleted. If a fixed-angle rotor, such as a Ti60, is used, more care must be taken to avoid mixing the DNA with the supernatant.

17. Concentrate the fractions in 1 M KCl before dialysis into 100 mM KCl, since this helps to prevent precipitation and loss of protein.

18. The amount of KCl or NaCl in a reaction should be 70 mM. The repair levels are considerably reduced at levels higher than this, and specificity for NER is reduced at lower salt. The extract dialysis buffer contains 100 mM KCl. When adding purified proteins to extracts, take into account the amount of salt in the protein buffer, and adjust the amount of KCl in premix B to make the final concentration in the 50-µL repair reaction 70 mM.
19. It is important to add enough total extract or CFII protein to a reaction. Do not add <50 µg of CFII or 100 µg of whole-cell extract protein to a 50-µL reaction, as these protein concentrations do not sustain significant NER.
20. Titrate the protein added to the mutant extract or CFII to find the optimal activity. Always add buffer alone to the reactions as a control.
21. A chase of unlabeled dATP can be added about 2 min after the PCNA pulse to reduce the background further if this is a problem. However, this can reduce the specific signal if added too soon, so time points must be taken to establish exactly the amount and when to pulse. Stop the reactions by immersing the tube in dry ice.
22. The film should be preflashed so that the response of the film will be as linear as possible for quantification.
23. Addition of pure proteins to CFIIs in repair reactions requires careful quantification in order to establish the contribution of the protein to the reaction as a whole. The first step is to determine the counts per minute in each band of interest. Scintillation counting of the gel can be carried to high efficiency without solubilizing it when ^{32}P has been used as a label. The bands remain stained with ethidium bromide in the dried gel, and guidelines for cutting the bands can be drawn while illuminating with UV-A or UV-B light (wear suitable eye protection and cover exposed skin). Under standard conditions described for the repair reactions, each 50-µL reaction mixture contains 2 µCi of [α-^{32}P] dATP at 3000 Ci/mmol and 8 µM cold dATP, so that there are 11 dpm/fmol of dATP. With a counting efficiency of 90%, this gives about 10 counts/min/fmol of dATP. Under these particular conditions, fmol of dAMP incorporated/reaction can be calculated by dividing the counts/min by 10 and by the correction factor for the activity date of the isotope as read from a ^{32}P decay table. This value should in turn be normalized for any differences in DNA recovery between the tracks.

Acknowledgments

We thank the past members of our laboratory for discussions and contributions to these procedures, and Takashi Yagi for organizing the procedure for purifying human PCNA.

References

1. Wood, R. D., Araújo, S. J., Ariza, R. R., et al. (2000) DNA damage recognition and nucleotide excision repair in mammalian cells. *Cold Spring Harbor Symp. Quant. Biol.* **65,** 173–182.

2. Friedberg, E. C., Walker, G. C., Siede, W., Wood, R. D., Schultz, R. A., and Ellenberger, T. (2005) *DNA Repair and Mutagenesis*, American Society for Microbiology Press, Washington, D.C.
3. Sugasawa, K., Okamoto, T., Shimizu, Y., Masutani, C., Iwai, S., and Hanaoka, F. (2001) A multistep damage recognition mechanism for global genomic nucleotide excision repair. *Genes Dev* **15**, 507–521.
4. Evans, E., Fellows, J., Coffer, A., and Wood, R. D. (1997) Open complex formation around a lesion during nucleotide excision repair provides a structure for cleavage by human XPG protein. *EMBO J.* **16**, 625–638.
5. Evans, E., Moggs, J. G., Hwang, J. R., Egly, J.-M., and Wood, R. D. (1997) Mechanism of open complex and dual incision formation by human nucleotide excision repair factors. *EMBO J.* **16**, 6559–6573.
6. O'Donovan, A., Davies, A. A., Moggs, J. G., West, S. C., and Wood, R. D. (1994) XPG endonuclease makes the 3' incision in human DNA nucleotide excision repair. *Nature* **371**, 432–435.
7. Clarkson, S. G. (2003) The XPG Story. *Biochimie* **85**, 1113–1121.
8. Mu, D., Wakasugi, M., Hsu, D. S., and Sancar, A. (1997) Characterization of reaction intermediates of human excision-repair nuclease. *J. Biol. Chem.* **272**, 28,971–28,979.
9. Matsunaga, T., Mu, D., Park, C. H., Reardon, J. T., and Sancar, A. (1995) Human DNA-repair excision nuclease - analysis of the roles of the subunits involved in dual incisions by using anti-XPG and anti-ERCC1 antibodies. *J. Biol. Chem.* **270**, 20,862–20,869.
10. Riedl, T., Hanaoka, F., and Egly, J. M. (2003) The comings and goings of nucleotide excision repair factors on damaged DNA. *EMBO J.* **22**, 5293–5303.
11. Araújo, S. J., Tirode, F., Coin, F., et al. (2000) Nucleotide excision repair of DNA with recombinant human proteins: definition of the minimal set of factors, active forms of TFIIH and modulation by CAK. *Genes Dev.* **14**, 349–359.
12. Manley, J. L., Fire, A., Samuels, M., and Sharp, P. A. (1983) In vitro transcription: whole cell extract. *Methods Enzymol.* **101**, 568–582.
13. Manley, J. L., Fire, A., Cano, A., Sharp, P. A., and Gefter, M. L. (1980) DNA-dependent transcription of adenovirus genes in a soluble whole-cell extract. *Proc. Natl. Acad. Sci. USA* **77**, 3855–3859.
14. Wood, R. D., Robins, P., and Lindahl, T. (1988) Complementation of the xeroderma pigmentosum DNA repair defect in cell-free extracts. *Cell* **53**, 97–106.
15. Sibghat-Ullah, Husain, I., Carlton, W., and Sancar, A. (1989) Human nucleotide excision repair *in vitro*: repair of pyrimidine dimers, psoralen and cisplatin adducts by HeLa cell-free extract. *Nucleic Acids Res.* **17**, 4471–4484.
16. Robins, P., Jones, C. J., Biggerstaff, M., Lindahl, T., and Wood, R. D. (1991) Complementation of DNA repair in xeroderma pigmentosum group A cell extracts by a protein with affinity for damaged DNA. *EMBO J.* **10**, 3913–3921.
17. O'Donovan, A. and Wood, R. D. (1993) Identical defects in DNA repair in xeroderma pigmentosum group G and rodent ERCC group 5. *Nature* **363**, 185–188.

18. Masutani, C., Sugasawa, K., Yanagisawa, J., et al. (1994) Purification and cloning of a nucleotide excision repair complex involving the xeroderma pigmentosum group C protein and a human homologue of yeast RAD23. *EMBO J.* **13,** 1831–1843.
19. Sijbers, A. M., de Laat, W. L., Ariza, R. R., et al. (1996) Xeroderma pigmentosum group F caused by a defect in a structure-specific DNA repair endonuclease. *Cell* **86,** 811–822.
20. Biggerstaff, M. and Wood, R. D. (1992) Requirement for *ERCC-1* and *ERCC-3* gene products in DNA excision repair *in vitro*: complementation using rodent and human cell extracts. *J. Biol. Chem.* **267,** 6879–6885.
21. van Vuuren, A. J., Appeldoorn, E., Odijk, H., Yasui, A., Jaspers, N. G. J., and Hoeijmakers, J. H. J. (1993) Evidence for a repair enzyme complex involving ERCC1, ERCC4, ERCC11 and the xeroderma pigmentosum group F proteins. *EMBO J.* **12,** 3693–3701.
22. Shivji, K. K., Kenny, M. K., and Wood, R. D. (1992) Proliferating cell nuclear antigen is required for DNA excision repair. *Cell* **69,** 367–374.
23. Wang, Z., Wu, X., and Friedberg, E. C. (1991) Nucleotide excision repair of DNA by human cell extracts is suppressed in reconstituted nucleosomes. *J. Biol. Chem.* **266,** 22,472–22,478.
24. Masutani, C., Sugasawa, K., Asahina, H., Tanaka, K., and Hanaoka, F. (1993) Cell-free repair of UV-damaged simian virus 40 chromosomes in human cell extracts 2. Defective-DNA repair synthesis by xeroderma pigmentosum cell extracts. *J. Biol. Chem.* **268,** 9105–9109.
25. Gaillard, P. H. L., Martini, E. M. D., Kaufman, P. D., Stillman, B., Moustacchi, E., and Almouzni, G. (1996) Chromatin assembly coupled to DNA-repair - a new role for chromatin assembly factor-I. *Cell* **86,** 887–896.
26. Sugasawa, K., Masutani, C., and Hanaoka, F. (1993) Cell-free repair of UV-damaged Simian virus-40 chromosomes in human cell-extracts 1. Development of a cell-free system detecting excision repair of UV-irradiated SV40 chromosomes. *J. Biol. Chem.* **268,** 9098–9104.
27. Wang, Z. G., Wu, X. H., and Friedberg, E. C. (1993) Nucleotide excision repair of DNA in cell-free-extracts of the yeast *Saccharomyces cerevisiae*. *Proc. Natl. Acad. Sci. USA* **90,** 4907–4911.
28. He, Z. G., Wong, J. M. S., Maniar, H. S., Brill, S. J., and Ingles, C. J. (1996) Assessing the requirements for nucleotide excision repair proteins of *Saccharomyces cerevisiae* in an in vitro system. *J. Biol. Chem.* **271,** 28,243–28,249.
29. Shivji, M. K. K., Grey, S. J., Strausfeld, U. P., Wood, R. D., and Blow, J. J. (1994) Cip1 inhibits DNA replication but not PCNA-dependent nucleotide excision repair. *Curr. Biol.* **4,** 1062–1068.
30. Gaillard, P.-H. L., Moggs, J. G., Roche, D. M. J., et al. (1997) Initiation and bidirectional propagation of chromatin assembly from a target site for nucleotide excision repair. *EMBO J.* **16,** 6282–6289.
31. Barret, J. M., Calsou, P., Laurent, G., and Salles, B. (1996) DNA-repair activity in protein extracts of fresh human-malignant lymphoid-cells. *Mol. Pharmacol.* **49,** 766–771.

32. Biggerstaff, M., Szymkowski, D. E., and Wood, R. D. (1993) Co-correction of the ERCC1, ERCC4 and xeroderma pigmentosum group F DNA repair defects *in vitro*. *EMBO J.* **12,** 3685–3692.
33. Biggerstaff, M. and Wood, R. D. (1999) Assay for nucleotide excision repair protein activity using fractionated cell extracts and UV-damaged plasmid DNA. *Methods Mol. Biol.* **113,** 357–372.
34. Wood, R. D., Biggerstaff, M., and Shivji, M. K. K. (1995) Detection and measurement of nucleotide excision repair synthesis by mammalian cell extracts in vitro. *Methods: A Companion to Methods in Enzymology* **7,** 163–175.
35. Asahara, H., Wistort, P. M., Bank, J. F., Bakerian, R. H., and Cunningham, R. P. (1989) Purification and characterization of *Escherichia coli* endonuclease III from the cloned *nth* gene. *Biochemistry* **28,** 4444–4449.
36. Henricksen, L., Umbricht, C., and Wold, M. (1994) Recombinant replication protein-A - expression, complex-formation, and functional-characterization. *J. Biol. Chem.* **269,** 11,121–11,132.
37. Fien, K., and Stillman, B. (1992) Identification of replication factor C from *Saccharomyces cerevisiae*: a component of the leading-strand DNA replication complex. *Mol. Cell. Biol.* **12,** 155–163.
38. Ausubel, F. M., Brent, R., Kingston, R. E., et al. (1989) *Current Protocols in Molecular Biology*, Greene Publishing Associates and Wiley-Interscience, New York, NY.
39. Manley, J. L. (1983) Transcription of eukaryotic genes in a whole-cell extract, in: *Transcription and Translation: A Practical Approach*, Vol. 101. (Hames, B. D., and Higgins, S. J., eds.), IRL Press, Oxford, pp. 71–88.
40. Biggerstaff, M., Robins, P., Coverley, D., and Wood, R. D. (1991) Effect of exogenous DNA fragments on human cell extract-mediated DNA repair synthesis. *Mutat. Res.* **254,** 217–224.

30

Assaying for the Dual Incisions of Nucleotide Excision Repair Using DNA with a Lesion at a Specific Site

Mahmud K. K. Shivji, Jonathan G. Moggs,
Isao Kuraoka, and Richard D. Wood

Summary

Analysis of the mechanism of nucleotide excision repair (NER) using cell-free extract systems and purified proteins requires DNA substrates containing chemically defined lesions that are placed at a unique site in a DNA duplex. In this way, NER can be readily and specifically measured by detecting the 24–32 nucleotide products of the dual-incision reaction. This chapter describes several methods for detection of repair of a specific lesion in closed-circular DNA. As a model lesion, we use the well-repaired 1,3-intrastrand d(GpTpG)-cisplatin crosslink. Three methods are given for analysis of repair. One is to incorporate a radioactive label internally near the lesion and measure excision by detecting radioactive excised oligomers. Two other methods use DNA that is not internally labeled so that it can be stored and used when convenient. The first method for detection of repair of such unlabeled DNA is to detect excision products with a labeled complementary oligonucleotide by Southern blot hybridization. The second method is to 3'-end-label the excised oligonucleotide directly with radiolabeled dNTP and a DNA polymerase, using a complementary oligonucleotide with a 5'-overhang that serves as a template. This protocol is fast and sensitive, but relies on accurate foreknowledge of the site of 3'-incision for the particular lesion being used.

Key Words: Cisplatin; end-labeling; gel electrophoresis; nucleotide excision repair; M13 DNA; xeroderma pigmentosum.

1. Introduction

Cells remove a wide array of potentially toxic and mutagenic lesions from their genomes by a major repair pathway called nucleotide excision repair (NER). This repair process involves a multiprotein nuclease complex that incises a damaged DNA strand on the 5'- and 3'-sides of a lesion *(1,2)*. In humans, defects in

From: *Methods in Molecular Biology: DNA Repair Protocols: Mammalian Systems, Second Edition*
Edited by: D. S. Henderson © Humana Press Inc., Totowa, NJ

the genes that participate in NER can lead to a rare recessive disorder, xeroderma pigmentosum (XP) conferring hypersensitivity to sunlight and a predisposition to skin cancer *(3)*. NER-defective XP cells belong to one of seven genetic complementation groups (A–G). Although the NER process in mammalian cells consistently leads to the excision of damaged DNA fragments 24–32 nucleotides in length, the exact positions of the 5′- and 3′-incisions depend on the lesion being repaired. For instance, during the removal of a thymine dimer, incisions are made at the 22nd–24th phosphodiester bond on the 5′-side of the lesion and at the 5th phosphodiester bond on the 3′-side of the lesion *(4)*. The main sites of incision during repair of DNA containing a 1,3-intrastrand d(GpTpG)-cisplatin crosslink are at the 16th–20th phosphodiester bond 5′- and the 8th-9th phosphodiester bond on the 3′-side of the lesion *(5)*.

The core factors that participate in NER in eukaryotes have been identified using a combination of biochemical and genetic approaches *(6–8)*. Analysis of the mechanism of NER using cell-free extract systems and purified proteins requires suitable DNA substrates containing characterized DNA lesions. For certain mutagens, sufficiently defined DNA substrates can be obtained by damaging DNA with UV light or using chemicals that make specific adducts, such as thymine-psoralen mono-adducts, cisplatin-DNA intrastrand crosslinks, or acetylaminofluorene–DNA adducts. In this case, repair DNA synthesis can be monitored as an end point by measuring incorporation of radiolabeled deoxynucleotides during repair of the damaged DNA as described in the preceding chapter.

In many instances, more detailed information is required. This can be obtained by using DNA substrates containing chemically defined lesions that are placed at a unique site in a DNA duplex. In this way, NER can be readily and specifically measured by detecting the 24- to 32-mer nucleotide products of the dual-incision reaction. This is a simpler reaction than the full repair process, since it does not involve a eukaryotic DNA polymerase holoenzyme. Moreover, the production of 24- to 32-mers is very specific for NER, so that there can be no confusion with signals arising from other repair pathways, such as base excision repair or mismatch repair. One direct method is to construct DNA substrates containing a single defined lesion in a linear duplex of sufficient size, usually in the middle of a 140–150-mer. This is obtained by ligating together a short oligonucleotide containing the lesion with a series of complementary and overlapping oligonucleotides. An internal radiolabel is placed near the lesion, so that excised fragments can be detected. This approach has been well described elsewhere *(9–11)*. An alternative procedure is to place the lesion at a specific site in a covalently closed-circular DNA molecule containing a lesion. Closed-circular plasmid molecules are useful for a number of purposes in biological systems, including studies of DNA repair and mutagenesis *(12)*. Circular duplexes also can be used to study the effects of nucleosome structure on NER *(13–17)*, and the coupling of chromatin assembly to NER, where chromatin assembly propagates from a site of repair *(18)*. Circular duplexes can be constructed with or without an internal radiolabel. The advantage of an internal label is that repair can be detected directly after electrophoresis, and quantification of repair is straightforward *(4,19,20)*. The disadvantage is that the labeled substrate must be used within 1 or 2 wk of construction because of radioactive decay.

This chapter describes several methods for detection of repair of a specific lesion in closed-circular DNA. As a model lesion, we use the 1,3-intrastrand d(GpTpG)-cisplatin crosslink. This adduct has two main advantages. First, it is exceptionally well repaired by the mammalian NER system. Second, with care it is possible for any laboratory to construct DNA modified in this way, without specialized tools or reagents. The adduct has been used in many studies of the mechanism of NER, for example *(5,17,21–24)*.

Three methods are given for analysis of repair. One is to incorporate a radioactive label internally near the lesion and measure excision by detecting radioactive excised oligomers *(4,20)*. Two other methods use DNA that is not internally labeled. The advantage of this approach is that a large amount of specifically modified DNA substrate can be prepared and stored frozen, and then used when convenient. The first method for detection of repair of such unlabeled DNA is to detect repair products with a labeled complementary oligonucleotide by Southern blot hybridization *(5)*. The second, more straightforward method is to 3'- end-label the excised oligonucleotide directly with radiolabeled deoxyribonucleoside triphosphate (dNTP) and a DNA polymerase, using a complementary oligonucleotide with a 5'-overhang that serves as a template *(21,22,25)*. This protocol has the advantage of being the fastest and probably the most sensitive method of detection, but it relies on accurate foreknowledge of the site of 3'-incision for the particular lesion being used.

2. Materials

2.1. Cisplatin-Adducted Oligonucleotide

1. *cis*-Platinum(II)diamminedichloride (cisplatin; Sigma).
2. 2X Platination buffer: 6 mM NaCl, 1.0 mM Na_2HPO_4, 1.0 mM NaH_2PO_4.
3. T4 Polynucleotide kinase (10,000 U/mL) and 10X kinase buffer (New England Biolabs; NEB).
4. 5 M NaCl.
5. Thin-layer chromatography (TLC) plates suitable for ultraviolet (UV) shadowing analysis (silica gel; Merck F254 or Polygram CEL 300 PEI/UV_{254}—20 × 20 cm^2).
6. Sephadex G25 (Amersham Biosciences).
7. [γ-^{32}P]ATP (>5000 Ci/mmol; Amersham).

8. Sequencing gel solutions: Concentrate, diluent, and buffer (Sequagel, National Diagnostics) are mixed according to the manufacturer's instructions to make 12–20% gels. To 100 mL of the above mixture add 0.32 mL of 25% ammonium persulfate and 40 µL of N,N,N',N'-tetramethylethylenediamine (TEMED) to the gel components prior to pouring the gel. Use appropriate spacers to form either 0.4- or 1.5-mm thick gels.
9. 10X TBE: 108 g of Tris base, 55 g of boric acid, 40 mL of 0.5 M EDTA, pH 8.0, and deionized water to a final volume of 1 L.
10. Sequencing gel-loading buffer: 0.9 mL of deionized formamide mixed with 0.1 mL of dye solution (10X TBE, 0.25% bromophenol blue and 0.25% xylene cyanol); 1.0-mL aliquots can be stored frozen (–20°C).
11. Intensifying screens for Kodak X-ray film (XOMAT AR or BioMax MS) or PhosphorImager plates (Molecular Dynamics).

2.2. Closed-Circular DNA Containing Cisplatin-Adduct: Radiolabeled and Nonradiolabeled

1. Single-stranded form (+ strand) of bacteriophage M13mp18 DNA (modified to accommodate a lesion as in **Subheading 3.2.**; *see* **Note 1**).
2. T4 DNA polymerase (5000 U/mL, HT Biotechnology; or 3000 U/mL, NEB; *see* **Note 2**).
3. T4 DNA ligase (400,000 cohesive end U/mL, NEB).
4. Recombinant T5 exonuclease *(26)* (Amersham Biosciences).
5. Sephacryl S-400 (Amersham Biosciences).
6. 10X annealing/complementary strand synthesis/ligation buffer: 100 mM Tris-HCl, pH 7.9, 500 mM NaCl, 100 mM MgCl$_2$, and 10 mM dithiothreitol (DTT) (*see* **Note 3**).
7. dNTPs and ribonucleotides (NTPs) (Amersham Biosciences).
8. Nuclease-free bovine serum albumin (BSA) (10 mg/mL) (Gibco BRL or NEB).
9. Restriction endonuclease *Apa*Ll and 10X restriction endonuclease buffer (NEB buffer 4).
10. Agarose (Gibco BRL).
11. Ethidium bromide (10 mg/mL, Bio-Rad).
12. Cesium chloride (CsCl, Gibco BRL).
13. Quick-Seal polyallomer tubes (2 mL, 11 × 32 mm, Beckman cat. no. 344625).
14. H$_2$O-saturated butanol.
15. Centricon 30 or Centricon 100 ultrafiltration units.
16. TE: 10 mM Tris-HCl, pH 8.0, 1 mM EDTA.

2.3. Southern Blot Method

1. 5X Repair reaction buffer: 200 mM N-2-hydroxyethylpiperazine-N'-2-ethanesulfonic acid (HEPES)–KOH, pH 7.8, 25 mM MgCl$_2$, 2.5 mM DTT (Sigma), 10 mM ATP, 110 mM phosphocreatine (di-Tris salt, Sigma), 1.8 mg/mL of BSA (nuclease-free, Gibco BRL). The pH of the final repair reaction mixture should be 7.8. Aliquots should be stored at –80°C. (*See* **Note 4**.)

Analysis of Nucleotide Excision Repair

2. Creatine phosphokinase (CPK, rabbit muscle; Sigma), 2.5 mg/mL in 10 mM glycine, pH 9.0, 50% glycerol. Store aliquots at –20°C.
3. 1 M KCl.
4. 0.5 M EDTA, pH 8.0.
5. Sodium dodecyl sulfate (SDS), 10 and 20% (w/v).
6. Proteinase K (2 mg/mL).
7. Phenol–chloroform–isoamyl alcohol (25:24:1).
8. 7.5 M Ammonium acetate.
9. Glycogen (20 mg/mL; Roche).
10. Yeast tRNA (1–2 mg/mL).
11. Absolute ethanol (–20°C).
12. 70% Ethanol (–20°C).
13. Restriction endonucleases: *Hin*dIII, *Xho*I, and l0X restriction endonuclease buffer (NEB buffer 2).
14. TLC plates: *see* **Subheading 2.1., item 5**.
15. Sequencing gel solutions, l0X TBE and sequencing gel loading buffer (*see* **Subheading 2.1., items 8–10**).
16. Hybond N+ DNA transfer membrane (Amersham Biosciences).
17. Whatman 3MM paper.
18. 20X SSC: 3 M sodium chloride, 0.3 M trisodium citrate, pH to 7.0 with 1 N HCl.
19. Hybridization buffer (per 200 mL): 70 mL of 20% SDS, 80 mL of 25% PEG 8000, 10 mL of 5 M NaCl, 26 mL of 1 M potassium phosphate buffer, pH 7.0, and 14 mL of H$_2$O.

2.4. End-Labeling Dual-Incision Products

1. 5X Repair reaction buffer: *see* **Subheading 2.3., item 1**.
2. CPK (*see* **Subheading 2.3., item 2**).
3. Restriction endonucleases (*Hin*dIII, *Xho*I) and l0X restriction endonuclease buffer (NEB buffer 2).
4. Sequencing gel solutions, l0X TBE and sequencing gel-loading buffer (*see* **Subheading 2.1., items 8–10**).
5. Sequenase enzyme (Sequenase™ v2.0; 13 U/µL; Amersham Biosciences) and Sequenase dilution buffer.
6. *Escherichia coli* DNA polymerase I (Klenow fragment) and 10X DNA polymerase buffer (Roche).
7. *Msp*I-digested pBR322 DNA (NEB).
8. [α-^{32}P]dCTP (3000 Ci/mmol; Amersham Biosciences).

3. Methods

3.1. Synthesis, Purification and Characterization of an Oligonucleotide Containing a Single 1,3-Intrastrand d(GpTpG)-Cisplatin Crosslink

1. Purified 24-mer oligonucleotide containing a unique GTG sequence (5′-TCTTC TTCTGTGCACTCTTCTTCT-3′) is allowed to react at a concentration of 1 mM

with a threefold molar excess of cisplatin (3 mM) for 16 h at 37°C in a buffer containing 3 mM NaCl, 0.5 mM Na$_2$HPO$_4$ and 0.5 mM NaH$_2$PO$_4$ *(27)*, as follows: Dissolve cisplatin in 2X platination buffer to a final concentration of 6 mM. Dissolve the 24-mer oligonucleotide at a concentration of 2 mM in water, and add an equal volume of this solution to the buffered cisplatin solution. Also perform a mock platination reaction using 2X platination buffer without cisplatin. Incubate in the dark at 37°C for 16 h. Use reaction volumes <100 µL to facilitate purification in **step 2**. (*See* **Note 5**.)

2. Stop the platination reaction by adding NaCl to 500 mM. Purify the platination reaction mixture on a 1-mL Sephadex G25 spin column equilibrated in H$_2$O.
3. Dilute a 1-µL aliquot of the purified platination reaction mixture (and also the mock-treated control oligonucleotide reaction mixture) in water to 20 pmol/µL (assume 100% recovery). Perform 5′-^{32}P-phosphorylation analysis by setting up a reaction mixture containing 1 µL (20 pmol) of platinated (or mock-treated) oligonucleotide, 1 µL of 10X T4 polynucleotide kinase buffer, 4 µL of 100 µM ATP, 0.2 µL of [γ-^{32}P]ATP (>5000 Ci/mmol), 1 µL of T4 polynucleotide kinase (10 U/µL), and 2.8 µL of H$_2$O. Incubate at 37°C for 1 h, and stop the reaction by incubating at 68°C for 15 min.
4. Add 8 µL of sequencing gel-loading buffer to each 10 µL of reaction mixture volume. Heat the samples at 95°C for 3 min, and place immediately on ice before loading on a 0.4-mm thick denaturing 20% polyacrylamide gel (40 cm long). Run the gel in 1X TBE buffer until the xylene cyanol dye-front is approx 28 cm from the wells (the dyes have mobilities on a 20% polyacrylamide gel such that xylene cyanol migrates as a 29-mer oligonucleotide and bromophenol dye migrates as a 9-mer oligonucleotide). Expose the gel (without fixing or drying) to X-ray film for 1 h. The 1,3-intrastrand crosslinked platinated oligonucleotide is identified as a single band whose mobility is retarded by approximately one nucleotide relative to the nonmodified oligonucleotide (**Fig. 1**; *see* **Notes 6** and **7**).
5. After analysis of a small aliquot of each platination reaction (as described in **steps 3** and **4**), a preparative denaturing polyacrylamide gel (20%; 1.5 mm thick; 40 cm long) is used to purify the remainder of each reaction. Mix each purified platination reaction mixture with 0.8 volume of sequencing gel-loading buffer (without dyes; *see* **Note 8**). Heat the samples at 95°C for 3 min, and place immediately on ice before loading the gel. Load sequencing gel-loading buffer with dyes in adjacent lanes to follow the migration during electrophoresis. Run the gel in 1X TBE buffer until the xylene cyanol dye front is approx 28 cm from the wells.
6. Place the appropriate region of the preparative gel onto a TLC plate. Visualize the oligonucleotides using a handheld UV lamp (254 nm; *see* **Note 9**), excise the desired platinated and nonplatinated oligonucleotides (use a clean scalpel for each sample) and place in a microcentrifuge tube (*see* **Note 10**). Crush or finely slice each polyacrylamide fragment before resuspending in 0.5–1.0 mL of H$_2$O. Incubate at 37°C for 16 h with agitation. A rapid freeze–thaw step may improve recovery. Centrifuge the samples in a microcentrifuge to pellet the polyacryla-

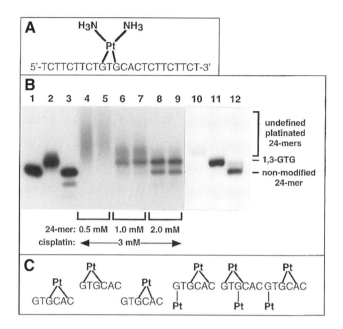

Fig. 1. (**A**) 24-mer oligonucleotide containing a single 1,3-intrastrand d(GpTpG)-cisplatin crosslink. (**B**) ^{32}P-labeled platinated and nonmodified oligonucleotides were separated in a denaturing 20% polyacrylamide gel. The 1,3-intrastrand crosslinked platinated oligonucleotide (lane 2) is observed as a single band whose mobility is retarded by approximately one nucleotide relative to the nonmodified oligonucleotide (lane 1). Lane 3 contains nonmodified 24-mer (HPLC-purified) used for the platination reactions in lanes 4–9. Platination reactions using a 3:1 molar ratio of cisplatin to oligonucleotide lead to the platination of almost all the available oligonucleotide (lanes 6 and 7), although the purest preparation of oligonucleotide containing a single 1,3-intrastrand d(GpTpG)-cisplatin crosslink was obtained using a 3:2 molar ratio of cisplatin to oligonucleotide (lanes 8 and 9), because overplatinated DNA products form a smear above the desired product (lanes 4 and 5). DNA was eluted from gel slices containing either the smear of undefined platinated 24-mers (lane 10), the defined 1,3-GTG platinated 24-mer (lane 11), or nonmodified 24-mer (lane 12). (**C**) The probable identity of platinated DNA products present in the smear includes an assortment of monoadducts, 1,3- and 1,4-intrastrand crosslinks.

mide fragments and recover the eluted oligonucleotide. It is convenient to lyophilize this solution to approx 100 µL for purification on a 1-mL Sephadex G25 spin column. Alternatively, the oligonucleotide can be recovered by ethanol precipitation or with a Microcon-30 concentrator. Quantify an aliquot of each sample by spectrophotometry. (*See* **Note 11**.)

7. Repeat the 5'-^{32}P-phosphorylation analysis described in **steps 3** and **4** for 20 pmol of each gel-purified oligonucleotide to determine the purity (*see* **Note 12**). Purified platinated oligonucleotides can be stored lyophilized or in TE buffer for several years at –80°C. Before use in the construction of closed-circular duplex DNA substrates (described in **Subheading 3.2.**), it is necessary to 5'-phosphorylate the platinated oligonucleotides from **step 6** with T4 polynucleotide kinase and ATP.

3.2. Construction of Closed-Circular Duplex DNA Containing a Single 1,3-Intrastrand d(GpTpG)-Cisplatin Crosslink

A 5- to 10-fold molar excess of 24-mer platinated oligonucleotide containing a single 1,3-intrastrand d(GpTpG)-cisplatin crosslink (described in **Subheading 3.1.**) is annealed to the single-stranded form (+ strand) of bacteriophage M13mp18 DNA modified to contain a sequence complementary to the platinated oligonucleotide within the polycloning site *(5,28)*. The 3'-terminus of the oligonucleotide acts as a primer for complementary strand synthesis by T4 DNA polymerase. The newly synthesized DNA strand is covalently closed with T4 DNA ligase resulting in a circular DNA duplex containing the single cisplatin-DNA adduct at a specific site (**Fig. 2A**).

1. For the annealing reaction, assemble a 75-µL reaction mixture containing 50 µL (25 µg) of single-stranded M13 DNA (500 ng/µL), 3.8 µL (380 ng) of 5'-phosphorylated 24-mer oligonucleotide containing a 1,3-intrastrand cisplatin crosslink (100 ng/µL), 7.5 µL of 10X annealing/complementary strand synthesis/ligation buffer, and 13.7 µL of H$_2$O. Incubate the mixture at 65°C for 5 min, 37°C for 30 min, 25°C for 20 min, and finally 4°C for 20 min.
2. For the complementary strand synthesis/ligation reaction prepare on ice a 200-µL reaction mixture containing the 75 µL of annealed DNA reaction mixture from **step 1**, 12.5 µL of 10X annealing/complementary strand synthesis/ligation buffer, 2 µL of 10 mg/mL of BSA, 4 µL of 100 m*M* ATP, 12 µL of 10 m*M* dNTP mixture (contains 10 m*M* each of dATP, dCTP, dGTP, dTTP), 15 µL of T4 DNA polymerase, 4 µL of T4 DNA ligase, and 75.5 µL of H$_2$O. Incubate at 37°C for 3 h (*see* **Note 13**).
3. Analyze the reaction products (remove 1-µL aliquots from 200-µL reaction mixtures) using a 0.8% agarose gel run in 1X TBE buffer at 50 V for 16 h (**Fig. 2B**). Add ethidium bromide (0.25 µg/mL) to the gel and running buffer. It is useful to load single-stranded DNA to assess the efficiency of complementary strand synthesis. Typical reactions result in the conversion of almost all single-stranded DNA into either closed-circular (approx 65% of molecules), linear (approx 15%), or nicked-circular DNA (approx 20%) (*see* **Fig. 2B**, lane 3; *see* **Note 14**).
4. Removal of nicked and linear DNA forms from closed-circular DNA can be achieved using CsCl/ethidium bromide density gradients with a final CsCl concentration of 1.55 g/mL. For 8 × 2.0 mL gradients, dissolve 15 g of CsCl in 9.5 mL of H$_2$O at 37°C for 1 h. Aliquot 1.352 mL of this solution into 2.2-mL

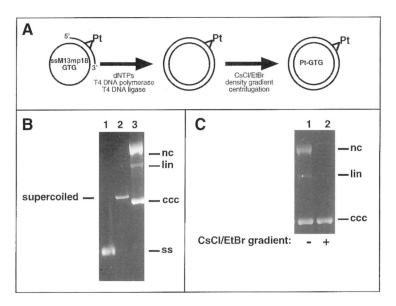

Fig. 2. (**A**) 24-mer platinated oligonucleotide containing a single 1,3-intrastrand d(GpTpG)-cisplatin crosslink (described in **Subheading 3.1.**) is annealed to the singlestranded form (+ strand) of bacteriophage M13mp18 DNA modified to contain a sequence complementary to the platinated oligonucleotide within the polycloning site (ssM13mp18GTG). The 3'-terminus of the oligonucleotide acts as a primer for complementary strand synthesis by T4 DNA polymerase. The newly synthesized DNA strand is covalently closed with T4 DNA ligase, resulting in a circular DNA duplex containing the single cisplatin-DNA adduct at a specific site. Closed-circular DNA (Pt-GTG) is then purified using CsCl–ethidium bromide gradient centrifugation. A control DNA substrate is prepared in the same way, except that a nonmodified 24-mer oligonucleotide is used as a primer. (**B**) Typical reactions (lane 3) result in the conversion of almost all single-stranded DNA (ss, lane 1) into either covalently closed-circular (ccc, approx 65% of molecules), linear (lin, approx 15%), or nicked-circular DNA (nc, approx 20%). Lane 2 contains the supercoiled replicative form of M13mp18GTG. The % yield of each form of DNA was quantified by densitometry correcting for the 1.6-fold increased fluorescence of linear and nicked-circular DNA. (**C**) More than 95% of the DNA substrate is in the closed-circular form after CsCl–ethidium bromide density gradient centrifugation (lane 2).

microcentrifuge tubes. Dilute the DNA samples, (+/– *Apa*LI digestion; *see* **Note 13**) to a final volume of 500 µL with TE buffer, and add to the 2.2-mL tubes containing CsCl solution. Transfer the DNA/CsCl solution to 2-mL Quick-Seal polyallomer tubes using a wide-bore needle and syringe. Cover each tube with foil, and add 148 µL of 10 mg/mL ethidium bromide solution to each Quick-Seal tube using a fine pipet tip. Heat-seal the polyallomer tubes under dim light.

Perform CsCl/ethidium bromide density gradient centrifugation at 315,000*g* (85,000 rpm) in a TLA100.2 (Beckman) ultracentrifuge rotor at 18°C for 24 h (*see* **Note 15**).

5. Unload the centrifuge tubes very carefully from the rotor in a dark room. Use a handheld UV lamp (312 nm) to visualize closed-circular DNA. Use a wide-bore needle and 1-mL syringe to remove this DNA in approx 500 µL of solution. It is important to keep the DNA away from light until all traces of ethidium bromide have been removed to avoid nicking. Add an equal volume of H_2O-saturated butanol, vortex for 5 s, and centrifuge for 1 min. Remove and discard the upper phase (pink owing to ethidium bromide). Repeat the extractions until no traces of ethidium bromide remain in either phase. Remove as much butanol as possible after the final extraction, and dilute the DNA solution to 2.0 mL using TE buffer.

6. Place each sample in a Centricon 100 or Centricon 30 ultrafiltration unit, and centrifuge at 1000*g* for 30–60 min at 4°C to concentrate the DNA to approx 50 µL. Add more TE buffer to 2.0 mL, and repeat this step four more times (*see* **Note 16**). Quantify the final DNA solution using a spectrophotometer and confirm the DNA purity using the gel electrophoresis conditions described in **step 3** (*see* **Note 17**). The purified DNA may be stored in aliquots at –80°C for more than 1 yr.

3.3. Construction of Internally Radiolabeled Closed-Circular Duplex DNA Containing a Single 1,3-Intrastrand d(GpTpG)-Cisplatin Crosslink

1. Radiolabel the gel-purified 24-mer oligonucleotide containing the 1,3-intrastrand d(GpTpG)-cisplatin crosslink (**Subheading 3.1.**) at its 5′-end by adding 30 pmol of oligonucleotide (1 µL; 0.25 µg/µL) to a microcentrifuge tube containing 1 µL of 10X T4 kinase buffer, 7 µL of [γ-^{32}P] ATP (>5000 Ci/mmol) and 1 µL of T4 polynucleotide kinase. Mix the contents and centrifuge. Incubate the reaction mixture at 37°C for 30 min. Supplement the reaction mixture with a cold-chase of 100 µM ATP (1 µL of 1 m*M* ATP) and 0.2 µL of T4 polynucleotide kinase. After mixing the contents, incubate the reaction mixture at 37°C for a further 30 min. Inactivate the kinase reaction by heating at 95°C for 5 min. The ^{32}P-labeled oligonucleotide can be used as a marker on sequencing gels (*see* **Note 18**).

2. For the annealing reaction, add 10 µL (10 µg; 4 pmol) of single-stranded M13 DNA and 2.2 µL of 10X annealing/complementary strand synthesis/ligation buffer to the ^{32}P-labeled 24-mer oligonucleotide (11 µL; 30 pmol). The ratio of oligonucleotide to M13 DNA should be titrated to find conditions that give the highest yield of closed-circular product. Incubate the reaction as described in **Subheading 3.2., step 1**.

3. For the complementary strand synthesis and ligation reaction, add the contents of the annealing mixture (23 µL) to a microcentrifuge tube containing 10 µL of 10X NEB buffer 2, 1 µL of BSA (10 mg/mL), 2 µL of 100 m*M* ATP, 30 µL of 2 m*M* dNTP mixture (contains 2 m*M* each of dATP, dGTP, dTTP, and dCTP), 10 µL of T4 DNA polymerase (3 U/µL), 2 µL of T4 DNA ligase, and 22 µL of H_2O. Mix

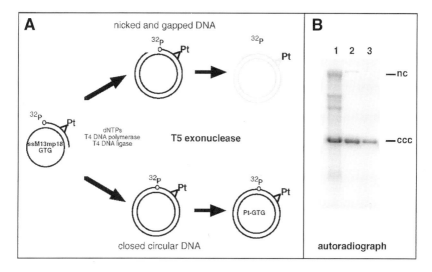

Fig. 3. Production of internally labeled closed-circular DNA containing a 1,3-intrastrand d(GpTpG)-cisplatin adduct. (**A**) Schematic. (**B**) Autoradiograph showing: lane 1, internally labeled products of a reaction with T4 DNA polymerase and T4 DNA ligase; lane 2, reaction mixture after treatment with T5 exonuclease to digest nicked and single-stranded DNA; lane 3, reaction mixture after purification on an S-400 spin column.

the contents of the tube, and incubate at 37°C for 4 h. Stop the reaction by heating at 75°C for 10 min.

4. The reaction now contains a mixture of complete and incomplete products of complementary strand synthesis (**Fig. 3A**). The latter products are degraded by T5 exonuclease *(26)*. Add 1 µL (2 µg) of T5 exonuclease, and incubate at 37°C for 2 h. Stop the T5 exonuclease activity by incubating the reaction mixture at 70°C for 5 min.
5. To remove the T5 exonuclease-degraded products, load the 100-µL reaction mixture onto a spin column containing Sephacryl S-400 equilibrated in TE (*see* **Note 19**). Centrifuge at 320*g* (2000 rpm) for 1 min, and collect the eluate (**Fig. 3B**).
6. Confirm the presence of the lesion by digesting the DNA with *Apa*L1 restriction endonuclease (*see* **Note 13**). Measure the radioactivity in the DNA by Cerenkov scintillation counting. Dilute the DNA to a concentration of 50 ng/µL, and store aliquots at –80°C (*see* **Note 20**).

3.4. Gel Purification of Oligonucleotides and Preparation of Radiolabeled Probes for Southern Blots

The 27-mer oligonucleotide (5′-GAAGAGTGCACAGAAGAAGAGGCCTGG-3′) that is used as a probe for Southern blotting should be gel purified.

1. Pour a denaturing 20% polyacrylamide gel (1.5 mm thick; 40 cm long) with wells of 2–3 cm width.
2. Prerun the gel at 50°C in 1X TBE.
3. Prepare the oligonucleotides (50 µg; 2.5 µL) diluted 1:1 in loading buffer (2.5 µL) without dyes (bromophenol blue or xylene cyanol; see **Note 8**).
4. Heat-denature the oligonucleotides at 95°C for 3 min and cool on ice. Load 50-µg aliquots of oligonucleotide/lane. To detect the oligonucleotide by UV shadowing, approx 0.5-2.0 OD units (approx 10 µg) must be loaded on the gel. Load a blank lane with sequencing gel-loading buffer containing bromophenol blue and xylene cyanol. Run the gel in 1X TBE at 50°C until the xylene cyanol has migrated two-thirds of the distance from the wells (see **Subheading 3.1., step 4**).
5. Transfer the gel to a TLC plate, and cover with Saran Wrap. The DNA will appear as a shadow under a short-wave UV lamp (see **Note 9**). Mark the desired band, and excise it from the bulk of the gel.
6. Crush or finely slice each polyacrylamide fragment before resuspending in 0.5–1.0 mL of H_2O. Incubate at 37°C for 16 h with agitation. A rapid freeze–thaw step may improve recovery. Centrifuge the samples in a microcentrifuge to pellet the polyacrylamide fragments and recover the eluted oligonucleotide. It is convenient to lyophilize this solution to approx 100 µL for purification on a 1-mL Sephadex G25 spin column. Quantify the DNA concentration by spectrophotometry, and store aliquots at –80°C.
7. To prepare radiolabeled probes, add 1.5 µL of 27-mer oligonucleotide (10 pmol) to 1 µL of 10X kinase buffer, 6.5 µL of [γ-^{32}P]ATP (>5000 Ci/mmol, 10 mCi/mL), and 1.0 µL of T4 polynucleotide kinase (10 U/µL). The radiolabel should be as fresh as possible. Incubate at 37°C for 60 min, and then inactivate the reaction by incubating at 65°C for 15 min. Increase the volume to 100 µL with TE, and purify on a Sephadex G25 spin column. Store shielded at –20°C until required.

3.5. Analysis of NER DNA Products by the Southern Blot Method

1. Reaction mixtures are set up in 50-µL aliquots. Add 10 µL of 5X repair reaction buffer, 1.0 µL of CPK, 1.5 µL of 1 M KCl, and 7.5 µL of H_2O to a 20-µL mixture comprising cell extract (100–300 pg of protein) and cell extract dialysis buffer (which contains 100 mM KCl). The total salt concentration (either KCl or NaCl) in the reactions is kept to a 70 mM final concentration.
2. Preincubate the reaction mixtures at 30°C for 5 min. Add 10 µL (0.25 µg) of DNA substrate to the reactions and incubate for a further 25–30 min.
3. Add 2 µL of 0.5 M EDTA, pH 8.0, 3 µL of 10% SDS, and 6 µL of Proteinase K (2 mg/mL) to the reaction tubes. Mix the contents thoroughly, and incubate at 37°C for 30 min or 56°C for 20 min.
4. Increase the volume to 100 µL with TE, and mix with an equal volume of phenol–chloroform–isoamyl alcohol. Mix thoroughly and centrifuge for 5–10 min at maximum speed.
5. Transfer 75 µL of the aqueous phase to a fresh tube containing 30 µL of 7.5 M ammonium acetate, and either 0.5 µL of glycogen (20 mg/mL) or yeast tRNA

(100 μg/mL). Mix well. Add 212 μL of ice-cold absolute ethanol. Mix thoroughly and place on dry ice for 20 min.

6. Centrifuge the samples at maximum speed in a chilled microcentrifuge for 30 min. Using a glass Pasteur pipet with a drawn-out tip, gently remove the alcohol (*see* **Note 21**).
7. Add 200 μL of chilled 70% ethanol, and centrifuge in a chilled microcentrifuge at maximum speed for 5–10 min. Carefully remove the ethanol and discard.
8. Dry the DNA pellet in a centrifuge under vacuum for 5 min. Add 8 μL of deionized water and dissolve the DNA by incubating the tubes at 37°C for 30–60 min.
9. Make a premixture of *Hin*dIII/*Xho*I enzyme such that each tube will contain 1.0 μL of 10X restriction endonuclease buffer, 0.4 μL of H_2O, and 0.3 μL each of *Hin*dIII and *Xho*I. Add 2 μL of this premix to the DNA, mix, and incubate at 37°C for 60 min (*see* **Note 22**).
10. Add 8 μL of loading buffer. The samples can be either stored frozen (–80 or –20°C) or loaded on a 12% sequencing gel.
11. Prerun the sequencing gel at a constant temperature of 50°C in 1X TBE. When the gel is at the appropriate temperature, heat the DNA and loading buffer mixture samples at 95°C for 5 min. Centrifuge and keep on ice until required. Wash the wells thoroughly to remove traces of urea that leached out during the prerun. Load the samples and run the gel at 50°C until the bromophenol blue dye has migrated approx 30 cm from the wells. As a marker, a 5′-phosphorylated 24-mer platinated oligonucleotide (150 pg) is also loaded in one of the lanes (*see* **Note 23**).
12. Cut the nylon membrane (Hybond N+; *see* **Note 24**) and four pieces of Whatman 3MM to the size of the gel area to be transferred. Presoak the nylon membrane and one piece of Whatman 3MM paper in 10X TBE.
13. Place the glass plate horizontally on the bench with the gel face up. Cover the gel from the wells to the bromophenol blue dye front with the nylon membrane followed by one piece of Whatman 3MM paper (*see* **Note 25**).
14. Lay three more pieces of dry Whatman 3MM paper on top followed by a glass plate. Place a modest weight on top of the plate, for example, two 500-mL bottles filled with water. Allow the DNA to transfer for 1.5–3 h (*see* **Note 26**).
15. Fix the DNA on the membrane by placing it (DNA side up) in 0.4 *M* NaOH for 20 min without submerging the membrane. The nylon membrane can be used immediately (or it can be covered with Saran Wrap and kept in a cold room until required). Wash the nylon membrane in 5X SSC for 2 min.
16. Roll the nylon membrane in a nylon mesh and place in a hybridization bottle (*see* **Note 27**). Pour 15 mL of hybridization buffer into the bottle, and coat the membrane in buffer by rotating the bottle (*see* **Note 28**). Prehybridize at 42°C for 10 min. Add the radiolabeled probe to the bottle, and incubate in an oven at 42°C for 16 h. The solution containing the radiolabeled probe can be decanted into a suitable container and reused if necessary.
17. Carefully remove the membrane from the hybridization bottle with forceps. Wash the membrane twice for 10 min each time with 1 L of 1X SSC containing 0.1% SDS (warm to >30°C; *see* **Note 29**).

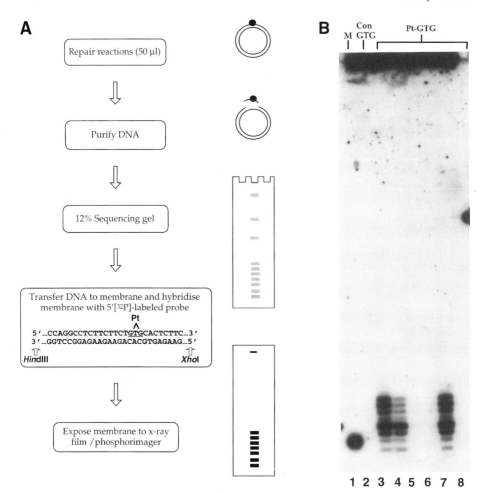

Fig. 4. Detection of excised fragments containing a 1,3-intrastrand d(GpTpG)-cisplatin adduct by Southern blot method. **(A)** Schematic. **(B)** Autoradiograph of the membrane containing the excised DNA fragments. Reaction mixtures (150 µL) contained replication protein A (RPA)-depleted HeLa cell extract (144 µg of protein) and either recombinant wild-type (lanes 2, 3, and 7) or mutant (lanes 4–6) forms of human RPA. The reaction mixture without RPA is shown in lane 8. The reaction mixtures included DNA containing the lesion (lanes 3–8) or control DNA (without the lesion; lane 2). The samples were treated as described (schematic and **Subheading 3.5.**). Lane 1 contains a 5′-phosphorylated 24-mer oligonucleotide marker.

18. Cover the membrane in Saran Wrap, and expose the membrane to Kodak X-ray film at −80°C or phosphorimager screens (*see* **Note 30**). An example of the type of image formed on the X-ray film or phosphorimager is shown in **Fig. 4B**.

3.6. Analysis of NER DNA Products by an End-Labeling Method

1. Each sample in this assay requires one-fifth of the components that are used for the assay described in **Subheading 3.5.** Therefore, 10-µL reaction mixtures can be used to analyze NER in cell extracts: combine 4.0 µL of cell extract plus buffer containing 0.1 M KCl, 2.0 µL of 5X repair reaction buffer, 0.2 µL of CPK, 0.3 µL of 1 M KCl, and 2.5 µL of H_2O. Incubate the reaction mixtures in the absence of DNA for 5 min at 30°C.
2. Add 1.0 µL of DNA substrate (50 ng). Incubate at 30°C for a further 25–30 min. At this point, the reaction mixtures can be stored frozen at –20°C.
3. To analyze the release of DNA containing the lesion, a 34-mer oligonucleotide is used (5'-pGGGGGAAGAGTGCACAGAAGAAGAGGCCTGGTCGp-3' with a phosphate group on the 3'-end to prevent priming). This oligonucleotide is complementary to the DNA fragment excised during NER (**Fig. 5A**) and the position of the major 3'-incision site *(5)* is such that the 34-mer oligonucleotide has a 5'-overhang. This 5'-overhang (shown in bold letters in **Fig. 5A**) is used as a template by Sequenase to incorporate radiolabeled dCMP on the 3'-end of the excised fragments. In the case of Pt-GTG-DNA substrate, add 1.0 µL of oligonucleotide (stock solution 6.0 µg/mL) to the 10-µL reaction mixture. It is recommended that the oligonucleotide is titrated to determine the right concentration for end-labeling.
4. Heat the tubes at 95°C for 5 min. Centrifuge the tubes to pull down any liquid that has evaporated and condensed on the side.
5. Allow the DNA to anneal by leaving the tubes at room temperature for 15 min.
6. Make up a Sequenase enzyme/[α-^{32}P]dCTP mixture such that each reaction contains 0.13 U of Sequenase enzyme and 2.0 µCi of [α-^{32}P]dCTP. These components are diluted in the Sequenase dilution buffer that is provided by the manufacturer. Add this mixture to the tubes, and incubate at 37°C for 3 min. Add 1.5 µL dNTP mixture (10 µM each of dATP, dGTP, dTTP, and 5 µM dCTP). Incubate at 37°C for a further 12 min.
7. Stop the reactions by adding 8.0 µL of loading buffer. Mix thoroughly, and heat the tubes at 95°C for 5 min. Centrifuge the tubes, and keep them on ice.
8. Prerun a 14% sequencing gel until it reaches a temperature of 50°C (*see* **Note 31**). Thoroughly wash the wells prior to loading one-third of the sample (6–7 µL) in the wells. Include appropriate end-labeled markers (in this case pBR322 DNA digested with *Msp*I and end-labeled with [α-^{32}P]dCTP; *see* **Note 32**). Electrophoresis should take place until the bromophenol blue dye migrates off the gel (which takes about 2 h).
9. Transfer the gel to 3MM paper, and dry the gel for 30–60 min. Expose the gel to Kodak BioMax film or PhosphorImager screens (*see* **Note 30**). An example of the type of image formed on the X-ray film or PhosphorImager is shown in **Fig. 5B**.

3.7. Analysis of NER Products Using Internally Radiolabeled DNA Substrates

1. Set up the repair reactions (10 µL) as described in **Subheading 3.6., step 1**.

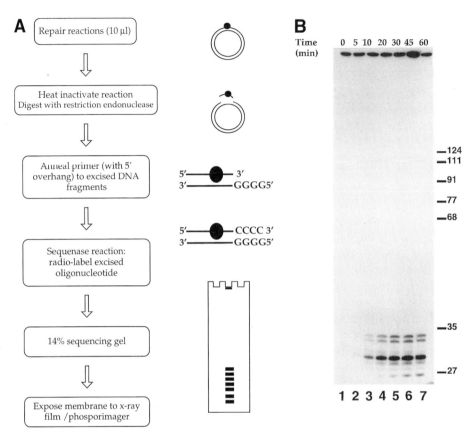

Fig. 5. Detection of excised fragments containing a 1,3-intrastrand d(GpTpG)-cisplatin adduct by the end-labeling method. (**A**) Schematic. (**B**) Autoradiograph of the dried sequencing gel containing the radiolabeled excised DNA fragments. A reaction mixture (80 µL) contained RPA-depleted HeLa cell extract (100 µg of protein) and recombinant human RPA (1 µg). This was incubated at 30°C for 5 min prior to addition of DNA containing the 1,3-intrastrand d(GpTpG)-cisplatin crosslink. Aliquots of 10 µL were removed at times 0, 5, 10, 20, 30, 45, and 60 min (lanes 1–7). The samples were treated as described (schematic and **Subheading 3.6.**). The positions of the radiolabeled *Msp*I-digested pBR322 DNA fragments are as shown (bold lines).

2. Add 1.0 µL of the internally radiolabeled DNA substrate (50 ng). Incubate at 30°C for a further 25–30 min. Stop the reactions by heating the reaction mixtures for approx 3 min at 95°C and add 8.0 µL of sequencing gel-loading buffer. At this point, the reaction mixtures may either be stored frozen at –20°C or kept on ice.
3. Prerun a 14% sequencing gel until it reaches a temperature of 50°C (*see* **Note 31**). Thoroughly wash the wells prior to loading most of the sample (approx 15 µL) in

Analysis of Nucleotide Excision Repair

the wells. Include appropriate end-labeled markers (in this case pBR322 DNA digested with *Msp*I and end-labeled with [α -^{32}P]dCTP; *see* **Note 32**). Electrophoresis should take place until the bromophenol blue dye migrates off the gel (which takes about 2 h).

4. Transfer the gel to 3MM paper, and dry the gel for 30–60 min. Expose the gel to Kodak BioMax film or phosphorimager screens (*see* **Note 30**).

4. Notes

1. These procedures use single stranded M13 phage (about 7200 bases) as the substrate for complementary strand synthesis. Shorter single-strand DNA has also been used with success, derived from the 2900 base pBluescript phagemid, and prepared with the aid of helper phage *(17,20,23)*. This may be a technical advantage as it requires the DNA polymerase to synthesize considerably less DNA.
2. We have found that T4 DNA polymerase or the closely related RB69 phage DNA polymerase are the most appropriate enzymes for complementary synthesis, as these enzymes can synthesize around a phage circle without displacing the primer. Recombinant RB69 polymerase can be efficiently prepared as described *(29)*. Briefly, plasmid pRB-C6His is grown in *E. coli* expression strain BL21 (DE3) and induced with 1 m*M* isopropyl-β-D-thiogalactopyranoside (IPTG) overnight at 16°C. The bacterial pellet is lysed by sonication in buffer consisting of PBS supplemented to 300 m*M* NaCl (final), 5 m*M* β-mercaptoethanol, 10% glycerol, 15 m*M* imidazole, and a commercial protease inhibitor tablet. The cleared lysate is loaded onto a Ni-NTA column equilibrated with the same buffer omitting β-mercaptoethanol, and then washed with this buffer without the additional NaCl. Protein is eluted with buffer containing 0.5X PBS, 200 m*M* imidazole, 10% glycerol and protease inhibitors, and dialyzed against buffer A containing 30 m*M* Tris-HCl, pH 8.0, 1 m*M* EDTA, 1 m*M* DTT, 1 m*M* sodium bisulfite, 0.1% Na azide, and 10% glycerol. The cleared dialysate is loaded onto a MonoQ FPLC column, washed with buffer A, and eluted with a 5–20% gradient of buffer B (the same as buffer A but with 2 *M* KCl). Pure fractions are combined and dialyzed against storage buffer containing 100 m*M* potassium phosphate (pH 6.5), 10 m*M* β-mercaptoethanol, and 50% glycerol. Aliquots are stored at –80 °C.
3. This is identical to 10X restriction endonuclease buffer (NEB buffer 2; NEB).
4. There are no dNTPs present in this buffer (*see* Chapter 29).
5. It is necessary to warm the solution to 37°C and stir for at least 1 h for complete dissolution of the yellow crystals. All cisplatin solutions should be protected from light.
6. Increasing the final concentration of oligonucleotide to 1.5 or 2.0 m*M* reduces the yield of platinated oligonucleotides, but gives a higher proportion of the oligonucleotide containing the desired 1,3-intrastrand crosslink relative to other platinated reaction products (*see* **Fig. 1**). This may be advantageous in obtaining a very pure preparation of platinated oligonucleotide.
7. This reduction in mobility results from the cisplatin-DNA adduct bending the DNA, adding a +2 charge, and increasing the molecular weight by 223 *(27)*.

8. The dyes interfere with subsequent recovery of DNA.
9. TLC silica gel contains a UV chromophore that is masked by the DNA. The DNA appears as a shadow *(30)*. Exposure to UV should be as brief as possible in order to avoid DNA damage.
10. This step determines the final purity of the platinated oligonucleotide, and thus, it is essential to be conservative in the amount of material excised from each band in order to avoid contamination from adjacent oligonucleotide species.
11. An alternative method for purification of the platinated oligonucleotide from preparative denaturing polyacrylamide gels avoids the use of any UV irradiation during visualization and excision of the desired reaction products. The platinated reaction products can be 5′-phosphorylated with T4 polynucleotide kinase and ATP prior to electrophoresis. ^{32}P-labeled oligonucleotide reaction products can be run in adjacent lanes to locate the desired 5′-phosphorylated platinated oligonucleotide products by autoradiography. The oligonucleotides excised from the gel must be dephosphorylated prior to any 5′-^{32}P-phosphorylation analysis for purity, and it is necessary to excise and analyze several gel fragments from each lane to ensure recovery of the desired species.
12. The identity of the 1,3-intrastrand d(GpTpG)-cisplatin crosslinks can be confirmed by enzymatic digestion of the platinated oligonucleotides to their component nucleosides using DNase I, P1 nuclease and alkaline phosphatase followed by reverse-phase HPLC analysis (K. J. Yarema and J. M. Essigmann, personal communication). It is generally easier, however, to confirm the presence of this lesion by restriction endonuclease or primer extension analysis after incorporation of the platinated oligonucleotide into closed-circular DNA (described in **Subheading 3.2.**). The platinum lesion can also be removed by treatment with sodium cyanide resulting in reversion of the oligonucleotide to the nonmodified form.
13. The 1,3-intrastrand d(GpTpG)-cisplatin crosslink is located within a unique *Apa*LI restriction site. Resistance to cleavage by this enzyme is diagnostic for the presence of the cisplatin–DNA adduct *(5,27)*. Platinated (but not control) DNA preparations may be cleaved with *Apa*LI after completion of complementary strand synthesis to linearize any molecules lacking the 1,3-intrastrand cisplatin crosslink. This is conveniently done by supplementing the 200-µL reaction mixtures with 50 µL of 10X NEB restriction digestion buffer 4, 5 µL of 10 mg/mL BSA, 10 µL of *Apa*LI (10 U/µL), and H$_2$O to 500 µL and incubating for a further 3 h at 37°C; 2.5-µL aliquots can be analyzed as described in **Subheading 3.2., step 3**.
14. The presence of gapped circular DNA molecules (these have a mobility in between closed-circular and linear DNA forms under the electrophoresis conditions described) indicates that insufficient amounts of dNTPs or DNA polymerase were present in the reaction.
15. The centrifugation conditions described give a good separation of closed-circular DNA from nicked-circular and linear forms.
16. DNA containing CsCl may be dialyzed against TE buffer rather than desalted in a Centricon 100 ultrafiltration unit, but a subsequent concentration step may then be required.

17. In addition to resistance to cleavage by *Apa*LI (*see* **Note 13**), the presence of a site-specific cisplatin–DNA crosslink in DNA substrates can be confirmed by primer extension analysis of the damaged DNA strand *(5)*.
18. The stock solution for the 24-mer oligonucleotide marker is made by adding 0.2 µL of the reaction mixture to 199.8 µL of TE. For a gel, use 1 µL of this plus 4 µL of TE and 4 µL of sequencing gel-loading buffer.
19. Sephacryl S-400 resin comes as a suspension in 20% ethanol. Spin columns (1 mL) are prepared by washing in TE six times (each 100-µL wash is centrifuged at 320g [2000 rpm] for 1 min). This step is very important in order to remove small DNA fragments produced by the T5 exonuclease digestion. These fragments inhibit the NER reaction.
20. The radiolabeled DNA must be used within 1–2 wk, because the substrate undergoes both radioactive decay and radiolytic decomposition.
21. Pipets with drawn-out tips are simply made, by heating the tip of the Pasteur pipet in a gas burner flame and pulling the molten tip in a curve with a pair of forceps until the glass has a very fine diameter. The glass is snapped near where the glass begins to taper. The bore should be approx 0.5–1.0 mm in diameter. When removing the alcohol, make sure that the tip of the drawn-out Pasteur pipet is facing away from the DNA pellet.
22. Digesting the DNA with *Hin*dIII and *Xho*I prior to gel electrophoresis allows detection of uncoupled incisions made either 3′ (*Hin*dIII) or 5′ (*Xho*I) to the lesion *(31)*.
23. As an aid to identifying the platinated oligonucleotides formed during the dual-incision process with Pt-GTG DNA substrates, the 5′-phosphorylated 24-mer oligonucleotide prepared in **Subheading 3.1.** can be loaded alongside the products of the repair reaction.
24. Other types of membrane, for example, Electran® positively charged nylon membrane (BDH), do not work as well with this procedure.
25. Remove any air bubbles between the membrane and the gel by rolling a glass pipet over the Whatman 3MM paper. The presence of air bubbles can interfere with the capillary transfer of DNA.
26. Always handle the nylon membrane by the edges.
27. Ensure that the leading edge formed by the membrane/mesh rotates into the buffer when placed in the hybridization oven.
28. SDS precipitates cause background spots and smears. It is important to avoid this. The solution can be stored at room temperature. If precipitates are seen, the solution should be filtered though 0.22-µm Nalgene filters. Warm the solution to 55°C before adding it to the hybridization bottle.
29. Keeping the wash buffer at >30°C prevents the SDS from precipitating out of solution. The stringency of wash depends on salt concentrations. Check for localized radioactivity after the second wash using a Geiger counter. If the membrane is to be reprobed, soak the membrane in 0.4 M NaOH for 30 min. Wash in 5X SSC for 10 min, or boil the blot in 0.5% SDS.
30. The exposure time to X-ray film will depend on the type of film used. Kodak XOMAT AR is good for overnight exposures if the Geiger counter readings from

the membrane (**Subheading 3.5.**) or dried gel (**Subheadings 3.6.** and **3.7.**) are in the range of approx 10–50 counts/s. Alternatively, another type of X-ray film (Kodak BioMax MS) or a PhosphorImager screen (Molecular Dynamics) is four times more sensitive compared to Kodak XOMAT AR. In the case of the more sensitive alternatives, the exposure times are approx 1–4 h where >50 counts/s are registered on the Geiger counter.

31. Using a 14% polyacrylamide gel allows separation of the 22–27 nucleotide fragments from the faster migrating salt front.
32. The DNA markers (*Msp*I-digested pBR322 plasmid) are end-labeled in Klenow buffer and [α-^{32}P]dCTP with DNA polymerase I (Klenow fragment). The reaction is incubated on ice for 30 min. The reaction is terminated by the addition of 0.5 M EDTA, pH 8.0, to a final concentration of 20 mM. Loading approx 1–2 ng of this marker on a sequencing gel gives an appropriate signal after a 4- to 5-h exposure to Kodak BioMax X-ray film. The marker is stable for several weeks even when repeatedly thawed and frozen at –20°C.

Acknowledgments

We thank the past members of our laboratory for discussions and contributions to these procedures, Kevin Yarema and John Essigmann for instruction in preparation of oligonucleotides modified with cisplatin, Jon Sayers for T5 exonuclease, and W. Konigsberg for procedures regarding preparation of RB69 DNA polymerase.

References

1. Lindahl, T. and Wood, R. D. (1999) Quality control by DNA repair. *Science* **286**, 1897–1905.
2. Friedberg, E. C., Walker, G. C., Siede, W., Wood, R. D., Schultz, R. A., and Ellenberger, T. (2005) *DNA Repair and Mutagenesis*, American Society for Microbiology Press, Washington, D.C.
3. Mitchell, J. R., Hoeijmakers, J. H., and Niedernhofer, L. J. (2003) Divide and conquer: nucleotide excision repair battles cancer and ageing. *Curr. Opin. Cell Biol.* **15**, 232–240.
4. Huang, J. C., Svoboda, D. L., Reardon, J. T., and Sancar, A. (1992) Human nucleotide excision nuclease removes thymine dimers from DNA by incising the 22nd phosphodiester bond 5' and the 6th phosphodiester bond 3' to the photodimer. *Proc. Natl. Acad. Sci. USA* **89**, 3664–3668.
5. Moggs, J. G., Yarema, K. J., Essigmann, J. M., and Wood, R. D. (1996) Analysis of incision sites produced by human cell extracts and purified proteins during nucleotide excision repair of a 1,3-intrastrand d(GpTpG)-cisplatin adduct. *J. Biol. Chem.* **271**, 7177–7186.
6. Wood, R. D., Araújo, S. J., Ariza, R. R., et al. (2000) DNA damage recognition and nucleotide excision repair in mammalian cells. *Cold Spring Harbor Symp. Quant. Biol.* **65**, 173–182.

7. de Boer, J. and Hoeijmakers, J. H. J. (2000) Nucleotide excision repair and human syndromes. *Carcinogenesis* **21,** 453–460.
8. Petit, C. and Sancar, A. (1999) Nucleotide excision repair: from *E. coli* to man. *Biochimie* **81,** 15–25.
9. Hess, M. T., Schwitter, U., Petretta, M., Giese, B., and Naegeli, H. (1997) Bipartite substrate discrimination by human nucleotide excision-repair. *Proc. Natl. Acad. Sci. USA* **94,** 6664–6669.
10. Huang, J. C. and Sancar, A. (1994) Determination of minimum substrate size for human excinuclease. *J. Biol. Chem.* **269,** 19,034–19,040.
11. Mu, D. and Sancar, A. (1997) DNA excision-repair assays. *Prog. Nucleic Acids Res. Mol. Biol.* **56,** 63–81.
12. Yarema, K. J. and Essigmann, J. M. (1995) Evaluation of the genetic effects of defined DNA lesions formed by DNA-damaging agents. *Methods* **7,** 133–146.
13. Sugasawa, K., Masutani, C., and Hanaoka, F. (1993) Cell-free repair of UV-damaged Simian virus-40 chromosomes in human cell-extracts 1. Development of a cell-free system detecting excision repair of UV-irradiated SV40 chromosomes. *J. Biol. Chem.* **268,** 9098–9104.
14. Wang, Z., Wu, X., and Friedberg, E. C. (1991) Nucleotide excision repair of DNA by human cell extracts is suppressed in reconstituted nucleosomes. *J. Biol. Chem.* **266,** 22,472–22,478.
15. Ura, K., Araki, M., Saeki, H., et al. (2001) ATP-dependent chromatin remodeling facilitates nucleotide excision repair of UV-induced DNA lesions in synthetic dinucleosomes. *EMBO J.* **20,** 2004–2014.
16. Hara, R. and Sancar, A. (2002) The SWI/SNF chromatin-remodeling factor stimulates repair by human excision nuclease in the mononucleosome core particle. *Mol. Cell Biol.* **22,** 6779–6787.
17. Frit, P., Kwon, K., Coin, F., et al. (2002) Transcriptional activators stimulate DNA repair. *Mol. Cell* **10,** 1391–1401.
18. Gaillard, P.-H. L., Moggs, J. G., Roche, D. M. J., et al. (1997) Initiation and bidirectional propagation of chromatin assembly from a target site for nucleotide excision repair. *EMBO J.* **16,** 6282–6289.
19. Kuraoka, I., Bender, C., Romieu, A., Cadet, J., Wood, R. D., and Lindahl, T. (2000) Removal of oxygen free-radical induced 5′,8 purine cyclodeoxynucleosides from DNA by the nucleotide excision repair pathway in human cells. *Proc. Natl. Acad. Sci. USA* **97,** 3832–3837.
20. Sugasawa, K., Okamoto, T., Shimizu, Y., Masutani, C., Iwai, S., and Hanaoka, F. (2001) A multistep damage recognition mechanism for global genomic nucleotide excision repair. *Genes Dev.* **15,** 507–521.
21. Araújo, S. J., Tirode, F., Coin, F., et al. (2000) Nucleotide excision repair of DNA with recombinant human proteins: definition of the minimal set of factors, active forms of TFIIH and modulation by CAK. *Genes Dev.* **14,** 349–359.
22. Araújo, S. J., Nigg, E. A., and Wood, R. D. (2001) Strong functional interactions of TFIIH with XPC and XPG in human DNA nucleotide excision repair, without a pre-assembled repairosome. *Mol. Cell Biol.* **21,** 2281–2291.

23. Riedl, T., Hanaoka, F., and Egly, J. M. (2003) The comings and goings of nucleotide excision repair factors on damaged DNA. *EMBO J.* **22,** 5293–5303.
24. Zamble, D. B., Mu, D., Reardon, J. T., Sancar, A., and Lippard, S. J. (1996) Repair of cisplatin-DNA adducts by the mammalian excision nuclease. *Biochemistry* **35,** 10,004–10,013.
25. Shivji, M. K. K., Ferrari, E., Ball, K., Hübscher, U., and Wood, R. D. (1998) Resistance of human nucleotide excision repair synthesis in vitro to p21^{Cdn1}. *Oncogene* **17,** 2827–2838.
26. Sayers, J. (1996) Viral polymerase-associated 5′→3′-exonucleases: expression, purification, and uses. *Methods Enzymol.* **275,** 227–238.
27. Yarema, K. J., Lippard, S. J., and Essigmann, J. M. (1995) Mutagenic and genotoxic effects of DNA-adducts formed by the anticancer drug *cis*-diamminedichloroplatinum(II). *Nucleic Acids Res.* **23,** 4066–4072.
28. O'Donovan, A., Davies, A. A., Moggs, J. G., West, S. C., and Wood, R. D. (1994) XPG endonuclease makes the 3′ incision in human DNA nucleotide excision repair. *Nature* **371,** 432–435.
29. Yang, G., Lin, T., Karam, J., and Konigsberg, W. H. (1999) Steady-state kinetic characterization of RB69 DNA polymerase mutants that affect dNTP incorporation. *Biochemistry* **38,** 8094–8101.
30. Ausubel, F. M., Brent, R., Kingston, R. E., et al. (1989) *Current Protocols in Molecular Biology*, Greene Publishing Associates and Wiley-Interscience, New York, NY.
31. Sijbers, A. M., de Laat, W. L., Ariza, R. R., et al. (1996) Xeroderma pigmentosum group F caused by a defect in a structure-specific DNA repair endonuclease. *Cell* **86,** 811–822.

31

Analysis of Proliferating Cell Nuclear Antigen (PCNA) Associated With DNA Excision Repair Sites in Mammalian Cells

A. Ivana Scovassi and Ennio Prosperi

Summary

Proliferating cell nuclear antigen (PCNA) is a homotrimeric protein adopting a ring structure that may encircle DNA. In this form, PCNA functions as a sliding platform to which different types of catalytic and regulatory proteins are tethered to perform DNA transactions such as replication and repair synthesis, methylation, chromatin assembly and remodeling, as well as sister chromatid cohesion. In addition, PCNA coordinates DNA metabolism with cell cycle progression by interacting with cyclins, cyclin-dependent kinases (CDK), and CDK inhibitors. PCNA participates in different pathways of DNA repair, including nucleotide and base excision repair, as well as mismatch repair, by interacting with proteins involved in these processes. A fundamental step in DNA repair involves the transition of PCNA from a freely soluble nucleoplasmic form, to a chromatin-bound form associated with repair sites. This chapter describes biochemical and immunofluorescence methods for detection of the chromatin-bound form of PCNA involved in DNA repair. Cellular fractionation and nuclear extraction procedures are provided for Western blot analysis, as well as for protein–protein interaction studies. An *in situ* extraction protocol is described for immunostaining, fluorescence microscopy and flow cytometric analyses of nuclear localization, and cell cycle distribution of PCNA associated with DNA repair sites.

Key Words: Chromatin-bound proteins; DNA damage; DNA repair; DNase I; proliferating cell nuclear antigen; UV-C radiation.

1. Introduction

Proliferating cell nuclear antigen (PCNA), first described as an autoantigen and then as a cofactor of DNA polymerases δ and ε, is today known to play a central role in several metabolic pathways, including DNA replication and

From: *Methods in Molecular Biology: DNA Repair Protocols: Mammalian Systems, Second Edition*
Edited by: D. S. Henderson © Humana Press Inc., Totowa, NJ

repair, chromatin remodeling, as well as cell cycle progression *(1–4)*. The importance of PCNA in all these processes is well explained by its crystal structure showing that it is a homotrimeric ring-shaped protein with a central hole accommodating DNA *(5)*. Thus, PCNA may function as a DNA clamping platform necessary to recruit to the relevant site of activity, proteins involved in different pathways of nucleic acid metabolism *(6)*. PCNA-interacting proteins include various DNA replication/repair enzymes, such as DNA helicases, endonucleases and glycosylases, several polymerases (for both replicative and translesion synthesis), and DNA ligase I *(7,8)*. In addition, other interactions are relevant to basic processes like DNA methylation *(9)*, protein poly-ADP ribosylation *(10)*, histone acetylation/deacetylation *(11,12)*, as well as chromatin assembly *(13)*, and sister chromatid cohesion *(14)*. Many of these proteins interact with a central region of PCNA, termed the interdomain connecting loop, through a conserved motif of eight amino acids (QxxL/I/MxxFF/FY), which is also present in the CDK inhibitor p21$^{waf1/cip1}$ *(15)*.

PCNA is present both in a freely soluble nucleoplasmic form and in a minor fraction resistant to detergent extraction, which is loaded by replication factor C (RFC) onto DNA for replication *(2)* and is referred to as the chromatin-bound form *(16)*. A similar recruitment process of PCNA to DNA repair sites is postulated, based on the observations that RFC is also required for nucleotide excision repair (NER) *(2)* and NER factors are assembled at the site of a DNA lesion *(17)*.

The involvement of PCNA in DNA repair was first documented at the cellular level on human amnion cells *(18)*, and in human fibroblasts exposed to ultraviolet (UV) light *(19)*, and later supported by molecular studies showing the steps at which PCNA is required for NER, and in general, for DNA excision repair *(20,21)*. Other reports have provided modified methods for detection of the chromatin-bound (actually the DNA-encircling) form of PCNA, and to analyze the mechanisms underlying the transition from the soluble to the detergent-resistant form of PCNA, in the NER process *(22,23)*. Recruitment of PCNA to chromatin after UV-induced DNA damage, was also investigated on xeroderma pigmentosum (XP) cells belonging to different complementation groups, which are defective in definite steps of the NER pathway. These studies clearly showed that the incision step is required for the efficient association of PCNA with DNA repair sites *(24–26)*. Further studies have investigated the chromatin association of PCNA induced by different types of DNA damage *(27–29)*, implying involvement of PCNA in other repair pathways, like base excision repair (BER). Specifically, studies at the molecular level have shown that PCNA is required for the long-patch BER sub-pathway *(30)*, although an interaction of PCNA with DNA polymerase β has been recently reported *(31)*. The involvement of PCNA in other repair pathways, such as mismatch repair

(MMR), has been supported by repair model studies and by the interaction of PCNA with proteins of this repair system *(3,7,32)*. In addition, colocalization and interaction of PCNA with DNA glycosylases and endonucleases at DNA replication foci during S phase, have demonstrated the involvement of PCNA in postreplication DNA repair *(33,34)*.

However, several aspects of PCNA functions in DNA repair processes still await to be clarified, the following among these: (1) The binding mode of different partners to the PCNA trimer (each molecule has three potential binding sites)—during DNA replication, for instance, PCNA interacts with DNA polymerase δ or DNA ligase I, in a mutually exclusive manner *(8)*; (2) The interaction dynamics of PCNA with proteins of different repair systems, that is, so-called partner exchange *(35,36)*; and (3) the association dynamics of PCNA with repair sites. At DNA replication sites, PCNA shows little turnover *(37)*, and recycling of PCNA during DNA repair has been suggested *(35,38)*.

In order to specifically analyze PCNA associated with DNA repair sites, a suitable extraction procedure is required to release the detergent-resistant form of PCNA. Cell fractionation is thus required for differential extraction of detergent-soluble and resistant forms, and for subsequent Western blot analysis and immunoprecipitation studies.

In this chapter, we describe a procedure that enables the release of these fractions of PCNA protein, without disruption of its interaction with relevant partners, thus making this protocol suitable for Western blot analysis and/or further immunoprecipitation studies. The procedure is based on a hypotonic lysis in the presence of a detergent, for the release of soluble proteins *(23,38)*. The detergent-resistant fraction, representing chromatin-bound proteins, is then solubilized by enzymatic digestion of DNA with DNase I *(38,39)*. The release of PCNA is made possible by its property to slide off a linear DNA molecule *(40)*, making this procedure particularly suited for co-immunoprecipitation of PCNA interacting proteins (*see* **Fig. 1A**).

An *in situ* extraction procedure *(29)*, necessary to remove the detergent-soluble proteins from cells adherent on cover slips, is also described for the immunofluorescence detection of chromatin-bound PCNA involved in the DNA repair process. A similar procedure applied to cells in suspension allows the flow cytometric analysis of cell cycle distribution of chromatin-bound PCNA *(23,29)*.

2. Materials

1. UV-C germicidal lamp (e.g., Philips T-UV9, emission peak at approx 254 nm).
2. UV-C radiometer (e.g., Spectronics, Westbury, NY).
3. Fluorescence microscope.
4. Flow cytometer.

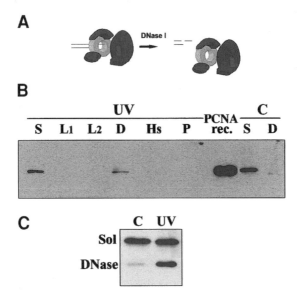

Fig. 1. **(A)** Schematic representation of PCNA complex release by DNA digestion. **(B)** Western blot analysis of PCNA in fractionated cell extracts obtained from quiescent human normal fibroblasts (3.5 × 10⁶) exposed to UV-C light (10 J/m^2) and collected 30 min later. The sample (UV) was fractionated as detergent-soluble (S), first and second low salt wash (L1 and L2), DNase I-released material (D), high salt extract (Hs), and final pellet (P). Recombinant PCNA (PCNA rec.) was loaded as a control. Detergent-soluble (S) and DNase-released (D) fractions from untreated control cells (C) are also shown. **(C)** Western blot analysis of detergent-soluble (Sol) and DNase-released fraction (DNase) obtained from untreated control (C) and UV-irradiated (10 J/m^2) proliferating human fibroblasts (UV). The samples were probed with PC10 mouse monoclonal antibody to PCNA, with secondary biotinylated antimouse antibody, and then with HRP-conjugated streptavidin.

5. Standard polyacrylamide gel electrophoresis (PAGE) and protein transfer equipment.
6. Culture media for human normal (e.g., fibroblasts) or tumor cells (e.g., HeLa).
7. Protease inhibitor cocktail (solution in dimethyl sulfoxide [DMSO]; Sigma or Roche) without EDTA. Store in aliquots at –20°C.
8. Phosphatase inhibitor cocktail (Sigma). Store in aliquots at –20°C.
9. Formaldehyde (supplied as 37% solution stabilized with methanol).
10. Hydrogen peroxide (supplied as 30% solution) is diluted in bidistilled H$_2$O to 50 mM prior to use.
11. Methylmethane sulfonate (MMS) is prepared as a 100 mM working solution in ethanol, and diluted in culture medium prior to use. N-Methyl-N'-nitro-N-nitro-

soguanidine (MNNG) dissolved in DMSO (stock is 10 mM, store in aliquots at −20°C).
12. Anti-PCNA antibodies: monoclonal antibody PC10 (Dako); rabbit polyclonal antibody FL-261 (Santa-Cruz Biotech.); goat polyclonal antibody C-20 (Santa-Cruz Biotech.).
13. Antimouse, antirabbit, antigoat secondary antibodies conjugated with fluorescein isothiocyanate (FITC), Alexa 594 (Molecular Probes), or biotin.
14. Streptavidin-FITC, streptavidin-Texas red, streptavidin-horseradish peroxidase (HRP).
15. Protein A- and Protein G–sepharose (Amersham Biosciences or Sigma).
16. Bovine serum albumin (BSA).
17. Peroxidase substrates for enhanced chemiluminescence (ECL) detection (Amersham Biosciences).
18. Phosphate-buffered saline (PBS): 137 mM NaCl, 2.7 mM KCl, 10.6 mM Na$_2$HPO$_4$, 1.4 mM KH$_2$PO$_4$, pH 7.4.
19. Physiological saline: 154 mM NaCl.
20. Hypotonic lysis buffer: 10 mM Tris-HCl, pH 7.4, 2.5 mM MgCl$_2$, 0.5% Nonidet P-40. Use freshly prepared solution, and immediately before use add 1 mM dithiothreitol (DTT; stock is 1 M, store in aliquots at −20°C), 1 mM phenylmethylsulfonyl fluoride (PMSF; stock is 200 mM in isopropanol, store at 4°C), 0.2 mM sodium vanadate (Na$_3$VO$_4$; stock is 100 mM; store in aliquots at −20°C), 0.5 µM okadaic acid (stock is 150 µM; store in aliquots at −20°C) and 100 µL/10^7 cells of protease inhibitor cocktail. Keep supplemented buffer on ice.
21. Washing buffer: 10 mM Tris-HCl, pH 7.4, 150 mM NaCl, 1 mM PMSF, and 50 µL/10^7 cells of protease inhibitor cocktail.
22. DNA digestion buffer: 2X solution is 20 mM Tris-HCl, pH 7.4, 20 mM NaCl, 10 mM MgCl$_2$, 20 µL/10^7 cells of protease inhibitor cocktail. Prepare fresh solution each time.
23. DNase I solution: DNase I (Sigma, D-4527, or equivalent) dissolved in 20 mM NaCl and 0.1 mM PMSF just prior to use.
24. SDS-loading buffer: 65 mM Tris-HCl, pH 7.4, 100 mM DTT, 1% sodium dodecyl sulfate (SDS), 10% glycerol, 0.02% bromophenol blue. Store at −20°C.
25. PBS-Tween-20 solution: PBS containing 0.2% Tween-20.
26. PBT solution: PBS containing 1% BSA and 0.2% Tween-20. Prepare fresh solution each time.
27. Ponceau S: 0.1% in 5% acetic acid solution. Store at room temperature (RT)
28. Blocking solution: 5% nonfat dry milk plus 0.2% Tween-20 in PBS. Prepare a fresh solution each time.
29. Propidium iodide (PI): stock solution is 500 µg/mL in bidistilled H$_2$O. Store at 4°C.
30. Bisbenzimide H 33258 (Hoechst 33258): stock solution is 1 mM in bidistilled H$_2$O. Store at 4°C.
31. 4′,6′-Diamidinophenylindole (DAPI): stock solution is 1 mM in bidistilled H$_2$O. Store at 4°C.

32. RNase A. Use freshly prepared at 1 mg/mL in PBS.
33. Mounting medium (e.g., Mowiol) containing 0.25% antifading agent (1,4-diazabicyclooctane).

3. Methods

The methods described here allow the investigation of the form of PCNA associated with DNA repair sites. DNA repair is triggered in response to different types of DNA damage. To obtain reliable results, treatment of cells with DNA damaging agents under accurately determined conditions are described first (*see* **Subheading 3.1.**). The transition of PCNA protein from a detergent-soluble to a detergent-resistant state is analyzed with biochemical techniques by using cell fractionation procedures (*see* **Subheading 3.2.**). The extraction of the detergent-soluble protein is obtained with a hypotonic solution in the presence of a nonionic detergent *(18,23,39)*. The subsequent release of chromatin-bound proteins is obtained by DNA digestion *(8,39)*. Detection of PCNA in each cell extract fraction is achieved by Western blot analysis. Immunoprecipitation studies of PCNA using different commercially available antibodies can also be easily performed *(8,39)*. Alternatively or concomitantly, chromatin-bound PCNA may be detected by means of immunocytochemical techniques and fluorescence microscopy (*see* **Subheading 3.3.**), which allow direct visualization of the protein associated with DNA repair sites. The same procedure can be also applied to cell suspensions for flow cytometric analysis of relative content and cell cycle distribution of chromatin-bound PCNA *(23,29)*.

3.1. Induction of DNA Damage With Chemical or Physical Agents

DNA damage and the consequent pathway of DNA repair greatly depends on the type of lesion induced. To detect chromatin-bound PCNA involved in the NER or BER processes, suitable procedures are based on the exposure of the cells to UV-C light, or to base-modifying agents, respectively. Best results are obtained on quiescent cells (e.g., fibroblasts growth-arrested by serum starvation) because they do not replicate, and hence PCNA is chromatin-bound only for DNA repair. However, studies with proliferating cells may be also performed to obtain information otherwise not available on quiescent cells. Studies may also be undertaken with transformed cells, but their repair capacity must be taken into account, depending particularly on the p53 status *(41)*.

3.1.1. UV Light Irradiation

UV irradiation (254 nm) is one of the most used methods to induce DNA damage repaired by the NER pathway. Typical lesions thus produced are the cyclobutane pyrimidine dimers and 6–4 photoproducts. Since the emission spectrum of commonly used UV lamps (often used are germicidal lamps) may

Analysis of PCNA Associated With Repair Sites

be different, an accurate determination of the energy supplied to the cells must be performed.

Particular care must be taken to avoid accidental exposure to the UV-C light of the lamp: Wear appropriate gloves and eye protection. As the optimal condition, the lamp is placed under a three-sided closed wooden box, with the open side covered with a black sheet of paper or cloth, which can be lifted to introduce the Petri dishes. If possible, perform all subsequent steps under a laminar flow hood to avoid contamination during exposure. The glass window of the hood will also protect against UV light reflected from the steel working level. To avoid reflection, a sheet of black paper may be used.

1. Monitor the energy of the lamp in the UV-C region by using a radiometer equipped with a detector sensitive in the 200–280 nm range.
2. Set the distance of the UV source from the working level so that the fluence of UV light will be 0.5–1 $J/m^2/s$. Check that the UV energy will be uniformly distributed throughout the surface to be exposed.
3. Use cells grown on Petri dishes (e.g., 10-cm diameter). For immunocytochemical analysis, grow the cells attached on cover slips (24 × 24 mm) that are placed into 35-mm diameter dishes. Cells in suspension may be also used, but care must be taken to layer into a Petri dish as a thin film in PBS, so that irradiation will be as uniform as possible (*see* **Note 1**).
4. Remove the culture medium and wash the cells twice with warm sterile PBS. Aspirate almost all PBS, leaving just a thin liquid layer on top of the cells.
5. Remove the dish lid and expose the cells to UV light for the period of time required to supply the chosen dose (e.g., 10 J/m^2).
6. Add back the culture medium and re-incubate cells at 37°C for the desired period of time (a suitable time course ranges from 30 min to 24 h).
7. Detach cells to be used for biochemical analysis with a standard trypsinization procedure. Alternatively, fix the samples for immunofluorescence analysis, as described in **Subheading 3.3.**

3.1.2. Exposure to Base-Modifying Agents

The BER pathway may be analyzed by treating cells with chemical agents inducing base oxidation or alkylation. Typical substances used to induce such lesions are hydrogen peroxide, or MMS and MNNG, respectively.

1. Dilute the drug in whole medium, except H_2O_2 which will be used in PBS to avoid oxidation of serum proteins.
2. Incubate the cells in the presence of the drug for 1 h, then discard the medium and reincubate in whole fresh medium for the desired period of time (e.g., 30 min–3 h).
3. Detach the cells or fix the samples, as described in the following.

3.2. Biochemical Analysis of PCNA Associated With DNA Repair Sites

The analysis of PCNA associated with DNA repair sites by biochemical techniques includes: (1) extraction of the detergent-soluble protein; (2) release of chromatin-bound protein by DNA digestion with DNase I; (3) immunoprecipitation of both detergent-soluble and DNase-released fractions with anti-PCNA antibodies.

3.2.1. Extraction of Detergent-Soluble PCNA Protein

Both fresh cells or frozen pellets in 2-mL Eppendorf tubes can be used with this extraction procedure.

1. Chill fresh cells, or thaw frozen samples to 0°C on ice and resuspend in 1 mL of hypotonic lysis buffer. The volume of lysis buffer (1 mL) is given for extraction from a cell pellet of about 10^7 cells, but it can be scaled down for smaller samples (*see* **Note 2**). However, for an efficient extraction of soluble nuclear proteins, add about 10 volumes of lysis buffer to the cell pellet.
2. Gently resuspend the cell pellet by pipetting the sample about three or four times, and allow lysis to occur for 10 min on ice (*see* **Note 3**).
3. Pellet the samples by low-speed centrifugation ($300g$, 1 min, 4°C). Collect the supernatant containing the detergent-soluble protein fraction, and measure protein content with the Bradford method. For normal human fibroblasts, about 2 mg/mL of soluble proteins are released from 10^7 cells. If the amount is lower, other soluble proteins may be released in the next washing step (*see* below), meaning that extraction has not been complete (*see* **Note 4**). Determine the protein concentration also in the fraction released in the washing buffer, and if necessary, add a further washing step. The detergent-soluble proteins may be analyzed by Western blotting or used for subsequent immunoprecipitation studies, to be compared to the chromatin-bound protein fraction.

3.2.2. Extraction of Detergent-Resistant PCNA by DNase I Digestion of DNA

After extraction of the detergent-soluble protein fraction, chromatin-bound proteins must be solubilized for Western blot analysis. The procedure described below, which is also suitable for immunoprecipitation studies *(8,39)*, allows the release of DNA-bound protein complexes, and it is based on DNA digestion with DNase I (*see* **Note 5**).

1. Resuspend pelleted permeabilized cells in washing buffer in order to remove as much as possible any trace of soluble proteins (*see* **Note 4**).
2. Centrifuge the samples ($300g$, 1 min) and then resuspend the pellets in half-volume of DNase I solution (e.g., 250 µL for a final digestion volume of 500 µL) containing 100–200 DNase I units/10^7 cells. After thorough resuspension, add the second-half volume of 2X digestion buffer (*see* **Note 6**). The amount of DNase I (*see* **Note 7**) to be added is dependent on the cell type (e.g., cells with a DNA content higher than diploidy, require proportionally higher amounts of DNase I, as compared with normal diploid cells).
3. Carry out the digestion for 15–30 min at 25° or 37°C, with constant agitation in order to avoid cell sedimentation and clumping (*see* **Note 8**).

4. Pellet the samples by high-speed centrifugation (14,000g, 1 min) and collect the resulting supernatants, containing the DNase-released proteins, for subsequent use.
 5. Load 30–40 µg of detergent-soluble proteins, and a similar or greater volume (see **Note 9**) of the DNase-released fraction, on a 12% polyacrylamide gel for standard Western blot analysis.
 6. Perform Western blot analysis with antibody against PCNA (see **Fig. 1B,C**).

3.2.3. Immunoprecipitation of Detergent-Soluble and DNase-Released Proteins

If fractionated extracts will be used for immunoprecipitation studies, perform the following steps:

 1. Take a sample of soluble cell lysate containing 1–1.5 mg total protein, and use the whole sample of DNase-released proteins. Set apart a small volume (30–50 µL) of each fraction to be loaded on the gel as a control for the protein input.
 2. Dilute both samples in washing buffer and add 4 µg of monoclonal antibody (PC10) to PCNA, or 20 µL of polyclonal antibody (C-20) to the C-terminus of PCNA, for the soluble fraction. Since the amount of target protein to be precipitated from the DNase-fraction is generally lower than the soluble one, the amount of antibody to be added may be reduced (e.g., 2 µg or 10 µL).
 3. Incubate for 30 min (RT), and then overnight at 4°C.
 4. Add 100 µL of Protein A- or Protein G–sepharose beads (10% v/v in 50 mM Tris-HCl buffer, pH 7.4) to the mixture. Incubate for 1 h at 4°C on a rocking platform (see **Note 10**).
 5. Precipitate the beads by centrifugation at 14,000g for 20 min at 4°C. Remove the supernatant by gentle aspiration.
 6. Wash the immune complexes on the beads three times with washing buffer. Each time, add 0.5 mL of washing buffer and resuspend the beads by gentle vortex-mixing. The final wash should be removed completely.
 7. For SDS-PAGE, add 60 µL of SDS-loading buffer to each sample.
 8. Denature the proteins in the sample by heating to 100°C for 5 min. Centrifuge for 15 s at 10,000g and load half of the supernatant (30 µL) on the gel.

3.2.4. Immunoblot Analysis of Fractionated Cell Extracts or Immunoprecipitated Proteins

 1. After SDS-PAGE, transfer the proteins to a nitrocellulose (or PVDF) membrane with a conventional apparatus (semidry or immersion).
 2. Check the protein transfer to the membrane by Ponceau S staining.
 3. Remove excess stain with several washes in double-distilled water (ddH$_2$O).
 4. Block the membranes for 30 min at RT with blocking solution.
 5. Incubate the membranes for 1 h at RT with monoclonal or polyclonal antibody to PCNA (diluted 1:1000, or 1:500 respectively, in PBS–Tween-20), then wash four to six times (10 min each) with PBS–Tween-20 under continuous agitation.

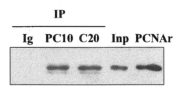

Fig. 2. Immunoprecipitation of chromatin-bound (detergent-resistant) PCNA protein released with DNase I from normal human fibroblasts. UV-irradiated (UV) samples were collected 30 min after irradiation (10 J/m^2), lysed in hypotonic solution, and then processed for enzymatic digestion to release chromatin-bound proteins. Immunoprecipitation of PCNA from the DNase-fraction was performed with PC10 monoclonal antibody, or with C-20 polyclonal antibody. Specificity was checked by immunoprecipitation with an irrelevant antibody (Ig). Protein input (1/20 of DNase fraction) before immunoprecipitation, was loaded (Inp), together with recombinant PCNA (PCNAr). Detection of PCNA in the immunoprecipitates was performed with PC10 antibody.

6. Incubate the membranes (30 min, RT) in a secondary biotinylated antibody (diluted 1:2000 in PBS–Tween-20), then repeat the washing as in the previous step.
7. Incubate the membranes (30 min, RT) with streptavidin–HRP (diluted 1:1000 in PBS–Tween-20) and then repeat the washing as before. This last step is needed to amplify the immunological reaction and enhance the visualization of PCNA in the DNase fraction. Samples obtained after immunoprecipitation may not need amplification; in this case, a two-step reaction with a secondary HRP-conjugated antibody is generally sufficient to detect the relevant signals.
8. Incubate the nitrocellulose (or PVDF) membranes with peroxidase substrates for ECL detection and expose to autoradiographic film (*see* **Fig. 2**).

3.3. Detection of PCNA Bound to DNA Repair Sites by Immunofluorescence Microscopy and Flow Cytometry

For the cytochemical determination of PCNA by immunofluorescence microscopy and flow cytometry, hypotonic extraction is necessary to remove the soluble protein not associated with DNA repair sites *(23,39)*. However, the two techniques require different cell preparation procedures, thus separate descriptions are given.

3.3.1. In Situ *Extraction for Fluorescence Microscopy Analysis of Chromatin-Bound PCNA*

A procedure similar to that given in **Subheading 3.2.1.** is also useful for extraction of detergent-soluble protein from cells adherent to cover slips *(29)* or microscope slides. However, it requires some modification, as follows:

1. Seed cells (about 5×10^4) on cover slips or microscope slides (previously cleaned with a 1:1 mixture ethanol–ether, and sterilized). Grow the cells to an 80% density before use.

Analysis of PCNA Associated With Repair Sites

2. Rinse the cover slips with PBS, then dip into cold ddH$_2$O for about 4 s *(29)*. For some cells (e.g., HeLa) dipping in cold physiological saline may be preferable (*see* **Note 11**).
3. Transfer the cover slips to a Petri dish containing cold hypotonic lysis buffer (containing 0.1 m*M* PMSF), in which the detergent (NP-40) concentration has been reduced to 0.1% to avoid detachment of cells (*see* **Note 12**). Keep the Petri dishes at 4°C for about 10 min, with gentle agitation every 3–4 min.
4. Remove the lysis solution and wash the cells carefully with cold PBS or physiological saline, then aspirate and replace with fresh PBS (e.g., 2 mL for a 35-mm dish).
5. Add an equal volume of 2% formaldehyde (*see* **Note 13**) to reach a final 1% concentration in PBS, and fix the permeabilized cells for 5 min at RT under continuous agitation (*see* **Note 14**).
6. Wash the cover slips again with PBS and further postfix in cold 70% ethanol. The samples can be stored in this solution at –20°C for 1–2 mo before further processing for immunocytochemical staining.

3.3.2. In Situ *Extraction for Flow Cytometric Analysis of Chromatin-Bound PCNA*

For flow cytometric analysis of chromatin-bound PCNA, again follow the procedure described in **Subheading 3.2.1.** for the extraction of detergent-soluble protein, except that about 1–2 × 10^6 cells are required for each sample to be examined *(23,29)*. Keep the lysis volume to 1 mL, as this will facilitate the complete extraction of soluble protein and reduce the number of subsequent passages involving several centrifugation steps. Use fresh cells in 15-mL plastic tubes.

1. After the hypotonic lysis, resuspend the cell pellet in 1 mL of cold PBS or physiological saline.
2. Fix the samples by adding 1 mL of 2% formaldehyde (the final concentration is 1%), incubate for 5 min at RT, then remove the fixative by centrifugation (300*g*, 5 min). Resuspend the cells in 1 mL of cold physiological saline and postfix by adding 2.7 mL of cold 95% ethanol (the final concentration is 70%), with gentle mixing of the two phases. The samples can be stored at –20°C up to 1–2 mo before further processing for immunofluorescence staining.

3.3.3. *Immunocytochemical Staining of PCNA Associated With DNA Repair Sites*

Described below is the method for immunostaining, which is essentially similar for lysed cells attached on cover slips or for cells in suspension. However, the procedures are described separately in order to highlight some relevant, different steps.

3.3.3.1. Cells on Cover Slips

1. Remove the fixative and wash the cells with PBS. All subsequent steps are done at RT.

2. Block the nonspecific staining sites by incubation for 15 min with PBT solution.
3. Incubate the cover slips with anti-PCNA monoclonal or polyclonal antibody (diluted 1:100 in PBT solution) for 1 h. Place the cover slips with cells facing upside-down onto a drop (50 µL) of antibody on a flat parafilm strip, so that a sandwich is formed (*see* **Note 15**).
4. Stop the incubation by returning the cover slips to Petri dishes with the cells facing up, and wash three times (10 min each) with PBS–Tween-20.
5. The following steps are performed in the dark to avoid photobleaching of fluorochromes. Incubate the cover slips for 30 min with appropriate secondary antibody (diluted 1:100 in PBT solution) by again placing the cells upside-down onto a drop of antibody, as described in **step 3**. Choose a secondary antibody conjugated with FITC (green fluorescence) or with Alexa 594 (red fluorescence) depending on the color emission required (*see* **Note 16**). Use a secondary antibody conjugated with biotin, if amplification of the signal is needed.
6. Remove the cover slips and wash them in Petri dishes three times (10 min each) with PBS–Tween-20.
7. If a step with a biotinylated secondary antibody was performed, then incubate for 30 min with streptavidin–FITC or streptavidin–Texas Red (both 1:100 dilution), and then wash three times with PBS–Tween-20, as in **step 6**.
8. Counterstain DNA with Hoechst 33258 or DAPI (0.2 µM in PBS) for 2 min, then remove the dye and wash the cover slips twice (5 min each) with PBS.
9. Mount the cover slips in aqueous mounting medium (e.g., Mowiol) containing an antifading agent.
10. After mounting, view slides with a fluorescence microscope (*see* **Fig. 3**) equipped with filter sets for UV, blue, and green excitation of fluorescence.
11. Seal the cover slips with transparent nail polish to prevent them from drying, and to allow storage of slides at –20°C.

3.3.3.2. Cell Suspensions

1. Bring samples stored at –20°C to RT and remove the fixative by centrifugation (300g, 5 min). Resuspend the cells in PBT solution and incubate the samples for 15–30 min at RT to block the nonspecific staining sites, as described for cover slips (**Subheading 3.3.3.1.**).
2. Remove the PBT solution by centrifugation, and add 100 µL of anti-PCNA monoclonal antibody (diluted 1:100 in PBT). If necessary, polyclonal antibodies (C-20 or FL-261) may be used with similar results.
3. Add a negative control sample to be incubated with an irrelevant IgG (e.g., Sigma M5409), to evaluate the background fluorescence *(29)*, and to select the baseline values on the flow cytometer.
4. Incubate for 1 h at RT with frequent mild agitation of the tubes to avoid cell sedimentation. This may lead to an incomplete staining of sedimented cells.
5. Add 1 mL of PBT solution and centrifuge (300g, 5 min). Remove the supernatant and wash the cells twice (10 min each) in 1 mL of PBT solution.

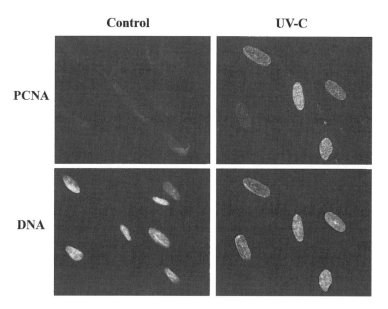

Fig. 3. Immunofluorescence staining of chromatin-bound PCNA in control (C) quiescent human fibroblasts, or in samples collected 30 min after UV-C irradiation (UV). Chromatin-bound protein was detected after *in situ* hypotonic lysis to remove the detergent-soluble fraction. After fixation, the samples were immunostained with PC10 monoclonal antibody, followed by incubation with FITC-conjugated secondary antibody. DNA was counterstained with Hoechst 33258 dye. Visualization was performed with an Olympus BX51 fluorescence microscope at ×1000 magnification.

6. Incubate the cells (30 min, RT) in the dark with 100 µL of FITC-conjugated antimouse secondary antibody (diluted 1:100 in PBT solution). Use frequent agitation, as in **step 4**.
7. Add 1 mL of PBT solution and centrifuge (300g, 5 min), wash the cells twice, as in **step 5**, except that after the last wash, centrifuge the cells again and resuspend in 1 mL of PBS containing 10 µg/mL of propidium iodide (PI) and 1 mg of RNase A to remove any RNA interfering with the PI staining of DNA (*see* **Note 17**).
8. Stain for 30 min at RT, and then overnight at 4°C (*see* **Note 18**), then measure the samples on a flow cytometer (*see* **Fig. 4**).

4. Notes

1. An alternative to whole cell UV-irradiation is the micropore irradiation technique (**42**) which permits the introduction of DNA damage at localized sites in the nucleus. The procedure uses polycarbonate isopore filters (different types are available, but we prefer those with 3- or 8-µm pore size) to mask the cells,

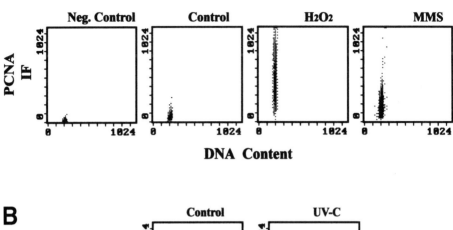

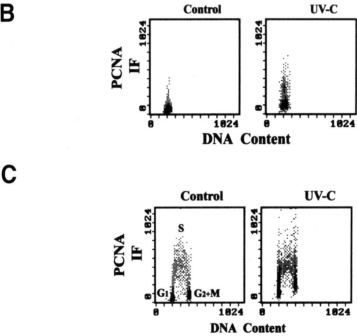

Fig. 4. Flow cytometric analysis of chromatin-bound PCNA immunofluorescence (IF). **(A)** Quiescent human fibroblasts, treated for 1 h with 100 μM hydrogen peroxide (H_2O_2), with 9.1 μM MMS, or mock-treated (Control), were further incubated in drug-free medium for 30 min. After hypotonic lysis and fixation, samples were immunostained with PC10 monoclonal antibody to PCNA, followed by a FITC-conjugated secondary antibody. The sample marked as negative control (neg. control) was incubated with an irrelevant mouse IgG antibody, instead of anti-PCNA antibody, to evaluate nonspecific staining. *(Continued on next page)*

and allow the irradiation only through the pores. Filters are placed on top of the cells after the PBS wash, just before irradiation. After UV exposure (energy may vary from 20 up to 100 J/m^2), the medium is added back, so that the filter will float and be removed without cell detachment.

2. It is worth using not fewer than 5×10^6 cells to easily detect the chromatin-bound fraction of PCNA, and consider about 3.5×10^6 cells as the lower limit for detection.
3. Resuspension of the cell pellet by pipetting may produce foam due to the presence of detergent in the hypotonic lysis buffer. However, this will not compromise extraction of soluble PCNA protein. Care must be taken if cells with small cytoplasm (e.g., lymphocytes) are used, since pipetting could result in nuclear damage and loss. In this case, pipet only two or three times.
4. When the extraction of detergent-soluble proteins is efficient, the amount of proteins released in the subsequent washing buffer should not be higher than 20% of the initial release.
5. An alternative procedure to release chromatin-bound proteins is based on the production of DNA strand breaks by sonication. This procedure is equally effective as DNA digestion in releasing DNA-bound proteins in a form suitable for immunoprecipitation experiments. This would not be otherwise possible with protocols using SDS to release nuclear-bound proteins. However, when using sonication, care must be taken to reduce protein degradation by using a "low" setting and short pulses, and by performing the entire procedure on ice.
6. Good resuspension of the cell pellet in the DNase solution is a requisite for efficient DNA digestion and consequent release of DNA-bound proteins. The addition of 2X digestion buffer after cell resuspension is performed because of the Mg^{2+} contained in the digestion buffer, which will favor cell clumping, thereby reducing the extent of DNA digestion.
7. The quality of DNase I is very important, since contaminants present in different preparations (e.g., chymotrypsin) may be responsible for proteolytic degrada-

---

Fig. 4. *(Continued)* (**B**) Human peripheral blood lymphocytes were suspended in warm PBS, layered onto a Petri dish, and exposed to UV-C irradiation (10 J/m^2). Cells were immediately collected with warm medium and incubated at 37°C for 30 min. After that, samples were processed for hypotonic lysis, fixed, and immunostained as before. (**C**) Proliferating human fibroblasts were exposed to UV-C light (10 J/m^2) and harvested 30 min later. After hypotonic lysis and fixation, cells were immunostained with PC10 antibody, followed by FITC-conjugated secondary antibody. All samples (**A**, **B**, and **C**) were stained with PI for DNA content determination and then measured with a Coulter Epics XL flow cytometer. In (**C**), the cell cycle compartments (G_1, S, and G_2+M) are indicated in the control panel to show the distribution of PCNA immunofluorescence. After UV exposure, the increase in PCNA immunofluorecence is detectable only in the G_1 and G_2 + M compartments.

tion. A chromatographically purified preparation (e.g., Sigma D-4527) is recommended.
8. Digestion conditions (e.g., time and temperature) should be determined depending on the experiment. For coimmunoprecipitation studies, possible degradation of PCNA-interacting proteins may be reduced by shortening both incubation time and temperature *(8)*. If cell clumping occurs during DNA digestion, a very brief sonication will help to disrupt cellular aggregates and facilitate DNA breakage.
9. Given that PCNA bound to DNA constitutes only a minor fraction of the total cellular amount of these proteins, loading of proportionally higher volumes of DNase extracts, as compared to the soluble fraction, may be required to improve the visualization of these proteins by Western blot. It is advisable to check the protein loading of different samples, by detecting another protein (e.g., actin or histone) as an internal standard.
10. It may be useful to prebind the antibody to the Protein A–Sepharose beads, so that the complex will be incubated with fractionated cell extracts only once. This step will reduce the total time of the procedure, thereby also avoiding possible protein degradation.
11. Dipping of the cover slips in cold ddH$_2$O is particularly useful with cells with large cytoplasm (e.g., fibroblasts) since this will favor the extraction of detergent-soluble proteins. However, when using cells with a small cytoplasm (e.g., some tumor cells, such as HeLa), the hypotonic stress is likely to result in a substantial detachment of the cells from the cover slip.
12. Avoid pipetting the hypotonic lysis solution directly onto the cells, since this will increase the risk of cell detachment from the cover slip.
13. Formaldehyde vapors are toxic. Solutions containing formaldehyde should be prepared in a chemical hood.
14. For some cell types (e.g., HeLa cells), the final formaldehyde concentration is increased up to 4% because it has been noticed that cells fixed with lower concentrations will lose their nuclear morphology after freezing–thawing of slides stored at –20°C.
15. The antibody sandwich formed with the coverslip on the parafilm strip will not dry under normal RT conditions, so that is not necessary to prepare a humid chamber. This has the advantage of reducing the incubation volume, thus saving antibody.
16. The choice of secondary antibody depends on the fluorescence emission to be detected under the microscope or with the flow cytometer. In the latter case, a FITC-conjugated antibody (or streptavidin-FITC) is preferred because the fluorescence emission can be measured in the green channel (515–540 nm), thus leaving the red channel for the DNA stain (PI) emission (590–620 nm).
17. Propidium iodide (like ethidium bromide) is an intercalating agent with mutagenic properties, and therefore accidental exposure of skin, or other body parts is to be avoided. Always wear gloves, especially when handling concentrated stock solutions.

18. Formaldehyde fixation reduces the ability of intercalating dyes, such as PI, to stain DNA. Thus, prolonged DNA staining is necessary to obtain optimal resolution of the DNA histogram, which greatly depends on the amount of fluorochrome bound to DNA.

Acknowledgments

We wish to thank our collaborators that have been involved in these investigations during the years. These studies have been supported by CNR Target and Special Projects, and by Italian MIUR (FIRB project RBNE0132MY).

References

1. Kelman, Z. (1997) PCNA: structure, functions and interactions. *Oncogene* **14,** 629–640.
2. Jónsson, Z. O. and Hübscher, U. (1997) Proliferating cell nuclear antigen: more than a clamp for DNA polymerases. *BioEssays* **19,** 967–975.
3. Prosperi, E. (1997) Multiple roles of proliferating cell nuclear antigen: DNA replication, repair and cell cycle control, in: *Progress in Cell Cycle Research*, Vol. 3 (Meijer, L., Guidet S., and Philippe, M., eds.), Plenum Press, New York, NY, pp. 193–210.
4. Tsurimoto, T. (1999) PCNA binding proteins. *Front. Biosci.* **4,** 849–858.
5. Krishna, T. S. R., Kong, X.-P., Gary, S., Burgers, P. M., and Kuriyan, J. (1994) Crystal structure of the eukaryotic DNA polymerase processivity factor PCNA. *Cell* **79,** 1233–1243.
6. Kelman, Z. and Hurwitz, J. (1998) Protein-PCNA interactions: a DNA scanning mechanism? *Trends Biochem. Sci.* **23,** 236–238
7. Maga, G. and Hübscher, U. (2003) Proliferating cell nuclear antigen (PCNA): a dancer with many partners. *J. Cell Sci.* **116,** 3051–3060.
8. Riva, F., Savio, M., Cazzalini, O., et al. (2004) Distinct pools of proliferating cell nuclear antigen associated to DNA replication sites interact with the p125 subunit of DNA polymerase δ or with DNA ligase I. *Exp. Cell Res.* **293,** 357–367.
9. Chuang, L. S.-H., Ian, H.-I., Koh, T.-W., Ng, H.-H., Xu, G., and Li, B. F. L. (1997) Human DNA-(Cytosine-5) methyltransferase-PCNA complex as a target for p21WAF1. *Science* **277,** 1996–2000.
10. Frouin, I., Maga, G., Denegri, M., et al. (2003) Human proliferating cell nuclear antigen, poly(ADP-ribose) polymerase 1, and p21$^{waf1/cip1}$. A dynamic exchange of partners. *J. Biol. Chem.* **278,** 39265–39,268.
11. Milutinovic, S., Zhuang, Q., and Szyf, M. (2002) Proliferating cell nuclear antigen associates with histone deacetylase activity, integrating DNA replication and chromatin modification. *J. Biol. Chem.* **277,** 20,974–20,978.
12. Hasan, S., Hassa, P. O. Imhof, R., and Hottiger, M. O. (2001) Transcription coactivator p300 binds PCNA and may have a role in DNA repair synthesis. *Nature* **410,** 387-391.
13. Shibahara K. and Stillman, B. (1999) Replication-dependent marking of DNA by PCNA facilitates CAF-1-coupled inheritance of chromatin. *Cell* **96,** 575–585.

14. Skibbens, R. V., Corson, L. B., Koshland, D., and Hieter, P. (1999) Ctf7p is essential for sister chromatid cohesion and links mitotic chromosome structure to the DNA replication machinery. *Genes Dev.* **13,** 307–319.
15. Warbrick, E. (1998) PCNA binding through a conserved motif. *BioEssays* **20,** 195–199.
16. Bravo, R. and Macdonald-Bravo, H. (1987). Existence of two populations of cyclin/proliferating cell nuclear antigen during the cell cycle: association with DNA replication sites. *J. Cell Biol.* **105,** 1549–1554.
17. Volker, M., Moné, M. J., Karmakar, P., et al. (2001) Sequential assembly of the nucleotide excision repair factors in vivo. *Mol. Cell* **8,** 213–224.
18. Celis, J. E. and Masden, P. (1986) Increased nuclear cyclin/PCNA antigen staining of non-S phase transformed human amnion cells engaged in nucleotide excision DNA repair. *FEBS Lett.* **209,** 277–283.
19. Toschi, L. and Bravo, R. (1988) Changes in cyclin/proliferating cell nuclear antigen distribution during DNA repair synthesis. *J. Cell Biol.* **107,** 1623–1628.
20. Shivji, M. K. K., Kenny, M. K., and Wood, R. D. (1992) Proliferating cell nuclear antigen is required for DNA excision repair. *Cell* **69,** 367–374.
21. Nichols, A. F. and Sancar, A. (1992) Purification of PCNA as a nucleotide excision repair protein. *Nucleic Acids Res.* **20,** 3559–3564.
22. Miura, M., Domon, M., Sasaki, T., Kondo, S., and Takasaki, Y. (1992) Two types of proliferating cell nuclear antigen (PCNA) complex formation in quiescent normal and xeroderma pigmentosum group A fibroblasts following ultraviolet light (UV) irradiation. *Exp. Cell Res.* **201,** 541–544.
23. Prosperi, E., Stivala, L. A., Sala, E., Scovassi, A. I., and Bianchi, L. (1993) Proliferating cell nuclear antigen complex-formation induced by ultraviolet irradiation in human quiescent fibroblasts as detected by immunostaining and flow cytometry. *Exp. Cell Res.* **205,** 320–325.
24. Aboussekhra, A. and Wood, R.D. (1995) Detection of nucleotide excision repair incisions in human fibroblasts by immunostaining for PCNA. *Exp. Cell Res.* **221,** 326-332.
25. Balajee, A. S., May, A., Dianova, I., and Bohr, V. A. (1998) Efficient PCNA complex formation is dependent upon both transcription coupled repair and genome overall repair. *Mutat Res.* **409,** 135–146.
26. Miura, M. (1999) Detection of chromatin-bound PCNA in mammalian cells and its use to study DNA excision repair. *J. Radiat. Res.* **40,** 1–12.
27. Miura, M., Domon, M., Sasaki, T., Kondo, S., and Takasaki, Y. (1992) Restoration of proliferating cell nuclear antigen (PCNA) complex formation in xeroderma pigmentosum group A cells following cis-diamminedichloroplatinum (II)-treatment by cell fusion with normal cells. *J. Cell Physiol.* **152,** 639–645.
28. Stivala, L. A., Prosperi, E., Rossi, R., and Bianchi, L. (1993) Involvement of proliferating cell nuclear antigen in DNA repair after damage induced by genotoxic agents in human fibroblasts. *Carcinogenesis* **14,** 2569–2573.
29. Savio, M., Stivala, L. A., Bianchi, L., Vannini, V., and Prosperi, E. (1998) Involvement of the proliferating cell nuclear antigen (PCNA) in DNA repair induced by

alkylating agents and oxidative damage in human fibroblasts. *Carcinogenesis* **19,** 591–596.

30. Fortini, P., Pascucci, B., Parlanti, E., D'Errico, M., Simonelli, V., and Dogliotti, E. (2003) The base excision repair: mechanisms and its relevance for cancer susceptibility. *Biochimie* **85,** 1053–1071.
31. Kedar, P. S., Kim, S. J., Robertson, A., et al. (2002). Direct interaction between mammalian DNA polymerase β and proliferating cell nuclear antigen. *J. Biol. Chem.* **277,** 31,115–31,123.
32. Kleczkowska, H. E., Marra, G., Lettieri, T., and Jiricny, J. (2001). hMSH3 and hMSH6 interact with PCNA and colocalize with it to replication foci. *Genes Dev.* **15,** 724–736.
33. Otterlei, M., Warbrick, E., Nagelhus, T. A., et al. (1999). Post-replicative base excision repair in replication foci. *EMBO J.* **18,** 3834–3844.
34. Tsuchimoto, D., Sakai, Y., Sakumi, K., Nishioka K., Sasaki, M., Fujiwara, T., and Nakabeppu, Y. (2001). Human APE2 protein is mostly localized in the nuclei and to some extent in the mitochondria, while nuclear APE2 is partly associated with proliferating cell nuclear antigen. *Nucleic Acids Res.* **29,** 2349–2360.
35. Cox, L. S. (1997) Who binds wins: competition for PCNA rings out cell-cycle changes. *Trends Cell Biol.* **7,** 493–498.
36. Warbrick, E. (2000) The puzzle of PCNA's many partners. *BioEssays* **22,** 997–1006.
37. Sporbert, A. Gahl, A., Ankerhold, R., Leonhardt, H., and Cardoso, M. C. (2002) DNA polymerase clamp shows little turnover at established replication sites but sequential de novo assembly at adjacent origin clusters. *Mol. Cell* **10,** 1355–1365.
38. Stivala, L. A., Riva, F., Cazzalini, O., Savio, M., and Prosperi, E. (2001) p21$^{waf1/cip1}$-null human fibroblasts are deficient in nucleotide excision repair downstream the recruitment of PCNA to DNA repair sites. *Oncogene* **20,** 563–570.
39. Savio, M., Stivala, L. A., Scovassi, A. I., Bianchi, L., and Prosperi, E. (1996) p21$^{waf1/cip1}$ protein associates with the detergent-resistant form of PCNA concomitantly with disassembly of PCNA at nucleotide repair sites. *Oncogene* **13,** 1591–1598.
40. Burgers, P. M. J. and Yoder, B. L. (1993). ATP-independent loading of the proliferating cell nuclear antigen requires DNA ends. *J. Biol. Chem.* **268,** 19,923–19,926.
41. Riva, F., Zuco, V., Supino, R., Vink, A. A., and Prosperi, E. (2001) UV-induced DNA incision and proliferating cell nuclear antigen recruitment to repair sites occur independently of p53-replication protein A interaction in p53 wild type and mutant ovarian carcinoma cells. *Carcinogenesis* **22,** 1971–1978.
42. Katsumi, S., Kobayashi, N., Imoto, K., et al. (2001) *In situ* visualization of ultraviolet-light-induced DNA damage repair in locally irradiated human fibroblasts. *J. Invest. Dermatol.* **117,** 1156–1161.

32

Analysis of DNA Repair and Chromatin Assembly In Vitro Using Immobilized Damaged DNA Substrates

Jill A. Mello, Jonathan G. Moggs, and Geneviève Almouzni

Summary

Significant advances have been made in identifying a complex network of proteins that could play a role in the repair of DNA damage in the context of chromatin. Insights into this process have been obtained by combining damaged DNA substrates with mammalian cell-free systems that contain both DNA repair and chromatin assembly activities. The methods described in this chapter provide a powerful approach for the detection of proteins recruited during the recognition and repair of DNA lesions, including repair proteins and chromatin associated factors. Substrates for the recruitment assay consist of DNA containing damage that is immobilized on magnetic beads. A human cell-free system that supports both DNA repair and chromatin assembly is incubated with the immobilized DNA-damaged substrates, and proteins associated with the DNA are then isolated and subjected to analysis. We present here protocols for preparing bead-linked DNA substrates containing different types of lesions, for the reaction of the damaged DNA with cell-free systems, and for the subsequent analysis of proteins that are recruited to the immobilized damaged DNA substrates.

Key Words: Assembly; chromatin; DNA damage; DNA linked to beads; histone chaperone; repair.

1. Introduction

The DNA of a eukaryotic cell experiences thousands of daily insults from both environmental and endogenous sources, resulting in numerous types of DNA damage that must be repaired in order to maintain genomic stability *(1)*. The recognition, removal, and repair of DNA lesions in the eukaryotic genome requires repair enzymes to operate within the complex chromatin environment of the nucleus. Chromatin is a nucleoprotein complex consisting of a basic repeating unit known as the nucleosome. A single nucleosome contains two turns of DNA wrapped around a core histone octamer comprising the histones H2A, H2B, H3, and H4 *(2)*. Nucleosomes represent the first level of compaction in chromatin, restricting access to enzymes involved in DNA metabolism *(3)*. In addition to these basic components, linker histones and a variety of nonhistone proteins are incorporated to generate a fully functional genome within a higher-order chromatin structure *(4)*. Maintenance of this complex chromatin structure is critical to ensure regulated DNA metabolism and proper epigenetic inheritance *(3)*. This position of chromatin at the crossroads of DNA metabolism places it as an ideal sensor to participate in checkpoint controls.

The repair of DNA damage in the context of chromatin has been proposed to involve the sequential steps of accessing the DNA lesion within chromatin, repairing the DNA lesion, and restoring the canonical chromatin structure, referred to as the "access, repair, restore" model *(5,6)*. In recent years, significant advances have been made in identifying a complex network of repair proteins, chromatin remodeling factors and chromatin assembly factors that could play a role in this process (*see* **refs. 6–8**). To coordinate the repair of DNA damage with the remodeling and maintenance of chromatin structures, a fine tuning of the functions of these protein factors is required. Significant mechanistic insights into how this process is operating at the nucleosomal level have been obtained by combining damaged DNA substrates with mammalian cell-free systems that contain both DNA repair and chromatin assembly activities *(9–11)*. The methods described in this chapter provide a powerful approach for the detection of proteins recruited during the recognition and repair of DNA lesions, including repair proteins and chromatin associated factors. Substrates for the recruitment assay consist of DNA containing damage that is immobilized on magnetic beads. DNA damage can be introduced with ultraviolet (UV)-C (UV-C) light (UV_{254nm}) to induce predominantly cyclobutane pyrimidine dimers and 6–4 photoproducts, DNA lesions that are repaired by the nucleotide excision repair pathway. Alternatively, DNase I can be used to enzymatically introduce single-strand DNA breaks that are repaired by the base excision repair pathway. In addition, ionizing radiation (^{137}Cs) can be used to induce more complex DNA damage in the DNA, including single-strand breaks. A human cell-free system that supports both DNA repair and chromatin assembly is incubated with the immobilized DNA-damaged substrates, and proteins associated with the DNA are then isolated and subjected to analysis. This experimental strategy is summarized (*see* **Fig. 1**). Discussed here are optimized conditions for preparing bead-linked DNA substrates containing these different types of lesions, for the reaction of the damaged DNA with cell-free systems, and for the subsequent analysis of proteins that are recruited to the immobilized damaged DNA substrates. A detailed description of the preparation of the human cell-free extracts is available elsewhere *(12)* and is therefore

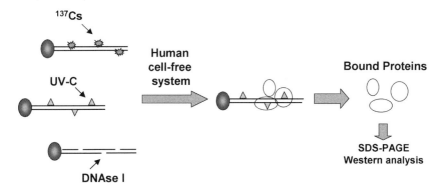

Fig. 1. Schematic representation of assay for DNA repair and chromatin assembly factors recruited during DNA damage processing. Magnetic bead-linked DNA substrates are prepared, and lesions in DNA created using UV-C, ^{137}Cs, or DNase I as indicated. These magnetic bead-linked DNA substrates (or undamaged duplex DNA as a control) are then incubated in a human cell-free system that is competent for both DNA repair and chromatin assembly. Specific proteins that bind to this DNA can be detected by SDS-PAGE, coupled to Western blotting or silver staining.

not included in these protocols. It should also be noted that this assay can be relatively easily adapted to additional DNA damaging agents and alternative cell-free systems.

Examples of experimental data obtained with the recruitment assay during chromatin assembly associated with a variety of DNA excision repair mechanisms *(11,13)* are shown (*see* **Fig. 2**). Proteins associated with the damaged DNA can be monitored over time to gain mechanistic insights into the order in which various proteins participate in DNA damage processing and chromatin remodeling. The approach described herein can be further adapted by using damaged DNA that has been preassembled into chromatin in order to analyze proteins potentially involved in chromatin rearrangements during the recognition of DNA lesions *(14)*.

2. Materials
2.1. Immobilizing Linearized DNA on Paramagnetic Beads

1. pUC19 plasmid (New England Biolabs [NEB], Beverly, MA) transformed into *E. coli* strain DH5α (Invitrogen).
2. Qiagen plasmid purification maxi-kit.
3. TE, pH 8.0: 10 mM Tris-HCl, pH 8.0, 1 mM EDTA pH 8.0.
4. 10X TBE stock solution: 108 g of Tris base, 55 g of boric acid, 40 mL of 0.5 M EDTA pH 8.0, dissolved in deionized water to a final volume of 1 L.

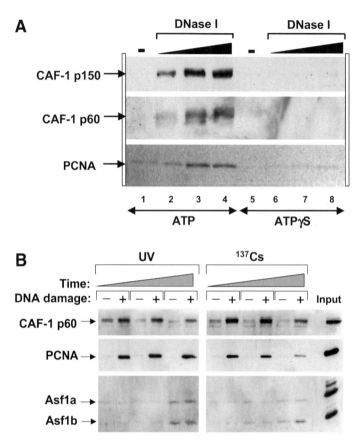

Fig. 2. **(A)** Co-recruitment of PCNA and chromatin assembly factor-1 (CAF-1) p150 and p60 subunits to bead-linked DNA substrates containing different amounts of DNase I-induced single-strand breaks (lanes 2–4). The replacement of ATP by ATPγS (a nonhydrolysable analog of ATP) inhibits DNA excision repair, and thus prevents the recruitment of both PCNA and the two largest subunits of CAF-1 (lanes 6–8). **(B)** Recruitment of PCNA and the CAF-1 p60 subunit to bead-linked DNA substrates containing DNA lesions induced by ionizing radiation (^{137}Cs) and UV$_{254nm}$ irradiation. By contrast, the Asf1 (antisilencing function) proteins, Asf1a and Asf1b, are not specifically recruited to the damaged DNA.

5. Agarose (ultrapure, Sigma) for a minigel: 0.8% agarose, 0.25 μg/mL of ethidium bromide, running buffer: 1X TBE, 0.25 μg/mL of ethidium bromide.
6. Restriction enzyme *Xma*I (NEB).
7. Restriction enzyme *Eco*RI (NEB).
8. Klenow polymerase (exo-free) (Roche).

9. Biotin-14-dATP (Roche), 0.4 mM stored at –20°C.
10. α-Thio-deoxynucleotide triphosphates (α-thio-dNTPs) (Roche), 10 mM stored at –20°C.
11. Sephadex G50 column (Roche).
12. 2X Wash/storage buffer: 10 mM Tris-HCl, pH 7.5, 2 M NaCl, 1 mM EDTA.
13. 10 mL Dynabeads M-280 in suspension at about 6–7 × 10^8 beads/mL (Dynal SA, Oslo, Norway).
14. Phosphate-buffered saline (PBS): 137 mM NaCl, 2.7 mM KCl, 1.4 mM KH$_2$PO$_4$, 0.01 M Na$_2$HPO$_4$, pH 7.4.
15. PBS containing 1 mg/mL of bovine serum albumin (BSA) and 0.05% Nonidet P-40 (NP-40).
16. 5 M Ammonium acetate.

2.2. Induction of DNA Lesions in Bead-Linked DNA

1. Bead-linked DNA.
2. Buffer A: 40 mM N-2-hydroxyethylpiperazine-N'-2-ethanesulfonic acid (HEPES), pH 7.8, 40 mM KCl, 0.05% NP-40.
3. UV-C 254-nm lamp.
4. Ice-cold steel block.
5. Vilber Lourmat VLX-3W UV dosimeter.
6. Siliconized Petri dishes.
7. 2X Wash/storage buffer: 10 mM Tris-HCl, pH 7.5, 2 M NaCl, 1 mM EDTA.
8. Buffer B: 10 mM HEPES, pH 7.6, 50 mM KCl, 1.5 mM MgCl$_2$, 0.5 mM EGTA, 10% glycerol.
9. DNase I (stock solution of 10 U/µL in buffer B; stored at –20°C) (Roche).
10. 0.5 M EDTA.
11. ^{137}Cs-source γ-irradiator (IBL 637 irradiator, CIS Biointernational).

2.3. Reaction of Bead-DNA Substrates With Human Cell-Free Extracts

1. Buffer B: 10 mM HEPES, pH 7.6, 50 mM KCl, 1.5 mM MgCl$_2$, 0.5 mM EGTA, 10% glycerol.
2. Cytosolic extract derived from HeLa cells (*see* **Note 1**).
3. Nuclear extract derived from HeLa cells (*see* **Note 1**).
4. Buffer A: 40 mM HEPES, pH 7.8, 40 mM KCl, 0.05% NP-40.
5. Buffer A containing 1 mg/mL of BSA.
6. 5X Reaction buffer: 25 mM MgCl$_2$, 200 mM HEPES–KOH, pH 7.8, 2.5 mM dithiothreitol (DTT, Sigma), 200 mM phosphocreatine (di-Tris salt, Sigma), 100 µM each of dGTP, dATP, dTTP, 40 µM dCTP (dNTPs, 100 mM lithium salt solutions, Roche). The 5X reaction buffer can be prepared in advance, aliquoted, and stored at –80°C.
7. 100 mM ATP (Amersham Biosciences), aliquoted and stored at –80°C. Note that each aliquot should be used only once.
8. Creatine phosphokinase (Roche): 2.5 mg/mL in H$_2$O aliquoted and stored at –80°C. Each aliquot is used only once.

9. 1X Laemmli buffer: 60 mM Tris-HCl, pH 6.8, 2% sodium dodecyl sulfate (SDS), 10% glycerol, 100 mM DTT, 0.001% bromophenol blue.
10. Reaction mix: 10 µL of 5X reaction buffer, 2 µL of ATP (final concentration of 4 mM), 2 µL of creatine phosphokinase (final concentration of 4 µg), 20 µL of cytosolic extract (200 µg of proteins), 4 µL of nuclear extract (20 µg), 12 µL of H_2O.

3. Methods

A scheme of the general experimental strategy is summarized (*see* **Fig. 1**). Substrates for the recruitment assay consist of DNA containing damage that is immobilized on magnetic beads. Distinct damaging agents can be used to obtain different types of lesions. The bead-linked DNA substrates can then be used to assay the binding of various proteins. As a source of protein, a human cell-free system that supports both DNA repair and chromatin assembly is incubated with the immobilized DNA-damaged substrates. Proteins associated with the DNA can then be isolated and subjected to analysis.

3.1. Immobilizing Linearized DNA on Paramagnetic Beads

We describe here how to immobilize linearized DNA on paramagnetic beads. The conditions are chosen in order to bind DNA at a density that should not be too high in order to avoid steric hindrance and remain accessible to the various DNA damaging agents. This is critical to obtain an equal distribution of lesions (on average) over all the DNA molecules and ensure reproducibility in the experiments.

3.1.1. Preparation of Biotinylated Linear DNA

1. Purify pUC19 DNA from bacteria using a Qiagen maxiplasmid purification kit. Resuspend the collected plasmid in TE, pH 8.0, and determine the DNA concentration by measuring the optical density at 260 nm (OD_{260} of 1 = 50 µg/mL). Check the quality of the DNA preparation to verify that it is nick-free DNA (<5% nicked form) by running a 200-ng aliquot on a 0.8% mini-agarose gel containing 0.25 µg/mL of ethidium bromide (*see* **Note 2**).
2. Linearize 40 µg of pUC19 with *Xma*I and verify that it is completely digested by running 200 ng on a 0.8% mini-agarose gel.
3. Add an equal volume 5 M Ammonium acetate and two volumes of cold ethanol to precipitate the DNA, place at –20°C for 30 min, centrifuge at maximum speed in a cooled microcentrifuge, wash the pellet with cold 70% ethanol, air-dry, and resuspend in 40 µL of TE.
4. Digest the linearized plasmid with *Eco*RI to generate a 5′-overhang at one end.
5. Precipitate and resuspend as in **step 3**.
6. Fill in the 5′-overhang with Klenow polymerase using a final concentration of 50 µM biotin-14-dATP, as well as 200 µM α-thio-dCTP, 200 µM α-thio-dTTP, and 200 µM α-thio-dGTP typically in a 60-µL volume reaction (*see* **Note 3**).

7. Remove the unincorporated dNTPs and the small restriction fragment by passing the DNA over a Sephadex G50 column. The eluted, biotinylated DNA (usually 60 µL can be recovered) can be used immediately or frozen and stored for later experiments.

3.1.2. Coupling Biotinylated Linear DNA to Dynal M-280 Beads

For the coupling reaction we usually work at a ratio of 0.0625 pg of DNA per bead.

1. Take a 960-µL aliquot of Dynabead M-280 suspension from stock solution and place in a 1.5-mL microcentrifuge tube. Concentrate the beads using a magnetic rack (*see* **Note 4**), then remove the liquid with a pipet and discard. Wash the Dynabeads by resuspending in 300 µL of PBS containing 1 mg/mL of BSA and 0.05% NP-40. Repeat this procedure two times with 300 µL of 2X wash/storage buffer (*see* **Note 4**).
2. The coupling of linearized DNA to Dynabeads is carried out here by adding 40 µg of DNA to the bead suspension (here in 300 µL). The reaction can be scaled up or down as required (*see* **Note 5**).
3. Dilute 40 µg of biotinylated DNA to a final volume of 1340 µL with 2X wash/storage buffer. Resuspend the washed, concentrated beads in this DNA solution. Incubate on a rotating wheel for 16 h at room temperature.
4. Concentrate the beads using the magnetic rack (*see* **Note 4**), discard the supernatant (save a 10-µL aliquot) and wash the beads twice with 2X wash/storage buffer. The bead-linked DNA can be stored up to one month in this same buffer at 4°C.
5. Run a 10-µL sample of the supernatant on an agarose gel alongside 300 ng of linearized pUC19 to verify efficient binding to the beads. No DNA should be visible in the supernatant if efficient binding was achieved.

3.2. Induction of DNA Lesions in Bead-Linked DNA

The DNA damage can be introduced on the immobilized DNA using different methods that will result in the formation of distinct type of lesions (*see* **Subheading 1.**). These different types of substrates can be useful to distinguish binding activities that can be lesion specific.

3.2.1. UV-C Radiation

1. Prewarm a UV 254-nm lamp for at least 15 min before irradiation. Place a steel block on ice under the lamp at a distance of approx 10 cm. Measure the UV fluence at the position of the block with a Vilber Lourmat VLX-3W dosimeter. A typically achieved UV fluence is 5–10 $J/m^2/s$. For UV Protection the manipulation is carried out in a chamber, and gloves and glasses should be used.
2. Resuspend the bead-linked DNA to be irradiated (typically 10 µg) in 500 µL of buffer A, and then transfer it to a siliconized Petri dish. Place the Petri dish on the ice-cold steel block and remove the lid to begin irradiation. Agitate the Petri dish every 20 min during the irradiation to resuspend the beads until the desired dose

is achieved. For example, to achieve a dose of 50,000 J/m² (5 J/cm²), irradiate the bead-linked DNA at 8 J/m²/s for 105 min. Treat the control, nonirradiated DNA similarly on ice, but without the irradiation.
2. Following irradiation, recover the bead-linked DNA from the Petri dish and transfer to a fresh 1.5-mL tube. Concentrate the beads using the magnetic rack as before, discard the supernatant. Wash the Petri dish with 500 µL of 2X wash/storage buffer and add this suspension to the recovered and concentrated beads. Concentrate the beads again, wash twice with 300 µL of 2X wash/storage buffer and store in the same buffer at 4°C.

3.2.2. DNase I

1. To determine the optimal conditions for DNase I treatment that will produce single-strand but not double-strand breaks, a titration should first be carried out on circular pUC19. Dilute the stock of DNase I (10 U/µL) to 0.1 U/µL in buffer B just before use. To titrate the DNase I digestion, dilute 1 µg of plasmid DNA to 65 µL in buffer B, add 0.65 U of DNase I, and incubate at 25°C over a range of time (for example 2, 4, and 6 min) or in the absence of DNase I as control. Remove 20-µL aliquots corresponding to each time point and transfer to a tube containing EDTA to 100 mM final concentration to stop the reaction. Analyze the products on an agarose gel. Choose conditions that yield the greatest number of nicked circular molecules but no linear molecules. This is usually obtained within the time range indicated above.
2. To induce single-strand breaks in immobilized DNA, concentrate 10 µg of bead-linked DNA (approx 25% from **step 3** in **Subheading 3.1.2.**) and resuspend in 650 µL of buffer B. Digest the DNA with 6.5 U of DNase I for the amount of time determined above. Stop the reactions with EDTA as described in above, wash the DNA with 2X wash/storage buffer, and store in this same buffer at 4°C.

3.2.3. Ionizing Radiation

1. Resuspend the bead-linked DNA to be irradiated (typically 10 µg) in 500 µL of buffer A, and then transfer to a siliconized Petri dish as described above for UV irradiation.
2. Place the plate in a ^{137}Cs-source γ-irradiator and irradiate at 2.4 Gy/min for 30 min to achieve a dose of 75 Gy. Treat the control DNA similarly but without irradiation during the same amount of time.
3. Recover bead-linked DNA as for UV-C irradiation and store at 4°C.

3.3. Reaction of Bead-DNA Substrates With Cell-Free Extracts

Set up reactions by mixing the bead-linked DNA with a human cell-free system derived from HeLa cells (*see* **Note 1**). Standard reactions of 50 µL contain the following: 600 ng of bead-linked DNA, 200 µg of proteins from the cytosolic extract, 20 µg of proteins from the nuclear extract, 40 mM HEPES, pH 7.8, 5 mM MgCl$_2$, 0.5 mM DTT, 4 mM ATP, 20 µM each of dGTP, dATP, dTTP, 8 µM dCTP, 40 mM phosphocreatine, and 5 µg of creatine phosphokinase.

1. Prior to the incubation, wash the bead-linked DNA once in buffer A containing 1 mg/mL of BSA, and once in buffer A alone.
2. Prepare the reaction mix and preheat at 37°C for 5 min.
3. Aliquot the bead-linked DNA to tubes, concentrate the beads, and remove the supernatant.
4. Add 50 µL of preheated reaction mix to each tube and incubate at 37°C.
5. Remove 15-µL aliquots at 2, 5, and 15 min. Terminate the reaction of the aliquots by concentrating the beads on a magnet and immediately removing the supernatant, and then washing the beads five successive times with 500 µL of buffer A.
6. Resuspend the concentrated beads in 30 µL 1X Laemmli buffer and elute the bound proteins by heating to 95°C for 5 min. Samples can then be stored at –20°C until ready to analyze.
7. Load samples onto an SDS-polyacrylamide gel of appropriate percentage for proteins of interest.
8. Detect bound proteins by silver staining, or transfer gel to a membrane and carry out Western blot analysis for proteins of interest (*see* **Fig. 2** and **Note 6**).

4. Notes

1. HeLa cell-free extracts usually used in our assays are prepared from adherent HeLa cells grown in plates as described (*12*). These extracts were largely characterized for their competence in repair of DNA damages and chromatin assembly assays (*12*). Other extract preparations can potentially be used, but their abilities to repair DNA damage and assemble chromatin should then be tested.
2. DNA produced by this method should migrate as a single band corresponding to the supercoiled form. If a significant amount of nicked form (which migrates more slowly than the supercoiled form *[15]*) is present, the DNA can be further purified on sucrose gradients (*16*; *see also* Chapter 29).
3. The incorporation of phosphothioate bonds using α-thio-dNTPs is carried out to protect the 3′-end from possible exonuclease activity during the incubation with cell-free extract (*see* **Subheading 3.3.**).
4. Follow the advice of Dynal for general handling of beads and for using the magnetic tube racks. See *Biomagnetic Techniques in Molecular Biology* for further information (*17*). Note that beads should never be allowed to dry out. Beads can be stored at 4°C, and should not be frozen.
5. These binding conditions, which do not yield a high quantity of DNA per bead, were specifically chosen to obtain an optimal induction of DNA lesions at this recommended DNA:bead ratio. Although higher levels could be achieved, we found these were not optimal for functional assays, owing to steric hindrance.
6. Several variations can be introduced into the experimental protocol in order to garner further mechanistic insight into the interplay between DNA repair and chromatin assembly/remodeling. For example, in one variation to this experiment, damaged DNA was reconstituted into nucleosomal organization by salt dialysis using pure histones prior to carrying out the recruitment assay (*14*). In another example of this application, the provided protocol can be extended to the identifi-

cation of novel repair/chromatin remodeling proteins by first separating recruited proteins by SDS-PAGE as described in **Subheading 3.3.**, and then carrying out tryptic digests of eluted polypeptides and mass spectrometry analysis for protein identification. Finally, given that other researchers have successfully used bead-linked DNA substrates to assay RNA polymerase II mediated transcription *(18)*, it may be possible to adapt the assay described here to gain insights into the interplay between DNA repair, chromatin remodeling, and transcription.

Acknowledgments

The authors thank Dr. J.-P. Quivy for advice regarding the use of Dynal beads and D. Roche for many insightful suggestions. This work was supported by la Ligue Nationale contre le Cancer (Equipe labellisée la Ligue), Euratom (FIGH-CT-1999-00010 and FIGH-CT-2002-00207), the Commissariat à l'Energie Atomique (LRC no. 26), and RTN (HPRN-CT-200-00078 and HPRN-CT-2002-00238).

References

1. Friedberg, E., Walker, G., and Siede, W. (1995) *DNA Repair and Mutagenesis*. American Society for Microbiology Press, Washington, DC.
2. Luger, K., Mäder, A. W., Richmond, R. K., Sargent, D. F., and Richmond, T. J. (1997) Crystal structure of the nucleosome core particle at 2.8 Angstrom resolution. *Nature* **389**, 251–260.
3. Wolffe, A. P. (1997) *Chromatin: Structure and Function*. Academic Press, New York, NY.
4. Vaquero, A., Loyola, A., and Reinberg, D. (2003) The constantly changing face of chromatin. Web-site: http://sageke.sciencemag.org/cgi/content/full/sageke;2003/14/re4.
5. Smerdon, M. J. (1991) DNA repair and the role of chromatin structure. *Curr. Opin. Cell Biol.* **3**, 422–428.
6. Green, C. M. and Almouzni, G. (2002) When repair meets chromatin. First in series on chromatin dynamics. *EMBO Rep.* **3**, 28–33.
7. Moggs, J. G. and Almouzni, G. (1999) Chromatin rearrangements during nucleotide excision repair. *Biochimie* **81**, 45–52.
8. Meijer, M. and Smerdon, M. J. (1999) Accessing DNA damage in chromatin: insights from transcription. *Bioessays* **21**, 596–603.
9. Gaillard, P.-H., Martini, E., Kaufman, P. D., Stillman, B., Moustacchi, E., and Almouzni, G. (1996) Chromatin assembly coupled to DNA repair: a new role for chromatin assembly factor I. *Cell* **86**, 887–896.
10. Gaillard, P.-H., Moggs, J. G., Roche, D. M. J., et al. (1997) Initiation and bidirectional propagation of chromatin assembly from a target site for nucleotide excision repair. *EMBO J.* **16**, 6281–6289.
11. Moggs, J. G, Grandi, P., Quivy, J.-.P., Jónsson, Z. O., Hübscher, U., Becker, P. B., and Almouzni, G. (2000) A CAF-1 - PCNA-mediated chromatin assembly pathway triggered by sensing DNA damage. *Mol. Cell. Biol.* **20**, 1206–1218.

12. Martini, E., Roche, D. M. J., Marheineke, K., Verrault, A., and Almouzni, G. (1998) Recruitment of phosphorylated chromatin assembly factor 1 to chromatin following UV irradiation of human cells. *J. Cell Biol.* **143,** 563–575.
13. Mello J. A., Sillje, H. H., Roche D.M., Kirschner, D.B., Nigg, E.A., and Almouzni G. (2002) Human Asf1 and CAF-1 interact and synergize in a repair-coupled nucleosome assembly pathway. *EMBO Rep.* **3,** 329–334.
14. Brand, M., Moggs, J. G., OuLad-Abdelghani, M., et al. (2001) A UV-damaged DNA binding protein in the TFTC complex links DNA damage recognition to nucleosome acetylation. *EMBO J.* **20,** 3187–3196.
15. Gaillard P. H., Roche, D., and Almouzni, G. (1999) Nucleotide excision repair coupled to chromatin assembly, in: *Methods in Molecular Biology*, Vol. 119, (P. Becker, Ed.), Humana Press, Totowa, NJ, pp. 231–243.
16. Wood, R. D., Biggerstaff, M., and Shivji, M. K. K. (1995) Detection and measurement of nucleotide excision repair synthesis by mammalian cell extracts in vitro. *Methods* **7,** 163–175.
17. Dynal Technical Handbook. *Biomagnetic Techniques in Molecular Biology*, 3rd Ed., Dynal, Oslo, Norway.
18. Nightingale, K. P., Wellinger, R. E., Sogo, J. M., and Becker, P. B. (1998) Histone acetylation facilitates RNA polymerase II transcription of the Drosophila hsp26 gene in chromatin. *EMBO J.* **17,** 2865–2876.

Index

A

Abasic site, *see* AP site
N-acetylaminofluorene (AAF), 161, 436
Acridine orange, 159, 161
Adenine phosphoribosyltransferase (APRT), 134ff
　gene targeting of, 135ff
Adenovirus-based vectors for gene transfer, 13, 14
Adriamycin, 56
Agarose-embedded cells, 96ff, 257–258
　irradiation of, 102–103
　preparation of "naked" DNA from, 98, 100, 101–102, 107
　preparation of nuclei from, 99, 102, 107
ALASA selection, 143, 147, 148
Alkaline agarose gel electrophoresis, 161, 172–173, 177, 208, 251ff
　calculation of lesion frequencies, 264, 268–269
　calculation of sizes and numbers of DNA molecules, 264–269
　troubleshooting, 259, 260–261
Alkaline filter elution (AFE), 288
2-Aminoanthracene, 295
AP (apurinic/ apyrimidinic)
　endonucleases, 157, 355, 356, 357, 358, 360, 366, 369, 374, 378, 390, 393
AP lyases, 157, 356, 357, 358, 360
AP (abasic) site, 156, 157, 270, 287, 302, 337, 341, 355, 356, 357, 358, 360, 365, 366, 369, 372, 373, 374, 378, 383, 390, 393
　AP site-containing oligonucleotide substrates, 356, 362
　　preparation of, 358–361

AP site-containing plasmid substrates, 366, 369–372, 373, 374, 390
　preparation of, 367–368, 373, 374, 390, 392
APE//Ref1, *see also* AP endonuclease, 356, 358, 359, 360
Apoptosis, 81, 82, 88, 91, 109, 110, 111, 112, 283, 284
　DNA fragmentation in, 74, 78, 81, 91
　detection by flow cytometry, 82, 85, 86
　detection by laser scanning cytometry, 82, 86, 87
　DNA strand break labeling, 81ff
Asf1, 480
Ataxia telangiectasia (AT), 1, 52, 53, 55, 62
Ataxia telangiectasia-like disorder (ATLD), 52
ATM, 52, 53, 55, 62, 73
ATR, 52, 53, 73

B

Bacteriophage M13 DNA, 438, 442, 444, 451
Baculovirus expression system, 64, 67
Base excision repair (BER), 27, 156, 157, 185, 355ff, 365ff, 377ff, 458, 462, 478
　assays of, 356–358, 362, 366, 372–373, 381–382, 386–390
　long patch repair (PCNA-dependent), 157, 366, 394, 458
　mammalian cell extracts, 366
　preparation, 372, 380–381, 384–386, 393
　short patch repair, 157, 366, 394

489

substrates, 356–361, 362, 366, 367–368, 369–372, 373, 374, 377–380, 382–384, 390, 392, 394
Biotinylated DNA, 482, 483
  coupling to paramagnetic beads, 483
  inducing lesions in bead-linked DNA, 483–484
Bleomycin, 3, 5, 26, 56
BLM, 53, 398
BND-cellulose chromatography, 348–349, 351, 352
Butadiene, 26

C

Camptothecin (CPT), 65, 87, 89, 90
Caspase-3, 78
Cell death
  delayed, 44
  effects on repair assays, 176, 283, 419
Cell freezing, 3, 21, 46, 48
Cell survival/ viability determinations, 5, 18, 20–21, 31, 48, 145, 146, 167
Cesium chloride gradient centrifugation, 349, 368, 370–371, 374, 392–393, 422–423, 429, 430, 442–444
  purification of parental DNA by, 171–172, 173
Challenge assay, see Cytogenetic challenge assay
Checkpoints, 52, 53, 56, 62, 478
Chernobyl nuclear contamination, 289, 298
Chesterton, G. K., 215
Chinese hamster cells (see also CHO cells), 2, 134ff
  electroporation of, 144–145
  freezing of, 3
  gene-targeting in, 134ff
  mutagenesis by ENU, 2, 3, 4
  ploidy, 5, 6, 135
  preparation of cell extract, 384
  V79, 2, 6, 56, 283
Chinese hamster ovary (CHO) cells, 2, 6, 56, 134ff, 384
  43-3B, 428

CHO-9, 56, 389, 393
CHO 27-1, 428
CHO-AA8-4 (AA8), 148, 327
CHO-AT3-2, 148
CHO-ATS49tg, 135, 139, 140, 147, 148, 149
CHO-K1, 135, 147
CHO UV41, 428
E1KO7-5, 140, 148
RMP41, 135
Chromatin assembly factor-1 (CAF-1), 480
Chromatin (nucleosome) assembly and DNA repair, 73, 419, 437, 478, 479, 482
  recruitment assay, 478, 479, 482, 484–485
  preparation of substrates, 482–484, 485
Chromosomal instability, 44, 49
Chromosome aberrations (CAs), 25, 26, 27, 32, 38
  chromatid breaks, 38
  chromatid exchanges, 38
  deletions, 35, 37, 38
  dicentrics, 35, 37, 38, 43, 44
  translocations, 35, 38, 43, 109, 110, 111, 112, 114
Chromosome painting, 49
Chromosome preparation (metaphase), 33–35, 45, 46, 47
  hypotonic solution, 33, 45, 46
Cisplatin, 5, 56, 161, 277, 326, 329, 331, 337, 341, 440
  DNA adducts, 156, 329, 436, 437, 440, 441, 443, 444, 445, 452, 453
  platination reaction, 337, 437, 439–440, 441, 451
Coblenz, W. W., see xvii
Cockayne syndrome (CS), 9, 10
Colcemid, 28, 33, 45, 46
Comet assay (single-cell gel electrophoresis), 27, 74, 158,

275ff, 288, 290, 302
Comet classification/ image analysis, 281–282, 283
electrophoresis and staining, 281, 283
preparation of cells, 280–281
preparation of slides, 280
statistical tests, 282
Complementation, 5, 10, 12, 13, 53
assays, 12, 13, 53
Crosslinks, *see* DNA crosslinks
Crystal violet, 45, 48
Cyclobutane pyrimidine dimers, *see* UV photoproducts
Cyclophosphamide, 277
Cytodex-1 microcarrier beads, 6
Cytogenetic challenge assay, 25ff
analysis of chromosomes, 35–40
collection of cell specimens, 30
isolation of lymphocytes, 30–31
irradiation of cells
ionizing radiation, 32, 35
UV radiation, 32–33
preparation of metaphase chromosomes, 33–35

D

DDB1 (*see also* UV-DDB), 332, 338
DDB2, 338
DEAE cellulose (*see also* BND-cellulose chromatography), 127
DEAE sepharose, 127, 129
DNA crosslinks, 52, 277, 329, 436, 437, 440, 441, 443, 444, 452, 453
DNA double-strand breaks (DSBs), 43, 73, 74, 95ff, 109, 123, 124, 148, 270, 277, 287, 327
γH2AX as a marker of, *see* Histone H2AX
measurement by pulsed-field gel electrophoresis, 104, 105
fraction of activity released (FAR), 105, 107

rare-cutting endonucleases, 103
DNA dyes
DAPI, 45, 47, 461, 468
ethidium bromide, 104, 105, 167, 172, 173, 176, 208, 253, 261, 263, 265, 279, 281, 288, 330, 366, 380, 438, 472
Giemsa, 28, 35, 45, 47, 48, 49
Hoechst-33258, 92, 461, 468
PicoGreen, 175, 187, 188, 189, 190, 288, 289, 290, 291, 300
OliGreen, 248
Propidium iodide (PI), 45, 74, 75, 76, 78, 84, 461, 469, 471, 472, 473
SYBR Gold, 128, 129
SYBR Green, 366, 369, 373
YOYO-1, 282, 292, 299
DNA electroblotting, 210
DNA electroelution, 330, 334
DNA glycosylases, 156, 204, 355, 356, 366, 374, 377
DNA helicases, *see* Helicases
DNA isolation, damage induced during, 107, 158, 186, 226, 303
DNA isolation (nuclear), 170–171, 186, 188, 207, 222, 226, 234–235, 243–244, 311
DNA labeling, 81, 82
BrdU, 53, 82, 83, 84, 88, 89, 92, 163, 173, 176
compared to other labels, 82, 83
FdU, 163, 173
IdU, 83, 88, 89, 92
Klenow polymerase, 330, 334–335, 341, 439, 454, 482–483
Klenow/ exonuclease III, 330, 335, 341
Sequenase, 449
T4 polynucleotide kinase (PNK), 308, 310, 312, 317, 319, 350, 359, 361, 368, 370, 379, 382, 384, 392, 393, 403, 404, 440, 442, 444, 446
Taq DNA polymerase, 210

Thymidine (radiolabeled), 17, 20, 53, 54, 98, 101, 105, 240, 243, 248
DNA ligase I, 157, 458, 459
DNA ligase III, 157
DNA ligase IV, 96, 124, 129
DNA-PK/ DNA-PKcs, 73, 96, 124, 324, 328, 329, 333, 340, 341, 342
DNA polymerase β, 366, 378, 394, 458
DNA polymerase δ, 418, 457, 459
DNA polymerase ε, 418, 457
DNA repair gene polymorphisms, 26, 27
DNA replication, inhibition by ionizing radiation, 51–53, 56, 61–62
DNA single-strand breaks, 276, 277, 287, 288, 289, 302, 303, 478, 484
  detection by Fast Micromethod, 288ff
DNA topoisomerase I, 87
DNase I, 452, 461, 464, 466, 471, 478, 481, 484
Dot blots, 174, 229ff
Drosophila S3 *N*-glycosylase/AP lyase, 356, 358, 360
DT40 cells (chicken), 134
dRPase, 157
Dual incision assays, *see* Nucleotide excision repair
Dynabeads, 481, 483, 485

E

Electrophoretic mobility shift assay (EMSA), 323ff
  cell extracts, preparation of, 330–331, 335–336, 341
  competitor DNAs, 324, 325, 327, 328, 329, 331, 333, 336, 337, 339, 340, 342
  nondenaturing gel, preparation of, 331, 336
  probes, 323, 324, 325, 327, 328, 329, 333, 334, 335, 336, 337, 339, 340
  reverse EMSA, 325, 326, 332, 337
  supershift assay, 327, 328, 333, 340
β-Elimination, 207, 356, 358, 363
δ-Elimination, 356, 358, 363
Electroelution, *see* DNA electroelution
Electroporation of mammalian cells, 144–145
ELISA, 218, 222–223, 224–226, 227
Endonuclease III/ Nth protein (*E. coli*), 204, 269, 278, 356, 357, 358, 360, 380, 383, 390, 391, 395, 420, 422, 424, 428, 429
Environmental toxicants, 26
ERCC1 (ERCC1), 134, 140, 418, 428
  gene-targeting of, 135ff
ERCC4, 428
Ethyl methanesulfonate (EMS), 2, 3, 4, 5
N-Ethyl-*N*-nitrosourea (ENU), 2, 3, 5
European Standards Committee on Oxidative DNA Damage (ESCODD), 157, 158
Exonuclease III (*E. coli*), 330, 335, 341, 356
Exonuclease V (*E. coli*), 352

F

Fanconi anemia, 1, 53
Fast Micromethod, 288ff
  strand scission factor (SSF), 295, 297, 298, 300, 301, 302, 303
FEN-1, 157, 410, 413
FIAU, 138, 142, 145, 146, 147
Flow cytometry, 74, 82, 85, 86, 88, 89, 459, 462, 466, 467, 468–469
  steps to minimize cell loss, 78, 90
Formaldehyde, 277
Formamidopyrimidine DNA glycosylase (Fpg), 161, 162, 165, 166, 169, 204, 269, 277, 356, 357, 358, 360
Fluorescence *in situ* hybridization (FISH), 45, 47–48, 49, 278
Fluorochromes, 82, 85, 86
Fluorometric analysis of DNA unwinding (FADU), 288, 301

G

G418 (Geneticin), 15, 17, 142, 145, 146
Gancyclovir, 138
Gel mobility shift assay (*see also* Electrophoretic mobility shift assay), 402, 408–409
Gene-specific damage and repair, 158ff, 184ff, 202ff
Gene targeting, 133ff
  homology requirements, 138
  recombinant classes, 139
  targeting vector design, 136–138
Gene therapy, 10, 13, 15
  virus-based vectors, 13, 14
Gene transfer/ transduction, 13, 14
Genomic instability, 43, 44, 48, 123, 399
Ghost cells, 283
Global genomic repair, 326
Glyoxal gels, 270
GM10115 human–hamster hybrid cells, 48, 49
GPT, 146, 148, 149

H

HAT selection, 143, 146, 149
Hedgehogs, 283
HeLa cells, 69, 100, 105, 129, 230, 237, 289, 294, 295
HeLa cell extracts, 69, 105, 129, 349, 391, 448, 450, 484, 485
  preparation of, 63, 65–66, 69, 98, 100, 105, 124–125, 126–127
Helicases, 53, 397ff
  ATP and Mg^{2+} requirements, 399, 412
  ATPase assay, 402–403, 409–411
  gel mobility shift assay, 402, 408–409
  inhibition by DNA lesions, 399
  kinetics, 400, 409, 411
  nitrocellulose filter binding assay, 402, 407–408
  processivity, 398
  strand displacement assays, 398
    fluorescence resonance energy transfer (FRET), 400
    radiometric, 398, 400, 401–402, 403–407
  substrates, 398, 399, 400, 406, 407, 408
    preparation of two- or three-stranded duplex, 403, 412
    preparation of four-stranded (Holliday) substrates, 404–405, 411
High-performance liquid chromatography (HPLC), 158, 216, 308, 310, 315–316, 319
Histone H2AX, 73, 76
  γH2AX and DNA double-strand breaks, 73, 74, 76, 78
  antibodies against γH2AX, 73, 75, 78
  apoptosis-associated γH2AX, 78
HN2, *see* Nitrogen mustard
Homology-directed repair (HDR), 96, 124
Host-cell reactivation, 12
Hot alkali treatment of DNA, 230, 233, 235–236, 358, 362
HSV-tk, 138, 147
Human embryonic cell line-293, 15
Human HL-60 cell line, 83, 87, 90, 91
Human lung carcinoma cell line A549, 101, 107
Human MM6 cell line, 114
Human TK6 cell line, 114, 118, 119, 283
Hydrogen peroxide (H_2O_2), 186, 277, 463, 470
8-Hydroxydeoxyguanosine (8-OHdG), 186, 319
Hypoxanthine phosphoribosyltransferase (HPRT), 136, 184
  HPRT mutants, 4, 6, 148, 149

I

Imaging cytometer (iCyte), 82
Immunoaffinity purification, *see* Protein-A sepharose chromatography

Immunofluorescence staining of cells in culture, 220–221, 466–468
Immunoprecipitation, 240, 242–243, 244, 459, 465, 472
Inverse PCR (IPCR), 110ff
Ionizing radiation, 4,43,45-46, 48, 54, 56, 65, 102, 124, 129, 270, 277, 278, 287, 327, 377
 delayed effects of, 43–44
 DNA synthesis inhibition, 51–53, 56, 61–62
 γ-rays, 18, 19, 26, 27, 32, 35, 38, 102, 289, 290, 294, 297, 301, 481, 484
 X-rays, 4, 5, 6, 27, 32, 35, 38, 48, 65, 102, 297

K

Keratinocytes, 10ff, 217ff
 culturing of, 15, 17, 18–19, 21, 22, 217, 220
 SV40-immortalized, 217, 220
 transduction of, 19
Ku, 56, 96, 124, 327, 328, 329, 333, 339, 340, 342

L

Laser scanning cytometry, 74, 82, 86, 87, 89
Leukemias, 89, 109
Ligation-mediated PCR (LM-PCR), 110, 111, 114–115, 202ff
 reactions, 117–118, 208–209
 sequencing gel analysis, 209–210
Light meter (radiometer/ photometer) (*see also* UV light meters), 174
Liver S9 fraction, 295

M

Metaphase chromosomes, preparation of, 33–35, 46
Methyl methanesulfonate (MMS), 3, 4, 5, 277, 278, 283, 460, 463, 470

N-Methyl-N'-nitro-N-nitrosoguanidine (MMNG), 331, 337, 461, 463
Methylene blue (MB), 3, 5, 157, 159, 161, 162, 165, 167, 168, 169, 171, 174
Micrococcal nuclease, 308, 309, 311, 319
Micrococcus luteus UV endonuclease, 252, 253
Micropore UV-irradiation technique, 469, 471
Mismatch repair (MMR), 134, 156, 157, 329, 345ff, 459
 analysis of excision intermediates, 350–351
 analysis of mismatch correction, 349–350
 MMR genes/ proteins, 124, 134, 136, 156, 329
 preparation of substrate, 347–349
Mitochondrial DNA damage, 156ff, 185–186
 associations with human diseases, 185
 estimates of levels of damage, 158, 175, 185
 induction of damage during isolation of mtDNA, 158, 186
Mitochondrial DNA repair, 156, 157, 175, 185
Mitomycin C (MMC), 3, 4, 6, 21, 56
MLL (mixed-lineage leukemia) gene, 109ff
Mouse cell lines
 A9, 56
 embryo stem (ES) cells, 134, 136
 L5178Y, 283, 291, 295, 301
 Swiss 3T3 J2 fibroblasts, 15, 18, 19, 20, 21
 3T3 cells, 372
MRE11, 52
Mutagen-sensitive mutants, 2, 6
Mytilus galloprovincialis, 301

N

NBS1, 52, 53

NEO, 13, 15, 138, 146, 147, 148
Nicking endonucleases, 346, 352
   N.*Alw*I, 347, 349, 352
   N.*Bst*NBI, 347, 351, 352
Nijmegen breakage syndrome (NBS), 1, 52, 53, 55
Nitrogen mustard (HN2), 277, 342
4-Nitroquinoline-*N*-oxide (4NQO), 161, 289, 290, 294, 295
Nonhomologous end-joining (NHEJ), 96, 111, 124, 141
Nth protein, *see* Endonuclease III
Nuclease P1, 308, 309, 311, 319, 452
Nucleotide excision repair (NER), 10, 12, 27, 134, 156, 185, 201, 202, 324, 325, 329, 399, 417ff, 435ff, 458, 462, 478, 479
   dual incision assay, 436
      analysis by end-labeling, 437, 439, 449
      analysis by Southern blot, 437, 438–439, 445–448
      analysis using internally labeled substrates, 437, 444–445, 449–451
      preparation of substrates, 436, 437–438, 439–446, 452, 453
   repair synthesis assay, 418, 419, 436
      CFII, 419, 425, 426, 427, 428, 431
      preparation of cell extract, 420–421, 425–426, 430
      preparation of plasmid substrate, 419–420, 421–423, 429
      repair reactions, 421, 426–428, 431

O

OGG1, 156
Oxidative DNA damage/stress, 155ff, 226, 227, 269, 270, 356, 377
   induced during DNA extraction/isolation, 158, 186, 226
8-Oxoguanine, 156, 158, 159, 161, 356, 358

P

^{32}P-Postlabeling, 216, 307ff
   enzymatic digestion of DNA sample, 308, 309, 311–312
   HPLC analysis, 308, 310, 315–317
   labeling adducted nucleotides, 310, 312
   PAGE analysis, 308, 310, 313–315, 316
   standards, 318, 319
   TLC analysis, 308, 310, 312–313, 315, 316
$p21^{waf1/cip1}$, 458
p53, 44, 52, 186
Paramagnetic beads (*see also* Dynabeads), 478ff
PCNA, *see* Proliferating cell nuclear antigen
Peripheral blood mononuclear cells (PBMCs), 289ff, 471
pH, control of, 54, 56, 105
Phosphodiesterases, 378
   spleen phosphodiesterase, 308, 309, 311, 319
Photolyases
   E. coli, 202, 205, 208, 230, 231, 232, 234
   yeast, 329, 341
Photolysis of halogenated bases in DNA, 83, 88–89, 91, 92
   enhancement by Hoechst-33258, 92
Photoreactivation/ photorepair, 230, 233, 257, 329
Photosensitizer, 247
   acetone, 242, 247
   acetophenone, 231, 233
PicoGreen DNA stain, *see* DNA dyes
Piperidine (*see also* Hot alkali treatment of DNA)
   cleavage of (6-4) photoproducts by, 202, 207
   cleavage of AP sites by, 357, 362
Plasmid-based end-joining assays, 96, 124ff

Polycyclic aromatic hydrocarbons (PAHs), 277
Postreplication repair, 459
Potato apyrase, 310, 312
Proliferating cell nuclear antigen (PCNA), 366, 378, 394, 418, 419, 426, 428, 429, 431, 457ff, 480
  anti-PCNA antibodies, 461, 465, 466, 468
  flow cytometric analysis, 459, 462, 466, 467, 468–469
  immunofluorescence detection, 459, 462, 466–468
  immunoprecipitation, 459, 465, 472
  soluble versus chromatin-bound forms, 458, 459, 462, 464, 465, 466, 467
    extraction of detergent-resistant PCNA, 464–465, 471
    extraction of detergent-soluble PCNA, 464, 467, 472
Protein-A sepharose chromatography, 64, 66–68, 70, 219
Pulsed-field gel electrophoresis, 104, 105, 270
  AFIGE, 104
  CHEF, 105

Q

Quantitative polymerase chain reaction (QPCR), 183ff
  long PCR, 184, 186, 188
  normalization for mtDNA copy number, 193

R

Rad1, 134, 140, 418
Rad2, 418
Rad3, 399, 418
Rad4, 418
Rad10, 134, 140, 418
Rad14, 418
Rad23, 418

RAD23B, 418
RAD50, 52
Radioimmunoassay (RIA), 239ff
Radioresistant DNA synthesis (RDS), 52, 53, 55
  assay of, 53–55
  controlling assay variability, 54, 55, 56
Rare-cutting endonucleases, 103
RB69 DNA polymerase, 451
Reactive oxygen species (ROS), 155, 156, 185, 278
Recombination, 96, 133, 134, 135, 138, 139, 140, 141, 148
RecQ, 398, 399
Replica plating, 2, 4
Replication factor C (RFC), 418, 458
Replication protein A (RPA), 398, 418, 419, 426, 428, 429, 448, 450
Replicative status of cultured cells
  30-h labeling index, 176
  population doubling level (PDL), 177
Retrovirus-mediated gene transfer, 13, 14
Rhodamine blue staining, 18, 21

S

Saccharomyces cerevisiae, 134, 140
Scintillation counting, 54, 55, 315
Single-cell gel test, *see* Comet assay
Single strand annealing, 134, 140
Skin biopsy, 12, 18–19, 218
Skin cancers, 10, 216, 436
Solar spectrum lamp, 294
Somatic cell fusion, 10
Southern blot gene-specific assay, 158ff
  Poisson calculations, 170
Ssl2, 418
Strand breaks induced by photolysis (SBIP), 83, 88–89, 92
Styrene, 26
Sucrose/glycerol fractionation, 380, 385
Sucrose gradient centrifugation, 420, 423, 424, 485
SV40 DNA replication system, 62, 64, 65, 68–69

Index

TAg, 62, 66, 69, 70
  purification of, 63–64, 66–68, 69, 70

T

T4 DNA polymerase, 380, 438, 442, 444, 451
T4 endonuclease V, 202, 205, 207, 208, 249, 252, 253
T4 gene 32 protein, 380, 382
T4 polynucleotide kinase, *see* DNA labeling
T5 exonuclease, 438, 445, 453
T7 RNA polymerase, 332, 338
Tamoxifen DNA adducts, 312, 316, 317, 318
Terminal deoxynucleotidyltransferase (TdT), 82, 84, 87, 91
Tetrahydrofuran, 357, 358, 360, 368
TFIIH, 418
Thin layer chromatography (TLC), 308, 312–313, 402, 403, 409–411
  TLC plates, 310, 312, 313, 402, 437, 440, 446, 452
6-Thioguanine, 4, 6
Transcription-coupled repair, 160, 202
Transcription/ translation in vitro, 325, 332, 338
Trichloroacetic acid (TCA) precipitation, 20, 55, 56–57
Trichothiodystrophy (TTD), 9, 10

U

Unscheduled DNA synthesis (UDS), 10, 12, 16, 17–18, 19–20
Uracil DNA glycosylase (UDG), 156, 357, 360, 361, 368, 372, 379, 380, 383, 388, 391, 395
Uranium, 26
Urine, DNA photoproducts in, 224–226
UV-DDB, 324, 325, 326, 329, 331, 337, 338, 341, 342
UV DNA cellulose affinity chromatography, 332, 338

UV endonucleases, 252, 253, 258
UV irradiation, 4, 5, 6, 10, 20, 27, 29, 32–33, 38, 41, 83, 88, 92, 206, 211, 215–216, 219, 220, 221–222, 234, 242, 243, 247, 257, 277, 324, 336, 341, 422
  through polycarbonate isopore filters, 469, 471
UV light meters (radiometers), 17, 18, 29, 84, 88, 92, 204, 217, 226, 247, 253–254, 331, 459, 481, 483
UV light sources (*see also* UV-A, UV-B or UV-C), 17, 18, 29, 40, 84, 88
UV photolysis, *see* Photolysis of halogenated bases in DNA
UV photoproducts, 201, 202, 203, 212, 216, 429
  antibodies against, 216, 220, 230, 232, 242, 247
  cyclobutane pyrimidine dimers (CPDs), 156, 161, 201, 202, 208, 211, 216, 226, 229, 230, 235, 237, 239, 242, 244, 245, 247, 252, 256, 258, 287, 326, 329, 341, 462, 478
  detection by dot blot, 229ff
  detection by ELISA, 222–223, 224–226
  detection by LMPCR, 202ff
  detection by RIA, 239ff
  detection in urine, 224–226
  immunocytochemical detection of, 220–221
  immunohistochemical detection of, 218, 223–224
  levels induced in DNA, 211, 249, 270, 430
  (6-4) photoproducts, 201, 202, 211, 229, 230, 235, 236, 237, 239, 242, 244, 245, 247, 288, 462, 478
  detection by dot blot, 229ff
  detection by RIA, 239ff

selective chemical destruction of, 202, 207, 230, 235–236
UV shadowing, 437, 440, 446, 452
UV survival test, 5, 18, 20–21
UV-A, 215, 226, 254
   definition of, xvii, 215
   radiation source, 205, 208, 330
UV-B, 20, 21, 211, 215, 226, 242, 243, 254
   definition of, xvii, 215
   irradiation and biopsy of skin, 218
   irradiation of cultured keratinocytes, 220, 221–222
   radiation source, 17, 18, 204, 211, 217, 218, 231, 247, 248, 294
UV-C, 21, 202, 215, 225, 226, 229, 232, 242, 243, 257, 289, 290, 430, 462–463, 478, 483–484
   definition of, xvii, 215
   irradiation of cells in culture, 206, 463, 471
   radiation source (germicidal lamp), 204, 206, 217, 247, 294, 331, 336, 459, 462, 463
UvrABC (*E. coli*), 204

V

V-79 cells, *see* Chinese hamster cells

V(D)J recombination, 74, 109, 110, 324, 327, 328

W

WRN, 53, 398, 399, 405, 406, 407, 408, 410, 413

X

X-rays, *see* Ionizing radiation
Xeroderma pigmentosum (XP), 1, 9, 10, 15, 21, 184, 202, 278, 436, 458
   complementation groups, 10, 12
XPA, 418
XPB, 10, 418, 428
XPC, 15, 418
XPD, 10, 27, 135, 139, 418
XPE, 325, 326
XPF, 134, 140, 418, 428
XPG, 10, 418
XRCC1, 27, 157
XRCC3, 27
XRCC4, 96, 124

Y

Yeast, 134, 229, 233, 329

Z

z-VAD-FMK caspase inhibitor, 78